T0345256

# Computer Vision in Robotics and Industrial Applications

# Series in Computer Vision

ISSN: 2010-2143

**Series Editor:** C H Chen *(University of Massachusetts Dartmouth, USA)*

---

*Published*

Series in Computer Vision - Vol. 3

# Computer Vision in Robotics and Industrial Applications

Editors

## Dominik Sankowski
## Jacek Nowakowski

Lodz University of Technology, Poland

**World Scientific**

NEW JERSEY · LONDON · SINGAPORE · BEIJING · SHANGHAI · HONG KONG · TAIPEI · CHENNAI

*Published by*

World Scientific Publishing Co. Pte. Ltd.

5 Toh Tuck Link, Singapore 596224

*USA office:* 27 Warren Street, Suite 401-402, Hackensack, NJ 07601

*UK office:* 57 Shelton Street, Covent Garden, London WC2H 9HE

**British Library Cataloguing-in-Publication Data**
A catalogue record for this book is available from the British Library.

**Series in Computer Vision — Vol. 3**
**COMPUTER VISION IN ROBOTICS AND INDUSTRIAL APPLICATIONS**

Copyright © 2014 by World Scientific Publishing Co. Pte. Ltd.

ISBN 978-981-4583-71-8

Printed in Singapore

# PREFACE

In recent years, a strong development of practical applications of image processing and analysis is easily noticeable. Different vision systems are widely used among others in the automotive industry, pharmacy, military and police equipment, automated production and measurement systems. In each of these fields of technology, digital image processing and analysis module is a critical part of the process of building this type of system. Without properly developed methods and algorithms it is almost impossible to design or operate any contemporary computer based machine vision.

Many companies are currently making a large financial investment in teaching young people the skills to work with video and multimedia applications. Hence the field of image processing is an increasingly important element in the education for professionals needed to shape the future of information society.

The authors are aware that the literature related to image processing and analysis methods, is very rich and easily accessible. However, the majority of published books focuses on theoretical issues. Books that discuss methods of image processing and analysis in the practical application are in a substantial minority, and among those, which are available, the majority covers medical or multimedia applications. Not many publications deal with industrial applications, especially industrial measurement applications. The authors of the book present mainly results of research work performed at the Institute of Applied Computer Science of the Lodz University of Technology.

This volume in general is divided into four main parts. Part one contains a theoretical introduction to image reconstruction and processing. Chapter 1 deals with important aspects of data classification, what is a very important element of image processing. The $k$ nearest neighbors algorithm ($k$-NN) is one of the most popular classification algorithms: simple to implement and intuitive. But in the standard $k$-NN a complete data set is used during the classification. This causes problems with large data sets resulting in high memory requirements and a decrease in the speed of classification. To eliminate these problems reduction and editing algorithms can be used, which remove a significant part of the data. This chapter presents reduction and editing algorithms, both of which use a measure of representativeness.

Chapter 2 presents some selected computer vision applications, for which image segmentation accuracy is essential. The chapter presents the development of effective dedicated image segmentation algorithms.

Chapter 3 presents challenging application of fractional order differential calculus in image processing.

The second part of the book - chapters 4 to 9 - deals with computer vision in robotics. Chapter 4 presents management software responsible for supervision of the complete mobile robot system. It presents architecture of the operating system and the Human-Machine Interface of the mobile robot. The following two chapters (5 and 6) present various systems like laser scanners, ultrasonic sensors, and stereovision systems used for detection and analysis of characteristic features of objects in robot environment. Last chapters of this part (7 and 8) deal with application of fractional order calculus for edge detection in mobile robot systems.

Human-computer interfacing has become one of the major scientific aspects in designing systems and machines that are capable of performing complex, possibly, autonomous tasks. Chapter 9 presents research on the use of speech or gestures as sources of control commands and auditory and visual feedback. It also addresses some aspects of increase in human-system communication simplicity, which will reduce the exclusion of groups that have insufficient background for using modern technology or groups that due to physical disabilities cannot operate conventional computer interfaces. Affective computing, with objectives to identify and properly respond to human emotional states, thus improving quality of human-machine interaction is also considered.

The third part of the book presents industrial applications of computer vision in process tomography, material science and temperature control. Chapter 10 describes use of the hybrid boundary element method for solving diffusion tomography problems. The next four chapters (11 to 14) present various application of data processing for 3D reconstruction and monitoring of industrial process phenomena, such as multiphase flow, pneumatic transport and storage of friable materials. Specialized 3D image processing methods for the analysis of X-ray tomography data are described in Chapter 15. At the end of this part of the book, computer vision based applications for temperature measurements and control are presented in chapters 16 to 17.

Medical, biological and other applications of computer vision are gathered in the fourth part of this book. Some examples of image processing for cultivation analysis of plants (chapter 18), immunoenzymatic visualization of secretory activity with ELISPOT method (chapter 19), diagnosis of selected brain diseases (chapter 20) and studying of dynamic properties of materials at high temperatures (chapter 21) are presented in this part.

Authors hope, that material presented in this book will help readers to enhance their knowledge on application of image processing in different areas and stimulate to develop new ideas.

Editors

Dominik Sankowski

Jacek Nowakowski

Łódź, July 2013.

# CONTENTS

# Part 1

# Theoretical Introduction to Image Reconstruction and Processing

# CHAPTER 1

# DATA SET PREPARATION FOR *K*-NN CLASSIFIER USING THE MEASURE OF REPRESENTATIVENESS

Marcin Raniszewski

*Institute of Applied Computer Science*
*Lodz University of Technology*
*90-924 Łódź, ul. Stefanowskiego 18/22*
*mranisz@kis.p.lodz.pl*

In data classification a decision is made on the basis of information of the data transfer (described by a set of features). Proper and rapid classification depends on the preparation of the data set, as well as the selection of classification algorithms. The *k* nearest neighbors algorithm (*k*-NN) is one of the most popular classification algorithms: simple to implement and intuitive. The *k*-NN has high classification accuracy in a wide range of applications. But in the standard *k*-NN a complete data set is used during the classification. This causes problems with large data sets: high memory requirements and the speed of classification decrease. To eliminate these problems reduction and editing algorithms can be used. Reduction algorithms remove a significant part of the data. The remaining samples must provide an acceptable level of classification accuracy. Editing algorithms filter a data set, removing noise and redundant data.

This chapter presents reduction and editing algorithms, both using a measure of representativeness. Each proper sample of a data set is a part of its class distribution. A measure of representativeness expresses the level of this information retained by a sample.

The presented algorithms were tested on seven real data sets well known in the literature. The results are promising in comparison with the results of popular reduction and editing algorithms.

## 1.1 Introduction

The Nearest Neighbor Rule (NN), the particular case of the *k* Nearest Neighbor Rule for *k* = 1, is still in use in many applications of Pattern Recognition (Duda, Hart and Stork 2001; Theodoridis and Koutroumbas 2006). This method of classification is intuitive, simple and effective: a new sample is classified to a class of its nearest neighbor (the closest sample) from a training set. In (Duda, Hart and Stork 2001) the authors proved that an NN classification error is never beyond a double classification error of the Bayesian classifier for sufficiently large training sets. NN has also other advantages: the lack of training phase and faster classification as opposed to $k$-NN with $k > 1$.

On the other hand, NN (like standard $k$-NN) uses a complete training set to classify new samples. It causes computational loads and increases memory requirements for large training sets (e.g. in image analysis). It is a serious problem in applications with large reference sets, where the speed of classification is crucial.

A well-known solution of the problem is the reduction of a reference set: the majority of samples are removed and only the most important samples are preserved (Wilson and Martinez 2000). Obviously, the classification accuracy of NN operating on the reduced reference set should be acceptable. After reduction the classification accuracy can even increase for certain data sets as a consequence of removal of a noise.

There are many popular and widely used reduction methods. The newer methods use heuristics such as Monte Carlo, Random Mutation Hill Climbing, Tabu Search or Genetic Algorithms (Skalak 1994, Kuncheva and Bezdek 1998, Cerveron and Ferri 2001) and despite good results they provide, all of them are random and have several parameters. The aim of presented research was to construct a non-random reduction algorithm providing similarly good results as heuristic procedures, with fewer parameters and a much shorter training phase than Tabu Search or genetic methods. The proposed reduction algorithm uses the idea of the most representative samples. They are selected according to the measure of representativeness.

The NN procedure is very vulnerable to a noise. All new samples for which the noise samples are nearest neighbors are misclassified. The editing algorithm tries to solve this problem: noise, mislabelled and atypical samples are recognized by the algorithm and removed from the reference set. The results of proposed reduction method based on the most representative samples was so promising that the idea of representativeness was also used to create an editing method.

Proposed in this paper reduction and editing algorithms were tested on seven real data sets and compared with the results of other well-known methods.

## 1.2 Well-known reduction techniques

Probably the most popular reduction algorithm is *Condensed Nearest Neighbor Rule* (CNN) described by Hart (Hart 1968). The algorithm produces a consistent reference set, which means that it correctly classifies (using NN) all samples from a training set. Hart called it a consistency criterion. This criterion was used for a long time in other reduction methods as a reliable stop condition in creating a reduced reference set.

Gates proposed *Reduced Nearest Neighbor Rule* (RNN) (Gates 1972). This method uses the result of the CNN algorithm and returns a consistent subset of the CNN reduced set.

Gowda and Krishna published a number of articles about different uses of the concept of *Mutual Nearest Neighborhood* (MNN) and *Mutual Neighborhood Value* (MNV). One of their papers concerns a reference set reduction method (GK) (Gowda and Krishna 1979). Gowda and Krishna's method also results in a consistent reduced set.

Dasarathy used the concept of *Nearest Unlike Neighbor* (NUN): the sample $x$ is NUN($y$) if it is from a different class than sample $y$ and is $y$'s nearest neighbor (Dasarathy 1994). Dasarathy believed his procedure (MCS) results in minimal consistent reduced sets. Kuncheva and Bezdek presented in (Kuncheva and Bezdek 1998) a counter-example.

Skalak proposed two heuristics: MC1 and RMHC-P (Skalak 1994). The former is based on Monte Carlo heuristic, while the latter on Random Mutation Hill Climbing. In RMHC-P initially $m$ samples are selected to a reduced set and $n$ replacements (mutations) are made: random samples from a reduced set are replaced with random samples (which is not actually in the reduced set) from a training set. If the replacement increases $k$-NN classification accuracy counted for all the samples from the training set with the actual reduced set as a reference set, then the replacement is accepted, otherwise rejected. After all $n$ replacements, the procedure finishes. The reduced set contains $m$ samples and is inconsistent.

Kuncheva described heuristics based on genetic algorithms (GA) (Kuncheva 1995, Kuncheva 1997, Kuncheva and Bezdek 1998). The proposed heuristics can be treated as editing or reduction procedures depending on the fitness function. In Kuncheva's GA every subset of a training set is represented as a binary string (chromosome) (the $i$-th bit is set to 1 if the $i$-th sample is in the reduced set, otherwise to 0). Every bit in a chromosome is initially set to 1 with a predefined probability, called the reduction rate. In a reproduction phase two offspring chromosomes are produced by every couple of parent chromosomes (each pair of corresponding bits of parent chromosomes are swapped at the crossover rate) and then each bit of each offspring chromosome alternates (mutates) at the defined mutation rate. Both, the crossover and the mutation rates are predefined probabilities. The fitness function $J$ is based on the number of correct classifications and uses the cardinality of the currently reduced set (the penalty term) (Kuncheva 1997):

$$J(Y) = \frac{n(Y)}{n} - \alpha \cdot card(Y),$$ (1.1)

where: $n(Y)$ denotes the number of correctly classified samples from the training set using only the reduced set $Y$ to find the $k$ nearest neighbors (sample $s$ is classified using $k$-NN with $Y$-$\{s\}$), $n$ – the number of all samples in the training set, $\alpha$ – the positive coefficient (the higher $\alpha$ is, the higher the penalty gets) and $card(Y)$ – the cardinality of $Y$. GA's reduced set is inconsistent.

Cerveron and Ferri proposed a heuristic based on Tabu Search (TS) (Cerveron and Ferri 2001). The method, similarly to GA, returns inconsistent sets and uses chromosome representations of subsets. All subsets that differ from the actual subset $S$ by only one element create a new neighborhood of $S$. The objective function is the same as (1.1). Two different methods of creating an initial subset $S$ are recommended by authors: condensed or constructive. The former uses a CNN procedure, the latter uses TS with disabled sample deletion (in creating the neighborhood of actual solution) starting from a subset with randomly selected single samples from each class. The constructive procedure is broken when a consistent set is obtained.

## 1.3 Well-known editing techniques

Wilson proposed *Edited Nearest Neighbor rule* (ENN) – the $k$-NN rule operating with an edited reference set (Wilson 1972). In the procedure each sample is reclassified using $k$-NN with a training set. After the reclassification misclassified samples are removed from the training set and the remaining samples constitute the edited reference set.

Tomek described two editing procedures based on Wilson's ENN rule: the Repeated ENN (RENN) rule and the All $k$-NN rule (Tomek 1976). Generally, RENN repeats ENN until no more samples are removed and the All $k$-NN rule executes ENN for all $k_0$ from 1 to $k$ and then removes all misclassified samples.

Devijver and Kittler proposed the MULTIEDIT algorithm (Devijver and Kittler 1980). It can be described in the following few steps (initially the training set is denoted as $S$):

1.  Diffusion: Make a random partition of samples from $S$ into $N$ separable subsets.
2.  Classification: Classify samples from each subset using NN rule operating with samples from the next subset (samples from the $N$-th subset are

classified using NN operating with samples from the first subset).

3. Editing: Remove all the samples misclassified in step 2.
4. Confusion: The remaining samples create a new set *S*.
5. Termination: If the last *I* iterations produced no editing, exit with the final solution *S*, otherwise go to step 1.

As mentioned in Section 1.2, Kuncheva proposed a heuristic method based on Genetic Algorithms (GA) for reference set edition (Kuncheva 1995). Kuncheva proposed a fitness function *J* based on the number of neighbors leading to the correct classification:

$$J(Y) = \frac{1}{n}\sum_{i=1}^{m} k_i,$$ (1.2)

where *n* denotes the number of all samples in the training set, *m* – denotes the number of correctly classified samples from the training set using only the current edited reference set *Y* to find the *k* nearest neighbors (sample *s* is classified using *k*-NN with *Y*-{*s*}) and $k_i$, $i = 1,2,\ldots,m$ – the number of neighbors leading to the correct classification of an *i*-th sample.

## 1.4 Methods proposed

In reduction methods one should choose only the most representative samples and preserve them in the reduced reference set. However, the definition of representatives is not clear. There is a group of algorithms which preserve samples from the class boundaries. Other methods add samples from the centers of homogeneous (one-class) areas to the reduced set.

### 1.4.1 *The measure of representativeness*

We proposed our own definition of representativeness (called the measure of representativeness): for any sample *x* it is defined as a number of samples from the same class as *x* (called voters) which lie closer to *x* than to any sample from the opposite class (Fig. 1.1). Let us denote a value of the measure of representativeness for sample *x* as RV(*x*). The higher RV(*x*) is, the more representative *x* is. Hence:

(1) if RV(*x*) > RV(*y*), then sample *x* is more representative than sample *y*;
(2) a sample with more voters is more representative for its class.
If RV(*x*) = RV(*y*), then a sample with the higher distance to its NUN (see Sec. 1.2) is more representative.

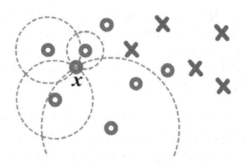

Fig. 1.1  RV(x) = 4.

### 1.4.2  *The reduction method proposed*

The proposed reduction algorithm, called the Reduction Algorithm based on Representativeness (RAR), has only one parameter $RV_{min} \geq 0$. The parameter denotes the minimal value of the measure of representativeness and should be established experimentally. Only a sample with RV (see Sec. 1.4) greater than $RV_{min}$ is added to the reduced set. Hence, smaller reduced sets for higher value of $RV_{min}$ are obtained.

RAR can be described in the following steps:

1.  Mark all the samples in the training set as "not included in the reduced set" and "available".
2.  Count RV for all samples marked as "available" and "not included in the reduced set". Samples marked as "unavailable" cannot be voters (see Sec. 1.4.1).
3.  Choose a sample $x$ with the greatest representative measure (break tie by choosing a sample with the higher distance to its NUN). If the greatest representative measure is less or equal to $RV_{min}$, go to step 5, otherwise mark $x$ as "included in the reduced set" and all $x$'s voters as "unavailable".
4.  If there is at least one sample marked as "not included in the reduced set" and "available", go to step 2.
5.  The samples marked as "included in the reduced set" constitute the reduced set.

After the construction of the reduced set for a specific value of $RV_{min} \geq 1$ the algorithm can simply be continued (from step 2) for $RV_{min} - 1$. In this way, all the reduced sets for decreasing values of $RV_{min}$ (from initial value to 0) can easily be generated.

As can be seen, samples are added to the reduced set according to their actual RV and the distance to their NUNs. Hence, the results obtained by RAR are repeatable.

### 1.4.3 *The proposed editing method*

In edited training sets samples that are not a noise should be preserved. Hence, one expects higher classification accuracy after the edition. The reduction level in editing techniques is not important.

The idea used in the proposed editing method is to use a reduced training set (which consists of the most representative samples) as a filter to recognize a noise. The consistency criterion can be used here: only the samples that are correctly classified by NN operating with the reduced training set are marked as "not noise". After the classification they constitute the edited training set.

One can describe the whole procedure, called the Editing Algorithm based on the Consistency Criterion (EACC), in the following steps:

1.  Reduce the training set (let us denote the reduced training set as $Y$).
2.  Classify each sample from the training set using NN operating with $Y$. All the correctly classified samples mark as "not noise".
3.  All the "not noise" samples constitute the edited training set.

Note that no particular technique of reduction is indicated. EACC can be run on any reduced training set. In this paper, we test EACC on reduced training sets obtained by RAR.

The EACC edited training set is consistent with set $Y$: all samples from the EACC edited set are correctly classified using NN operating on $Y$.

### 1.5 Test results

Seven real data sets were used in the tests: Liver Disorders (BUPA), GLASS Identification, IRIS, PHONEME, PIMA Indians Diabetes, Wisconsin Diagnostic Breast Cancer (WDBC) (Diagnostic) and YEAST (Frank and Asuncion 2010; The ELENA Project Real Databases).

The number of class labels, attributes and samples in the data sets are summarized in Table 1.1.

*M. Raniszewski*

Table 1.1  A brief summary of all data sets used in tests.

| Data set | No. of classes | No. of attributes | No. of samples |
|---|---|---|---|
| BUPA | 2 | 6 | 345 |
| GLASS | 6 | 9 | 214 |
| IRIS | 3 | 4 | 150 |
| PHONEME | 2 | 5 | 5404 |
| PIMA | 2 | 8 | 768 |
| WDBC | 2 | 30 | 569 |
| YEAST | 10 | 8 | 1484 |

Stratified ten-fold cross-validation (with NN classification) was used for each experiment.

We used an Intel Core i3-2310M, 2.10 Ghz processor with 4 GB RAM for the tests. We implemented all algorithms in Java.

### 1.5.1  Test results of reduction methods

We reduced data sets with seven well-known algorithms described in Sec. 1.2: CNN, RNN, GK, MCS, GA, TS, RMHC-P and the proposed RAR (Sec. 1.4.2).

In GA we set the number of nearest neighbors to 1, the number of iterations to 200, the reduction rate to 0.1, the number of chromosomes to 20, the crossover rate to 0.5, the mutation rate to 0.025 and fitness function (1.1) with $\alpha = 0.01$. The implementation of GA is equivalent to that proposed in (Kuncheva and Bezdek 1998).

We set TS parameters to the values proposed in (Cerveron and Ferri 2001) with one difference: we used the condensed methods of creating the initial subset for PHONEME data set due to very long phase of constructive initialization for that training set.

In RMHC-P: $k = 1$, $m$ was equal to the reduced set size obtained by the best (according to classification accuracy for the specified data set) reduction algorithm from the group of GA, TS and RAR (methods creating inconsistent reduced set) and we set $n$ to 200 for IRIS, 300 for GLASS, 400 for BUPA, 600 for WDBC, 700 for PIMA, 1000 for YEAST and 2000 for PHONEME.

We applied RAR with five decreasing values of $RV_{min}$: 4, 3, 2, 1 and 0 (and named them as RAR4, RAR3, RAR2, RAR1 and RAR0, respectively). As pointed in Sec. 1.4.2, we obtained all five reduced sets in a single RAR run.

The results of classification accuracy were also compared with NN operating on complete training sets.

We highlighted in bold the best result of the reduction method for each data set (Table 1.2 and Table 1.3). For clarity, we divided tables into three parts:

(1) older reduction algorithms (which create consistent reference sets): CNN, RNN, GK, MCS;
(2) newer reduction algorithms (heuristics which create inconsistent reference sets): GA, TS, RMHC-P;
(3) proposed reduction method RAR with different values of $RV_{min}$.

Table 1.2  Classification accuracies (in percentages) of reduction methods.

|  | BUPA | GLASS | IRIS | PHONEME | PIMA | WDBC | YEAST |
|---|---|---|---|---|---|---|---|
| NN | 62,89 | 71,56 | 96,00 | 90,82 | 67,20 | 91,19 | 53,04 |
| CNN | 60,30 | **69,36** | 93,33 | **88,79** | 63,95 | 90,84 | 50,90 |
| RNN | 58,55 | 66,12 | 92,00 | 88,16 | 62,91 | 91,19 | 49,88 |
| GK | 56,51 | 66,14 | 94,00 | 88,34 | 63,69 | 91,55 | 50,29 |
| MCS | 55,05 | 65,44 | 92,67 | 87,45 | 64,73 | 91,36 | 49,19 |
| GA | 67,51 | 64,98 | **97,33** | 80,90 | 69,29 | 92,42 | 46,00 |
| TS | 68,13 | 68,83 | 96,00 | 84,53 | 69,54 | 93,14 | 57,36 |
| RMHC-P | 63,43 | 67,66 | 94,00 | 84,57 | 70,20 | 92,62 | 56,73 |
| RAR0 | 63,22 | 68,83 | 93,33 | 86,66 | 66,55 | 92,41 | 54,06 |
| RAR1 | 66,07 | 64,12 | 94,00 | 85,84 | 71,88 | **93,67** | 56,06 |
| RAR2 | 67,86 | 60,23 | 96,00 | 85,79 | 73,31 | 93,32 | 57,15 |
| RAR3 | **68,37** | 58,31 | 96,00 | 85,47 | **74,49** | 93,14 | **57,69** |
| RAR4 | 66,34 | 55,46 | 96,00 | 84,29 | 74,23 | 92,79 | 55,74 |

Table 1.3  Reduction levels (in percentages) of reduction methods.

|  | BUPA | GLASS | IRIS | PHONEME | PIMA | WDBC | YEAST |
|---|---|---|---|---|---|---|---|
| CNN | 40,97 | 51,87 | 87,70 | 76,11 | 46,66 | 82,97 | 33,54 |
| RNN | 48,24 | 57,89 | 88,59 | 80,28 | 55,22 | 86,70 | 39,53 |
| GK | 43,51 | 55,76 | 88,52 | 78,69 | 53,30 | 85,98 | 37,20 |
| MCS | 49,50 | 60,69 | 89,93 | 81,67 | 57,47 | 86,27 | 40,90 |
| GA | 94,94 | 93,82 | **96,67** | 90,79 | 92,48 | 93,85 | 91,35 |
| TS | 87,57 | 77,05 | 94,00 | **99,37** | 94,21 | **98,52** | 96,89 |
| RMHC-P | 96,13 | 78,35 | **96,67** | 86,86 | 95,40 | 95,59 | 97,12 |
| RAR0 | 76,59 | 78,35 | 92,44 | 86,86 | 78,43 | 92,33 | 78,44 |
| RAR1 | 87,34 | 89,87 | 94,44 | 91,27 | 89,89 | 95,59 | 90,99 |
| RAR2 | 92,59 | 93,61 | 95,48 | 93,91 | 93,95 | 96,49 | 95,24 |
| RAR3 | 96,13 | 95,69 | 95,93 | 95,40 | 95,40 | 97,13 | 97,12 |
| RAR4 | **98,13** | **96,83** | 96,37 | 96,38 | **96,56** | 97,62 | **98,28** |

*M. Raniszewski*

## 1.5.2 *Test results of editing methods*

We edited all the data sets with five well-known editing algorithms described in Sec.1.3: ENN, RENN, All k-NN, MULTIEDIT, GA and the proposed EACC (Sec. 1.4.3).

We tested ENN, RENN and GA with $k = 1$ (ENN1, RENN1, GA1) and $k = 3$ (ENN3, RENN3, GA3) and All $k$-NN with $k = 3$ (All3NN).

In MULTIEDIT we set the parameter $N$ to 3 and the parameter $I$ to 5 for small data sets: BUPA, GLASS, IRIS, PIMA, WDBC and YEAST. For PHONEME data set we set $N$ to 9 and the parameter $I$ to 10.

In GA (in accordance with (Kuncheva 1995)) for all the data sets we set the number of iterations to 200, the reduction rate to 0.8, the number of chromosomes to 50 and the mutation rate to 0.05. Additionally, we set the number of iterations to 200 and the number of chromosomes to 50 for all the data sets except PHONEME. For PHONEME data set, due to a very long phase of edition, we decreased the number of iterations to 50 and the number of chromosomes to 10.

We used the proposed RAR with $RV_{min}$ parameter value from 0 to 4 as the EACC initial reduction method. Hence, we obtained 5 reduced training sets for each data set and then we ran the EACC method on each of them. We denoted the results as EACC0, EACC1 EACC2, EACC3 and EACC4, where EACC$i$ is the EACC result for the data set reduced by RAR$i$ (Sec. 1.4.1).

We highlighted the best result of the editing method for each data set in bold (Table 1.4). We added the reduction level results in Table 1.5 for better comparison of the algorithms tested. For clarity, we divided tables into two parts:

(1) well-known algorithms: ENN, RENN, All $k$-NN, MULTIEDIT and GA;
(2) proposed editing method EACC, which builds edited reference set on the basis of RAR reduction method with different values of $RV_{min}$.

Table 1.4 Classification accuracies (in percentages) of editing methods.

|  | BUPA | GLASS | IRIS | PHONEME | PIMA | WDBC | YEAST |
|---|---|---|---|---|---|---|---|
| NN | 62,89 | 71,56 | 96,00 | 90,82 | 67,20 | 91,19 | 53,04 |
| ENN1 | 63,73 | 70,15 | 96,67 | 89,45 | 68,49 | 92,25 | 56,27 |
| ENN3 | 65,50 | 65,91 | 96,67 | 88,92 | 71,37 | 91,72 | 56,13 |
| RENN1 | 65,18 | 66,35 | 96,67 | 89,06 | 68,10 | 92,77 | 56,87 |
| RENN3 | 67,54 | 63,44 | 96,67 | 88,32 | 71,77 | 92,41 | 56,73 |
| ALL3NN | 67,21 | 67,28 | 96,67 | 88,42 | 72,79 | 93,14 | 57,00 |
| MULTIEDIT | 61,46 | 52,79 | 95,33 | 78,31 | 69,80 | 90,68 | 51,97 |
| GA1 | 64,64 | **71,31** | 96,67 | 89,27 | 68,76 | 91,53 | 51,39 |
| GA3 | 66,94 | 71,04 | **97,33** | 88,75 | 70,20 | **93,49** | 53,04 |

| EACC0 | 64,34 | 71,14 | 96,67 | **89,97** | 66,81 | 91,54 | 54,50 |
| EACC1 | 66,39 | 64,90 | 96,67 | 88,77 | 71,63 | 93,14 | 56,59 |
| EACC2 | 69,84 | 63,93 | 96,67 | 87,95 | 73,19 | 93,31 | 57,75 |
| EACC3 | **71,26** | 63,45 | 96,67 | 87,32 | 73,58 | 92,43 | **58,03** |
| EACC4 | 68,92 | 60,27 | 96,00 | 86,18 | **74,10** | 92,08 | 57,02 |

Table 1.5  Reduction levels (in percentages) of editing methods.

|           | BUPA  | GLASS | IRIS  | PHONEME | PIMA  | WDBC  | YEAST |
|-----------|-------|-------|-------|---------|-------|-------|-------|
| ENN1      | 37,78 | 27,31 | 4,15  | 9,29    | 32,16 | 8,36  | 47,84 |
| ENN3      | 36,23 | 32,71 | 3,85  | 10,85   | 30,54 | 7,30  | 45,24 |
| RENN1     | 40,93 | 31,00 | 4,15  | 10,22   | 35,26 | 8,77  | 51,47 |
| RENN3     | 42,09 | 39,09 | 3,85  | 12,53   | 34,66 | 8,05  | 51,09 |
| ALL3NN    | 54,46 | 40,19 | 5,41  | 15,16   | 44,27 | 11,44 | 60,32 |
| MULTIEDIT | 73,04 | 63,00 | 12,74 | 40,36   | 54,54 | 14,49 | 76,19 |
| GA1       | 50,85 | 44,96 | 51,04 | 38,16   | 49,05 | 48,23 | 48,52 |
| GA3       | 51,05 | 45,12 | 48,37 | 37,17   | 50,12 | 50,56 | 49,85 |
| EACC0     | 18,74 | 14,96 | 3,19  | 4,99    | 14,53 | 4,45  | 28,72 |
| EACC1     | 24,73 | 26,01 | 3,26  | 8,32    | 21,09 | 5,84  | 37,17 |
| EACC2     | 28,08 | 32,34 | 3,93  | 10,54   | 23,32 | 5,78  | 40,83 |
| EACC3     | 29,92 | 37,43 | 4,00  | 12,17   | 24,03 | 6,27  | 42,77 |
| EACC4     | 33,72 | 41,95 | 4,37  | 13,56   | 25,22 | 6,80  | 44,22 |

## 1.6  Discussion

In this section we discuss the test results of reduction and editing methods. For better comparison between the proposed and well-known methods, we summarized the test results on box and whisker plots in Fig. 1.2, 1.3, 1.4 and 1.5 (the median, quartiles, minimum and maximum of classification accuracies and reduction levels for each method).

### 1.6.1  *Discussion of reduction methods test results*

The most important thing in the reduction method is the high reduction level. Obviously, satisfactory classification accuracy is also important, but the main goal we want to achieve is the high increase in NN speed. Hence, let us focus on reduction levels.

As we can see in Table 1.3 and Fig. 1.2 all of the older reduction methods have the reduction level of approx. between 40% and 85%. This is not a good result in comparison with the newer algorithms, such as GA, TS, RMHC-P and the proposed RAR, which reduce the training set to the level almost always above 90%. The main reason for such results is the consistency criterion used in older methods as a stop condition. This criterion ensures that all the samples from the original training set are properly classified using NN with the reduced set. For real data sets with a noise (as the one used in the tests) the consistency criterion is not appropriate. It causes the reduction method to add a noise to the reduced set and all the samples from a neighborhood of the noise. It increases the size of the reduced set and causes a decrease in classification accuracy (the noise is still in the reduced set).

Hence, let us focus on newer reduction methods: GA, TS, RMHC-P and the proposed RAR. They create an inconsistent reduced set, which allows them to reject a noise and choose only those samples which correctly and fully describe the distribution of their classes. As mentioned above, for almost all data sets the algorithms (excluding RAR0) achieved high averaged reduction levels (above 90%).

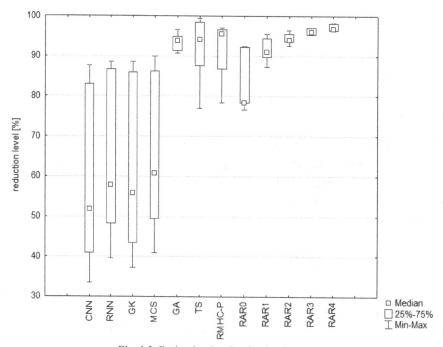

Fig. 1.2 Reduction levels of reduction methods.

For the proposed RAR methods the reduction level is obviously higher for the higher values of parameter $RV_{min}$. Higher values of parameters $RV_{min}$ cause only samples with a higher RV (Sec. 1.4.1) to be added to the reduced set.

We should confront reduction levels with classification accuracies to check whether the higher reduction level does not reduce classification accuracy.

In Table 1.2 and Fig. 1.3 we can see that despite the high reduction level the newer methods maintain classification accuracy at a satisfactory level, often higher than the classification accuracy achieved with the original (unreduced) data set.

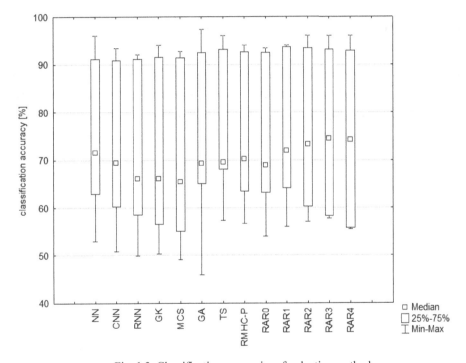

Fig. 1.3 Classification accuracies of reduction methods.

For four data sets RAR achieved the highest classification accuracies (but for different values of parameter $RV_{min}$), and for BUPA and PIMA they were significantly higher than the classification accuracies obtained on complete data sets.

The classification accuracies obtained by CNN for GLASS and PHONEME data sets are surprisingly high in comparison with the results of newer reduction methods. This can be explained by the specific form of the data sets which, for higher reduction levels, are not able to maintain the classification accuracy at a satisfactorily high level.

*M. Raniszewski*

One can notice that the smaller RAR reduced sets (obtained with a higher value of $RV_{min}$) do not provide higher classification accuracy. For all the tested data sets the best classification accuracies were achieved for $RV_{min}$ between 0 and 3.

GA, RMHC-P and TS are heuristics with many parameters and their results are not unique (because of randomization). It is possible that their results would be better or worse with other values of the parameters.

The time of the reduction phase of most of the tested algorithms for average size data sets (like PHONEME) is sufficiently small (several seconds or minutes) in contrast to GA and TS (about 2 and 4 hours, respectively).

### 1.6.2 *Discussion of editing method test results*

Unlike reduction methods, we used editing methods not to reduce the size of the reference set but to remove a noise and improve classification accuracy. The reduction level in editing methods is not important. However, for better comparison of the editing methods, in addition to the classification accuracies we also summarized the reduction levels.

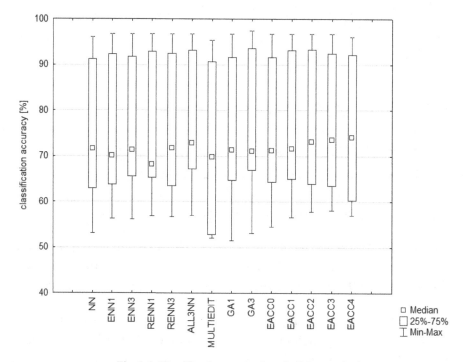

Fig. 1.4 Classification accuracies of editing methods.

The classification accuracies of all the tested methods are very similar, except MULTIEDIT (Table 1.4 and Fig. 1.4). MULTIEDIT results are worse than the results on the original data set for almost all data sets (except PIMA). This can be the consequence of the inadequate choice of parameter values.

The best results are within two algorithms: GA and the proposed EACC. EACC achieved the highest classification accuracies with four data sets while GA with three. We can observe a significant improvement in classification accuracy for the EACC procedure with BUPA, PIMA and YEAST data sets (EACC for RAR3 and RAR4).

We can also discuss the results of reduction levels (Table 1.5 and Fig. 1.5). In an editing method it is important to correctly recognize a noise. If the reduction level is higher than 20%-30%, then the editing method is likely to remove more than only a noise.

GA and MULTIEDIT reduction levels are very high (from 40% to 50% for GA and from 15% to as much as 75% for MULTIEDIT). Considering that reduction levels are so high, one should answer the question whether GA and MULTIEDIT are really editing methods.

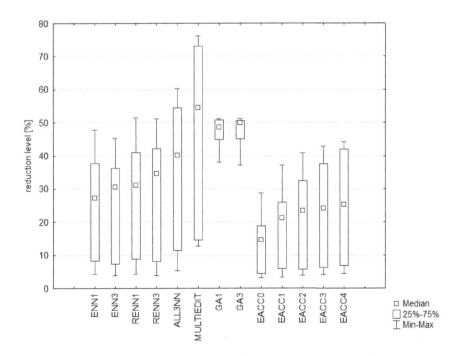

Fig. 1.5  Reduction levels of editing methods.

The EACC reduction levels are very promising. They are basically lower than 40% and in most cases lower than the reduction levels achieved by ENN, RENN and All 3-NN. Obviously, for the higher values of $RV_{min}$, we have higher reduction levels for RAR and consequently higher reduction levels of EACC.

We can use the EACC procedure with any method of reduction. In this paper we used the EACC with RAR results. It is possible that for data sets such as GLASS, IRIS, PHONEME, we should use the CNN or GA reduction set (see Table 1.2).

The training phase of all algorithms was very short (a few milliseconds or seconds), except GA (about 2 hours for the PHONEME data set).

### 1.6.3 *Choice of the best RAR*

Since we can easily (in one run) achieve RAR results for decreasing values of $RV_{min}$ (in our tests from 4 to 0), we can quickly choose the reduced set which is adequate for our purpose. If we want strong reduction, we can choose RAR with a higher value of $RV_{min}$. If we focus on high classification accuracy, we can choose the best reduced set according to this criterion.

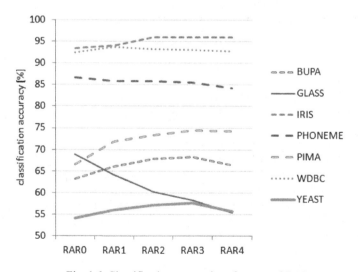

Fig. 1.6 Classification accuracies of proposed RAR.

As can be seen in Table 1.2 and Fig 1.6 the range of $RV_{min}$ from 0 to 4 was sufficient to obtain the reduced set with the highest classification accuracy for all the data sets tested. We achieved the maximum classification accuracy for almost all the data sets (except IRIS) for $RV_{min} < 4$.

For higher values of $RV_{min}$ the reduced sets are smaller (we get stronger reduction levels – see Fig. 1.7) but the classification accuracies are lower. Hence, the range of $RV_{min}$ from 0 to 4 should be sufficient for most real data sets.

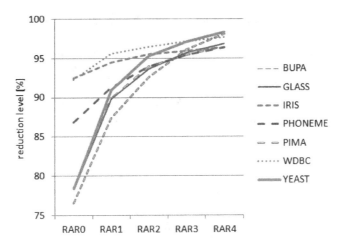

Fig. 1.7  Reduction levels of proposed RAR.

## 1.7  Conclusions

In this chapter we presented two methods of data set preparation for NN classifications: the Reduction Algorithm based on Representativeness (RAR) and the Editing Algorithm based on the Consistency Criterion (EACC).

In RAR only the representative samples are added to the reduced set, that is to say, samples with RV higher than the initially fixed $RV_{min}$.

RAR has the following advantages:

(1) a high reduction level for $RV_{min} \geq 1$ (approx. above 90 %);
(2) a high classification accuracy (often higher than accuracies obtained on a complete training set);
(3) a unique solution (lack of randomization);
(4) a short time of the reduction phase;
(5) only one parameter $RV_{min}$ (with only five values: {0, 1, 2, 3, 4} which should be considered);
(6) a possibility to obtain reduced sets for all the considered values of $RV_{min}$ in a single RAR run.

However, there is also a disadvantage of the method presented: an additional phase of validation is necessary to choose the best reduced set from all the sets generated.

The EACC is a very simple method and it can be run on any reduced data set. In this paper we used RAR results for the EACC.

The EACC has the following advantages:

(1) a high classification accuracy (mostly higher than accuracies obtained by NN operating on a complete training set);
(2) a very short time of the training phase (not counting the time of reduction);
(3) no parameter (the number of parameters of EACC is the number of parameters of the reduction method);
(4) a unique solution as opposed to MULTIEDIT and GA editing methods;
(5) low reduction levels (most likely noise is not added to the edited reference set).

## References

Cerveron, V., Ferri, F.J. (2001). Another move towards the minimum consistent subset: A tabu search approach to the condensed nearest neighbor rule, IEEE Trans. on Systems, Man and Cybernetics, Part B: Cybernetics, Vol. 31(3) pp. 408-413.

Dasarathy, B.V. (1994). Minimal consistent set (MCS) identification for optimal nearest neighbor decision systems design, IEEE Transactions on Systems, Man, and Cybernetics 24(3), pp. 511-517.

Devijver, P.A. and Kittler, J. (1980). On the edited nearest neighbor rule. Proc. 5th International Conf. on Pattern Recognition, pp. 72-80.

Duda, R.O., Hart, P.E. and Stork, D.G. (2001). Pattern Classification – Second Edition, John Wiley & Sons, Inc.

Frank, A. and Asuncion, A. (2010). UCI Machine Learning Repository [http://archive.ics.uci.edu/ml]. Irvine, CA: University of California, School of Information and Computer Science.

Gates, G.W. (1972). The reduced nearest neighbor rule, IEEE Transactions on Information Theory, Vol. IT 18, No. 5, pp. 431–433.

Gowda, K. C. and Krishna, G. (1979). The condensed nearest neighbor rule using the concept of mutual nearest neighborhood, IEEE Transaction on Information Theory, v. IT-25, 4, pp. 488-490.

Hart, P.E. (1968). The condensed nearest neighbor rule, IEEE Transactions on Information Theory, vol. IT-14, 3, pp. 515-516.

Kuncheva, L.I. (1995). Editing for the k-nearest neighbors rule by a genetic algorithm, Pattern Recognition Letters, Vol. 16, pp. 809-814.

Kuncheva, L.I. (1997). Fitness functions in editing k-NN reference set by genetic algorithms, Pattern Recognition, Vol. 30, No. 6, pp. 1041-1049.

Kuncheva, L.I. and Bezdek, J.C. (1998). Nearest prototype classification: clustering, genetic algorithms, or random search?, IEEE Transactions on Systems, Man, and Cybernetics, Part C: Applications and Reviews, Vol. 28(1), pp. 160-164.

Skalak, D.B. (1994). Prototype and feature selection by sampling and random mutation hill climbing algorithms, 11th International Conference on Machine Learning, New Brunswick, NJ, USA, pp. 293–301.

The ELENA Project Real Databases [http://www.dice.ucl.ac.be/neural-nets/Research/Projects/ELENA/databases/REAL/].

Theodoridis, S. and Koutroumbas, K. (2006). Pattern Recognition – Third Edition, Academic Press - Elsevier, USA.

Tomek, I. (1976). An Experiment with the edited nearest-neighbor rule. IEEE Transactions on Systems, Man, and Cybernetics, Vol. SMC-6, No. 6, pp. 448-452.

Wilson, D.L. (1972). Asymptotic properties of nearest neighbor rules using edited data. IEEE Transactions On Systems, Man and Cybernetics, Vol. 2, pp. 408-421.

Wilson, D.R. and Martinez, T.R. (2000). Reduction techniques for instance-based learning algorithms, Machine Learning, Vol. 38, No. 3, pp. 257–286.

# CHAPTER 2

# SEGMENTATION METHODS IN THE SELECTED INDUSTRIAL COMPUTER VISION APPLICATION

Anna Fabijańska and Dominik Sankowski

*Institute of Applied Computer Science*
*Lodz University of Technology*
*90-924 Łódź, ul. Stefanowskiego 18/22*
*{an_fab, dsan}@kis.p.lodz.pl*

Image segmentation is an essential task in all applications of machine vision. It addresses the problem of partitioning an image into disjoint regions of interest according to their specific features (gray levels, texture etc.) (Materka, 1991; Tadeusiewicz, 1997; Wiatr, 2003; Gonzalez and Woods, 2008; Petrou and Petrou, 2010). The accuracy of image segmentation significantly influences the results of image analysis performed in the following steps. Therefore, for many vision systems the development of an effective and dedicated image segmentation algorithm is an important task.

## 2.1 Introduction

Computer vision systems have become very popular nowadays. They are of great importance in almost every field of science, engineering and industry (Ranky, 2003; Obinata and Dutta, 2007). The most significant applications of machine vision are: the analysis of biomedical and industrial images, the control and management of manufacturing processes, computer thermography, process tomography, remote sensing, biometry and many others (Batchelor and Whelan, 2002; Steger et al., 2008; Gocławski et al., 2009). In all these applications vision systems model the real world or recognize objects and describe them based on digital images. The images are acquired using video, digital cameras, radars or other specialized sensors.

Recently, modern vision systems more and more often obtain information that is normally invisible for humans. This is possible due to the application of dedicated algorithms which analyze specific features of the process or object under investigation. Characteristic features of the processed and analyzed images differ depending on the application.

One of the biggest challenges of modern computer vision is the development of vision systems imitating the behavior of human sense of sight. These systems should identify, describe, analyze and understand objects that are relevant to a given image-based task. This requires extensive background knowledge and interplay between two related disciplines, i.e. image processing (for image content discovery) and image analysis (for image information content interpretation).

However, it should be remembered that the visual representation of information contained within a digital image is characterized by a high level of redundancy. Therefore, after converting the image into its digital representation, a stage of detailed image analysis is carried out, in order to separate information significant to the user or process from the entire information reaching the observer or the detector. In order to extract interesting information for further processing (such as description or recognition) image segmentation algorithms are applied in this step.

Image segmentation is an important issue during image analysis (Gonzalez and Woods, 2008; Petrou and Petrou, 2010; Jähne, 2012). It divides an input image into regions matching separate objects, visible in the image by finding in the image analyzed cohesive regions which are characterized by similar values of some attribute (e.g. lightness) or a set of features (e.g. texture). In this way, objects of interests are extracted for further processing such as description or recognition. The result of image segmentation should allow one to define the geometric features of objects placed in a scene as accurately as possible with the minimum computational complexity. In many applications the accuracy and the effectiveness of a segmentation algorithm is the most important criterion to be considered during the development of dedicated image processing and analysis algorithms.

In the field of digital image processing, image segmentation is one of the most often deliberated issues. Although the literature abounds with different approaches to the problem of image segmentation, there is no general theory of it. There is still a necessity to develop unambiguous criteria for image segmentation and classification so that image interpretation processes (such as medical or industrial image analysis) could be performed automatically.

## 2.2 Preliminaries of Image Segmentation

The aim of image segmentation is to separate an object (or objects) of interest from the background which is usually understood as non-significant (redundant) information contained within an image. It is equivalent to finding in the analyzed images cohesive regions, characterized by similar values of the certain attribute (e.g. gray level, color, gradient direction) or a set of features (for example, texture).

Every digital image $L(x, y)$ of the scene $D$ which consists of objects placed in this scene and the background can be described by the following equation (Putiatin, Averin, 1990; Strzecha, 2002):

$$L(x, y) = H_1(x, y) + H_2(x, y) + \ldots + H_s(x, y) + H_\varphi(x, y) \tag{2.1}$$

where:

$\quad x, y$   - the pixel coordinates,

$\quad\quad s$   - the number of objects in the scene,

$\quad H_k(x, y)$   - the image of $k$-th object, $k = \{1, 2, \ldots, s\}$,

$\quad H_\varphi(x, y)$   - the image of the background.

Moreover, conditions given by Equation (2.2) are fulfilled.

$$\left. \begin{array}{ll} H_k(x, y) = 0 & \text{for} \quad (x, y) \notin D_k \\ H_\varphi(x, y) = 0 & \text{for} \quad (x, y) \notin D_\varphi \end{array} \right\} \tag{2.2}$$

where:

$\quad D_k \subset D$   - the region of $k$-*th* object,

$\quad D_\varphi \subset D$   - the background.

The relationship between regions $D_k$ and $D_\varphi$ is given by Equations (2.3) and (2.4).

$$D_1 \cup \ldots \cup D_s \cup D_\varphi = D \tag{2.3}$$

$$D_i \cap D_j = 0 \text{ for } i \neq j \tag{2.4}$$

Image segmentation can be then defined as the decomposition of image $L(x, y)$ into images of:

- objects $H_k(x, y)$ placed in the scene, $k = \{1, 2, \ldots, s\}$;
- the background $H_\varphi(x, y)$.

The decomposition can be performed by labeling each pixel of the image ($(x, y) \in D$) with the label $\pi(x, y)$ in accordance with the transformation given by Equation (2.5).

$$\pi : D \rightarrow \{0, 1, \ldots, s\} \tag{2.5}$$

Labels $\pi(x, y)$ define each pixel affinity to a certain image component. Different labels are assigned to separate objects. The background is usually assigned the value of '0' (see Eq. (2.6)).

$$\pi(x, y) = \begin{cases} l & \text{for} \quad (x, y) \in D_l \\ 0 & \text{for} \quad (x, y) \in D_\varphi \end{cases}, \quad \text{where} \quad l \in \{1, 2, \ldots, s\} \tag{2.6}$$

The result of image segmentation is the decomposition of the scene $D$ into regions of objects $D_k$ and the background $D_\varphi$. In most practical applications, image segmentation algorithms divide the scene $D$ into two regions, namely: the object of interest $D_1$ ($k = 1$) and the background $D_\varphi$. In this case, the result of image segmentation is a binary image in which the value 0 is assigned to the pixels belonging to the background, and the value 1 is assigned to the pixels belonging to the object (or vice versa). Depending on the application, the result of image segmentation can be significantly different.

In order to meet the demands of different applications and effectively handle the segmentation of images characterized by different properties, numerous approaches to image segmentation have hitherto been proposed. They are briefly characterized in the following subsection.

## 2.3 Taxonomy of Approaches to Image Segmentation

Over the last forty years, numerous distinctly different approaches dedicated to the problem of image segmentation have been proposed in the literature (Weszka, 1978; Fu and Mui, 1981; Zenzo, 1983; Haralick and Shapiro, 1985; Sahoo et al., 1988; Jiang et al., 1993; Pal and Pal, 1993; Zhang, 1997; Pham et al., 200; Shi and Malik, 2001; Boykov and Jolly, 2001; Freixenet et al., 2002; Strzecha, 2002; Bieniecki et al., 2003; Muñoz et al., 2003; Felzenszwalb and Huttenlocher, 2004; Sezgin and Sankur, 2004; Yoo, 2004; Huart and Bertolino, 2005; Pantofaru and Hebert, 2005; Chang et al., 2006; Grady, 2006; Zhang, 2006; Maška et al., 2007; Viitaniemi and Laaksonen, 2008; Fabijańska, 2011; Ilea and Whelan, 2011; Mobahi et al., 2011; Gupta et al., 2012). Their increasing diversity and interpenetration of the ideas used significantly impedes the clear classification of the existing solutions to the problem considered. However, in general, the existing approaches to image segmentation can be qualified into one of the following groups:

- pixel based methods;
- edge based methods;
- region based methods;
- model based methods;
- graph based methods;

The general taxonomy of existing approaches to image segmentation is shown in Figure 2.1.

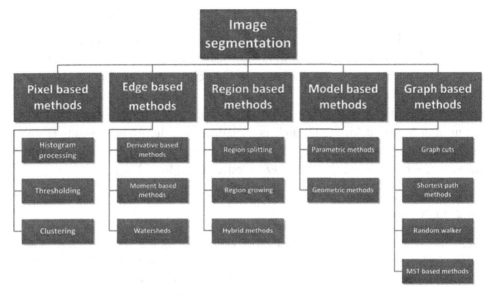

Fig. 2.1 A general taxonomy of approaches to image segmentation.

### 2.3.1 *Pixel based methods*

Pixel based methods perform image segmentation only with regard to attributes of the individual pixels, disregarding their relationship with the region under segmentation and properties of the background. These methods include mainly image thresholding and image clustering. Image thresholding is usually dedicated for images where pixels are described by a single attribute (e.g. intensity). In this case, the classification of pixels to the object and the background is performed by comparing the regarded attribute of each pixel with some threshold.

Pixels with the attribute higher than the threshold are included into the object. The remaining pixels are included into the background. This process is often referred to as binarization. The threshold can be set globally for the entire image (global thresholding), or locally – depending on the local characteristics of the image. In addition, the threshold can be static (static thresholding) or change adaptively, depending on image properties (Ridler and Calvard, 1978; Rosenfeld, 1983; Boukharouba et al., 1985; Kittler and Illingworth, 1986; Whatmough, 1991; Sahasrabudhe and Gupta, 1992; Li and Lee, 1993; Shanbag, 1994; Fan et al., 1996; Carlotto, 1997; Li and Tam, 1998; Ramar et al., 2000; Sezgin and Sankur, 2004; Fabijańska, 2009; Tajima and Kato, 2011). In the case of images in which the individual pixels are described by a set of features (e.g. color components), thresholding can be generalized to multi-dimensional thresholding. However, in this case clustering is usually applied for image segmentation. The clustering algorithms assign pixels to clusters corresponding to the object and the background. The resulting clusters are as homogeneous as possible (with regard to the attributes considered) and simultaneously the highest difference between the clusters is ensured. There are plenty image clustering algorithms. However, the most popular include: the c-means algorithm (MacQueen, 1967; Duda and Hart, 1973; Hartigan and Wong, 1979; Duda et al., 2000; Ostrovsky et al., 2006; Arthur and Vassilvitskii, 2007) and its variations e.g. the fuzzy c-means (Dunn, 1973; Bezdek, 1981; Bezdek et al., 1999; Wu and Yang, 2002) and the mean-shift approach (Comaniciu, Meer, 1997; Comaniciu, Meer, 2002).

### 2.3.2 *Edge based methods*

Edge based methods are the basic approaches to image segmentation. They localize objects by detecting their edges, i.e. pixels located on the boundary between the background and objects. These are usually accompanied with abrupt changes in image intensity (Davis, 1975; Torre and Poggio, 1984; Haddon and Boyce, 1990; Ziou and Tabbone, 1998; Sankowski et al., 2006; Gonzalez and Woods 2008; Pratt, 2007; Fabijańska, Sankowski, 2008c; Nadernejad et al., 2008; Senthilkumaran and Rajesh, 2009; Al-Kofahi et al., 2010; Petrou and Petrou, 2010; Fabijańska, 2011). According to the information used during boundary detection, edge-based approaches to image segmentation can be divided into two main groups:

- derivative-based methods;
- moment-based methods;

Image derivative based approaches approximate image derivatives using special masks. In particular, gradient methods (e.g. Sobel operator or Canny edge detector) utilize the image first derivative (Roberts, 1965; Prewitt, 1970; Frei and Chen, 1977; Canny, 1986), while zero-crossing methods (e.g. LoG) use the image second derivative (Marr, 1976; Marr and Hildreth, 1980; Haralick, 1984). In the case of gradient methods the edge position is defined by local extremes of the image first derivative. Most commonly they utilize local directional maxima of the gradient magnitude. Zero-crossing approaches locate edges with respect to zero-crossing in the image second derivative. The behavior of image derivatives at the edge is shown in Figure 2.2. The first derivative (gradient) is in the maximum in the center of the edge while the second derivative crosses zero.

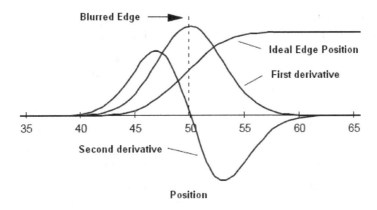

Fig. 2.2 Behavior of image derivatives at the edge.

Image derivative based methods are fast in locating edges. However, their accuracy is low since they provide wide edges, specifically in the case of blurred images. Moreover, they often fail when detecting edges in low contrast images.

Moment-based approaches to edge detection utilize image intensity moments or spatial moments to determine edge position. In particular, moments are expressed in terms of edge parameters (i.e. edge location, edge direction, background intensity, edge contrast) and then solved to determine edge position. Geometrical moments (Machuca and Gilbert, 1981; Lyvers et al., 1989), Zernike moments (Ghosal and Mehrotra, 1993, 1994) or orthogonal Fourier-Mellin moments (Bin et al., 2008) are utilized by the most popular moment-based approaches to edge detection. Moment based approaches are computationally complex but they are characterized by high precision of edge location. However, they often fail in the case of significantly blurred edges.

### 2.3.3 *Region based methods*

Region based approaches perform image segmentation by the separation of coherent regions $D_i$, $i = 1, 2, ..., n$, homogenous in terms of a given feature or a set of features (e.g. color, texture) where $D = \sum D_i$ and $D_i \cap D_j = \emptyset$ for $i \neq j$ (Brice and Fennema, 1970; Yachida and Tsuji, 1971; Feldman and Yakimovsky, 1974; Adams and Bischof, 1994; Haralick and Shapiro, 1985; 1994; Gonzalez and Woods, 2008; Petrou and Petrou, 2010). These methods can be divided into two main groups:

- agglomerative (bottom-up) approaches;
- divisible (top-down) approaches;

The representative of agglomerative methods is region growing approaches. Most commonly, these methods start from a seed point and iteratively expand the region by comparing all the unlabeled neighboring pixels to the pixels already qualified into the output regions. The pixel is assigned to the region containing the most similar pixels. This process continues until all the pixels are qualified into regions (Adams and Bischof, 1994; Mehnert and Jackway, 1997; Hijjatoleslami and Kittler, 1998; Revol-Muller et al., 2002; Wan and Higgins, 2003; Fan et al., 2005; Shih and Cheng, 2005; Gonzalez and Woods, 2008).

Divisible algorithms are represented by region splitting approaches, which divide the image into a set of disjoint regions. The splitting is usually based on quadtree image decomposition which starts at the root representing an input image. Then the image is divided recursively into four parts as long as the resulting regions are not coherent within themselves (Ohlander et al., 1979; Gevers and Smeulders, 1997; Gonzalez and Woods, 2008; Xiong et al., 2010; Gupta et al., 2012).

There are also hybrid solutions, which alternately split and merge regions within a digital image (Horowitz and Pavlidis, 1974; Yakimovsky and 1974; Horowitz and Pavlidis, 1976; Haralick and Shapiro, 1985; Beveridge et al., 1989; Chang and Li, 1994; Tremeau and Borel, 1997; Kelkar and Gupta, 2008).

The practical realization of the region based methods depends to a large extent on the data structure used to represent the image. However, in all the cases, the growth and division of the regions is performed with regard to some measure of similarity. In the simplest case the measure takes into account the statistics of pixel intensity within the region. However, there are also measures which regard region homogeneity, gradient or the presence of the repeating structural patterns (textons).

### 2.3.4 *Model based methods*

Model based approaches to image segmentation describe boundaries of an object using a mathematical deformable model which adapts its shape to the geometry of the object under segmentation. In this case, the model should be understood as an elastic contour (curve or plane) which, under the influence of the internal and the external forces, can change its shape and location. The source of internal forces is the shape of the contour while external forces come from contour location in the image (Terzopoulos, 1986). The internal forces smooth the contour while the external forces pull the contour to the object boundaries.

There are two main groups of model based approaches:

- parametric active contours;
- geometric active contours;

Parametric methods define the contour explicitly in the form of a parametric curve, plane or nodal contour points. The fitting of the contour into the boundary is performed by the iterative minimization of the energy associated with the contour.

By fitting the contour into structures within the image the boundary of the object of interest is found. The most popular parametric solutions include the snake, balloon, dual active contour and gradient vector flow (Kass et al., 1988; Cohen, 1989; Cohen and Cohen, 1993; Gunn and Nixon, 1994; Durikovic et al., 1995; McInerney and Terzopoulos, 1995; Gunn and Nixon, 1997; Xu and Prince, 1998; Piotrowski and Szczepaniak, 2000; Zimmer, Olivo-Marin, 2005; Tomczyk and Szczepaniak, 2008).

In contrast to the parametric approaches, the geometric methods represent the contour implicitly – in the form of a function. They are implemented based on the curve evolution theory and the level set theory. These allow more flexibility in contour deformation and evolution. Additionally, the contour can change both its shape and topology. This eliminates the necessity of the explicit topology changes, as in the case of parametric solutions (Osher and Sethian, 1988; Alvarez et al., 1993; Caselles et al., 1993, 1997; Sapiro and Tannenbaum, 1993; Malladi et al., 1995; Sethian, 1999; Osher and Fedkiw, 2001; Osher and Fedkiw, 2003).

### 2.3.5 *Graph based methods*

Graph based approaches represent a digital image as a weighted graph $G = (V, E)$ where $V$ stands for a set of nodes and $E$ stands for a set of edges. In particular, in the graph $G$ every pixel is represented by node $v_i \in V$ and every pair of neighboring nodes $v_i$ and $v_j$ is connected by an edge $e_{ij} = \{v_i, v_j\} \in E \subseteq V \times V$. Weights $w_{ij}$ assigned to edges $e_{ij}$ describe the relation between the neighboring nodes $v_i$ and $v_j$. Having the image represented by a graph $G$, image segmentation is obtained by partitioning the graph into two disjoint subgroups $G_A$ and $G_B$ where $A \cup B = V$ and $A \cap B = \emptyset$. The partitioning is performed according to some criterion and aims at the removal of edges connecting sub-graphs $A$ and $B$. This idea is explained in Figure 2.3.

*A. Fabijańska & D. Sankowski*

**G=(V,E)**

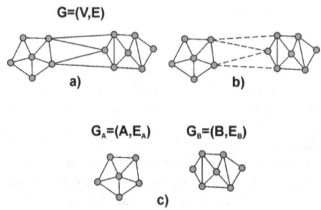

**G$_A$=(A,E$_A$)**      **G$_B$=(B,E$_B$)**

Fig. 2.3  Graph partitioning; (a) input graph; (b) edges to be removed are denoted by dashed lines; (c) disjoint sub-graphs.

The most challenging problem of graph based segmentation is finding edges to be removed. Several solutions to this problem have already been proposed. In general, they can be divided into one of the following groups:

- minimum spanning tree based methods;
- graph cut based methods;
- the shortest path based methods.

**Minimum spanning tree based methods** partition the minimal spanning tree of the graph representing an input image in order to find sub-trees whose nodes have similar properties. These sub-trees correspond with the homogeneous regions being segmented (Zahn, 1971; Urquhart, 1982, Morris et al., 1986; Kwok and Constantinides, 1997; Felzenszwalb and Huttenlocher, 2004).

**Graph-cut based methods** try to get the optimal bipartition of the graph representation of an image. The total weight of edges removed during graph partitioning is called a cut and is defined by Equation (2.7).

$$cut(A, B) = \sum_{u \in A, v \in B} w(u, v) \qquad (2.7)$$

According to the graph theory, graph bipartition is optimal when the cut is minimal. In this case the resulting subsets of nodes should contain the nodes the most different from each other while maintaining the highest possible homogeneity within each of the subsets. Recently, the most representative graph-cut methods are: **spectral graph partitioning** (Wang and Siskind, 2001; Hagen and Kahng, 1992; Cox et al., 1996; Wang and Siskind, 2003) and **combinatorial graph cuts** (Boykov and Jolly, 2001; Boykov, Kolmogorov, 2004). The first group of methods uses the eigenvectors of the graph Laplacian to find the optimal partition of the graph (Shi and Malik, 2000). This however requires solving a generalized eigenvalue problem, which is very time consuming (NP-complexity) and renders the method useless in practical applications. Combinatorial graph cuts perform segmentation by solving the min-cut/max-flow problem (Boykov and Jolly, 2001; Boykov, Kolmogorov, 2004). The method represents an image as a graph where a set of nodes contains two additional terminal nodes, the source $S$ and the sink $T$ which represent the foreground and the background, respectively. Each pixel has up to four edges (*n-links*) to the neighboring pixels and two terminal edges (*t-links*). Weights define edge capacities. In particular, weights assigned to *n-links* describe similarities between the nodes and weights assigned to *t-links* define the individual penalties for assigning a pixel to the object and the background. Image segmentation is next defined by the edges which get saturated when the maximum flow is sent from the source $S$ to the sink $T$. This method requires the user to impose conditions on the object and the background, respectively. However, it is also fast and efficient and as a result often applied in many applications.

**Shortest path methods** are dedicated to the graphs where the edge weights describe the costs of connecting incident nodes. These methods utilize the fact that in the graph representation of an image the shortest path is usually defined by edges connecting consecutive pixels which belong to the borders. In this case the shortest path is the path with the lowest total cost equal to the sum of the weights assigned to all the edges in the path. Most commonly, shortest path methods require the user to indicate pixels which belong to the border. The complete outline of the object is then defined as the shortest path between the indicated edge pixels and the shortest path between the first and last indicated pixel over all the indicated pixels (Barrett and Mortensen, 1995; Barret and Mortensen, 1996; Mortensen and Barret, 1998; Falcão et al., 1998). For the shortest path determination, most commonly Dijkstra's algorithm is used (Dijkstra, 1959; Cormen et al., 2009).

## 2.4  Image Segmentation in Thermo-Wet System

Well established methods for image segmentation often fail when applied to images from industrial measurement systems. This happens, for example, in the case of a Thermo-Wet system for high-temperature measurements of surface properties. The system determines the contact angles and surface tension of metals and alloys based on images of their molten, heat-emitting specimens. The considered surface properties are estimated from characteristic dimensions that are obtained on the basis of the specimen edges. However, due to specific measurement conditions (high temperature, intense thermal radiation, the application of protective gasses etc.) the specimen edge looks blurry and unsharp, which causes problems during image segmentation.

### 2.4.1 *The measurement process*

The system Thermo-Wet regarded in this paper determines surface properties of metals and alloys based on images of their molten specimens. In particular, image processing and analysis algorithms are applied to measure the surface tension and contact angles of the material under investigation. Measurements are performed at temperatures up to 1800°C and consider images presenting heat-emitting drops of the molten material. Exemplary images obtained from Thermo-Wet system are presented in Figure 2.4. In particular, Figure 2.4a presents a specimen of silver at 1134°C, in Figure 2.4b a specimen of steel at 1399°C can be seen and Figure 2.4c presents a specimen of gold at 1199°C.

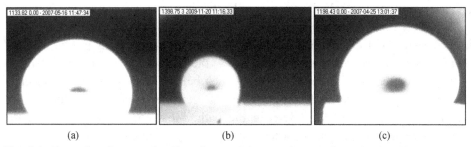

| (a) | (b) | (c) |

Fig. 2.4 Exemplary images of molten, heat-emitting specimens of metals; (a) silver, 1134°C; (b) steel, 1399°C; (c) gold, 1199°C.

The vision system considered determines surface properties of metals and alloys using the sessile drop method. The method uses characteristic dimensions of a drop of molten material to obtain values of surface tension and contact angles. Specifically, dimensions $X$, $Y$, $H$ (Fig. 2.5) are related to surface tension through Equation (2.8).

$$\gamma = g\Delta\rho\alpha^2 \qquad\qquad (2.8)$$

where:

$\gamma$       - the surface tension;

$g$       - the gravity acceleration;

$\Delta\rho$       - the gradient of densities;

$\alpha^2$       - the parameter determined using Equation (2.9).

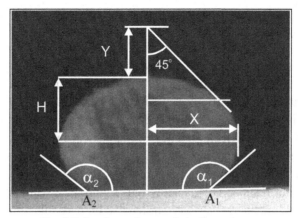

Fig. 2.5 Specimen characteristic dimensions.

$$\frac{\alpha^2}{X^2} = \left(\frac{H}{X}\right)^2 - 0{,}660\left(\frac{H}{X}\right)^3\left[1 - 4{,}05\left(\frac{H}{X}\right)^2\right]$$ (2.9)

Contact angles $\alpha_1$ and $\alpha_2$ are defined as angles between the upper edge of the base plate and tangents to the specimen profile in the contact of three phases $A_1$ and $A_2$ (see Fig. 2.5).

More detailed information on the measurement system considered can be found in the following papers and conference reports (Sankowski et al., 2000, 2001; Strzecha, 2002; Jeżewski, 2006; Fabijańska, 2007; Fabijańska and Sankowski, 2008ab, 2009ab; Koszmider, 2009; Strzecha et al., 2010, Koszmider et al., 2011).

### 2.4.2 *Problem definition*

Looking at Equation (2.9) it can be easily concluded that the more precisely the specimen shape is defined, the more accurate values of surface parameters are obtained. However, the proper determination of the specimen shape is hindered by an *aura*, i.e. glow which, due to high temperatures, forms around the specimen and blurs the boundary between the specimen and the background. As the boundary line is supposed to be located somewhere in the *aura* region, many approaches of image segmentation fail to locate the border with appropriate accuracy. In particular, they either join the *aura* with the object, which increases its dimensions or completely removes the *aura*, which in turn results in a decrease in the object dimensions.

### 2.4.3 *Selection of image segmentation strategy*

Images obtained from the Thermo-Wet system contain bright objects with homogenous surfaces seen on the contrasting background (see Fig. 2.4). Therefore, it would be desirable to perform image segmentation using the pixel based approaches, in particular, image thresholding. The key problem in thresholding is the choice of the threshold. There are a number of different methods for threshold selection. They use statistical information in the image as the basis for the threshold determination. However, in the case of the images under consideration, due to the *aura* presence, different thresholding approaches produce different results. This is shown in Figure 2.6, which presents results of global thresholding using various strategies. In particular, Ostu (Otsu, 1979), MaxEntropy (Kapur et al., 1985), Intermodes (Prewitt and Mendelsohn, 1966), IsoData (Ridler and Calvard, 1978), Minimum (Prewitt and Mendelsohn, 1966), Percentile (Doyle, 1962), Yen (Ten et al., 1995), Triangle (Zack et al., 1977) and Shanbhang (Shanbhag, 1994) algorithms were applied. Different segmentation results obtained by global thresholding cause the method to be insufficient as it conditions accuracy of surface properties by a threshold selection procedure.

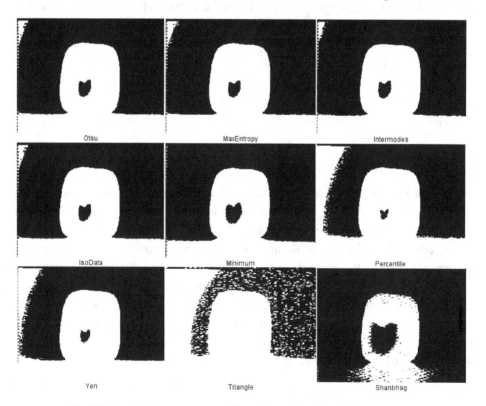

Fig. 2.6 Result of image segmentation using various thresholding approaches.

In order to overcome problems with global thresholding, an adaptive technique was developed for the segmentation of the class of images considered. In particular, a *"divide and conquer"* approach was proposed for image segmentation. The method works recursively by breaking down the problem of image segmentation into sub-problems of the same type, until these become sufficiently simple to be solved directly.

In particular, the image being segmented is divided into four smaller and equal sub-images for which segmentation is performed. The idea of splitting of the image into smaller regions during segmentation is presented illustratively in Figure 2.7.

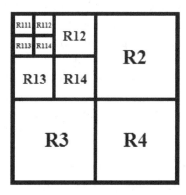

Fig. 2.7 The idea of recursive splitting of the image into the smaller regions.

The algorithm introduced starts by dividing the image into four equal regions. For each region the standard deviation of the gray level (contrast) is computed. If it is higher than the global standard deviation (i.e. computed for the whole image), the current region is split into for equal sub-regions for which the procedure of local and global contrast comparison is repeated. Otherwise, the current sub-region of the image is binarized with the threshold determined based on image local properties using the IsoData algorithm.

The pseudocode of the proposed algorithm is given below.

```
1. Compute standard deviation for the image
   (global stdev)
2.    IF (image dimensions large enough) THEN
   split image into four sub-images of equal size
3.    FOR EACH sub-image:
      a.    compute standard deviation (local
         stdev)
      b.    IF(local stdev>global stdev)
         GOTO step 2
```

> **ELSE** perform ISODATA thresholding for
> current sub-image.

In the authors' implementation of the algorithm proposed, regions are split into smaller ones as long as all the dimensions of the current sub-region are bigger than 32 pixels. Otherwise, the algorithm leads to over-segmentation.

The results of the application of the proposed adaptive segmentation algorithm to exemplary images from the Thermo-Wet vision system are presented in Figure 2.8. Moreover, a comparison with the results of global IsoData thresholding is given. In particular, the first row presents original images (from left to right: steel, 1311°C; gold, 761°C; steel, 798°C; copper, 948°C).

The results presented in Figure 2.8 clearly show that the proposed adaptive approach to image segmentation provides high quality results. Objects after segmentation have well defined, regular borders which are free from artifacts. Moreover, the specimen shape is properly defined – details of the specimen contour are extracted with high accuracy. The regions after segmentation accurately match the shape of the original objects. The results obtained by the proposed adaptive approach are more accurate than those obtained using the global IsoData thresholding. The new algorithm much more effectively segments regions characterized by very non-uniform background illumination. The differences in image segmentation qualities can especially be observed in the case of images presenting objects emitting very intense thermal radiation. Moreover, the *aura*, which could not be properly segmented by the global thresholding was effectively separated from the object by the proposed method.

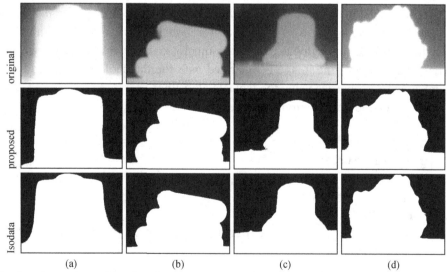

Fig. 2.8 Comparison of results of image thresholding obtained using the proposed method with results of global IsoData thresholding.

Another attempt was to perform image segmentation using edge based approaches. However, proper segmentation was again hindered by the aura effect. More precisely, the border provided by classical edge detectors has a width of several pixels, which is insufficiently accurate from the point of view of determination of surface properties. The effect is presented in Figure 2.9, which shows the magnified edge of an exemplary specimen provided by the Laplace operator (Fig. 2.9a) and the Sobel edge detector (Fig. 2.9b).

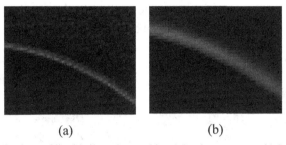

(a)                                                      (b)

Fig. 2.9  Edge of a drop of liquid silver detected by a) Laplace operator b) Roberts operator.

The results of thresholding edges provided by the Sobel operator are presented in Figure 2.10. In particular, the top panel presents input images (from left to right: Palladium 1550°C, Glass 770°C, Gold 1066°C, Silver 1073°C) while the corresponding images after edge thresholding are shown in the bottom panel.

From Figure 2.10 it can easily be seen that the resulting edges do not match the specimen shape sufficiently accurate for further quantitative analysis. The specimen profile is incomplete and not continuous, the upper edge of the base plate is not determined properly. Moreover, the border of the specimen profile is rather wide, whcih hinders proper edge localization. The profile round-offs at the base (particularly important from the point of view of determination of wetting angles) are strongly deformed.

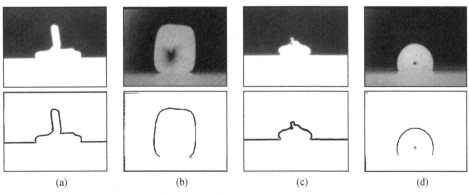

(a)                  (b)                  (c)                  (d)

Fig. 2.10  Results of edge detection in exemplary images; a)  Palladium 1550°C; b) Glass 770°C; c) Gold 1066°C; d) Silver 1073°C.

*A. Fabijańska & D. Sankowski*

Additionally, when applied to the considered class of images the traditional edge detection algorithms exhibit sensitivity to distortions and no uniform intensity distribution. These effects are presented in Figure 2.11.

Since the results of image segmentation using gradient operators are not satisfactory, a new approach to edge detection was developed. The proposed algorithm works in three stages. In the first stage the Sobel operator is applied for preliminary segmentation. The use of the gradient operator allows determining regions free from edges. An approximate outline of the specimen is found. The region of the further analysis is narrowed down.

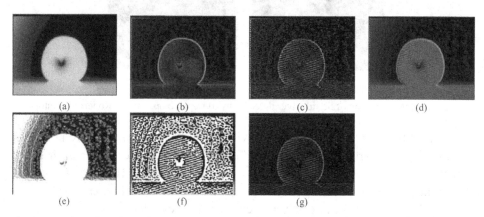

Fig. 2.11 Edges detection using gradient operators; a) original image (Glass, 860°C); b) Sobel's masks ; c) Laplace operator; d) Krish's masks; e) Frei-Chen's masks; f) Marr-Hildreth's masks; g) operator LoG.

In the next step of the algorithm, regions supposed to contain edges are subjected to proper segmentation. In this stage local thresholding with recursive threshold selection (with IsoData algorithm) is used. The results of the second stage of the algorithm (applied to the images presented in Fig. 2.11), are presented in Fig. 2.12. It can easily be seen that the quality of the specimen shape determination is much higher than in the case of the traditional approach.

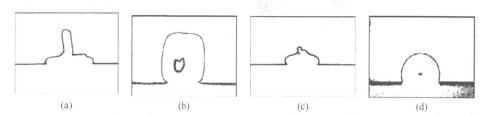

Fig. 2.12 Results of thresholding effects of preliminary segmentation carried out using gradient operator (Sobel operator).

In the last step of the algorithm a decision about border line qualification is taken. For the images characterized by the average intensity higher than the half of the intensity range, the border (which is connected with lightness value differences) is included in the background. The decision is based on the assumption that for such a class of images, intensity distribution is homogenous in the object region only. In the case of the average intensity value lower than the half of the intensity range, the algorithm qualifies the border to the object, assuming uniform intensity distribution in the background.

The final results of the algorithm described are shown in Fig. 2.12. For better presentation of the results, colors were applied. As can be seen, all the presented resultant images obtained from the segmentation process contain smooth, free of artefacts, boundaries. They provide good estimates of the contours of objects in the original image. The upper edge of the base-plate is clear and well defined. The obtained images can then be successfully used for the further quantity image analysis algorithms.

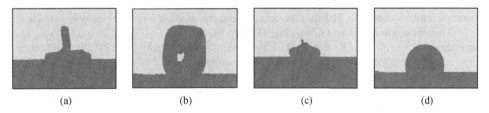

    (a)                (b)                (c)                (d)

Fig. 2.12 Final segmentation results (with grey scale applied for better results presentation).

## 2.5 Conclusions

As shown in this chapter, image segmentation is a very challenging task. Although there are a variety of image segmentation algorithms, in many practical applications of machine vision the selection and the development of the image segmentation algorithm often causes serious difficulties.

In the case of the system for high temperature measurements of surface properties of metals and alloys considered in our work, the main problem is the uncertainty of image segmentation results. Owing to specific properties of images obtained from the system (in particular the *aura* effect), different segmentation algorithms produce different results. Unfortunately, due to lack of ground truth images, it is impossible to determine which algorithm provides the most accurate results and the assessment of image segmentation quality can only be performed visually.

42                        A. *Fabijańska & D. Sankowski*

Having in mind the properties of the images under consideration for the segmentation of heat-emitting specimens, threshold-based and edge-based algorithms were used. However, in their original form they failed. Therefore, improvements to these approaches were proposed. As a result, the subjective accuracy of image segmentation was improved. In particular, the edges after segmentation were smooth and provided good estimates of the specimen contours.

## References

Adams, R. and Bischof, L. (1994). Seeded region growing. *IEEE T. Pattern Anal.,* 16(6), pp. 641-647

Al-Kofahi, Y., Lassoued, W., Lee, W. and Roysam B. (2010). Improved automatic detection and segmentation of cell nuclei in histopathology images. *IEEE T. Bio-Med. Eng.*, 57(4), pp. 841-852

Alvarez, L., Guichard, F., Lions, P. L. and Morel J. M. (1993). Axioms and fundamental equations of image processing. *Archive for Rational Mechanics and Analysis*, 123(3), pp. 199-257

Arthur, D. and Vassilvitskii S. (2007). K-means++: The advantages of careful seeding. *Proc. 18th Annual ACM-SIAM Symposium on Discrete Algorithms*, pp. 1027-1035

Batchelor, B. and Whelan, P. (2002). Intelligent Vision Systems for Industry. Springer, Germany.

Barret, W. and Mortensen, E. (1996). Fast, accurate and reproducible live-wire boundary extraction. *Visualization in Biomedical Computing*, 1131, pp.183-192

Beveridge, J. R., Griffith, J., Kohler, R. R., Hanson, A. R. and Riseman, E. M. (1989). Segmenting images using localizing histograms and region merging. *Int. J. of Comput. Vision*, 2(3), pp. 311-347

Bieniecki, W., Grabowski, S., Sekulska, J., Turant, M. and Kałużyński, A. (2003). Automatic segmentation and recognition of pathomorphological microscopic images. *Proc. the 7th Int. Conf. : The Experience of Designing and Application of CAD Systems in Microelectronics*, pp. 461-464

Bin, T. J., Lei A., Jiwen, C., Wenjing, K. and Dandan, L. (2008). Subpixel edge location based on orthogonal Fourier-Mellin moments. *Image Vision Comput.*, 26(4), pp. 563-569

Boukharouba, S., Rebordao, J. M. and Wendel, P. L. (1985). An amplitude segmentation method based on the distribution function of an image. *Graph. Model. Im. Proc.*, 29, pp. 47-59

Boykov, Y. and Jolly, M.-P. (2001). Interactive graph cuts for optimal boundary & region segmentation of objects in N-D images. *Proc. Int. Conf. on Comput. Vision*, pp. 105-112

Brice, C. R. and Fennema, C. L. (1970). *Scene analysis using regions*. Artificial Intelligence, 1(3-4), pp. 205-226

Carlotto, M. J., (1997). Histogram analysis using a scale-space approach. *IEEE T. Pattern Anal.*, 9, pp. 121-129

Canny, J. (1986). A computational approach to edge detection. *IEEE T. Pattern Anal.*, 8, pp. 679-698

Caselles, V., Catte, F., Coll, T. and Dibos, F. (1993). A geometric model for active contours. *Numerische Mathematik*, 66, pp. 1-31

Caselles, V., Kimmel, R., and Sapiro, G. (1997). Geodesic active contours. *Int. J. of Comput. Vision*, 22(1), pp. 61-79

Chang, C. I., Du, Y., Wang, J., Guo S. M. and Thouin P. D. (2006). Survey and comparative analysis of entropy and relative entropy thresholding techniques. *IEE Proceedings-Vision, Image and Signal Processing*, 153(6), pp. 837-850

Chang, Y. L. and Li, X. (1994). Adaptive image region-growing. *IEEE T. Image Process.*, 3(6), pp. 868-872

Cohen, L. D., (1989). On active contour models and balloons. *Computer Graphics and Image Processing*, 53(2), pp. 211-218

Cohen, L. D. and Cohen, I. (1993). Finite element methods for active contour models and balloons for 2d and 3d images. *IEEE T. Pattern Anal.*, 15(11), pp. 1131-1147

Comaniciu, D. and Meer, P. (1997). Robust analysis of feature spaces: color image segmentation. *Proc. IEEE Conf. on Comput. Vision and Pattern Recogn.*, pp. 750-755

Comaniciu, D. and Meer, P. (2002). Mean shift: a robust approach toward feature space analysis. *IEEE T. Pattern Anal.*, 24(5), pp. 603-619

Cormen, T. H., Leiserson, C. E., Rivest, R. L. and Stein, C. (2009). Introduction to Algorithms. MIT Press

Cox, I. J., Rao, S. B. and Zhong, Y. (1996). Ratio regions: a technique for image segmentation. *Proc. the Int. Conf. on Pattern Recogn.*, pp. 557-564

Davis, L. S. (1975). A survey of edge detection techniques. *Computer Graphics and Image Processing*, 4, pp. 248-270

Dijkstra, E. W. (1959). A note on two problems in connexion with graphs. *Numerische Mathematik*, 1, pp. 269-271

Doyle, W. (1962). Operation useful for similarity-invariant Pattern Recogn., *J. of the Association for Computing Machinery*, 9, pp. 259-267

Duda, R. O. and Hart P. E. (1973). Pattern Classification and Scene Analysis. Wiley

Duda, R. O., Hart P. E. and Stork D. G. (2000). Pattern Classification. Wiley-Interscience

Durikovic, R., Kaneda, K. and Yamashita, H. (1995). Dynamic contour: a texture approach and contour operations. *The Visual Computer*, 11, pp. 277-289

Fabijańska, A. (2007). Image enhancement algorithms for high temperature measurements of surface properties of metals and alloys. PhD Thesis, Lodz University of Technology, Poland

Fabijańska, A. and Sankowski, D. (2008a). Algorithm of optical filter self-acting change for high temperature applications of vision systems. *Proc. IEEE Int. Conf. on Signals and Electronic Systems*, pp. 363-366

Fabijańska, A. and Sankowski, D. (2008b). Preprocessing of images obtained from high temperature vision system. *Proc. IEEE Int. Workshop on Imaging Systems and Techniques*, pp. 204-207

Fabijańska, A. and Sankowski, D. (2008c). Edge detection in brain images. *Proc. IEEE 4th Int. Conf. on Perspective Technologies and Methods in MEMS Design*, pp. 60-62

Fabijańska, A. and Sankowski, D. (2009a). Comput. Vision system for high temperature measurements of surface properties. *Mach. Vision Appl.*, 20(6), pp. 411-421

Fabijańska, A. and Sankowski, D. (2009b). Improvement of image quality of high-temperature vision system. *Meas. Sci. Technol.*, 20, 104018, 9 pp.

Fabijańska, A. (2009c). The recursive approach to image segmentation, *Proc. IEEE 5th Int. Conf. Perspective Technologies and Methods in MEMS Design*, pp. 53-55

Fabijańska, A. (2011). Yarn image segmentation using the region growing algorithm. *Meas. Sci. Technol.*, 22, 114024, 9pp

Feldman, J. and Yakimovsky, Y. (1974). Decision theory and artificial intelligence: I. A semantics based region analyzer. *Artif. Int.*, 5(4), pp. 349-371

Felzenszwalb, P. F. and Huttenlocher, D. P. (2004). Efficient graph based image segmentation. *Int. J. Comput. Vision*, 59(2), pp. 167-181

Frei, W. and Chen, C. C. (1977). Fast boundary detection: a generalization and a new algorithm. *IEEE T. Comput.*, 26(10), pp. 988-998

Freixenet, J., Muñoz X., Raba, D., Martí, J. and Cufí, X. (2002). Yet another survey on image segmentation: region and boundary information integration. *Lect. Notes Comput. Sc.*, 2352, pp. 21-25

Fu, K. S. and Mui, J. K. (1981). A survey on image segmentation. *Pattern Recogn.*, 13(1), pp. 3-16

Gevers, T. and Smeulders, A. W. M. (1997). Combining region splitting and edge detection through guided delaunay image subdivision. *Proc. the IEEE Conf. on Comput. Vision and Pattern Recogn.*, pp. 1021-1026

Ghosal, S. and Mehrotra, R. (1993). Orthogonal moment operators for subpixel edge detection. *Pattern Recogn. Lett.*, 26(2), pp. 295-305

Ghosal, S. and Mehrotra, R. (1994). Zernike moment-based feature detectors. *Proc. Int. IEEE Conf. Image Processing*, 1, pp. 934-938.

Gocławski, J., Sekulska-Nalewajko, J., Gajewska, E. and Wielanek M. (2009). An automatic segmentation method for scanned images of wheat root systems with dark discolourations. *Int. J. Appl. Math. Comp.*, 19 (4), pp. 679-689

Gonzalez, R. C. and Woods, R. E. (2008). Digital Image Processing, Pearson-Prentice Hall, USA

Grady, L. (2006). Random walks for image segmentation. *IEEE T. Pattern Anal.*, 28(11), pp. 1768-1783

Gunn, S. R. and Nixon M. S. (1994). A model based dual active contour. *Proc. British Mach. Vis. Conf.*, pp. 305-314

Gunn, S. R. and Nixon, M. S. (1997). A robust snake implementation: a dual active contour. *IEEE T. Pattern Anal.*, 19(1), pp. 63-68

Gupta, G., Psarrou, A. and Angelopoulou, A. (2012). Image segmentation based on semi-greedy region merging. *Proc. IET Im. Proc. Conf.*, pp.1-4

Falcão, A. X., Udupa, J. K., Samarasekera, S., Sharma, S., Elliot B. H. and de A. Lotufo R. (1998). User-steered image segmentation paradigms: live wire and live lane. *Graph. Model. Im. Proc.*, 60, pp. 233-260

Fan, J., Wang, R., Zhang, L., Xing, D. and Gan, F. (1996). Image sequence segmentation based on 2D temporal entropy. *Pattern Recogn. Lett.*, 17, pp. 1101-1107

Fan, J., Zeng, G., Body, M. and Hacid, M. S. (2005). Seeded region growing: an extensive and comparative study. *Pattern Recogn. Lett.*, 26(8), pp. 1139-1156

Haddon, J. F. and Boyce, J. F. (1990). Image segmentation by unifying region and boundary information. *IEEE T. Pattern Anal.*, 12(10), pp. 929-948

Hagen, L. and Kahng, A. (1992). New spectral methods for ratio cut partitioning and clustering. *IEEE T. Comput. Aid. D.*, 11(9), 1074-1085

Haralick, R. M. and Shapiro L. G., (1985). Image segmentation techniques. *Comput. Vis. Grap. Im. Proc.*, 29(1), pp. 100-132

Hartigan, J. A. and Wong, M. A. (1979). A K-means clustering algorithm. *Appl. Stat.*, 28(1), pp. 100-108

Hijjatoleslami, S. A. and Kittler, J., (1998). Region growing: a new approach. *IEEE T. Image Process.*, 7(7), pp. 1079-1084

Horowitz, S. L. and Pavlidis, T. (1974). Picture segmentation by a directed split and merge procedure. *Proc. Int. Conf. on Pattern Recogn.*, pp. 424-433

Horowitz, S. L. and Pavlidis, T. (1976). Picture segmentation by a tree traversal algorithm. *J. ACM*, 23, pp. 368-388

Huart, J. and Bertolino, P. (2005). Similarity-based and perception-based image segmentation. *Proc. IEEE Int. Conf. Image Processing*, pp. 1148-1151

Ilea, D. E. and Whelan, P. F. (2011). Image segmentation based on the integration of colour-texture descriptors-a review. *Pattern Recogn.*, 44(10-11), pp. 2479-2501

Jähne, B. (2012). Digital Image Processing. Springer

Jeżewski, S., (2006). High temperature lighting model in issues of image processing of specimens in contact of solid-liquid phases. PhD Thesis. AGH University of Science and Technology, Cracow, Poland

Jiang, H., Toriwaki, J. and Suzuki, H. (1993). Comparative performance evaluation of segmentation methods based on region growing and division. *Systems and Computers in Japan*, 24(13), pp. 28-42

Kapur, J. N., Sahoo, P. K. and Wong, A. K. C. (1985). A new method for gray-level picture thresholding using the entropy of the histogram. *Graph. Model. Im. Proc.*, 29, pp. 273-285

Kass, M., Witkin, A. and Terzopoulos D. (1988). Snakes: active contour models. *Int. J. of Comput. Vision*, 1(4), pp. 321-331

Kittler, J. and Illingworth, J. (1986). Minimum error thresholding. *Pattern Recogn.*, 19, pp. 41-47

Kelkar, D. and Gupta, S. (2008). Improved quadtree method for split merge image segmentation. *Proc. First Int. Conf. on Emerging Trends in Engineering and Technology*, pp. 44-47

Koszmider, T. (2009). The integrated computer system for determination of geometrical parameters of specimens of metals and alloys in high temperatures. PhD Thesis, Lodz University of Technology, Lodz, Poland

Koszmider, T., Strzecha, K., Fabijańska, A. and Bąkała M. (2011). Algorithm for accurate determination of contact angles in vision system for high temperature measurements of metals and alloys surface properties. *Computer Recognition Systems 4, Advances in Intelligent and Soft Computing*, 95, pp. 441-448

Kwok, S. H. and Constantinides, A. G. (1997). A fast recursive shortest spanning tree for image segmentation and edge detection. *IEEE T. Image Process.*, 6(2), pp. 328-332

Li, C. H. and Lee, C. K. (1993). Minimum cross-entropy thresholding. *Pattern Recogn.*, 26, pp. 617-625

Li, C. H. and Tam, P. K. S. (1998). An iterative algorithm for minimum cross-entropy thresholding. *Pattern Recogn. Lett.*, 19, pp. 771-776

Lyvers, E. P., Mitchell, O. R., Akey, M.L. and Reeves A. P. (1989). Subpixel measurements using a moment-based-edge operator. *IEEE T. Pattern Anal.*, 11(12), pp. 1293-1309

Machuca, R. and Gilbert, A. L.(1981). Finding edges in noisy scenes. *IEEE T. Pattern Anal.*, 3, pp. 103-111

MacQueen, J. B. (1967). Some methods for classification and analysis of multivariate observations. *Proc. 5th Berkeley Symposium on Mathematical Statistics and Probability*. University of California Press, pp. 281-297

Malladi, R., Sethian, J. A. and Vemuri, B. C. (1995). Shape modeling with front propagation: a level set approach. *IEEE T. Pattern Anal.*, 17(2), pp. 158-175

Maška, M., Hubený, J., Svoboda, D. and Kozubek M., (2007). A comparison of fast level set-like algorithms for image segmentation in fluorescence microscopy. *Lect. Notes Comput. Sc.*, 4842, pp. 571-581

Materka, A. (1991). Elements of Digital Image Processing and Analysis. PWN, Poland (in Polish).

McInerney, T. and Terzopoulos D. (1995). Topologically adaptable snakes. *Proc. Int. Conf. on Comput. Vision*, pp. 840-845

Mehnert, A. and Jackway, P. (1997). An improved seeded region growing algorithm. *Pattern Recogn. Lett.*, 18(10), pp. 1065-1071

Mobahi, H., Rao, S., Yang, A., Sastry, S. and Ma, Y. (2011). Segmentation of natural images by texture and boundary compression. *Int. J. of Comput. Vision*, 95(1), pp. 86-98

Morris, O. J., Lee, M. J. and Constantinides, A. G., (1986). Graph theory for image analysis: an approach based on the shortest spanning tree. *IEE Proceedings-F. Communications Radar & Signal Processing*, 133, pp. 146-152

Mortensen, E. N. and Barrett, W. A. (1995). Intelligent scissors for image composition. *Proc. the 22nd Annual Conf. on Computer Graphics and Interactive Techniques*, pp. 191-198

Mortensen, E. N. and Barrett, W. A. (1998). Interactive segmentation with intelligent scissors. *Graph. Model. Im. Proc.*, 60, pp. 349-384

Muñoz, X., Freixenet, J., Cufí, X. and Martí, J., (2003). Strategies for image segmentation combining region and boundary information. *Pattern Recogn. Lett.*, 24,(1-3), pp. 375-392

Nadernejad, E., Sharifzadeh, S. and Hassanpour, H. (2008). Edge detection techniques: evaluations and comparisons. *Appl. Math. Sci.*, 2(31), pp. 1507-1520

Obinata, G. and Dutta, A. (2007). Vision Systems: Applications. ITech Education and Publishing, Vienna, Austria

Ohlander, R., Price, K. and Reddy, D. R. (1979). Picture segmentation using a recursive region splitting method. *Comput. Grap. Im. Process.*, 8(3), pp. 313-333

Osher, S. and Sethian, J. (1988). Fronts propagating with curvature dependent speed: algorithms based on the hamilton-jacobi formulation. *J. Comput. Phy.*, 79, pp. 12-49

Osher, S. and Fedkiw, R. (2001). Level *set methods: an overview and some recent results. J. Comput. Phy.*, 169, pp. 463-502

Osher, S. and Fedkiw, R. (2003). Level Set Methods and Dynamic Implicit Surfaces. Springer, New York, USA

Ostrovsky, R., Rabani, Y., Schulman, L. and Swamy, C. (2006). The effectiveness of Lloyd-type methods for the k-means problem. *Proc. 47th IEEE Symp. Foundations of Comput. Sci.*, pp. 165-174

Otsu, N. (1979). A threshold selection method from gray-level histograms. *IEEE T. Syst. Man Cyb.*, 9, pp. 62-66

Pal, N. R. and Pal, S. K. (1993). A review on image segmentation techniques. *Pattern Recogn.*, 26(9), pp. 1277-1294

Pantofaru C., Hebert M., (2005). A Comparison of Image Segmentation Algorithms. tech. report CMU-RI-TR-05-40, Robotics Institute, Carnegie Mellon University

Petrou, M. and Petrou, C. (2010). Image Processing. The Fundamentals. Willey, USA

Pham D. L., Xu C., Prince J., (2000). Current Methods in Medical Image Segmentation. Annual Review of Biomedical Engineering, 2, pp. 315-337

Piotrowski M., Szczepaniak P. S., (2000). Active Contour Based Segmentation of Low-Contrast Medical Images. *Proc. 1st Int. Conf. on Advances in Medical Signal and Information Processing*, pp. 104-109

Pratt, W. K. (2007). Digital Image Processing. Wiley-Interscience

Prewitt, J. M. S. and Mendelsohn, M. L. (1966). The analysis of cell images. Annals of the New York Academy of Sciences, 128, pp. 1035-1053

Prewitt J., (1970). Object Enhancement and Extraction. Lipkin B., Rosenfeld A. (Eds). Picture Processing and Psychopictorics, New York: Academic, pp. 75-149

Putiatin E., Averin S., (1990). Przetwarzanie Obrazów w Robotyce. Mashinostrojenie (Обработка изображений в робототехнике. Машиностроение) (w języku rosyjskim)

Ramar K., Arunigam S., Sivanandam S. N., Ganesan L., Manimegalai D., (2000). Quantitative Fuzzy Measures for Threshold Selection. Pattern Recogn. Lett., 21, pp. 1-7

Ranky, P. (2003). Advanced Machine Vision Systems and Application Examples, Sensor Review, 23(3), pp. 242–245

Revol-Muller C., Peyrin F., Carrillon Y., Odet C., (2002). Automated 3D Region Growing Algorithm Based on an Assessment Function. Pattern Recogn. Lett., 23 (1-3), pp. 137-150

Ridler T. W., Calvard S., (1978). Picture Thresholding Using an Iterative Selection Method. IEEE T. Syst. Man Cyb., SMC-8, pp. 630-632

Roberts L. G., (1965). Machine Perception of Three-Dimensional Solids. Tippett J. T., et al., (Eds). Optical and Electro-Optical Information Processing, Cambridge, MA: MIT Press

Rosenfeld A., De la Torre P., (1983). Histogram Concavity Analysis as an Aid in Threshold Selection. IEEE T. Syst. Man Cyb., 13, pp. 231-235

Sahasrabudhe S. C., Gupta K. S. D., (1992). A Valley-Seeking Threshold Selection Technique. Comput. Vision and Image Understanding, 56, pp. 55-65

Sahoo P. K., Soltani S., Wong A. K. C., Chen Y. C., (1988). A Survey of Thresholding Techniques. Comput. Vision Graphics and Image Processing, 41(2), pp. 233-260

Sankowski D., Strzecha K., Jeżewski S., (2000). Digital Image Analysis in Measurement of Surface Tension and Wettability Angle. Proc. IEEE Int. Conf. on Modern Problems in Telecommunication, Computer Science and Engineers Training, pp. 129-130

Sankowski D., Senkara J., Strzecha K., Jeżewski S., (2001). Automatic Investigation of Surface Phenomena in High Temperature Solid and Liquid Contacts. Proc. 18th IEEE Instrumentation and Measurement Technology Conf., pp. 346-249

Sankowski D., Strzecha K., Fabijańska A., (2006). Edge Detection Algorithm-The New Approach. SiS'(2006, XIV Konferencja Sieci i Systemy Informatyczne, pp. 195-197

Sankur B., Sezgin M., (2004). A Survey Over Image Thresholding Techniques and Quantitative Performance Evaluation. J. of Electronic Imaging, 13(1), pp. 146-165

Sapiro G., Tannenbaum A., (1993). Affine Invariant Scale-Space. *Int. J. Comput. Vision*, 11(1), pp. 25-44

Senthilkumaran, N. and Rajesh R. (2009). Edge detection techniques for image segmentation-a survey of soft computing approaches. *Int. J. of Recent Trends in Engineering*, 1(2), pp. 250-254

Sethian, J. A. (1999). Level set methods and fast marching methods. Cambridge University Press, UK

Shanbag, A. G. (1994). Utilization of information measure as a means of image thresholding. *Comput. Vis. Graph. Im. Proc.*, 56, pp. 414-419

Shi, J. and Malik J. (2000). Normalized cuts and image segmentation. *IEEE T. Pattern Anal.*, 22(8), pp. 888-905

Shih, F.Y. and Cheng S. (2005). Automatic seeded region growing for color image segmentation. *Image Vision Comput.*, 23(10), pp. 877-886

Steger, C., Ulrich, M. and Wiedemann, C. (2008) Machine vision algorithms and applications, Wiley-VCH, Denmark

Strzecha, K. (2002). Application of image processing and analysis in high temperature measurements of surface properties of selected materials. PhD Thesis, Lodz University of Technology, Lodz, Poland

Strzecha, K., Bąkała M., Fabijańska A. and Koszmider T. (2010b): The evolution of Thermo-Wet-the computerized system for high temperature measurements of surface properties. *Automatics Scientific Bulletin of AGH Academy of Science and Technology*, 14(3.1), pp. 525-535

Tadeusiewicz, R. and Korohoda P. (1997). Computer image processing and analysis. Wydawnictwo Fundacji Postępu Telekomunikacji, Cracow, Poland (in Polish)

Tajimaa, R. and Kato Y. (2011). Comparison of threshold algorithms for automatic image processing of rice roots using freeware image. *J. Field Crops Research*, 121(3), pp. 460-463

Terzopoulos, D. (1986). On matching deformable models to images. Technical Report 60, Schlumberger Palo Alto Research, November

Tomczyk, A. and Szczepaniak P. S. (2008). Active contour segmentation of disjoint objects applied to medical images. *J. of Medical Informatics & Technologies*, 12, pp. 163-168

Torre, V. and Poggio, T. (1984). On edge detection. *IEEE T. Pattern Anal.*, 8(2), pp. 147-163

Tremeau, A. and Borel N. (1997). A region growing and merging algorithm to color segmentation. *Pattern Recogn.*, 30(7), pp. 1191-1203

Urquhart, R. (1982). Graph theoretical clustering based on limited neighborhood sets. *Pattern Recogn.*, 15(3), pp. 173-187

Viitaniemi, V. and Laaksonen J. T. (2008). Techniques for image classification, object detection and object segmentation. *Proc. the 10th Int. Conf. on Visual Information Systems: Web-Based Visual Information Search and Management*, pp. 231-234

Wan, S. Y. and Higgins, W. E. (2003). Symmetric region growing. *IEEE T. Image Process.*, 12(9), pp.1007-1015

Wang, S. and Siskind J. M. (2001). Image segmentation with minimum mean cut. *Proc. Int. Conf. on Comput. Vision*, 1, pp. 517-524

Wang, S. and Siskind, J. M. (2003). Image segmentation with ratio cut, *IEEE T. Pattern Anal.*, 25(6), pp. 675-690

Weszka, J. S. (1978). A Survey of threshold selection techniques. *Comp. Graph. Im. Proc.*, 2, pp. 259-265

Whatmough, R. J. (1991). Automatic Threshold selection from a histogram using the exponential hull. *Graph. Model. Im. Proc.*, 53, pp. 592-600

Wiatr, K. (2003). Acceleration of calculations in vision systems. WNT, Warsaw, Poland (in Polish)

Xiong, W., Ong, S.H. and Lim, J.H. (2010). A recursive and model-constrained region splitting algorithm for cell clump decomposition. *Proc. Int. Conf. Pattern Recogn.*, pp. 4416-4419

Xu, C. and Prince, J. L. (1998). Snakes, shapes and gradient vector flow. *IEEE T. Image Process.*, 7(3), pp. 359-369

Yakimovsky, Y. (1974). Boundary and object detection in real world images. *Proc. IEEE Conf. on Decision and Control including the 13th Symposium on Adaptive Processes*, pp. 460-467

Yachida, M. and Tsuji, S. (1971). Application of color information to visual perception. *Pattern Recogn.*, 3(3), pp. 307-318

Yen, J. C., Chang, F. J. and Chang, S (1995), A new criterion for automatic multilevel thresholding, *IEEE Trans. Image Process.* 4(3), pp. 370-378

Yoo, S. T. (2004). Insight into images: principles and practice for segmentation, registration, and image analysis. A K Peters, Ltd.

Zack, G.W., Rogers, W.E. and Latt, S.A. (1977): Automatic measurement of sister chromatid exchange frequency, *J. Histochem. Cytochem.* 25(7), pp. 741–53

Zahn, C. T. (1971). Graph-theoretic methods for detecting and describing gestalt clusters. *IEEE T. Comput.*, 20, pp. 68-86

Zenzo, S. D. (1983). Advances in image segmentation. *Image Vision Comput.*, 1(4), pp. 196-210

Zhang, Y. J. (1997). Evaluation and comparison of different segmentation algorithms. *Pattern Recogn. Lett.*, 18(10), pp. 963-974

Zhang, Y. J. (2006). Advances in image and video segmentation. IRM Press

Zimmer, C. and Olivo-Marin, J.C. (2005). Coupled parametric active contours. *IEEE T. Pattern Anal.*, 27(11), pp. 1838-1842

Ziou, D. and Tabbone, S. (1998). Edge detection techniques: an overview. *Int. J. Pattern Recogn. Image Anal.*, 8(4), pp. 537-559

# CHAPTER 3

# LINE FRACTIONAL-ORDER DIFFERENCE/SUM, ITS PROPERTIES AND AN APPLICATION IN IMAGE PROCESSING

Piotr Ostalczyk

*Institute of Applied Computer Science*
*Lodz University of Technology*
*90-924 Lodz, ul. Stefanowskiego 18/22*
*piotr.ostalczyk@p.lodz.pl*

In this Chapter the fractional-order line difference/sum (FOLD/S) is defined. The main properties of the FOLD/S are derived. The mentioned mathematical notions are illustrated by numerous numerical examples. The proposed mathematical tool - FOLD/S may be useful in edge detection in image processing.

## 3.1 Introduction

The ever-increasing requirements for digital one- and two-dimensional signal processing quality are forcing the enhancement of commonly used techniques and the invention of new methods. Among the fundamental processing methods for analyzing digital images, treated as two-dimensional signals which may be described as two-dimensional scalar fields, the image edge detection and image noise rejection can be mentioned (Bai and Feng, 2007; Chen *et al.*, 2012; Chen and Fei, 2012; Cuesta *et al.*, 2012; Gan and Yang, 2010; Gao *et al.*, 2011b,a; Ghamisi *et al.*, 2012; Hu *et al.*, 2011b,a; Jalab and Ibrahim, 2012; Oustaloup *et al.*, 1991a,b; Pu *et al.*, 2010; Wang *et al.*, 2013). Widely known and commonly used edge detection methods, such as the Sobel, Prewitt, Canny, Laplacian (Watkins *et al.*, 1993) and Zero-Cross (Tadmor and Zou, 2008), now seem to be unsatisfactory due to the performance of image transforming computer processors. One of the solutions of the problem mentioned is the application of a new mathematical tool known as the Fractional Calculus.

For over forty years the Fractional Calculus has been a subject of growing interest (Kaczorek, 2011; Miller and Ross, 1993; Oldham and Spanier, 1974; Podlubny, 1999; Samko *et al.*, 1993). This is caused by its successful application in many scientific and technical fields. Mathematical modeling of real-time physical processes (Carpinteri and Mainardi, 1997) by linear or non-linear, time variant or time invariant fractional-order differential equations can be mentioned here. The interest ranges from physics and chemistry by technical as mechanical and electrical to economical and biological processes. analysis and control strategies synthesis.

The Fractional Calculus has its discrete counterpart where the fractional-order left derivative is replaced by a fractional-order backward difference (FOBD), whereas the fractional-order integral by a fractional-order sum (FOBS), respectively. There are four equivalent forms of the FOVD/S: Grünwald-Letnikov, Horner, Riemann-Liouville and Caputo (Lubich, 1986; Oustaloup, 1995). Nowadays in all the mentioned areas of the Fractional Calculus applications the digital images are used for storing, supplying and processing information.

The Chapter is organized as follows. In Sec. 3.2 some basic definitions concerning the fractional-order backward differences and sums are given. Section Sec. 3.3 contains a definition of the fractional-order line backward difference and sum (FOLBD/S). It is supported by a numerical example. Then, in Section Sec. 3.4 the FPLBD/S of two discrete-variable functions are considered. The last Section is devoted to the FOLBD/S application to edge detection.

### 3.1.1 *Notation*

Integer numbers will be denoted as $k,l,i,j,n,m$ and a set of integers will be denoted as $\mathbb{Z}$ whereas their subset containing all non-negative integers as $\mathbb{Z}_+$. Similarly, real numbers and non-negative real numbers will be denoted by $f,g,x,y$ with appropriate sets by $\mathbb{R}$ and $\mathbb{R}_+$, respectively. A finite elements subset of $\mathbb{Z}_+$ and $\mathbb{Z}_+ \times \mathbb{Z}_+$, where $\times$ denotes the Cartesian product, will be further denoted as $\mathbb{Z}_+ \supseteq K_{k_{a,b}} = \{k_a, k_{a+1} \cdots, k_{b-1}, k_b\}$ and $\mathbb{Z}_+ \times \mathbb{Z}_+ \supseteq K_{k_{1,a,b}, k_{2,a,b}} = \{(k_{1,a}, k_{2,a}), (k_{1,a}, k_{2,a+1}), \cdots, (k_{1,b}, k_{2,b})\}$.

A one discrete-variable $k$ function $y$ such that $y : \mathbb{Z}_+ \to \mathbb{R}_+$ using additional conditions $y(k) = 0$ for $k < k_a (k_a \geq 0)$ and $k > k_b$ is defined. The values related to the discrete variable set $K_{k_a, k_b}$ will be further collected in a column vector

$$a_n = \begin{bmatrix} y_{k_a} & y_{k_a+1} & \cdots & y_{k_b-1} & y_{k_b} \end{bmatrix}^T \tag{3.1}$$

where $T$ denotes transposition.

Next, two discrete-variable $k_1, k_2$ function $Y : \mathbb{Z}_+ \times \mathbb{Z}_+ \to \mathbb{R}$ described by a equation $z = y(k_1, k_2)$ is defined. A condition, forced by practical applications, that $y(k_1, k_2) = 0$ for $k_1 < k_{1,a}, k_2 < k_{2,a} (k_{1,a}, k_{1,2} \geq 0)$ and $k_1 > k_{1,b}, k_2 > k_{2,b}$ is also imposed. The values of interest related to the set $K_{k_{1,a,b}, k_{2,a,b}}$ will be further collected in the matrix

$$\mathbf{Z} = \mathbf{Y}_{k_1, k_2} \begin{bmatrix} k_{1a}k_{2a} & \cdots & k_{1a}k_{2b} \\ \vdots & \ddots & \vdots \\ k_{1b}k_{2a} & \cdots & k_{1b}k_{2b} \end{bmatrix} \tag{3.2}$$

Function (3.1) values can be treated as consecutive values of the discrete-time signal, for instance, voltage, measured speech signal data, whereas function (3.2) can represent a discrete image or measured scalar two dimensional temperature field. It can be visualized by a surface defined by a matrix (3.2) over the set $K_{k_{1,a,b},k_{2,a,b}}$.

On functions defined by equations (3.1) and (3.2) such operations as addition and multiplication by a real scalar can be performed. To function (3.1) such operations as scalar multiplications, discrete differentiation and integration can also be applied, which means evaluating the $n$-th order difference and the $n$-fold sum. In this article the main subject operations are the fractional-order differentiation and integration. To distinguish them from the commonly known differentiation and summation of integer orders, the fractional operation orders will further be denoted by Greek letters $\nu, \mu \in \mathbb{R}$ (contrary to their integer counterparts denoted usually by $n, m \in \mathbb{Z}$).

## 3.2  Fractional-order backward difference/sum

The Grünwald-Letnikov fractional-order $\nu \in \mathbb{R}$ backward-difference/sum (FOBD/S) is defined as a finite sum (Ostalczyk, 2000)

$$\begin{matrix} GL \\ k_a \end{matrix} \Delta_k^{(\nu)} f(k) = \sum_{i=k_a}^{k} a_{i-k_a}^{(\nu)} f_{k+k_a-i} \tag{3.3}$$

where $[k_a, k]$ defines the FOBD/S evaluation range $k \leq k_b$ and

$$a_i^{(\nu)} = \begin{cases} 0 & \text{for} \quad i = -1, -2, -3, \cdots \\ 1 & \text{for} \quad i = 0 \\ (-1)^i \frac{\nu(\nu-1)\cdots(\nu-i+1)}{i!} & \text{for} \quad i = 1, 2, 3, \cdots \end{cases} \tag{3.4}$$

For positive orders eq. (3.4) defines fractional-order backward difference (FOBD) and for negative orders fractional-order backward sum (FOBS). One can easily prove that for $0 < \nu \leq 1$:

$$\sum_{i=0}^{\infty} a_i^{(\nu)} = 0 \tag{3.5}$$

and

$$\sum_{i=1}^{\infty} a_i^{(\nu)} = 1 \tag{3.6}$$

Moreover, the operator (3.3) is linear

$$\begin{matrix} GL \\ k_a \end{matrix} \Delta_k^{(\nu)} [af(k)] = a \begin{matrix} GL \\ k_a \end{matrix} \Delta_k^{(\nu)} f(k) \quad \text{for} \quad a \in \mathbb{R} \tag{3.7}$$

and additive

$$\begin{matrix} GL \\ k_a \end{matrix} \Delta_k^{(\nu)} [f(k) + g(k)] = \begin{matrix} GL \\ k_a \end{matrix} \Delta_k^{(\nu)} f(k) + \begin{matrix} GL \\ k_a \end{matrix} \Delta_k^{(\nu)} g(k) \tag{3.8}$$

It should be noted that equation (3.3) is valid for integer orders $\nu = n \in \mathbb{Z}$ and defines commonly known backward differences and sums.

$$\,_{k_a}^{GL}\Delta_k^{(1)}f(k) = \,_{k-1}^{GL}\Delta_k^{(1)}f(k) = \Delta f(k) \tag{3.9}$$

$$\,_{k_a}^{GL}\Delta_k^{(2)}f(k) = \,_{k-2}^{GL}\Delta_k^{(2)}f(k) = f(k) - 2f(k-1) + f(k-2) \tag{3.10}$$

$$\,_{k_a}^{GL}\Delta_k^{(-1)}f(k) = \sum_{i=k_a}^{k} f(i) \tag{3.11}$$

$$\,_{k_a}^{GL}\Delta_k^{(n)}f(k) = \sum_{i=k_a}^{k} a_{i-k_a}^{(n)} f_{k+k_a-i} \tag{3.12}$$

Without the loss of generality in image processing one can assume that $k_a = 0$. Here one should note that the FOBD as well as the FOBS evaluates over an interval $[k_a, k]$ of growing length.

## 3.3   Fractional-order line backward difference/sum

In this Section a generalization of the FOBD/S is presented.

### 3.3.1   *Scalar fields and a two-dimesnional curve*

For the set $\mathbb{Z}_+ \times \mathbb{Z}_+$ or $K_{k_{1,a,b}, k_{2,a,b}} \subseteq \mathbb{Z}_+ \times \mathbb{Z}_+$ which can describe a scalar field one can define a curve C as a set of points $(k_1, k_2) \in K_{k_{1,a,b}, k_{2,a,b}} \subseteq \mathbb{Z}_+ \times \mathbb{Z}_+$ satisfying the condition

$$\mathbf{C} = \{(k_1, k_2) \in K_{k_{1,a,b}, k_{2,a,b}} : c(k_1, k_2) = 0\} \tag{3.13}$$

The curve $C$ can be parameterized by any bijective parameterization

$$\mathbf{c} : K_{k_{a,b}} \to \mathbf{C} \tag{3.14}$$

such that $c(k_a)$ and $c(k_b)$ are the endpoints of $C$ for $0 \leq k_a < k_b$. Hence eq. (3.13) can be expressed as

$$\mathbf{C} = \left\{ (k_1, k_2) \in K_{k_{1,a,b}, k_{2,a,b}} : \begin{bmatrix} k_1(k) \\ k_2(k) \end{bmatrix} = \begin{bmatrix} c_1(k) \\ c_2(k) \end{bmatrix} = \mathbf{c}(k), k \in K_{k_{a,b}} \right\} \tag{3.15}$$

The notions defined above are illustrated by the following numerical example, which will be used in further investigations.

**Example 3.1:**

One considers a scalar field described by the two discrete-variable function

$$Y(k_1, k_2) = 10 sin(k_1) cos(k_2) e^{-0.0025\sqrt{k_1^2 + k_2^2}} \tag{3.16}$$

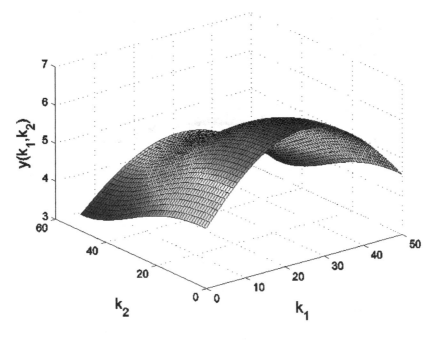

Fig. 3.1 The surface defined by Eq. (3.16).

for $K_{k_{1,a,b},k_{2,a,b}}$ with $a = 0, b = 50$. The surface defined by eq. (3.16) is presented in Fig. 3.1. Next one defines a curve on the two dimensional plane

$$\mathbf{C} = \left\{ (k_1, k_2) \in K_{k_{1,a,b},k_{2,a,b}} : (k_1 - k_{1,c})^2 + (k_2 - k_{2,c})^2 - \left( \frac{r_c}{h} \right)^2 = 0 \right\} \quad (3.17)$$

where $a = 0, b = 50$, $k_{1,c} = k_{2,c} = 0.5$, $r_c = 0.25$, $h = \pi r_c / a$. The curve (3.17) can also be presented in the parameterized form

$$\mathbf{C} = \left\{ (k_1, k_2) \in K_{k_{1,a,b},k_{2,a,b}} : \begin{bmatrix} k_1(k) \\ k_2(k) \end{bmatrix} = \begin{bmatrix} \frac{k_{1c} + r_c sin(kh)}{h} \\ \frac{k_{2c} + r_c sin(kh)}{h} \end{bmatrix}, k \in K_{k_{a,b}} \right\} \quad (3.18)$$

In Fig. 3.2 two curves are presented. The first one (2D) is defined by equation (3.18) whereas the second one is described by the function $Y[k_1(k), k_2(k)]$ for $k \in K_{k_{a,b}}$. This means that from surface (3.16) a 3D curve is extracted, related to variables $k_1(k), k_2(k)$.

Below, in Fig. 3.3 a discrete curve (3.18) (in black) and an orthogonal projection of the surface (3.16) on a $k_1, k_2$ plane with indicated constant value lines (with colours scale related to appropriate values due to) are presented.

The curve $C$ will further be treated as the domain of a discrete differentiation/integration.

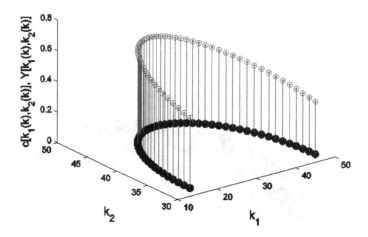

Fig. 3.2   Curves defined by Eq. (3.18) (black) and $Y[k_1(k), k_2(k)]$ (gray o).

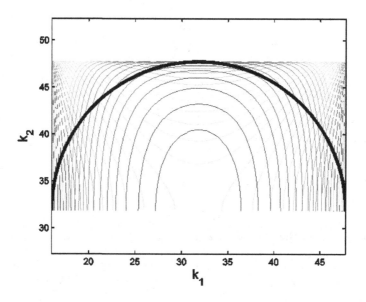

Fig. 3.3   Curve defined by Eq. (3.16) (black) and a constant values line of $Y[k_1(k), k_2(k)]$.

### 3.3.2   *Fractional-order line backward difference/sum definition*

The fractional-order $\nu \in \mathbb{R}$ line backward-difference/sum (FOLBD/S) of two discrete-variable function (Ostalczyk, 2001) is defined as a finite sum

$$
{}^{L}_{k_a}\Delta^{(\nu)}_{k_b} y[\mathbf{c}(k)]
$$

$$
= \sum_{i=k_a}^{k_b} a^{(\nu)}_{i-k_a} y_{c_1(k+k_a-i),c_2(k+k_a-i)} |\mathbf{c}(k+k_a-i) - \mathbf{c}(k+k_a-i-1)|^{\nu}
$$

(3.19)

where $[k_a, k_b]$ defines the FOBD/S evaluation range, $c(k)$ is a parametrized curve (line, path) function (3.11) and a terms $|\mathbf{c}(k + k_a - i) - \mathbf{c}(k + k_a - i - 1)|$ should be interpreted as a distance between subsequent points on the curve (3.11). This distance can be expressed as

$$s_i = |\mathbf{c}(k + k_a - i) - \mathbf{c}(k + k_a - i - 1)|$$

$$= \sqrt{[\mathbf{c}(k + k_a - i) - \mathbf{c}(k + k_a - i - 1)]^T [\mathbf{c}(k + k_a - i) - \mathbf{c}(k + k_a - i - 1)]}$$

$$= \sqrt{[c_1(k + k_a - i) - c_1(k + k_a - i - 1)]^2 + [c_2(k + k_a - i) - c_2(k + k_a - i - 1)]^2}$$

$$= \sqrt{[\Delta c_1(k + k_a - i)]^2 + [\Delta c_2(k + k_a - i)]^2}$$

$$(3.20)$$

Since, as mentioned above, in practical applications often $k_a = 0$, henceforth it will be assumed that $k_a = 0$. For a special case, when

$$\mathbf{C} = \left\{ (k_1, k_2) \in K_{k_{1,0,b}, k_{2,0,b}} : \begin{bmatrix} k_1(k) \\ k_2(k) \end{bmatrix} = \begin{bmatrix} c_1(k) \\ 0 \end{bmatrix} = \mathbf{c}(k), k \in K_{k_{0,b}} \right\} \quad (3.21)$$

the elementary arc distance equals

$$|\mathbf{c}(k + k_a - i) - \mathbf{c}(k + k_a - i - 1)| = 1 \quad (3.22)$$

and

$$_0^L \Delta_k^{(\nu)} y[c_1(k)] = _0^L \Delta_k^{(\nu)} y(k) = _0^{GL} \Delta_k^{(\nu)} y(k) \quad (3.23)$$

The FOLBD/S evaluation procedure will be illustrated by the following numerical example.

**Example 3.2:**

For function (3.16) and curve (3.18) (considered in Example 4.1) evaluate the FOLBD and FOLBS of orders: $n = -1$, $\nu = -0.5$, $\nu = 0.5$ and $n = 1$, respectively. In Fig. 3.4 a plot of function $y[c(k)]$ vs. the parameter $k$ is presented. Here the curve (3.18) is straightened.

In Fig. 3.3 the classical line sum along curve (3.18) is plotted.

The final value (related to $k = 50$) equals to a sum of elementary fields created by rectangles characterized by $y[c(k)]$ and elementary arches $s_k$ treated as high and width, respectively. This is an area represented as a curtain created by curve (3.18) and function $y[c(k)]$ presented in Fig. 3.2. In Fig. 3.6 the FOLBS of order $\nu = -0.5$ is given. In Fig. 3.7 and Fig. 3.8 the FOLBDs of orders $\nu = 0.5$ and $n = 1$ are presented, respectively. One should realize that the plot given in Fig. 3.8 presents the classical first-order backward difference of function (3.16) evaluated over an interval $[0, 50]$. Here one can connect the position of the maximum value of the function (presented in Fig. 3.4) with the zero value of the FOLBD function.

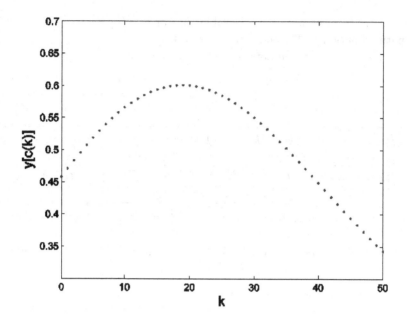

Fig. 3.4 The parametrized function $y[c(k)]$ vs. $k$.

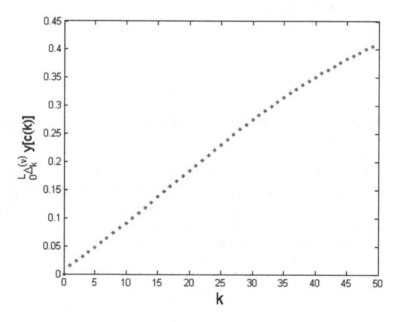

Fig. 3.5 The FOLBS of order $n = -1$ vs. $k$.

### 3.3.3 *Fractional-order line backward difference/sum properties*

The FOLBD/S fundamental properties will be presented in the form of the following theorems.

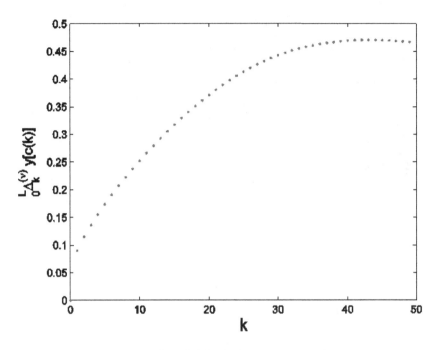

Fig. 3.6    The FOLBS of order $\nu = -0.5$ vs. $k$.

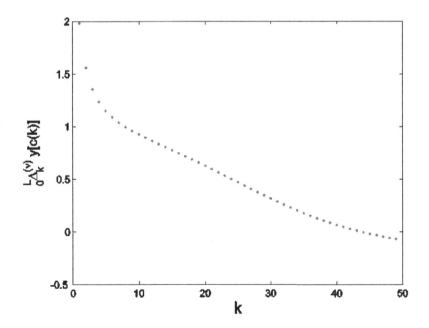

Fig. 3.7    The FOLBS of order $\nu = 0.5$ vs. $k$.

*P. Ostalczyk*

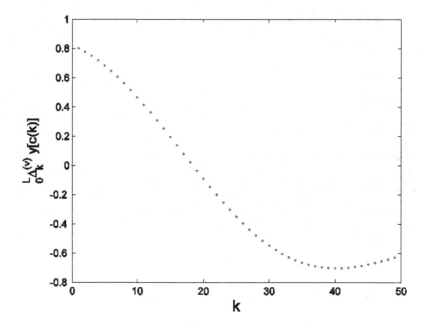

Fig. 3.8   The FOLBS of order $\nu = 1$ vs. $k$.

**Theorem 3.1.** *For a two discrete-variable function* (3.2) *defined over a curve* (3.13) *and a scalar $e \in \mathbb{R}$ one has*

$$\substack{L\\k_a}\Delta_k^{(\nu)}ey[\mathbf{c}(k)] = e\substack{L\\k_a}\Delta_k^{(\nu)}y[\mathbf{c}(k)] \tag{3.24}$$

***Proof.***   The proof is immediate by realizing the form of the FOLBD/S definition sum.   □

**Theorem 3.2.** *For a two discrete-variable functions $y$, $z$ defined by* (3.2) *over the same curve* (3.13) *one has*

$$\substack{L\\k_a}\Delta_k^{(\nu)}\{y[\mathbf{c}(k)] + z[\mathbf{c}(k)]\} = \substack{L\\k_a}\Delta_k^{(\nu)}y[\mathbf{c}(k)] + \substack{L\\k_a}\Delta_k^{(\nu)}z[\mathbf{c}(k)] \tag{3.25}$$

***Proof.***   The proof is also immediate by examining the form of the FOLBD/S definition sum.   □

**Theorem 3.3.** *In general*

$$\substack{L\\k_a}\Delta_k^{(\nu)}y[\mathbf{c}(k)] \neq \substack{L\\k_b}\Delta_k^{(\nu)}y[\mathbf{c}(k)] \tag{3.26}$$

**Proof.** According to equation (3.19) the FOLBD/S may be expressed as

$$\substack{L\\k_a}\Delta_k^{(\nu)}y[\mathbf{c}(k)]$$

$$= [a_0^{(\nu)}s_0^\nu \quad a_1^{(\nu)}s_1^\nu \quad \cdots \quad a_{k-1-k_a}^{(\nu)}s_{k-1-k_a}^\nu \quad a_{k-k_a}^{(\nu)}s_{k-k_a}^\nu] \begin{bmatrix} y_{c(k)} \\ y_{c(k-1)} \\ \vdots \\ y_{c(k_a+1)} \\ y_{c(k_a)} \end{bmatrix} \tag{3.27}$$

In general

$$\substack{L\\k}\Delta_{k_a}^{(\nu)}y[\mathbf{c}(k)]$$

$$= [a_0^{(\nu)}s_0^\nu \quad a_1^{(\nu)}s_1^\nu \quad \cdots \quad a_{k-1-k_a}^{(\nu)}s_{k-1-k_a}^\nu \quad a_{k-k_a}^{(\nu)}s_{k-k_a}^\nu] \begin{bmatrix} y_{c(k_a)} \\ y_{c(k_a+1)} \\ \vdots \\ y_{c(k-1)} \\ y_{c(k)} \end{bmatrix}. \tag{3.28}$$

$$\neq \substack{L\\k_a}\Delta_k^{(\nu)}y[\mathbf{c}(k)] \qquad \qquad \square$$

**Theorem 3.4.** *Consider two discrete-variable functions (3.2) defined over two curves (3.13) satisfying the relation*

$$\mathbf{c}_1(k_b) = \mathbf{c}_2(k_b) \tag{3.29}$$

*Then, for a new curve defined as*

$$\mathbf{c}_k = \mathbf{c}_1(k) \cup \mathbf{c}_2(k) \quad for \quad 0 \le k_a < k_b < k_c \tag{3.30}$$

*one has*

$$\substack{L\\k_a}\Delta_{k_b}^{(\nu)}y[\mathbf{c}_1(k)] + \substack{L\\k_b}\Delta_{k_c}^{(\nu)}y[\mathbf{c}_2(k)] = \substack{L\\k_a}\Delta_{k_c}^{(\nu)}y[\mathbf{c}(k)] \tag{3.31}$$

**Proof.** The proof is immediate by realizing the form of the FOLBD/S definition sum. $\square$

The notions defined above are illustrated by the following numerical example, which will be used in further investigations.

## 3.4 Fractional-order line backward difference/sum of elementary functions

In this Section the FOLBD/S of some discrete two-dimensional fundamental functions are evaluated. These are: a discrete Dirac pulse, a discrete unit step function, a linear function (a plane) and two-dimensional symmetric and anti-symmetric sine functions.

### 3.4.1  FOLBD/S of the discrete Dirac Pulse

The two discrete variable Dirac pulse function is defined as

$$\delta(k_1, k_2) = \begin{cases} 1 & \text{for} \quad k_1 = k_2 = 0 \\ 0 & \text{for} \quad k_1 \neq k_2 \neq 0 \end{cases} \tag{3.32}$$

A plot of the two discrete variable Dirac pulse is presented in Fig. 3.9.
Now it is assumed that a set $K_{k_{1,0,b}, k_{2,0,b}} \subseteq \mathbb{Z}_+ \times \mathbb{Z}_+$, over which a scalar field described by eq. (3.32) is defined, contains a curve (3.14). Moreover,

$$\begin{bmatrix} k_1(0) \\ k_2(0) \end{bmatrix} = \begin{bmatrix} c_1(0) \\ c_2(0) \end{bmatrix} = \begin{bmatrix} 0 \\ 0 \end{bmatrix} = \mathbf{c}(0) \tag{3.33}$$

The discrete curve (3.18) (in black) and the function $Y[k_1(k), k_2(k)]$ (grey o) are presented in Fig. 3.10. The FOLBD/S for $0 \leq k \leq k_b$ equals

$$_0^L\Delta_k^{(\nu)} y[\mathbf{c}(k)] = \sum_{i=0}^{k} a_i^{(\nu)} y_{c_1(k-i), c_2(k-i)} s_i^{\nu} \tag{3.34}$$

where for curve (3.18) one obtains

$$s_i = |\mathbf{c}(k-i) - \mathbf{c}(k-i-1)|$$

$$= \frac{r_c}{h} \sqrt{[sin(k-1) - sin(k-i-1)]^2 + [cos(k-1) - cos(k-i-1)]^2} \tag{3.35}$$

$$= \frac{2r_c sin\left(\frac{h}{2}\right)}{h}$$

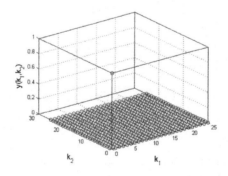

Fig. 3.9  Plot of two discrete-variable Dirac pulse function.

In Fig. 3.11 a plot of function $y[c(k)]$ vs. the parameter $k$ is presented (the curve (3.18) is straightened). Taking into account (3.32) and (3.35) the FOLBD/S finally equals

$$_0^L\Delta_k^{(\nu)} y[\mathbf{c}(k)] = a_k^{(\nu)} \frac{2r_c \sin\left(\frac{h}{2}\right)}{h} \tag{3.36}$$

Plots of function (3.36) for four different orders $\nu = -1, -0.5, 0, 5, 1$ are given in Fig. 3.12(a)(b)(c)(d).

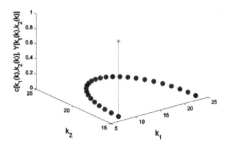

Fig. 3.10   Plot of parametrized curve (3.18) and function $Y[k_1(k), k_2(k)]$.

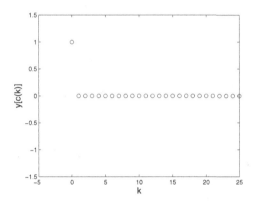

Fig. 3.11   The parametrized function $y[c(k)]$ vs. $k$.

### 3.4.2   *The FOLBD/S of the discrete unit step function*

The two discrete variable unit step function is defined as

$$\mathbf{1}(k_1, k_2) = \begin{cases} 0 & \text{for} \quad k_1 < k_2 < 0 \\ 1 & \text{for} \quad k_1 \geq k_2 \geq 0 \end{cases} \tag{3.37}$$

A plot of the two discrete variable unit step function pulse is presented in Fig. 3.13.

Now it is assumed that a set $K_{k_{1,0,b}, k_{2,0,b}} \subseteq \mathbb{Z}_+ \times \mathbb{Z}_+$, over which a scalar field described by eq. (3.32) is defined, contains a curve (3.14). The discrete curve (3.18) (in black) and the function $Y[k_1(k), k_2(k)]$ (grey o) is presented in Fig. 3.14. The FOLBD/S for $0 \leq k \leq k_b$ equals

$$_0^L\Delta_k^{(\nu)} y[\mathbf{c}(k)] = \sum_{i=0}^{k} a_i^{(\nu)} s_i^\nu = \sum_{i=0}^{k} a_i^{(\nu)} \tag{3.38}$$

with $s_i$ defined by equation (3.37). One can prove that

$$_0^L\Delta_k^{(\nu)} y[\mathbf{c}(k)] = \sum_{i=0}^{k} a_i^{(\nu)} = a_k^{(\nu+1)} \tag{3.39}$$

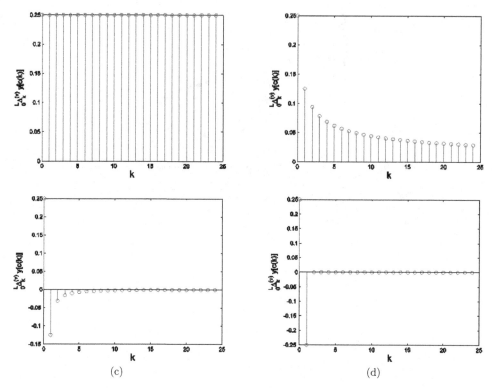

(c)                                                      (d)

Fig. 3.12   Plots of (3.36) for $\nu = -1, -0.5, 0, 5, 1$.

Plot of $Y[k_1(k), k_2(k)]$ along the curve (3.18) is presented in Fig. 3.14

In Fig. 3.15 a plot of function $y[\mathbf{c}(k)]$ vs. the parameter $k$ is presented (the curve (3.18) is straightened). The plots of function (3.37) for four different orders $\nu = -1, -0.5, 0, 5, 1$ are given in Fig. 3.16 (a)(b)(c)(d).

### 3.4.3   The FOLBD/S of the discrete linear function

The two discrete variable unit step function is defined as

$$1(k_1, k_2) = \begin{cases} 0 & \text{for} \quad k_1 < 0, k_2 < 0 \\ k_1 + k_2 & \text{for} \quad k_1 \geq 0, k_2 \geq 0 \end{cases} \tag{3.40}$$

A plot of the two discrete variable linear function is presented in Fig. 3.17.

Now it is assumed that a set $K_{k_{1,0,b}, k_{2,0,b}} \subseteq \mathbb{Z}_+ \times \mathbb{Z}_+$, over which a scalar field described by eq. (3.40) is defined, contains a curve (3.14). The discrete curve (3.18) (in black) and the function $Y[k_1(k), k_2(k)]$ (grey o) are presented in Fig. 3.18.

In Fig. 3.19 a plot of function $y[\mathbf{c}(k)]$ vs. the parameter $k$ is presented (the curve (3.18) is straightened).

The FOLBD/S evaluated numerically for $0 \leq k \leq k_b = 30$ is given in Fig. 3.20 (a), (b), (c), (d) for four different orders $\nu = -1, -0.5, 0, 5, 1$.

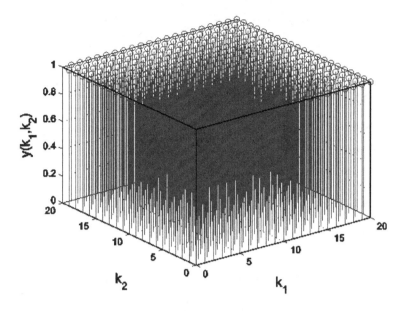

Fig. 3.13   Plot of two discrete-variable unit step function.

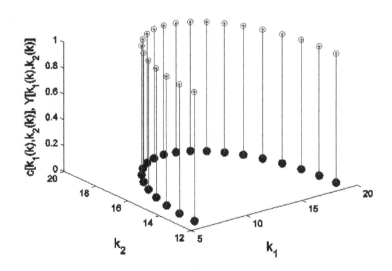

Fig. 3.14   Plot of parametrized curve (3.18) and function $Y[k_1(k), k_2(k)]$.

## 3.5   Fractional-order line backward difference/sum edge detector

The fundamental idea of edge detection in image processing is based on the calculation of the difference between adjacent pixels. This can be generalized by the

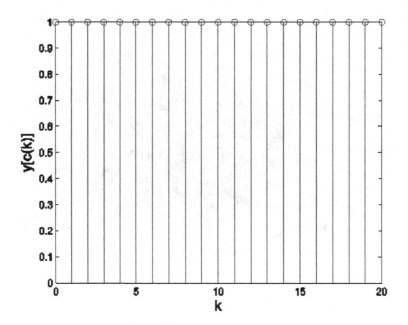

Fig. 3.15  The parametrized function $y[\mathbf{c}(k)]$ vs. $k$.

FOLBD/S application. As the fractional-order edge detector a weighted sum of the FOLBD/Ss is proposed. First, 8 lines described by sets of points are defined

$$\mathbf{c}_1(k_1, k_2, L_1) = \{(k_1, k_2) : k_1, k_1 - 1, \cdots, k_1 - L_1, k_2 = const\}$$

$$\mathbf{c}_2(k_1, k_2, L_2) = \{(k_1, k_2) : k_1 = k_2, k_1, k_1 - 1, \cdots, k_1 - L_2\}$$

$$\mathbf{c}_3(k_1, k_2, L_3) = \{(k_1, k_2) : k_1 = const, k_2, k_2 - 1, \cdots, k_1 - L_3\}$$

$$\mathbf{c}_4(k_1, k_2, L_4) = \left\{ \begin{matrix} (k_1, k_2) : & k_1, k_1 - 1, \cdots, k_1 - L_4 \\ & k_2, k_2 + 1, \cdots, k_1 + L_4 \end{matrix} \right\}$$

$$\mathbf{c}_5(k_1, k_2, L_5) = \{(k_1, k_2) : k_1, k_1 + 1, \cdots, k_1 + L_5, k_2 = const\}$$

$$\mathbf{c}_6(k_1, k_2, L_6) = \{(k_1, k_2) : k_1 = k_2, k_1, k_1 + 1, \cdots, k_1 + L_6\}$$

$$\mathbf{c}_7(k_1, k_2, L_7) = \{(k_1, k_2) : k_1 = const, k_2, k_2 + 1, \cdots, k_1 + L_7\}$$

$$\mathbf{c}_8(k_1, k_2, L_8) = \left\{ \begin{matrix} (k_1, k_2) : & k_1, k_1 + 1, \cdots, k_1 + L_8 \\ & k_2, k_2 - 1, \cdots, k_1 - L_8 \end{matrix} \right\}.$$

(3.41)

where $L_i, i = 1, 2, \cdots, 8$ denotes a line length. For the above lines the FOLBD/S is defined

$$_0^L \Delta_{L_i}^{(\nu_i)} y[\mathbf{c}_1(k)]$$

(3.42)

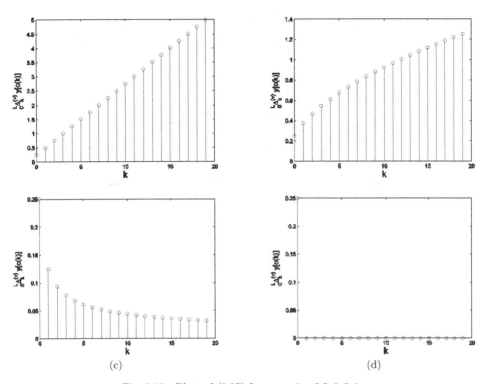

Fig. 3.16   Plots of (3.37) for $\nu = -1, -0.5, 0, 5, 1$.

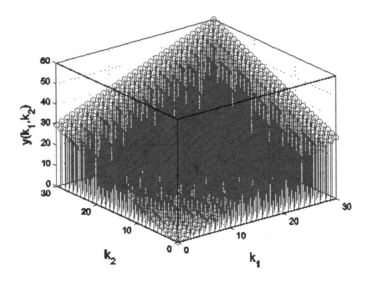

Fig. 3.17   Plot of two discrete-variable linear function.

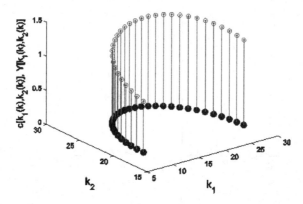

Fig. 3.18    Plot of parametrized curve (3.14) and function $Y[k_1(k), k_2(k)]$.

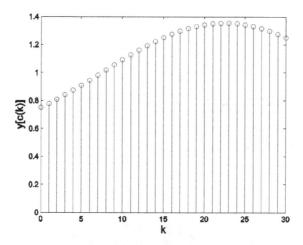

Fig. 3.19 · The parametrized function $y[\mathbf{c}(k)]$ vs. $k$.

where $y[\mathbf{c}_i(k)]$ is a two dimensional function and $\nu_i$ is a fractional or integer order (positive for differentiation and negative for integration). This function may be defined by equation (3.2) and may represent a discrete image. Now a fractional-order edge detector is defined as follows

$$p(k_1, k_2) = \frac{1}{8} \sum_{i=1}^{8} w_{i0}^{L} \Delta_{L_i}^{(\nu_i)} y[\mathbf{c}_1(k)] \tag{3.43}$$

where $w_i$ denote appropriate weight.


## 3.6    Conclusions

It can be shown that the fractional-order edge detector based on the proposed line fractional-order backward difference is a generalization of commonly used masks.

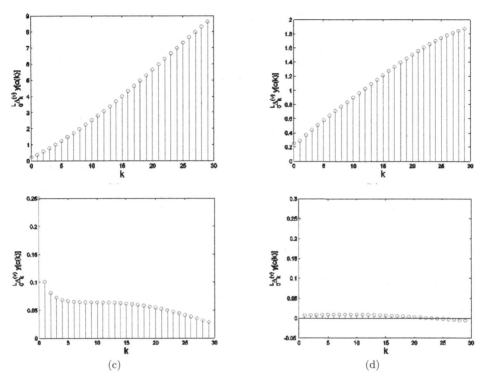

Fig. 3.20 Plots of the FOLBD/Ss for $\nu = -1, -0.5, 0, 5, 1$ of function (3.40).

Pixel value evaluated by equation (3.43) depends on eight coefficients $w_i$, lengths $L_i$ and orders $\nu_i$ what significantly increases the possibilities of mask shape forming. At the same time one must admit that the image processing with the use of equation (3.43) takes more time in a real-time processing. In the face of rapidly increasing computing performance processors, this shortcoming becomes irrelevant. For equal lengths and orders equation (3.43) simplifies essentially. It is well known that a differentiation operation in signal processing amplifies noise. Hence additional low pass filters are necessary. Applications of the proposed solution in edge detection in robotics are presented in Chap. 7.

## References

Bai, J. and Feng, X.-C. (2007). Fractional - order anisotropic diffusion for image denoising, *IEEE Transactions on Image Processing* **16**, 10, pp. 2492–2502.

Carpinteri, A. and Mainardi, F. (eds.) (1997). *Fractals and Fractional Calculus in Continuum Mechanics* (Springer-Verlag Wien and New York).

Chen, Q.-l., Huang, G. and Zhang, X.-q. (2012). A fractional differential approach to low contrast image enhancement, *International Journal of Knowledge and Language Processing* **3**, 2, pp. 20–29.

Chen, X. and Fei, X. (2012). Improving edge-detection algorithm based on fractional differential approach, in *2012 International Conference on Image, Vision and Computing (ICIVC 2012)*, Vol. 50 (Singapore).

Cuesta, E., Kirane, M. and Malik, S. A. (2012). Image structure preserving denoising using generalized fractional time integrals, *Signal Processing* **92**, 2, pp. 553–563.

Gan, Z. and Yang, H. (2010). Texture enhancement through multiscale mask based on rl fractional differential, in *International Conference of Information Networking and Automation*, Vol. 1 (Kunming, China), pp. 333–337.

Gao, C., Zhou, J., Hu, J. R. and Lang, F., Lang (2011a). Edge detection of colour image based on quaternion fractional differential, *IETImageProcessing* **5**, 3, pp. 261–272.

Gao, C., Zhou, J., Zheng, X. and Lang, F. (2011b). Image enhancement based on improved fractional differentiation, *Journal of Computational Information Systems* **7**, 1, pp. 257–264.

Ghamisi, P., Couceiro, M. S., Benediktsson, J. A. and Ferreira, N. M. F. (2012). An efficient method for segmentation of images based on fractional calculus and natural selection, *Expert Systems with Applications* **39**, pp. 12407–12417.

Hu, J., Pu, Y. and Zhou, J. (2011a). Fractional integral denoising algorithm and implementation of fractional integral filter, *Journal of Computational Information System* **7**, 3, pp. 729–736.

Hu, J., Pu, Y. and Zhou, J. (2011b). A novel image denoising algorithm based on riemann-liouville definition, *Journal of Computers* **6**, 7, pp. 1332–1338.

Jalab, H. A. and Ibrahim, R. W. (2012). Denoising algorithm based on generalized fractional integral operator with two parameters, *Discrete Dynamics in Nature and Society* **2012**, pp. 1–15.

Kaczorek, T. (2011). *Selected Problems of Fractional Systems Theory* (Springer-Verlag, Berlin).

Lubich, C. (1986). Discretized fractional calculus, *SIAM Journal on Mathematical Analysis* **17**, 3, pp. 704–719.

Miller, K. S. and Ross, B. (1993). *An Introduction to the Fractional Calculus and Fractional Differential Equations* (John Wiley & Sons Inc., New York, USA).

Oldham, K. and Spanier, J. (1974). *The Fractional Calculus* (Academic Press, New York).

Ostalczyk, P. (2000). The non-integer difference of the discrete-time function and its application to the control system synthesis, *International Journal of System Science* **31**, 12, pp. 1551–1561.

Ostalczyk, P. (2001). *Discrete-Variable Functions* (A Series of Monographs No 1018, Technical University of d, Poland).

Oustaloup, A. (1995). *La dérivation non entiére* (ditions Hermés, Paris, France).

Oustaloup, A., Mathieu, B. and Melchior, P. (1991a). Edge detection using non integer derivation, in *Presented at the IEEE European Conference on Circuit Theory and Design (ECCTD91).* (Copenhagen, Denmark).

Oustaloup, A., Mathieu, B. and Melchior, P. (1991b). Robust edge detector of non integer order: the CRONE detector, in *Presented at the 8th Congres de Cytometrie en Flux et dAnalyse dImage (ACF91).* (Mons, Belgium).

Podlubny, I. (1999). *Fractional differential equations* (Academic Press, San Diego, USA).

Pu, Y.-F., Zhou, J.-L. and Yuan, X. (2010). Fractional differential mask: A fractional differential-based approach for multiscale texture enhancement, *IEEE Transactions on Image Processing* **19**, 2, pp. 491–511.

Samko, S. G., Kilbas, A. A. and Marichev, O. I. (1993). *Fractional Integrals and Derivatives: Theory and Applications* (Gordon and Breach, London).

Tadmor, E. and Zou, J. (2008). Three novel edge detection methods for incomplete and noisy spectral data, *Journal Fourier Analysis and Applications* **14**, pp. 744–763.

Wang, C., Lan, L. and Zhou, S. (2013). Grunwald-letnikov based adaptive fractional differential algorithm on image texture enhancing, *Journal of Computational Information Systems* **9**, 2, pp. 445–454.

Watkins, C. D., Sadun, A. and Marenka, S. (1993). *Modern Image Processing* (Academic Press, Inc).

# Part 2

# Computer Vision in Robotics

# CHAPTER 4

# MANAGEMENT SOFTWARE FOR DISTRIBUTED MOBILE ROBOT SYSTEM

Maciej Łaski, Sylwester Błaszczyk, Piotr Duch, Rafał Jachowicz, Adam Wulkiewicz, Dominik Sankowski, and Piotr Ostalczyk

*Institute of Applied Computer Science*
*Lodz University of Technology*
*90-924 Lodz, ul. Stefanowskiego 18/22*
*{mlaski, sblaszc, pduch, rjachowicz, awulkie, dsan}@kis.p.lodz.pl,*
*piotr.ostalczyk@p.lodz.pl*

Robotic vehicles autonomy (Jeżewski *et al.*, 2009) is the subject of many studies conducted by various institutions. A growing interest in this discipline of engeneering may be seen e.g. in universities, companies and the military. This is because of an understandable tendency to replace humans for tasks that are deemed too dangerous, where life may be in danger. This is particularly true of tasks performed by the military as well as the police, mining or firefighter units.

Therefore, robots should perform as many autonomous tasks as it is possible. To accomplish this, different sensors are used to gather information about the surounding environment. Those can be distance measurement units or camera based vision systems. But synthetic information received from those sensors is only a piece of whole robotic system that allows autonomous tasks to be performed by the vechicle.

In this chapter the complete mobile platform managment system (Łaski *et al.*, 2012) is presented. This is the basis for algorithms that use computer vision to detect obstacles, measure distances and receive attributes of objects in the environment.

## 4.1 Architecture of mobile robot management system

This chapter presents a concept of a distributed management system for a mobile platform, which was designed and implemented at the Institute of Applied Computer Science (formerly the Computer Engineering Department) within the framework of the: Autonomous robot for surveillance and mine detection tasks.

The main task of the system is to encapsulate the functionality of the robotic platform as independent modules. They cover: the physical layer control, acquisition and fusion of data from different sensors, communication between the modules, analysis of image data. The aforementioned mechanisms are also designed to perform autonomous behaviours which currently are the subject of many researches performed by various institutions. A growing interest in this discipline of engeneering may be seen e.g. in universities, companies and the military. This is due to an understandable tendency to replace humans for tasks that are deemed too dangerous. This is particularly true of tasks performed by the military as well as the police, mining or firefighter units. The authors of this article focus on small mobile platforms which may be used in dangerous tasks of reconnaissance and mine detection. They define a set of mobile platform navigation funcionalities and present a concept of a modularized subsystem which implements these funcionalities in order to make navigation in the unknown environment possible.

The autonomy of mobile robots is currently very extensively developed. Environment perception and sophisticated processing systems are fundamental to realize platform with autonomous behaviors. For this reason this paper focuses on the specification and realization of necessary modules, communication protocols and inter-module data transfer. They will carry out the basic and complex functionality of these platforms. Each module has its own dedicated microprocessor to provide easy replacement or service. The main purpose of this system is to provide abstraction layer for hardware and to encapsulate functionality in modules.

### 4.1.1 *Implemented modules*

The mobile platform management system is responsible for many tasks performed by on-board systems. It covers power management, the stearing of physical devices such as engines, communicates with with different sensors and realizes algorithms that perform autonomous behaviors. All these tasks can be encapsulated so that the whole system will have small modules, each responsible for specific task.

Currently implemented modules are shown in Fig. 4.1. They can be divided into two categories: basic and complex modules. The first includes modules that are responsible for handling the physical layer of the mobile platform. Those modules are: Operator console, communication, image processing, sensor, driving and observation arm. Their tasks are: to control the DC (direct current) motors, to gather data from sensors. They also provide reliable communication between the modules and between the platform and the operator console. The second category consists of complex modules that use information generated by basic modules to perform sophisticated autonomous tasks. Their tasks are: autonomous driving with obstacle avoidance, operator tracking using computer vision systems and autonomous driving to a specified position/direction.

Each module is responsible for single, specific functionality and provides a clearly specified communication interface. This allows modules to be replaced easily and fast. In addition, the portability of the management system is increased when these

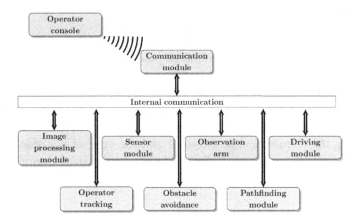

Fig. 4.1   Implemented and verified modules of mobile robot management system.

preconditions are applied. For example, the engine module performs strict orders and their implementation is in its internal mechanisms. It can be replaced with another module that executes the same commands, but in a different way or on a different platform. For higher-level algorithms such as obstacle avoidance it should not be relevant whether they control a small mobile platform or a large and heavy transporter. The only requirement is to provide the same communication interface for the drive unit.

### 4.1.2   *Driving module*

Designed robot is equipped with six drive components and one central unit that performs control and supervisory functions. Each drive component consists of a DC motor, a quadrature encoder, an H bridge driver and a logical part that implements a speed regulation loop [8] and communicates with the central unit. The central unit serves an interface between the other modules and separates the physical and the logical layer. Thermal and current sensors are used for security reasons.

The main concept of encapsulation of the drive module is the scalability and portability of the system to other platforms, regardless of their dimensions and traction capabilities. Higher level modules such as obstacle avoidance or operator tracking produce a list of orders which the drive component is able to execute. It does not matter how those orders will be executed. Other modules will retain their full functionality as long as the communication interface and the list of orders will not change.

A list of orders and protocols have been defined during the tests performed on the robot constructed at the Institute of Applied Computer Science. These are the most basic commands but they are sufficient to perform all tasks ordered by other modules. By joining them in sequences the platform is capable of performing very complicated tasks.

### 4.1.3  *Sensor module*

The presented autonomous mobile platform is equipped with a range of sensors which are required to locate the robot in position on the Earth. This module (Fig. 4.2) is responsible for receiving and processing data from them. Syntethic understanable data determined by this module is sent on the internal communication bus. Other modules can use it for navigation. The module can encapsulate arbitrary number of sensors of various types. They're not visible for the rest of the system. Only the communication protocol and the type of information that this module provides is fully defined. Other modules may use this information for their internal purposes.

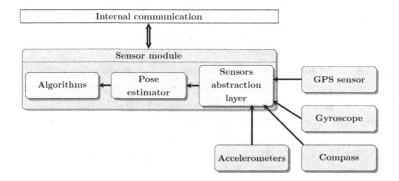

Fig. 4.2   Implemented and verified sensor module.

Currently this module supports the following sensors:

- GPS receiver — a global positioning system sensor allows the robot to read the location with an accuracy of 3 meters. It is used by various modules requiring information about the exact position of the robot when performing the task / command.
- Accelerometer — a triaxial acceleration sensor allows you to read the platform orientation in space.
- Magnetic field sensor — a triaxial compass to calculate the orientation of the platform in the global coordinate system, which is the basis for algorithms to maintain an autonomous direction.
- Gyroscope — a triaxial sensor to determine the orientation of the allowable swing to the platform.

### 4.1.4  *Vision module*

In the context of the autonomous management of the mobile platform, in particular referring to the terrain movement ability, the perfect mechanism to imitate is the human vision system. Human's stereovision system is the source of desired or even essential information. Using this kind of system, one can determine a given location and track objects in the environment. Moreover the distance between these objects can be calculated and even some of their important features (e.g. temperature) can be discovered using a proper type of cameras. In addition, all the functionalities mentioned above are available in real time execution mode.

In order to make implementation of all mentioned above tasks possible, the image processing and analysis module shown in Fig. 4.3 is equipped with cameras of three different types. These cameras are situated in a set that is adapted to work in a stereovision mode and are connected to a PC class computer installed on the mobile platform. The first camera is a basic daylight, high-resolution camera, which provides a simple real-time preview. The second camera is a device dedicated to infrared image acquisition, namely a thermo-vision camera. The third camera is a high-sensitivity device, which provides vision in low lighting conditions. It can be successfully used instead of military, noctovision solutions. The layout of all cameras is fixed and therefore known a priori at every moment. Imaging data gathered by different cameras can be integrated together using stereovision algorithms in order to determine and model the surrounding environment. The algorithms mentioned use the disparition effect, which occurs in stereovision systems. It is a shift of the same objects on images presenting the same scene, but acquired from different cameras (perspectives).

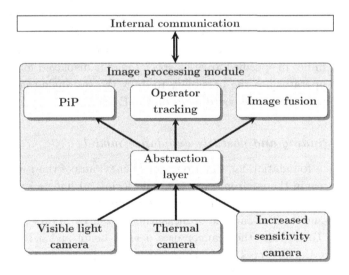

Fig. 4.3   Implemented and verified vision module.

The vision module is designed in such a way that its between-platform portability is possible irrespective of the types of tasks it will have to perform. Therefore, it does not issue the commands to other modules, but only shares essential information. The type of this information is of course dependent on the execution mode. Thus, if the system performs the scout task on the output of the image processing module it will find a properly prepared image. However, if the system ordered more complex task performance, e.g. object detection and tracking on the output, it will find information which refers to the location, velocity, distance etc.

### 4.1.5 *Operator console*

Respectively to the functionalities of the robot, a special operator console has been developed. It allows one to control the robot wirelessly up to a distance of three kilometers. The heart of this system is a laptop that is responsible for displaying information about the internal system status, showing images from the robot cameras and providing a convenient user interface. It visualizes the maps and environment generated from the data gathered from the robot sensors. It is capable of sending simple orders, such as: drive forward, turn, look left. In addition, the mechanism designed offers the possibility to order complicated tasks to the mobile platform, such as:

(1) Go to GPS position – a special map is displayed for the operator, who can choose the destination point for the robot by touching a specified position on the map. The robot will go to the position using the obstacle avoidance module.
(2) Go in a specified direction – using the data received from the robot, the console displays the compass in the coordinate system of robot. The operator may choose the direction that the robot will follow using the sensor module. The obstacle avoidance module can be turned off if such a configuration is needed.
(3) Search for mines in specified area – a special map is displayed for the operator. Then, the operator has the possibility to draw the rectangle that will be send to the robot in the global coordinates on the Earth.

### 4.1.6 *Path finding and obstacle avoidance module*

This module is a foundation for all autonomous behaviours of the robot. It detects obstacles that are in the environment of the platform and finds a safe path to the destination avoiding collisions. To provide this functionality a two dimensional laser scanner is used. It is capable of measuring distances to objects surrounding the platform. To do this, the scanner uses a laser beam and a rotating mirror. Output data are the distances to obstacles in the plane perpendicular to the mirror rotation axis. Used sensor is HOKUYO UTM-30LX (Fig. 4.5) which allows ane to perform 1081 distance measurements on 270 degrees during 25 milliseconds. It guarantees 30 meters range with 1 millimetre precision. The algorithm of path planning builds a map of the environment based on the measurements gathered from the laser scanner. The data is stored in a regular grid which has cells of the same size. They store information if there is an object in the space which occupies them. The current implementation uses A* (A-star) and Lee's algorithm.

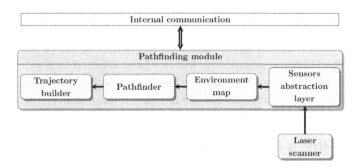

Fig. 4.4  Implemented and verified pathfinding module.

Fig. 4.5  2-dimensional HOKUYO UTM-30LX laser scanner.

This module awaits an order from the other modules to perform a drive with obstacle avoidance. When this command is received, it starts communicating with the laser and discovers a secure path to reach the target. According to its calculations, it sends orders to the drive module which performs a specified movement. During the platform movement the environment changes so the path planning module continue updating of the internal map and reestablishes a safe trajectory. When a new obstacle appears or platform re-discovers new place, which was not in area available for laser scanner, the safe path will be calculated again. In that case the moving and appearing objects are not the issue and will not produce a collision.

### 4.1.7  *Internal communication*

The entire robotic system implemented consists of independent modules which share some module specific information that can be used by the other parts of the system. This is the reason why the communication module has been implemented - to provide reliable and fast communication. All the modules use it to give orders and provide preprocessed data (from sensors, for example).

The robot is equipped with several electrical motors which are driven using pulse width modulation. This method causes a lot of noise in the sourounding electrical environment. Therefore, the physical medium has to be resistant to any interference. That's why, the CAN communication bus has been selected as a primary medium for data transmission. It is sufficient to transmit any sensor data from the onboard devices and resend all commands that the modules require from the other modules.

## 4.2   Human-Robot Interface for six wheeled mobile robot

### 4.2.1   *Human – Robot Interaction*

Human Robot Interaction (HRI) is an area of science dedicated to understanding and developing robot and human interaction techniques. Goodrich and Scholtz (2007) defined the main problem which has to be solved by the HRI as the pursuit of knowledge and development of interaction methods between one or more humans with one or more robots. On the one hand, the HRI studies the issues related to the robot such as designing semi-autonomous behaviours that allow one to perform required tasks with a minimum humans intervention and the problems associated with those behaviours (Adams and Skubic, 2005; Crandall and Cummings, 2007; Sheridan and Parasuraman, 2006; R. Parasuraman and Wickens, 2000). On the other hand, it also includes researches connected with capabilities and limitations of the robot operator (Adams, 2002), his role in the interaction with the robot (Scholtz, 2003) as well as designing a user interface (Yanco and Drury, 2007). Furthermore, HRI deals with the evaluation of the final system including the robot and the team of humans behind it (Scholtz *et al.*, 2004; Endsley, 1988). At the same time Goodrich and Scholtz (2007) focus their the attention on the fact that the process of interaction requires also conducting researches on methods of communication. Those methods depend on the relative position between the human and the robot. One can distinguish two general categories of interaction and communication:

- remote - the human and the robot are separated spatially and even temporally (Opportunity Mars rover)
- proximate - the human and the robot are collocated (service robots e.g. Roomba)

In this chapter the authors focused on remote interaction with mobile robots, often referred to as teleoperation (Sayers, 1998). According to the definition of HRI, the user interface requires communication media between the human and the robot.

### 4.2.2   *Operator Console Unit (OCU)*

In the process of designing a user interface, crucial parts are those responsible for collecting and processing data from users and those responsible for presenting processed information gathered from the robot sensors to the users (Input, Displays)(de Barros and Lindeman, 2008). They are usually in the form of Operator Concole Unit (OCU) particularly used in military solutions. Fig. 4.6 presents an operator console developed for the control of the mobile robot built at the Institute of Applied Computer Science, Lodz University of Technology in cooperation with the company Prexer Sp. z o.o. [Ltd].

Fig. 4.6   Operator console for mobile robot.

The user through the intuitive graphical interface is able to control all of the mobile platform modules such as, the driving module, the observation arm and the head module. Orders given by the user are sent by the wireless communication module to the interpretation module located in mobile platform. Next, the processed orders are delegated to the module responsible for their realization. The flow diagram of data transmission between the hardware modules is presented in Fig. 4.7.

### 4.2.3   *Realized control system of mobile platform for OCU*

The development of framework was necessary for providing the desired level of cooperation beetwen all the parts of OCU. Based on researches presented in Collett *et al.* (2005) the most relevant goals in the process of framework development were chosen by the authors: platform independence, scalability, development process simplification, real-time performance and software reusability.

Fig. 4.8 presents a block diagram of the developed and realized framework for the operator console unit shown in Fig. 4.6. One can distinguish three logical layers of software in the designed system architecture:

- DEVICES - procedures for handling devices and retriving data from them
- COMMUNICATION - data management in shared memory

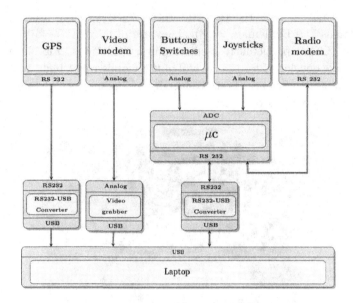

Fig. 4.7   Block diagram of communication between the hardware modules.

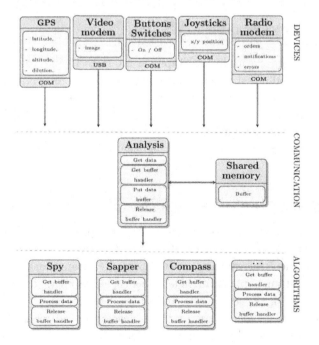

Fig. 4.8   Block diagram of communication between software modules in the operator console.

- ALGORITHMS - algorithms management

### 4.2.4 *Functional elements of mobile robot management system*

In Table 4.1 functionalities available in particular modes are presented.

Table 4.1   Functionalities of user interface.

| Mode | Functionalities |
|------|-----------------|
| Spy | • manual control of mobile platform <br> • manual control of observational head <br> • video |
| Sapper | • search for mines in the given area <br> • displaying map with marked position of detected mines and robot position |
| Compass | • autonomous drive with desired speed and in required geographical direction <br> • manual drive with desired speed and required in geographical direction <br> • view from selected camera <br> • local map view <br> • view of calculated robot path |
| Pattern | • autonomous detection and tracking of operator <br> • view from selected camera |
| Arm | • manual control of observational head <br> • manual control of arm <br> • view from selected camera |
| Comeback | • autonomous drive to operator <br> • displaying map with position of robot and operator |
| GoToGps | • autonomous drive to GPS coordinates <br> • displaying map with position of robot and operator |
| Settings | • presentation of system status <br> • vision system settings <br> • notification and error message settings |

Fig. 4.9   Execution of control system in Spy mode

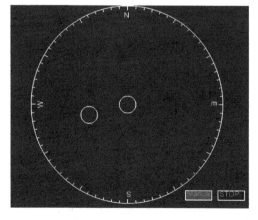

Fig. 4.10   Execution of control system in Compass mode

## 4.3   Laser scanner based obstacle avoidance system

One of key features of the modern autonomous robots is the ability to detect and avoid obstacles during navigation in an unknown environment. Robots gather spatial information using various distance measuring sensors such as ultrasound sonars, 2- and 3-dimensional laser scanners or stereo cameras which may be used to calculate distances to objects in the robot's field of view. This information is integrated in the data structure which often provides functionalities such as accelerated spatial searching. Those may be used during the integration process, e.g. when SLAM techniques (Montemerlo and Thrun, 2007; Nüchter, 2009; Stachniss, 2009) are implemented or by upper-level navigation algorithms using the map to find a safe path to the target.

The obstacle avoidance system implemented in the mobile robot which is being described in this chapter consists of three modules:

- pose module,
- pathfinding module,
- driving module.

Each module is intended to provide some functionalities to the rest of the system. They have well-defined interfaces and enclose implementation details therefore they may be easily moddified or replaced.

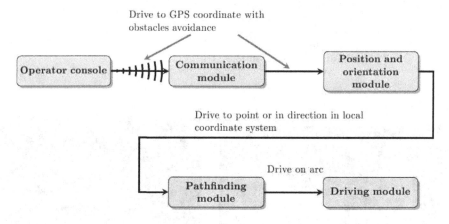

Fig. 4.11   The communication between modules during driving to GPS coordinates.

The modules communicate with each other by sending commands. For instance, during the task of driving to GPS coordinates (Fig. 4.11) the "drive to GPS with obstacle avoidance" command is sent from the operator console and received by the pose module. This module, which gathers information from the GPS sensor and the electronic gyroscope calculates the position or direction to the target in the robot's local coordinate system and sends the "drive to point or direction in local coordinate system" to the pathfinding module. The purpose of the latter is to build a map using data gathered from the laser scanner and to find a safe route. After the analysis of the map and the calculation of the path it sends the "drive on arc" command to the driving module.

### 4.3.1 *Environment map*

In order to find a safe path to the target robots build a map of the environment using spatial data gathered from the sensors. Typically the map is a data structure which allows performing spatial queries (Bentley, 1975; Finkel and Bentley, 1974; Stachniss, 2009). Depending on the type of data and the purpose of the map various data structures may be used.

The purpose of the robot which is being described is to operate without the need to exchange large amounts of data with the operator console for a long time. This implies:

- implementing all the necessary algorithms in the embedded system,
- using energy-efficient components.

Therefore at the current stage of development, the robot uses a 2-dimensional laser scanner and integrates spatial data in the occupancy grid. The occupancy grid is a regular grid in which probabilities of existance of objects in the robot's environment are stored. The map in Fig. 4.12 was created using data gathered by the laser scanner. Safe areas are white, obstacles black. Not discovered areas are grey.

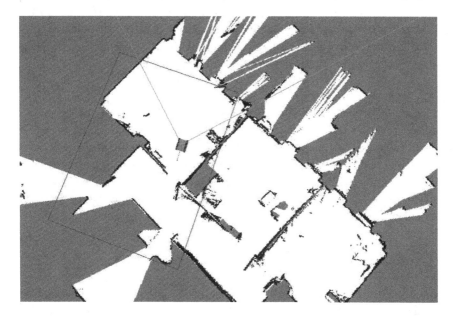

Fig. 4.12   2-dimensional map of the environment.

### 4.3.2   *Pathfinding algorithm*

In the literature one may find various examples of pathfinding algorithms (Dijkstra, 1959; Lee, 1961; Hadlock, 1977; Soukup, 1978). We have choosen the well-known A* algorithm (Hart *et al.*, 1968) which allows one to use a heuristic in the searching process. It employs a knowledge-plus-heuristic cost function of node to determine the order in which the search visits nodes of the map. A cost function (4.1) is the sum of two functions: the past path cost function, which is the distance from the starting point and fututre path cost function, which is estimated.

$$f(x) = g(x) + h(x) \tag{4.1}$$

The past path function and heuristic that were used may take into account the following factors:

- distance to target,
- angles of turns,
- obstacle proximity,
- known/unknown space.

The past path cost (4.2) is the sum of costs of transitions between subsequent neighbours on a path from the starting to the currently analyzed point.

$$g(x) = \sum_{i=1}^{n_x} |x_{i-1}, x_i| + P_o(x_i) + P_t(x_i) \tag{4.2}$$

, where $n_x$ is the number of nodes on the past path, $P_o(x)$ is the penalty related to the proximity of the obstacle at point $x$ and $P_t(x)$ is a penalty related to the turn angle in point $x$. The esstimate of the future path is made by the heuristic. In this case, we have used a well known distance heuristic (4.3) calculating the distance between the currently analyzed point $x$ and the target $x_t$.

$$h(x) = |x, x_t| \tag{4.3}$$

The result of the pathfinding algorithm (Fig. 4.13) is a set of the map cells connected together that form a path from the starting point to the target. This raw path consists of segments. Each path fragment is at an angle of the value equal to a multiple of 45 deg with respect to the previous segment. Therefore, we use a postprocessing method which smoothes the raw path. The actual arc radius along which the robot is going to move is proportional to the distance to the first turn on the calculated path.

## 4.4   Automatic control systems of mobile robot drives

The mobile platfrom management system is divided into small modules resopnsible for their own tasks. According to this, the mobile robot drives control system has been isolated from the whole mobile robot management system. Other modules can

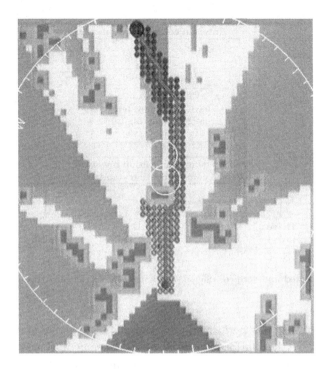

Fig. 4.13 Path found in the unknown environment.

give drive orders and ask about the current state of the drives using the communication interface with specified commands. That allows one to change the base of the robot without touching the rest of system. The only requirement is that the new drive system will perform and react in the same manner to identical requests. The rest of the system is not related to specific physical capabilities of the platfrom. This chapter presents the realized and verified engine management system for a six wheel mobile platform.

### 4.4.1 *Centralized subsystem for drives management and control*

The robot constructed is equipped with six drives. Each of them includes: a direct current motor, a quadrature encoder, an H bridge and a steering unit connected to the wheel that it powers. They are placed in a group of three on both sides of the mobile platform. All drives are controlled by a central unit which ensures communication interface with the rest of system. This central unit is the last element in the chain that has knowledge about the physical aspects of platform, such as properties of the wheels, their placement and spacing, the maximum speed and braking possibilites. A schema of the mobile robot drives control system is shown in Fig. 4.14.

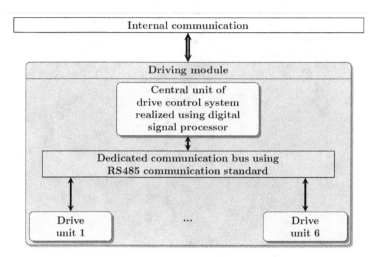

Fig. 4.14   Implemented and verified schema of driving module that controls six wheeled mobile platform.

As a result of the tests performed on a real platform, the minimum set of commands required by the platform to perform was specified. These commands are sufficient to perform all tasks associated with moving in an unfamiliar environment:

(1) Drive at a speed $V$ – after receiving this order the CPU sends the job to maintain the speed $V$ for all the six modules. The platform moves in a straight line at a specified speed.
(2) Follow the curve of radius $r$ at a speed $V$ – after receiving this order, the CPU calculates an appropriate speed for each of the six drive units to ensure that the platform is moving around a predetermined arc.
(3) Stop – This order stops the platform immediately. Usually the speed control should stop the platform by setting the speed to zero but in some hazardous situations there is a need for using the fast breaking system.

The central unit reacts to the received command immediately and updates the state of the drives that it controls. This allows the motion of the platform to be controlled by other modules. According to thier internal calculations the parameters for speed and turning radius are send to central unit which continuosly updates orders to all six drives it controls. For example, the module which performs the calculation of the collision-free motion path, first waits for data from the laser scanner. Then, the collision-free path is calculated and the appropriate command is sent to the central unit of the mobile robot drive control system. The robot starts to move and the surrounding environment changes. As a result, new data is received from the laser scanner, a new safe motion path is determined and new motion parameters are sent to the drives. This process is executed until the destination point is reached.

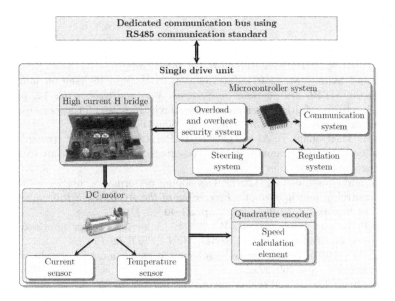

Fig. 4.15   Elements of single robot drive system.

The above example presents flexible capabilites of the system. Only a few commands that the platform base realizes are sufficient to follow even very complicated paths. Another thing is that that the module that calculates the safe path is not involved in aspects of controlling the physical elements of the robot, such as engines. So for this module there is no difference between a small six wheeled mobile platform and a bigger four wheeled truck until the base system can perform and react to the same commands.

### 4.4.2   *Dedicated module for single engine management*

A single drive unit, as shown in Fig. 4.15 can be isolated from the whole mobile platform drive control system. It is divided into four basic elements:

(1) Engine – A direct current motor with a gearhead produced by PITTMAN. It is connected directly to the wheel. Also the current and temperature sensors are connected to the engine for security reasons.
(2) H bridge – A High Current drive for the electric motor. It allows the motor rotation direction and the motor torque to be controlled using Pulse Width Modulation (PWM).
(3) Quadrature encoder – a sensor that allows the motor speed.
(4) Microcontroller system – It is the heart of the drive system. It controls speed and performs a speed regulation loop using a PID regulator. It is also responsible for communication with the central unit and for controling current and the temperature of the engine to prevent damage.

# References

Adams, J. (2002). Critical considerations for human-robot interface development, *2002 AAAI Fall Symposium on Human-Robot Interaction* .

Adams, J. and Skubic, M. (2005). Introduction to the special issue on human-robotic interaction, *IEEE Transactions on Systems, Man and Cybernetics* **12**, pp. 27–30.

Bentley, J. L. (1975). Multidimensional binary search trees used for associative searching, *Communications of the ACM, 18(9)* , pp. 509–517.

Collett, T. H., MacDonald, B. A. and Gerkey, B. P. (2005). Player 2.0: Toward a practical robot programming framework, *In Proceedings of the Australasian Conference on Robotics and Automation* .

Crandall, J. and Cummings, M. (2007). Developing performance metrics for the supervisory control of multiple robots, *Proceeding of the ACM/IEEE international conference on Human-robot interaction* , pp. 33–40.

de Barros, P. G. and Lindeman, R. W. (2008). A survey of user interfaces for robot teleoperation, *Technical Report Series, WPI-CS-TR-09-12* .

Dijkstra, E. W. (1959). A note on two problems in connexion with graphs, *Numerische Mathematik 1* , pp. 269–271.

Endsley, M. (1988). Design and evaluation of situation awareness enhancement, *Proceedings of the Human Factors Society 32nd Annual Meeting* , pp. 97–101.

Finkel, R. and Bentley, J. (1974). Quad trees: A data structure for retrieval on composite keys, *Acta Informatica 4 (1)* , pp. 1–9.

Goodrich, M. A. and Scholtz, A. C. (2007). Human-robot interaction: a survey, *Foundations and Trends in Human-Computer Interaction* , pp. 203–275.

Hadlock, F. O. (1977). A shortest path algorithm for grid graphs, *Networks , vol. 7, no. 4* , pp. 323–334.

Hart, P. E., Nilsson, N. J. and Raphael, B. (1968). A formal basis for the heuristic determination of minimum cost paths, *IEEE Transactions on Systems Science and Cybernetics SSC4 4 (2)* , pp. 100–107.

Jeżewski, S., Sankowski, D. and Dadan, W. (2009). Koncepcja autonomicznego robota pola walki przeznaczonego do zada zwiadu i wykrywania min, *Zeszyty naukowe AGH ISSN: 1429-3447* .

Łaski, M., Baszczyk, S., Duch, P., Jachowicz, R., Wulkiewicz, A. and Sankowski, D. (2012). Distributed mobile platform management system, *Zeszyty naukowe AGH ISSN: 1429-3447* .

Lee, C. Y. (1961). An algorithm for path connections and its applications, *IRE Transactions on Electronic Computers EC-10 (2)* , pp. 346–365.

Montemerlo, M. and Thrun, S. (2007). *FastSLAM, Springer Tracts in Advanced Robotics*, Vol. 27 (Springer), ISBN 978-3-540-46399-3.

Nüchter, A. (2009). *3D Robotic Mapping, Springer Tracts in Advanced Robotics*, Vol. 52 (Springer), ISBN 978-3-540-89883-2.

R. Parasuraman, T. S. and Wickens, C. (2000). A model for types and levels of human interaction with automation, *IEEE Transactions on Systems, Man and Cybernetics* , pp. 286–297.

Sayers, G. (1998). Remote control robotics, *Springer-Verlag* .

Scholtz, J. (2003). Theory and evaluation of human robot interactions, *In Proceedings of the 36th Annual Hawaii International Conference on System Sciences* .

Scholtz, J., Antonishek, B. and Young, J. (2004). Evaluation of a human-robot interface: Development of a situational awareness methodology, *In System Sciences, 2004. Proceedings of the 37th Annual Hawaii International Conference on* , pp. 3212–3217.

Sheridan, T. B. and Parasuraman, R. (2006). Human-automation interaction, *Reviews of Human Factors and Ergonomics* , pp. 89–129.

Soukup, J. (1978). Fast maze router, *DAC '78 Proceedings of the 15th Design Automation Conference* , pp. 100–102.

Stachniss, C. (2009). *Robotic Mapping and Exploration, Springer Tracts in Advanced Robotics*, Vol. 55 (Springer), ISBN 978-3-642-01096-5.

Yanco, H. A. and Drury, J. L. (2007). Rescuing interfaces: A multi-year study of human-robot interaction at the aaai robot rescue competition, *Autonomous Robots* , pp. 333–352.

# CHAPTER 5

# ADVANCED VISION SYSTEMS IN DETECTION AND ANALYSIS OF CHARACTERISTIC FEATURES OF OBJECTS

Adam Wulkiewicz, Rafał Jachowicz, Sylwester Błaszczyk, Piotr Duch, Maciej Łaski, Dominik Sankowski, and Piotr Ostalczyk

*Institute of Applied Computer Science*
*Lodz University of Technology*
*90-924 Łódź, ul. Stefanowskiego 18/22*
*{awulkie, rjachowicz, sblaszc, pduch, mlaski, dsan}@kis.p.lodz.pl,*
*piotr.ostalczyk@p.lodz.pl*

## 5.1 Introduction

Vision systems, as a basic technology, are one of the most complicated and, at the same time, essential when it comes to building autonomous or semiautonomous vehicles. Vision systems are designed to eliminate or to limit the necessity of human observation and the probability of an occurrence of mistakes. From many minor vision systems categories there are three main which include all other. These are: measurement, inspection and guidance systems. The purpose of a measurement system is to calculate dimensions of an object by analysis of its digital image. Inspection systems are ordered to check or to examine a given area and to detect any irregularities there. Guidance systems, in turn, are designed to perform orders based on the perception system of the machine. The vision system dedicated to vehicles which are to perform autonomous or semiautonomous tasks is included in the last category: guidance systems.

Nowadays, image processing and analysis is a rapidly developing branch of science, which increasingly often finds its application in the industry. More and more device control systems are based on information acquired by image data processing. Furthermore, those systems repeatedly acknowledge image data as the only reliable source of information (Zhang *et al.*, 2011; Liu *et al.*, 2009; Arora and Banga, 2012; Choudekar *at al.*, 2011). Due to the fact that image data is often ambiguous, the entrustment of making key decisions in any technological process to a system that is based on such data needs aid of additional image processing and analysis support techniques. In addition, in cases where an image analysis system is not a decision one, but only an image data display mechanism, some valuable image processing support techniques can be applied. These are tasks in which the image processing and the analysis system are only an informer, not the manager, e.g. surveillance and scout.

Advanced vision systems are increasingly complicated and most of them are based on the detection of characteristic features of objects, tracking and analysis. Therefore, the best exposition of key features of objects is essential. This approach has a special significance wherever the analysis of information acquired from a single camera, or even from many cameras, but analyzed separately, is insufficient. A human being is not capable of infallible information integration from many different sources. Therefore, the fusion of images acquired from two or more different sources to form a single image in order to enhance the quality of analysis of an observed scene has a significant practical value and is a continuously developed branch of image processing and analysis. Algorithms of such a type find their application in medicine (registering and combining magnetic resonance (MR), positron emission tomography (PET), computer tomography (CT) into composites to aid surgery (Derek *et al.*, 1994)), military and surveillance (object detecting and tracking (Snidaro *et al.*, 2009; Zin *et al.*, 2011) or in the industry (non-destructive evaluation techniques for inspecting parts (Blum and Liu, 2005)). Potential advantages of image fusion include:

(1) Information can be read more precisely, in a shorter time and at a lower cost.
(2) Image fusion allows distinguishing features that are impossible to perceive with any individual sensor.
(3) Compact representation of information.
(4) Extended spatial and temporal coverage.
(5) Exposure of all the objects present on the scene by the integration of images taken from the same viewpoint under different local settings.

For example, one can imagine a human hidden in the woody area at night. While using a Night Vision camera the observer is able to see details of the environment and exact shapes of objects, but will hardly detect a camouflaged human figure. In turn, on the Thermo Vision camera image the silhouette of a hidden man will be clearly visible, but it will be hard to determine its location in space, because the temperature of the surrounding elements of the environment is nearly identical. On the integrated image both features would be presented: the environment details and the temperature of the objects.

However, not only object exposition support is important, but also the aid of a computer system wherever a human operator has to simultaneously observe the scene and accurately control the robot's actuators. In such cases, the help of the system is a significant enhancement. Such solutions are already applied in the automotive industry, e.g. semi-automatic parking systems.

Fig. 5.1 IBIS robot and its operator console (PIAP, Warsaw, www.antyterroryzm.com).

In Fig 5.1 the pyrotechnic robot IBIS constructed by PIAP from Warsaw is presented. When operating such advanced devices, the operator is forced to maintain ceaseless concentration, especially when performing dangerous target tasks such as the examination of explosives. There are only a few available solutions which offer advanced aid systems in such situations and even more, a vast number of similar devices do not support even basic enhancement mechanics such as inverse kinematics (Sun *et al.*, 2012). Due to the necessity of operator support during accurate mobile platform maneuvers, the authors of the Mobile Robot operating system from the Institute of Applied Computer Science developed a support mechanism of suspicious object observation by the robot's head.

In this paper two operator work aid algorithms are presented. The algorithms use the robot's head as a perception system and are based only on image processing and analysis data. The algorithms are: Fusion of image data acquired from different types of cameras (among others Thermo Vision and Night Vision) and mobile platform positioning according to objects indicated by the operator. The result of the output of the first algorithm is the exposition of chosen features (often impossible to determine by using only one camera) of objects located on the scene, e.g. size, temperature, etc. The result of the output of the second algorithm reaches the optimal position of the robot's head. The optimal position is determined by the combination of safety and accessibility level to a suspicious object placed on the ground.

## 5.2 Image fusion algorithm

Fusion techniques of images acquired from different types of cameras can be classified into one of the following categories (Zhang, 2010; Al-Wassai *et al.*, 2011) (Figs 5.2, 5.3 and 5.4):

(1) The pixel/data level – the integration of raw data acquired from multiple sources into one image which includes more information (presented in a synthetic way) than each of the input images analyzed separately. The other possibility is to expose differences between data collections acquired at different times.
(2) The feature level – on this level, the fusion of images acquired from different sources into one image which can be used for later processing and analysis is achieved by the extraction and combination of many types of features such as edges, corners, lines or textures.
(3) The decision level – this fusion combines the results from multiple algorithms into a field of the final fused decision.

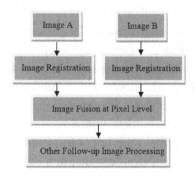

Fig. 5.2  Structure of image fusion at pixel level.

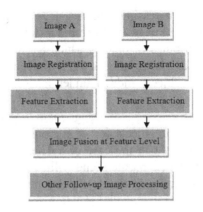

Fig. 5.3  Structure of image fusion at feature level.

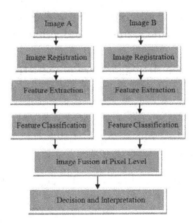

Fig. 5.4  Structure of decision-level image fusion.

The issue of fusion of images acquired from different cameras in order to enhance some interesting features is a complex task. In a common approach one can find four major stages:

(1) The first stage of preprocessing – in this stage it is necessary to implement mechanisms which allow one to geometrically fit many images acquired from different cameras to each other. Those images usually differ from each other in type and resolution, so special attention should be paid to the fitting mechanism. The most important steps to proceed in this stage are: the intersection of two or more input images determination, disparity calculation (disparity is a shift between two images, caused by not identical location of the cameras) and object deformation (caused by different perspectives). The technique which includes such steps is commonly called multi-sensor image registration.

(2) The second stage of preprocessing – in this stage a segmentation of the region of interest of the acquired images is performed. In the case of infrared cameras in such regions the attributes of objects which are not visible by normal optic sensors can be included. The L-band SAR (Synthetic Aperture Radar) can, for example, detect metal objects hidden under leaves, a tent, clothes or those which are painted.

(3) The third stage of preprocessing – this stage main functionality is the extraction of attributes or features of objects (or of regions of interest) and their valid description. This means that in the case of a Thermo Vision camera a proper temperature to proper objects on the scene will be assigned. In turn, in the case of SAR objects, the information of metal detection is assigned, etc.

(4) Image integration – in the last stage information about individual objects (which was extracted in the second and third stage of preprocessing) on the separate images is fused. Not only can individual objects be analyzed and fused, but also regions of interest. The result of this stage is nothing more than an output image which is a combination of all component images with clear and unequivocal enhancement of characteristic features of individual objects or regions.

It may seem that the three first stages are not strictly connected with the issue of image fusion but they are essential and necessary to perform a valid image data fusion process.

### 5.2.1 *Image fusion algorithm description*

The presented algorithm is depicted on the following block diagram (Fig. 5.5):

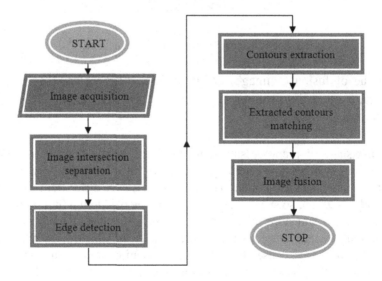

Fig. 5.5 Block diagram of the proposed algorithm.

### 5.2.2 *Image intersection determination*

There are two major types of differences between images acquired from different types of cameras. The first type is image spatial mismatch, which occurs very often and is caused by using two (or more) cameras with different resolutions and/or objective focal lengths. Differences between such images can be repaired by performing proper spatial transformation. The second type of differences is connected with the occurrence of such factors as lighting changes, using different sensors, moving objects on the analyzed scene, etc. These differences, which do not come off after image spatial mismatch, cannot be eliminated in the image registration process stage.

The difference between sizes of images acquired from different sources is caused by the use of different types of cameras. Due to that fact, in the end there will be two (or more) different resolutions of images of the scene analyzed. Sometimes it is even impossible to avoid that kind of situation, because some types of cameras have originally a lower resolution (e.g. Thermo Vision cameras) than others (similar quality but different sensor type, e.g. Night Vision or Day Light cameras). In addition, using different types of cameras (as in the approach presented) causes a necessity of using objectives with significantly different parameters. Therefore, the calibration stage is very important.

Due to the relation between viewing angles of camera sets and focals of their objectives, it is necessary to preliminarily calculate the final size (in pixels) of the analyzed scene. This operation is essential because in the next steps it will be necessary to have the same objects on both images (when using two cameras observing the scene). Additionally, to support performance speed boost these elements which are not present on all the images of the analyzed scene will be omitted. The issue described is illustrated in Fig. 5.6:

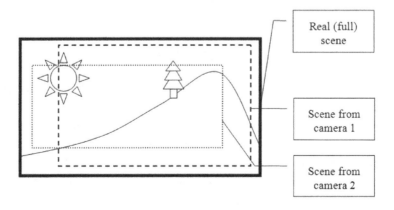

Fig. 5.6 Difference between acquired scenes from cameras with different parameters.

Concluding the figure 5.6 the intersection of two acquired scenes is the fragment presented in the figure 5.7:

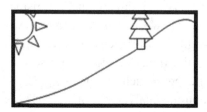

Fig. 5.7  Intersection of acquired scenes.

For the preliminary mathematical determination of the intersection, based on the knowledge about objective focals, the lens focal length and the relative position of cameras, an estimated viewing angle of both cameras should be calculated. Additionally, after final viewing angle estimation, the region of analysis should be limited by ignoring those parts of the scene which are inaccessible by any of the cameras. The mentioned region is shown in Fig. 5.8:

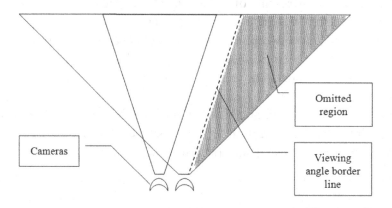

Fig. 5.8  Camera set region of analysis.

If two cameras do not have the same viewing angles (which is a usual case), there will always be an area in one of the images which can be omitted during the analysis. It can be determined by ignoring pixels which are cut off by the border line of the viewing angle (Fig. 5.8). This line is set by drawing a smaller viewing angle at the camera with a greater viewing angle.

The individual viewing angles of camera sets can be calculated by following the approach presented below (Fig. 5.9):

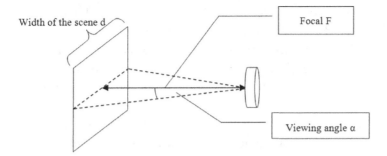

Fig. 5.9  Relation between the camera viewing angle sets and its focal.

In Fig. 5.9 the vertical viewing angle is deliberately omitted, because cameras are set in such a way that they lie on the surface which is parallel to the ground, which eliminates a vertical perspective shift. In order to calculate the viewing angle, a dashed triangle should be analyzed (Figs 5.9 and 5.10):

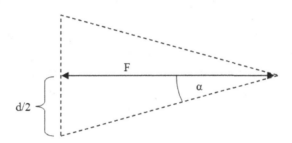

Fig. 5.10  Horizontal surface of the viewing angle.

Based on the trigonometric relation between the viewing angle α, focal F and half of the scene width, one can calculate:

$$\alpha = arctg(\tfrac{d}{2F})$$

(5.1)

The determined and calculated viewing angle of the camera sets is used to determine which pixels can be omitted during analysis.

### 5.2.3  *Image fusion on the feature level*

The next issue connected with image integration is matching the particular scene elements. The effect which is problematic is disparity, heavily used in stereovision systems. Disparity is a shift between the same objects in the separate images acquired from cameras which have different locations in a stereovision set. The relative shift between objects in the analyzed images is related to their distance from the stereovision set. Due to that fact, in the issue presented, it is impossible to fuse images on the pixel level, because the algorithm is not supported by information about the scene depth (distance to individual objects).

In the case of fusing images acquired from different types of cameras, presented by the authors, a characteristic feature detection based approach is used to determine the terms of image integration. Characteristic features are first detected in the first camera and then searched and matched to the same features in the image acquired from the second one. Those features are: contours of warm objects (acquired by a Thermo Vision camera) and contours of objects in rough lighting conditions (acquired by a Night Vision camera). Due to the fact that one of the major tasks of image fusion is to maintain characteristic features of the objects present on the analyzed images, the methods which are based on feature level integration yield subjectively better results than those which are pixel level based (Samadzadegan, 2004).

The main step of this image fusion stage is the analyzed scene decomposition to its characteristic features, e.g. edges. Next, the detected edges are connected in groups and each group represents one object located on the scene. Matching objects between the image from one camera to their equivalents in the image from another camera is done by their shape analysis based on comparing the extracted contours.

In the first step, on both images, the edges of all the objects are determined. In order to perform such a task, one of the many edge detection methods can be used, e.g. the Sobel Operator (Sobel and Feldman, 1973), the Prewitt Operator, the Laplacian of Gaussian or the Canny Edge Detection algorithm (Xin *et al.*, 2012). Even more sophisticated methods of edge detection based on a fractional order derivative can be used (Mathieu *et al.*, 2003; Yang *et al.*, 2011; Gan and Yang, 2010). Due to the fact that the authors' main target was to develop an algorithm which will be implemented in the Mobile Robot, the most important factor in choosing an algorithm was resistance to noise, which is a common problem in mobile vision systems. The second important factor was precision and detection quality. In addition, the authors wanted to choose the fastest algorithm and they planned to place the main emphasis on its optimization.

Therefore, the authors chose the Canny algorithm, which is very precise and resistant to noise. The complex computation requirements of its internal mechanisms were highly optimized to fit the needs of the project presented. To conclude features of the chosen Canny algorithm, it is very important to: gain a low error level, ensure that none of the existing edges is omitted during detection, ensure that none of the false edges, which do not really exist, is detected, properly locate the detected edges to their real position and avoid the multi-detection of one edge (each edge should be detected only once). Based on the above-mentioned criteria, the Canny algorithm in the first step that smoothes the image analyzed in order to remove noise, then finds the image gradient to highlight regions with high spatial derivatives. Finally, the threshold operation is used to choose only the strongest edges.

### 5.2.4 *Matching features detected on particular images*

In the image, an edge refers to an occurrence of a rapid image intensity transition. Edge detection algorithms can find these transitions but usually the edge map, which is the result of using the algorithms mentioned, consists of unconnected short parts of real edges. In order to use such an edge map in high-level algorithms in which objects (extracted from two images) matching algorithms are included, the edge parts present on the map provided have to be grouped in contours.

In order to connect the edges found, two operations are performed on the image: dilation and erosion. For the purpose of contour extraction, the algorithm introduced by S. Suzuki and K. Abe (Suzuki and Abe, 1985) is used. To simplify calculations, the authors decided to choose only the outline contours of all sets of the edges. Due to that simplification, smaller contours that lie inside the larger ones are omitted.

In the next stage, all the contours from the thermo picture are compared and matched with those from the high sensitivity picture. The comparison process can be implemented by using a simple square mean difference or more sophisticated and advanced methods, e.g. Pearson's correlation coefficient, which is widely used and optimized (Loğoğlu and Ateş, 2010). This method is described in detail in 7.2.2. section (equation 7.2).

The presented in 7.2.2. section solution of determining the correlation between two images is used also in the camera set positioning algorithm. This calibration process is performed based on the relation between the camera set position and the object position lying underneath. By means of Pearson's correlation coefficient, succeeding fragments of the indicated object, which are acquired from the previous frames, are located. Then, the proper command is sent to the mobile platform in order to robot's head reach the desired end position. The camera set positioning algorithm mentioned above is presented in chapter 5.3.

### 5.2.5 *Image fusion*

In this stage a new image is created based on data integration, which comes from images acquired from each of available cameras. Based on the objects detected in the process of the second stage and information about their correlation (similarity in both images), it is possible to create a result image which consists of imposed analyzed images. The final result of image fusion depends on the user's preferences, and it should preliminarily be determined which features and how should be enhanced and marked. The main advantage of the presented solution of image fusion is resistance to shift errors which may occur during the image integration process. This is done because each object on the scene is analyzed separately according to its own disparity. This process is described in more detail in the following chapter.

Due to the fact that authors place the main emphasis on Thermo-Vision and Night-Vision image integration, in the solution presented these two types of cameras are used. In the integration process the contours (calculated from edge positions) from a Thermo-Vision image are matched to those from a Night-Vision image. This priority allows one to preserve all the elements of the Nigh-Vision image with special enhancement of objects with a temperature significantly different from that of the environment.

### 5.3  Head (camera set) positioning algorithm

Due to the nature of tasks to which the Mobile Robot made at the Institute of Applied Computer Science is dedicated, its main aid subsystem is a vision based camera set positioning mechanism. The main functionality of this mechanism is to accurately drive towards a suspicious object lying on the ground in front of the robot and to not ram or overrun it. The need of such a functionality results from the fact that there is a big distance between the robot and the operator during the task performance. The only available source of vision information is a camera set installed in the robot's head. Its position above the suspicious object is vital and should be as optimal as possible. A situation in which such a vision aid subsystem is useful is presented in Fig. 5.11:

Fig. 5.11 Ground observation.

Nearly all inspection robots equipped with a camera installed on an arm manipulator are capable of performing close observation of an object keeping a safe distance between this object and the front of the drive platform. Unfortunately, such task performance needs the presence of a highly qualified and experienced operator. The assumption of the positioning algorithm project is the application of image data acquired from the robot's head as the only source of orientation information in a user-friendly aid subsystem which does not require any qualified personnel. Such ground observation mechanisms have already been researched and developed, and are in use (Kawamata *et al.*, 2002).

Besides the image sequence, the algorithm is supported only by the operator mark i.e. a spot on the screen of the console, which was pointed by the operator in order to determine a suspicious object. In Fig. 5.12 an example view from the robot's head observing the ground (inside the building) is presented:

Fig. 5.12 View from the robot's head observing the ground.

Owing to the fact that the algorithm cannot assume that the operator will mark the location of a suspicious object sufficiently well, the nearest neighborhood of the pointed spot should be analyzed in order to find places of reference which will be used for the camera set positioning process. The authors of the algorithm presented decided to use template matching mechanisms to compare the analyzed image (frame) and fragments of the previous frames which include the places of reference mentioned. To compare such a pair of images the authors used Pearson's correlation coefficient (Loğoğlu and Ateş, 2010).

It is assumed that the implemented functionality (camera set positioning) is performed in the forward-backward direction of the drive platform of the Mobile Robot. This means that, in the case of marking a suspicious object near the left or the right border of the console screen, the rotation of the drive platform will not be necessary. Such an approach can be adopted due to the fact that the robot's head is capable of rotating in the spherical system, so it will be able to face towards the marked object.

Owing to the moving direction limit of the mobile platform during the camera set positioning process, the algorithm can skip the deformation of objects placed on the analyzed scene caused by different perspectives of the camera set in different positions of the platform. Those deformations would be much more intense if the platform could rotate during the positioning process. Nevertheless, to ensure the minimization of incorrect detections of searched objects, the authors decided that the searched template is a strip of pixels cut from the previous frame which includes the marked object. The width of this strip is the same as the width of the analyzed image. In the succeeding frames the place where the correlation coefficient will be the highest (between the cut-off strip of pixels from the previous frame and the current frame) is a place where the marked object is located (only in the top-bottom image dimension, because the algorithm does not analyze the left-right variation of the object location). In the case of the first frame, the strip of pixels to match on the next frames will include the spot marked by the operator. This kind of approach implies that not only the marked object is being tracked but also any other objects that lie on the same vertical position. Due to that fact, the algorithm is often supported by more than one place of reference during performance. In Fig. 5.13 a case of its standard use is presented. The operator marked a suspicious object with a slight error (the place where the operator touched the console screen is marked as a cross with two circles).

Fig. 5.13 Ground observation with suspicious object marking performed by the operator.

For the image acquired from a ground observation camera depicted in Fig. 5.13, a fragment which will be searched for on the next frame is a strip of pixels with the determined width (the same as the width of the image depicted in Fig. 5.13) and height. The above-mentioned strip of pixels is shown in Fig. 5.14 and its vertical zoom in Fig. 5.15. The height of the pixel strip depends on the computing power of the computer executing the implementation of the algorithm presented. Due to the fact that the presented functionality is crucial from the viewpoint of the robot's safety, incorrect detections caused by slow computing are not allowed to happen. During the tests as well as in the target implementation, the authors decided to use a constant height value of the pixel strip which is 50 pixels. The proposed value is connected with a number of pixels necessary to fully display an object with sizes similar to an antipersonnel mine in the image acquired from cameras dedicated to the Mobile Robot project. To such objects sizes (pixel number needed to display) research, the authors took into consideration the robot's head position, which was tilted over the test object at about 50 cm.

Fig. 5.14 A fragment of the image (presented in Fig. 5.13) chosen to search for on the next frame.

Fig. 5.15 Vertical zoom of a fragment of the image (presented in Fig. 5.13) chosen to search for on the next frame.

While comparing images (current frame and the strip of pixels cut from previous frame) a Pearson's correlation coefficient is calculated repeatedly for each vertical position on the current frame. The result of such comparing process is a vertical vector of normed values of Pearson's correlation coefficient and the highest value index indicates a vertical position of a place where most likely lies the marked object. Such position can be calculated from the following equation:

$$i_{max} = index[\max \begin{pmatrix} p_1 \\ p_2 \\ p_3 \\ \vdots \\ p_{n-1} \\ p_n \end{pmatrix}] \qquad (5.2)$$

where:

$i_{max}$ – is the vector index referring to the position of the highest value of Pearson's correlation coefficient,

index() – is the function that returns an index to the given element,

max() – is the function that returns the maximum value of the given vector,

$p_1, p_2, ..., p_n$ – are the values of Pearson's correlation coefficient calculated by comparing the process between the chosen strip of pixels cut from the previous frame and the corresponding strip of pixels of the current frame (which is the same size). To calculate those values, equation 7.2 described in 7.2.2. section is used:

n – the size of the result vector which differs from the height of the current image. The difference is caused by the necessity of comparing the whole strip of pixels to the corresponding pixels of the current image. Therefore, comparing the edges on the current image (the middle row of pixels of the pixel strip is on the edge pixels of the current image) is impossible and comparison must be performed according to the determined range of the current image pixel indexes. The height of the result vector can be calculated from the following equation:

$$n = h_c - h_t + 1 \qquad (5.3)$$

where:

$h_c$ – is the current image height,

$h_t$ – is the height of pixel strip cut from the previous frame (in the tests and in target implementation this value was set to 50 pixels, as discussed earlier).

After conducting tests, the authors of the presented algorithm noticed a slight dislocation of the final indication of the marked object when the camera set reached the final position (above the target). This dislocation of indication and real position of the marked object is caused by a slight deformation of objects in the image. In turn, this deformation is caused by the change of perspective on the succeeding frames during the drive platform movement. Due to that deformation the highest value of Pearson's correlation coefficient location was slightly shifted from the real position of the marked object. The change in the perspective was obviously caused by the movement of the robot. Despite the fact that the final error was insignificantly small, the authors of the presented algorithm, due to the crucial nature of the working conditions of the algorithm, implemented additional security mechanisms. Those security systems are mainly based on the Region of Interest of the current frame control. Reducing the region of analysis aims to eliminate false indications which could occur if the whole image were analyzed. If the highest value of Pearson's correlation coefficient is located beyond the neighborhood of the marked object, the drive platform will move towards that location despite the fact that the object lies somewhere else. That kind of situation can be caused by the uniformity of the observed ground. Then, the differences between the correlation coefficients calculated for each vertical position of the current frame image are minimal. What is more, when on the scene there is an object similar to the marked one, the indication on the succeeding frames can be drastically different from that of its real location. Therefore, the authors of the presented algorithm decided to reduce the region of analysis of the succeeding frame to the following range: from the current position of the suspicious object to the center of the frame. In standard use case of the algorithm presented, the final position will never be located beyond this area. For the purpose of reduction of the region of analysis, the authors used the Region of Interest (ROI) mechanism. During the algorithm performance and the platform movement, ROI is progressively reduced. The ROI reduction described above and the validity of its use are illustrated in Figs. 5.16, 5.17 and 5.18:

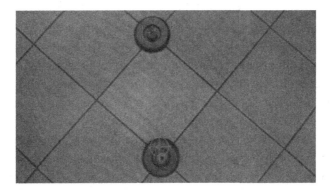

Fig. 5.16 Ground observation when more than one suspicious object lies on the scene. Marking of the operator is clearly visible.

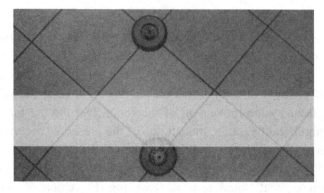

Fig. 5.17 Introduction of the region of analysis reduction. In the image an area between the vertical center of the image and the current position of the pointed object is marked with a bright color.

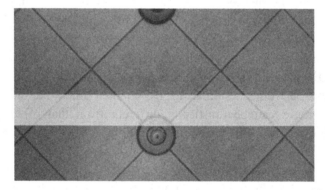

Fig. 5.18 Region of analysis reduction during the camera set positioning algorithm performance (the drive platform heads towards the marked object).

The bright area marked in Figs 5.17 and 5.18 is reduced while the drive platform is approaching the position in which the camera set is tilted exactly over the pointed object. The succeeding reduction of ROI illustrated secures the algorithm from any unexpected behavior, especially when any of the objects present on the scene is similar to the marked one. These situations can take place especially when the ground is uniform and lacks characteristic features while the marked object is sufficiently well camouflaged. Obviously, the ROI reduction in the range from the current object position to the center of the current frame does not secure one from false detections which could be located inside the already reduced ROI. Then, in the worst case the algorithm will finish its task earlier than the operator assumed but the platform will not run over the suspicious object which was marked by the operator.

Besides the increased security against unexpected behavior provided by the ROI control, the reduction of the region of analysis offers one more major advantage. This is a performance speed increase caused by excluding significant parts of the image from analysis. Additionally, the increase is greater when the camera set is located closer to the final position. And that is when the performance speed has the utmost importance in terms of accuracy.

## 5.4 Implementation

The solutions presented in this article were tested in the Mobile Robot built at the Institute of Applied Computer Science. Due to the necessity of increased efficiency and adaptations of the algorithm to work in real-time conditions, the methods for optimization in computation time were applied. The algorithms proposed were implemented on a computer with an Intel Atom Z530 (1,6 GHz) processor which is installed in the Mobile Robot platform. The technologies applied are: C++ with the use of the OpenCV library.

## 5.5 Results

During experiments conducted, the authors used two types of cameras: a Night Vision camera and a Thermo Vision camera. In the case of the first algorithm (image fusion) both cameras were necessary. In turn, to the second algorithm (camera set positioning) only a Night Vision camera or even a Day Light camera, which is also installed in the Mobile Robot's head, was needed. To test quality, both algorithms were implemented on a PC with an Intel(R) Core(TM)2 Quad (2,83 GHz) processor. In the case of image fusion algorithm, the result of enhancement of desired features (temperature) is presented on the following images: Figs. 5.19, 5.20, 5.21 and 5.22 (selected frames from the image sequence, displayed in real-time).

Fig. 5.19 Image acquired from a Night Vision camera.

Fig. 5.20  Image acquired from a Thermo Vision camera.

Fig. 5.21  Edges detected from the picture presented in Fig 5.20.

Fig. 5.22  Fused image. Enhancement of areas with a higher temperature.

The results of the quality tests of the camera set positioning algorithm in laboratory working conditions were satisfactory because no false detections and marked object run overs were recorded.

The target implementation of the image fusion algorithm done on an Intel Atom Z530 reached approximately 15 fps performance speed. In the case of the camera set positioning algorithm, the authors managed to reach the human eye processing fluency, which is approximately 25 fps. The speed was higher when the marked object was close to the center of the image. The results of the quantity tests of implementations of both vision aid algorithms are shown in the following table (Table 5.1):

Table 5.1 Performance tests results of presented algorithms.

|  | Intel(R) Core(TM)2 Quad (2,83GHz) [fps] | Intel Atom Z530 (1,6Ghz) [fps] |
| --- | --- | --- |
| Image fusion algorithm | over 25 | 15 |
| Camera set positioning algorithm | ~50 | ~25 |

## 5.6 Conclusions

Considering the image fusion algorithm, objects of distinguished temperatures are marked in the result image and can be quickly located on the scene. As can be noticed, the algorithm is resistant to disparity caused by different object distances from the camera set. In addition, due to the use of a high sensitivity camera, the feature enhancement can be performed in rough lighting conditions.

The results of the algorithm presented demonstrate immense possibilities of the use of such a solution in surveillance tasks. It can be a powerful enhancement of an observer workstation in many fields of interest.

As far as the camera positioning algorithm is concerned, a significant enhancement of the operator console was obtained. After the use of safety mechanisms the operator can easily and safely position the camera set in order to better observe a suspicious object.

## 5.7 Further research

The authors of the image fusion algorithm presented plan the following:

(1) to adapt the solution and its implementation to enhance more object features which can be detected by different types of cameras;
(2) develop the multiple (not only two) image source fusion;
(3) use numerous upgrades of the detection or comparison techniques which are now available.

As for camera positioning algorithm, the authors plan to implement not only one direction (forward-backward) calibration but also the rotation of the platform to fully support object centering on the image.

# References

Al-Wassai, F. A., Kalyankar, N. V. and Al-Zuky, A. A. (2011). Arithmetic and Frequency Filtering Methods of Pixel-Based Image Fusion Techniques, *International Journal of Computer Science Issues* (IJCSI), Vol. 8 Issue 3, p. 113.

Arora, M. and Banga, V. K. (2012). Intelligent Traffic Light Control System using Morphological Edge Detection and Fuzzy Logic, *International Conference on Intelligent Computational Systems*, Planetary Scientific Research Centre, Dubai.

Blum, R. S. and Liu, Z. (2005). Multi-sensor image fusion and its applications, *special series on Signal Processing and Communications*, Taylor & Francis, CRC Press.

Choudekar, P., Banerjee, S. and Muju, M. K. (2011). Real time traffic light control using image processing, *Indian Journal of Computer Science and Engineering*. Vol. 2, No. 1, pp. 6-10.

Derek, L.G. et al. (1994). Accurate Frameless Registration of MR and CT Images of the Head: Applications in Surgery a Radiotherapy Planning Dept. of Neurology, *United Medical and Dental Schools of Guy's and St. Thomas's Hospitals*, London, SEt 9R, U.K.

Gan, Z. and Yang, H. (2010). Texture Enhancement though Multiscale Mask Based on RL Fractional Differential, *International Conference of Information Networking and Automation*, Vol. 1, pp. 333-337.

Kawamata, S., Ito, N., Katahara, S. and Aoki, M. (2002). Precise position and speed detection from slit camera image of road surface Marks. *Intelligent Vehicle Symposium*, Versailles, France.

Liu, X., Wang, Y., Liu, Y., Weng, D. and Hu, X. (2009). A remote control system based on real-time image processing, *Fifth International Conference on Image and Graphics*, IEEE Computer Society, Xi'an, Shanxi, China.

Loğoğlu, K. B. and Ateş, T. K. (2010). Speeding-Up Pearson Correlation Coefficient Calculation On Graphical Processing Units, *Signal Processing and Communications Applications Conference (SIU)*, Diyarbakir.

Mathieu, B., Melchior, P., Oustaloup, A. and Ceyral, C. (2003). Fractional differentiation for edge detection, *Signal Processing*, Vol. 83, No. 3, pp. 2421-2432.

Samadzadegan, F. (2004). Data Integration Related to Sensors, Data and Models, *International Archives Of Photogrammetry Remote Sensing And Spatial Information Sciences*, Vol. 35, pp. 569-574.

Snidaro, L., Visentini, I. and Foresti, G. L. (2009) Multi-sensor Multi-cue Fusion for Object Detection in Video Surveillance, *AVSS '09 Proceedings of the 2009 Sixth IEEE International Conference on Advanced Video and Signal Based Surveillance*, IEEE Computer Society Washington, DC, USA, pp. 364-369.

Sobel, I. and Feldman, G. (1973). A 3x3 Isotropic Gradient Operator for Image Processing, in *R. Duda and P. Hart (Eds.), Pattern Classification and Scene Analysis*, pp. 271-272.

Sun, Z., He, D. and Zhang, W. J. (2012) A systematic approach to inverse kinematics of hybrid actuation robots. *Advanced Intelligent Mechatronics (AIM)*, Kachsiung.

Suzuki, S. and Abe, K. (1985). Topological Structural Analysis of Digitized Binary Images by Border Following. *Computer Vision, Graphics, And Image Processing 30*, pp. 32-46.

Xin, G., Ke, C. and Xiaoguang, H. (2012). An improved Canny edge detection algorithm for color image, *Industrial Informatics (INDIN)*, Beijing.

Yang, Z., Lang, F., Yu, X. and Zhang, Y. (2011). The Construction of Fractional Differential Gradient Operator, *Journal of Computational Information Systems 7:12,* pp. 4328-4342.

Zhang, J. (2010). Multi-source remote sensing data fusion: status and trends, *International Journal of Image and Data Fusion*, Vol. 1, Issue 1, pp. 5-24.

Zhang, Y., Zhao, G. and Zhang Y. (2011). Design of a remote image monitoring system based on GPRS, *International Conference on Machine Learning and Computing*, IACSIT Press, Singapore.

Zin, T. T., Takahashi, H., Toriu, T. and Hama, H. (2011). Fusion of Infrared and Visible Images for Robust Person Detection, *Image Fusion, Osamu Ukimura (Ed.)*, InTech.

# CHAPTER 6

# PATTERN RECOGNITION ALGORITHMS FOR THE NAVIGATION OF MOBILE PLATFORM

Rafał Jachowicz, Sylwester Błaszczyk, Piotr Duch, Maciej Łaski,
Adam Wulkiewicz, Dominik Sankowski, and Piotr Ostalczyk

*Institute of Applied Computer Science*
*Lodz University of Technology*
*90-924 Łódź, ul. Stefanowskiego 18/22*
*{rjachowicz, sblaszc, pduch, mlaski, awulkie, dsan}@kis.p.lodz.pl,*
*piotr.ostalczyk@p.lodz.pl*

Modern industrial robots and those dedicated for special operations, in particular, multi-tasking, more frequently require mobility to properly carry out their tasks. The mobility of such robots gives their operators wider flexibility than in the case of dealing with stationary robots. Mobile robots can also be equipped with some autonomous behaviors, which can lead to performing a given task independently. Regardless of the type of tasks which the robot has to carry out, it requires the system to recognize its surroundings. The process of recognition of the surroundings is very important, because the output data of this algorithm are further used as input data for other algorithms, and the correctness of those data is a prerequisite to the proper functioning of the algorithms. There are many systems for studying and modeling of the surrounding environment which works in real time. Among others, they are based on the following: laser scanners (Karakaya *et al.*, 2012; Isenburg and Shewchuk, 2009), ultrasonic sensors (Gomez *et al.*, 2003; Iida *et al.*, 2004) and stereovision systems (Wang *et al.*, 2011; Hu *et al.*, 2005).

## 6.1 Robot vision system

During the work on a mobile robot platform for reconnaissance and mine detection tasks, described in chapter 5, a subsystem for detecting and tracking objects located in the close surroundings of the robot, based only on image data was developed. Those data are received from three different types of cameras, namely: a high resolution vision camera, a high-sensitivity camera, which can be successfully used instead of military night vision cameras, and a thermal camera. All the cameras are mounted in the robot head, placed on a four-joint arm.

The aim of the designed video subsystem for the mobile robot was to use the image data as a source of information about the surrounding environment. The image data provides much more information about the observed scene in comparison to other sources of data, e.g. laser scanners or ultrasonic sensors. Stereo cameras can provide information about the image depth and spatial orientation of the robot in the same manner as lasers with additional information about visible objects. Fazl-Ersi and Tsotsosa (Fazl-Ersi and Tsotsos, 2012) present a survey on robot localization methods in the open air based only on image data.

We still do not know the mechanisms of extracting all the properties of the observed scene, based only on the images even if each of us can do this almost completely at "first sight". We can identify colors, shapes, sizes and distances to objects which appear in our field of vision. Based on this knowledge we are even able to recognize objects which are partially obscured or hidden.

In the case of image data analysis algorithms, the main challenge is the constantly changing background and the possibility of sudden changes in lightning. The latter can occur when a mobile robot enters or exits the room. The first problem disqualifies algorithms based on the subtraction of two consecutive frames, which is commonly used in algorithms for object detection and tracking (Wang *et al.*, 1993). Additionally, it also excludes algorithms based on background extraction (Chen and Yang, 2011), used in static monitoring systems. Such algorithms remember what the background looks like at different times of day, and then those remembered backgrounds are subtracted from the current frame and all the changes on the scene are detected. The methods of background analysis and segmentation mentioned above can rarely be used in mobile systems, and even in cases when those methods can be employed, there is no guarantee of sufficient and accurate results. However, a background analysis module can be used as a supporting module, which improves the efficiency of the algorithm. Thus, for the proper extraction, localization and tracking of objects, the algorithm should be focused on the analysis of characteristic features, the properties which make the object recognizable (He *et al.*, 2011). Just as we are able to recognize people or objects based on their specific features, the algorithm should choose from the many characteristic points those which best match the object searched for. It should be noted that we are doing this on the basis of our expert knowledge. The recognition of characteristic features is possible due to the prior knowledge and memorization of them. The more we know about these objects, the easier it is to notice and recognize them, even if they are partially hidden.

In order to investigate the possibility of the mobile platform navigation based only on image data, during the work on mobile robot platform for reconnaissance and mine detection tasks, the algorithm to find a specific marker in the image acquired from a light vision camera was developed. In the first step the marker is found on the image and the distance between the camera and the marker is calculated based on its size, which is known a-priori. Next, the designed system, based on the location of the marker on the image and the calculated distance, sends an order to the platform to move in an appropriate direction (toward the marker). The speed of this movement is calculated in such a way as to maintain an initially set distance between the marker and the platform. This functionality was implemented in order to release the operator from controlling the mobile platform during the march of the unit if another kind of transport for the mobile platform was unavailable.

## 6.2 Robot vision system

In this chapter a novel algorithm for the realization of the functionality (marker detection, confirmation and tracking) described in the previous section is presented. The solution is based on a pattern matching algorithm using the Pearson correlation coefficient. This method is commonly used in image recognition and analysis algorithms, e.g. to determine the similarity between two images (Neto *et al.*, 2007). The crucial step of the presented algorithm is the possibility of restoration of a mathematical shape of the marker at any time when the algorithm is running. The marker should be placed on the operator's back, so that the robot can follow him.

### 6.2.1 *Marker*

The marker used during the experiment carried out in a laboratory and later in the conditions close to real ones was a black circle of a diameter of 12 cm placed on white paper of A4 format. The size of the marker was determined on the basis of the conducted experiments that led to the determination of the minimal size of an object which can be detected on the images from a distance of 10 meters. In this case, the authors used a visible light camera with a resolution of 1000x1000 pixels with the lens with a viewing angle of 55°.

When the marker is found for the first time on an image from a sequence of images, the part of the currently analyzed image which contains the marker found is treated as a new pattern, which will be searched for on the following images. Additionally, the pattern found is validated, calibrated and its deformation and motion trajectory are examined.

### 6.2.2 *Pattern matching*

The basis of the algorithm is pattern comparison with all the areas of the analyzed frame where this pattern may appear. The authors, after testing many options, decided to choose the Pearson correlation coefficient as the best method to determine the similarity of the pattern and the analyzed part of the image. In the case of image analysis algorithms, a two-dimensional version of the Pearson correlation coefficient can be used:

$$R(x,y) = \sum_{x',y'} (T'(x',y') \cdot I'(x + x', y + y')) \tag{6.1}$$

where:

$T$–thepattern (the pattern dimension has to be equal or smaller than the dimension of image $I$).

$I$–the image where the pattern $T$ is searched.

$x', y', x'', y''$–the vertical and horizontal coordinates of the pixel in pattern $T$ in a range from 0 to the width and height of the pattern

$x, y$–the vertical and horizontal coordinates of the pixel in image $I$.

$w, h$ - the height and width of the pattern $T$

$$T'(x',y') = T(x',y') - \frac{1}{w \cdot h} \sum_{x'',y''} T(x'',y'')$$

$$I'(x + x', y + y') = I(x + x', y + y') - \frac{1}{w \cdot h} \sum_{x'',y''} I(x + x'', y + y'')$$

In order to obtain the value of the Pearson coefficient within the range of -1 – 1, an additional mathematical operation is necessary:

$$R(x,y) = \frac{\sum_{x',y'} (T'(x',y') \cdot I'(x + x', y + y'))}{\sqrt{\sum_{x',y'} T'(x',y')^2 \cdot \sum_{x',y'} I'(x + x', y + y')^2}} \tag{6.2}$$

Equation (6.2) returns normalized values of the Pearson coefficient, and 1 means that both of the images analyzed are identical. And value -1 indicates that the analyzed images are their negatives. Values close to 0 mean that images are completely different from each other. It should be noted that the presented method not only analyze the brightness of color of each pixel but also takes into account the average brightness of the analyzed image. Unlike the methods which are based on histogram analysis and which require additional methods of object shape and texture analysis (Zhang *et al.*, 2005), the Pearson correlation coefficient includes all the aspects necessary to determine the similarity of two analyzed images (which is an important part of the algorithm presented).

### 6.2.3 *Pattern tracking*

In the approach presented, an algorithm is prepared to track object which is marked with the symbol known a-priori. The most difficult part in this approach is to separate the object being tracked from the background. The authors are aware that such a separation may not always be possible, which means that in the analyzed part of the current frame some parts of the background may be included. For this reason, the authors decided to use the fragment of the current image which contains a pattern as a new pattern for the following frame. This is possible because of the assumed speed of the presented algorithm which is approximately 25 fps (frames per second), and is close to the capabilities of the human eye. The authors also assumed that the speed of the tracked object is smaller than 15 km/h (the speed of a running man), which is also limited by the maximum speed which can be achieved by the robot used, described in chapter 5. Due to such an assumption the difference between the consecutive frames in the area where the marker appears is insignificant, so the part of the previous frame can be used as a pattern for the next frame. Additionally, those differences are taken into account during the analysis of the Pearson correlation coefficient; namely, the value of the threshold for this coefficient can be changed. The threshold is the value of the coefficient from which the analyzed part of the image can be considered as a potential location of the marker. The value of this threshold can be also changed due to the appearance of various factors in the environment, such as changes in the lighting conditions. Although the lighting conditions are compensated by different image processing algorithms, e.g. histogram equalization, they still may make changes in the image in such a way that the results of calculation of the correlation coefficient will be disqualifying.

Regardless of the fact that the value of the threshold for the correlation coefficient has to be continuously controlled to avoid false results, the authors successfully implemented the detection and tracking algorithm based only on the search of the part of the previous frame which contains the marker on the current frame. The marker used by the authors was a black circle painted on a white rectangle. This marker can easily painted and restored. Figs. 6.1–6.6 show the results of applying the presented algorithm in real conditions. During the test the

*R. Jachowicz et al.*

mobile platform with a day vision camera was controlled manually by the robot operator, and the algorithm was run on the onboard computer. The task of the algorithm was to detect the marker in real time.

Fig. 6.1 and Fig. 6.2  The Frame from a video sequence with a correctly detected marker.

Fig. 6.3 and Fig. 6.4  Frame from a video sequence in which the marker was found with a visible error and the error is growing while processing succeeding frames.

Fig. 6.5 and Fig. 6.6 Frame from a video sequence in which the marker was lost. Instead to the marker some part of the background was detected and treated as a pattern for the next frame.

The black points in the figures above indicate the places where the value of the correlation coefficient, between the pattern (part of the previous frame where the marker was found) and the part of the image was the highest. The black circle is the estimated size of the marker, determined using the Hough transform (Xu and Velastin, 1993). The Hough transform was originally developed to detect lines on the image, but later its use was also extended to detect circles. An example of this algorithm is a version presented by Chengpinga and Velastina, based on the weighted Hough transform with the use of the Mehalanobis distance (Xu and Velastin, 1993):

$$M(\hat{a}_i) = \sum_{k=1}^{M} I[d_{ik}(\hat{Z}_k, \hat{a}_i)] \cdot [1 - \omega C(\hat{a}_i)] \tag{6.3}$$

for i = 1, ..., N.
where:
$d_{ik}$ – the Mehalanobis distance,
$\varepsilon$ – the chi-square variable,
$\omega$ – the weight coefficient for the cost function $C(\hat{a}_i)$ defined by:

$$C(\hat{a}_i) = \frac{\sqrt{\varepsilon}}{\sqrt{d_{ik}}} \tag{6.4}$$

Function I [$d_{ik}$] is defined by:

$$I[d_{ik}] = \begin{cases} 1 & d_{ik} \leq \varepsilon \\ 0 & d_{ik} > \varepsilon \end{cases} \tag{6.5}$$

The algorithm presented was implemented on the Intel® Core™ 2 Duo T6600 (2.2 GHz). The algorithm works with approximately 50 frames per second, using a day light camera with a resolution of 1002x1002 px.

### 6.2.3 Pattern confirmation

One can easily notice that with each consecutive frame the part of the image which is treated as a pattern for the following frame moves away from the real position of the marker in the image. Due to this fact, after a few frames, the marker was completely lost and was replaced in the pattern by part of the white background. This happened because of differences in the neighborhood of the template between the frames which were mentioned earlier and were decided to omit. In fact, if only the Pearson correlation coefficient is considered on one single frame, differences are insignificant because the maximum of its value lies in the close neighborhood of the real location of the marker. Unfortunately, in the case of exact detection of the marker on the sequence of frames, where the error on each frame can be multiplied, the differences between these frames imply a significant error on the final Distribution of the Correlation Coefficient (DCC), which is not less than the smallest possible unit of the whole system which refers to the granulation of this system. In the case of digital vision systems this unit is of course one pixel. Therefore, a shift between the prepared template indication and the real location of the marker in the image, even equal to this one pixel, is inevitable. In addition, one can notice that this shift is not random and is characterized by some tendency which is connected with the movement of the tracked object in the image. The effect is similar to the situation where the object escapes from under the magnifying glass. In response to the issue that arose, the authors of the presented algorithm used, a well-known solution found in common sources: the solution of smoothing the result image, which represents Distribution of the Correlation Coefficient, with an adaptive Gauss smoothing algorithm (Yan *et al.*, 2005). After this mathematical operation in the result image, which represents the DCC, the maxima the values of which were the highest, were slightly shifted towards their real location, because for each aggregation of high correlation coefficient, the values of the final local maximum were determined based on all of its closest values. The authors hoped that this solution would limit the effect of the error mentioned on the final shift of the template detection. Unfortunately, the attempts of DCC correction did not bring satisfactory results because the final shift was still inevitable, although it was significantly delayed.

To solve the problem presented in figures 6.1–6.6, the authors decided to add an additional step of the algorithm, namely pattern correction. In the proposed solution after the initial determination of the place where the marker may appear, the procedure of pattern confirmation and centralization is run. The procedure presented can be divided into three stages.

(1) In the first step, the shift of the point indicated by DCC and the real center of the figure is determined. To simplify the algorithm, the authors decided to introduce an additional marker with a known shape which is easy to restore. As a marker the authors used a black circle painted on a white rectangle, as mentioned in chapter 6.2.1. The dimensions of the circle and the rectangle were known a-priori. In the case of a marker with a circular or elliptical shape, the determination of its shift and center is relatively simple. This procedure is reduced to the calculation of the circle diameter or the ellipse axis length in pixels, and then the center of the marker can be easily found. In the case of non-smooth shapes, such as rectangular ones, in order to determine the center of the figure, one has to refer to diagonals. To find the center of the figure, the intersections of the diagonals have to be determined. In the case of non-symmetrical shapes, such as triangular ones, the axis of symmetry has to be used. In general, it can be assumed that if the figure used as a marker can be described mathematically, one can find the center of such figures using simple mathematical operations. In the worst case, the determination of the average values of the length and width of the figure, which can be found based on the color of the figure, will be necessary. The center of the figure calculated in such a way is close to the center of mass of the figure.

(2) In the second step, on the basis of the knowledge about the center of the figure, the area for further analysis, determined in the previous step, is limited to the place around the central point of the figure.

(3) The third step is the confirmation of the found marker. The mathematical description of the marker is known a priori, therefore it is possible to confirm whether this marker actually appears in the place of the highest possibility of occurrence of the marker. During the first two steps of the algorithm it is assumed that the marker was found correctly and its parameters (shape, size, color) are measured. In the third step this information are verified through the reconstruction of the marker and the comparison with the part of the image determined in the second step.

Checking the correctness of indication of the Pearson correlation coefficient is possible without the determination of the shape and color of the figures. However, it is computationally expensive. A better solution is to assume that the indication of the Pearson correlation coefficient is correct so that the marker is at this point. Then, its parameters can be determined, and based on them, the marker can be restored. Then, the comparison of the marker thus created with the part of the image is used. This solution eliminates the possibilities of incorrect detection of the marker.

### 6.2.3.1 *Image pre-processing*

To determine the shape and color of the figure, the authors used binary thresholding and methods for finding contours. In the first step, it is necessary to find the center of the figure which represents the marker, and set the area of the image so that the center of the figure will also be the center of the analyzed part of the image. This situation in presented in Fig. 6.7:

Fig. 6.7 Location of the indication of the highest value of the correlation coefficient.

The first part of the confirmation and centralization procedure is the separation of the background pixels from those of the figure pixels. This is done by the simple operation of thresholding. First, the image is converted to grayscale and next, the pixels with a value higher than the threshold are set to the maximum, while those with a value lower than the threshold are set to zero. As a result, one obtains a binary image, where the black color corresponds to the pixels belonging to the figure, whereas the white color corresponds to those belonging to the background. In Fig. 6.8 the image after applying those operations is presented.

Fig. 6.8 Figure 6.7 after applying the thresholding operation.

One can easily notice that after the thresholding operation applied with appropriate values of the threshold, the figure and the background are clearly separated. The value of the threshold is set during the analysis of the current frame. This threshold is calculated based on the average value of the pixels belonging to the figure, and those belonging to the background. In order to avoid the inclusion of the pixels belonging to other objects which may appear on the image in calculations, the background is defined as an area in the close vicinity of the figure. A different method for determination of the initial threshold value is used. In this case, the authors assumed that in the image the value of the pixels of the black color of the figure is close to the value of the darkest pixel in the image. This assumption was supported by the experiments conducted. In the same way, the value of the pixels belonging to the white color of the marker background is close to the brightest pixel on the analyzed image. Based on that assumption, the value of the threshold is determined by calculating the average of the brightest and the darkest pixels on the image.

### 6.2.3.2 *Determination of the center of the circle*

After applying the operation of thresholding on the image acquired from a camera, the next step is the calculation of the center of the figure, in this case, a circle. The center of the circle is determined based on the calculation of number of the black pixels belonging to the figure in each of four directions. Considering that the figure used as a marker is symmetric, it is easy to determine the shift between the point which is indicated by the DCC and the real center of the figure.

*R. Jachowicz et al.*

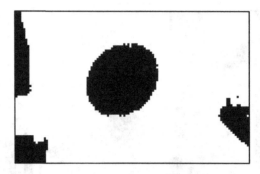

Fig. 6.9 Indication of the highest DCC value Shift determination using a pixel counting procedure.

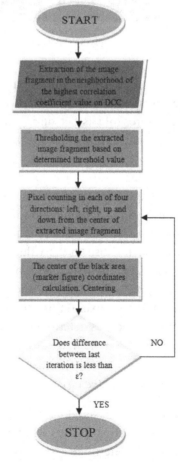

Fig. 6.10 Block diagram of DCC indication centering algorithm, and the determination of approximate marker parameters. ε - accuracy threshold, below which the algorithm acknowledges the determined shift of DCC indication as precise.

Fig. 6.9 presents a situation in which the point indicated by the highest value of the Pearson correlation coefficient is shifted from the real center of the figure. The gray point indicates the highest value of the Pearson coefficient while the white one the center of the figure – the place where the highest value of the coefficient should be found. The gray arrow represents the directions in which the number of black pixels is accumulated. One can easily notice that after applying the above-mentioned operation on the image, the calculated center of the figure is close to its real center point.

In Fig. 6.10 a complete algorithm for the correction of the DCC indication and calculation of the marker parameter is presented.

When the algorithm presented in Fig. 6.10 finishes (that is, when the difference between the coordinates of the center of the marker figure on the following frames is less than $\varepsilon$, for example, is equal to one pixel), it can be assumed that in the case of a circle as a marker figure, the average number of pixels in each direction is the radius of this circle.

However, there are situations where such an approximation can be insufficient. Those situations occur when the marker, due to its movement and rotation, is spatially deformed on a two-dimensional image. Such a situation is presented in Fig. 6.11:

Fig. 6.11 Spatial deformation of objects in two-dimensional image (a circle becomes an ellipse).

In Fig. 6.11 the marker, which in reality is a circle, in the two-dimensional image representation becomes an ellipse. In the case of other figures there can be even greater deformations, such as a square becoming a parallelogram or a rectangle becoming a square. In situations where the deformation mentioned occurs, two solutions can be applied. The first is to invoke a recursively centering procedure until satisfactory results are achieved. A satisfactory result of the recursive method can be DCC indication correction which is less than the assumed $\varepsilon$ error, as described earlier. The second is to consider analyzing mathematical characteristic features of figure which is used to create a marker. In this approach, mathematical characteristic features of the figure used to create a marker have to be known a priori. For a circle, this is a radius which, regardless of the spatial position of the surface on which it lies, will always be equal to the longest distance (observed on a two-dimensional image) between the center of the deformed figure and its edge. It follows that the diameter of real (undeformed) circle will be the longest distance between two points on the edge of the figure.

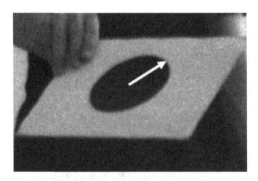

Fig. 6.12 Finding mathematical characteristic features of marker figures. In the image the circle radius is presented.

In Fig. 6.12 a marker the base figure of which is a circle is presented. The mathematical characteristic feature to be found in order to determine the marker parameters is marked with a white arrow. As has been described above, to find this characteristic feature, the authors used the fact that, in every spatial deformation of a circle presented on a two-dimensional image, always the longest distance between the center of the observed ellipse (which really is a circle) and its edge is the radius of this circle. This is the simplest case. In a situation where the marker figure is a square, the longest distance between the center of the deformed figure and its edge will be a value between half of the real side of the square and half of its real diagonal. A situation is similar when the marker figure is a rectangle but the lower limit of the range in which the value described can be put is half of its shorter side. In order to use such figures as markers, there is a need for an additional mathematical analysis of their characteristic features if the parameters of these figures are to be determined. The features mentioned can be the current field of the deformed figure or its current circuit which changes according to a spatial rotation, and many others. For laboratory tests as well as for the target implementation in the Mobile Robot built at Institute of Applied Computer Science, the authors used the first figure described (a circle) as a marker which was to be placed on the operator clothes (on his back).

As in the case of the tests conducted on the marker detection and the tracking algorithm based only on template matching, now the authors also succeeded in achieving the approximate performance speed of about 50 frames per second with a working image resolution of 1002x1002 px. The algorithm of template detection and tracking, enhanced by marker centering according to mathematical characteristic features of the base figure was implemented on the Intel® Core™ 2 Duo T6600 (2.2 GHz) processor.

However, despite the fact that the continuity of tracking was successfully achieved and the area selected to detect on the next frame almost always stayed in the neighborhood of the center of the real position of the marker, the authors noticed that sometimes the algorithm chose a false object for tracking. Those objects were very similar to the marker but were located far from it and were clearly recognizable for the operator. A good example of such false detection is the operator's head when he stands on a bright background (e.g. the sky). Such a situation is illustrated in Fig. 6.13:

Despite the fact that the operator observing the image presented in Fig. 6.13 will instantly recognize the difference between a human head and the marker, the algorithm will acknowledge those objects as very similar when using only the correlation coefficient. It often happens that the correlation coefficient between the template recently chosen to track and similar to the marker element will be greater than when compared to the real current location of the marker. There are many possible reasons for the occurrence of such situations. In particular, during the tests, the authors noticed that a large percentage of the above-mentioned situations are those in which the marker is slightly blurred because of its movement. This is because of the acquisition time which is set in the camera properties. This time has to be variable because in heavily underexposed rooms or at dusk, the time of acquisition of a single frame should be as long as possible (of course to maintain good image quality). On such a slowly acquired frame, objects which move very fast are blurred. In this case, when the marker moves and some other element of the scene, similar to the marker, remains still for a moment required to acquire a frame, it will appear that it will be the best current detection. Such a situation which, in an exaggerated manner, brings out the issue described is illustrated in Fig. 6.14:

*R. Jachowicz et al.*

Fig. 6.13 Finding mathematical characteristic features of marker figures. In the image the circle radius is presented.

Fig. 6.14 Detection of a marker blurred by movement. Target implementation execution.

Fig. 6.14 illustrates the result image of the final algorithm of template detection and tracking on an example of a circle as a marker figure. One can notice that the tracked marker is heavily blurred when its movement speed increases. It should be mentioned that movement of the marker on the surface which is perpendicular to the camera axis is the most blurring factor. The movement towards or from the camera set gives negligibly insignificant blur. The situations described can take place and they will happen in target working conditions of the robot, e.g. while tracking on the road bend at dusk, etc. To illustrate the solution of the issue described above, the authors deliberately used a frame (Fig. 6.14) from the algorithm target implementation done on the Intel(R) Atom Z530 processor with a CPU performance speed of 1.6 GHz. On the frame presented there are a number of elements which can easily help to understand the enhancements applied. These elements come from the debugging mode of the target application. These elements are:

(1) The indication of the marker pointed by a small spot with its approximate shape marked with an ellipse.
(2) The information panel, also called a clock panel, located in the bottom left corner of the image. It shows an approximate distance of the detected marker from the camera set expressed in meters (3.672 m), rough movement speed measured on the surface perpendicular to the camera axis (12 m/s), the brightness of the scene expressed as an average value of pixels of a monochromatic image (136/256) and performance speed (14.2 fps).
(3) The movement path, i.e. information about the locations of the marker on the previous frames expressed as a polyline drawn after indications locations of the marker.
(4) The Region of Analysis, also called the Region of Interest, marked with a rectangle in which the probability of marker detection is the highest. The size and location of this area depends on locations and speed analysis of the marker on the previous frames. In Fig. 6.14 the Region of Analysis (marked with a rectangle) is determined based on the mentioned locations and speed data in addition to the current position of the marker, and it will be used to track the template on the next frame. Thus, it shows the area in which the marker will be detected on the next frame, not on the current one.

The first idea which allowed one to limit false detections and, at the same time, improved performance speed was the introduction of the Region of Interest described above. The technique based on using such a tool is to limit analyzed area of the image. Using this technique the authors assumed that the area in which the probability of the marker detection is the highest can be determined based on the available movement path of this marker. Such a mechanism allowed one to automatically reject all objects which are similar to the marker, but do not lie inside the Region of Interest. As already mentioned, additionally the authors managed to gain a performance speed increase due to the reduction of the analysis area. The application module that is responsible for determining the location and size of the Region of Interest is called a Movement Prediction Module. During its operation it uses data from a certain number of previous frames in order to determine parameters necessary to estimate the next location of the marker on the next frame. These parameters are most of all pixel acceleration and pixel speed. The quantities mentioned give information about how the marker moves on a two-dimensional image and are not directly related to its real absolute acceleration and speed in a space. The final location of the Region of Interest is calculated by adding the distance determined based on the current pixel speed corrected by the current pixel acceleration to the current position of the marker. In Fig.14the Region of Interest (marked with a rectangle)is set according to the rules listed above and is prepared to detect the marker on the way which the marker will most probably follow. The position of the center of the Region of Interest can be calculated from the following equation (6.6):

$$p = p_c + (v_c + a_c t_a) t_a \qquad (6.6)$$

where:

$p_c$ – is the current position of the marker,

$v_c$ – is the current pixel speed of the marker (estimated by analyzing positions of the marker on the previous frames),

$a_c$ – is the current pixel acceleration of the marker (estimated by analyzing positions of the marker on the previous frames),

$t_a$ – is the time of the current frame acquisition.

The second idea which significantly improved the efficiency of the algorithm (the indicator of the rejected false detections has risen to over 95%) is the enhancement of the third part of the procedure of pattern confirmation and centralization mentioned earlier. The enhancement is to confirm the presence of the marker in the place of indication. By the artificial recreation of the marker (based on the quantities extracted in first and second phase of pattern confirmation and centralization procedure)and its comparison with the analyzed part of the image, the algorithm checks the correctness of the indication. Once again the Pearson correlation coefficient is used, and this time it is calculated only for two images whose sizes are similar to the current pixel size of the marker. To give an example, the third step of the pattern confirmation and centralization procedure performed for Fig. 6.13 based on the comparison of the following images:

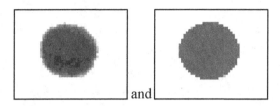

for confirming the presence of the marker in the neighborhood of its real location and:

for confirming the marker in the neighborhood of the operator's head.

Fig. 6.15 Zoomed fragments of an image presented in the figure 6.13 used to perform third part of pattern confirmation and centralization procedure. They were extracted from places with the highest correlation coefficient value.

## 6.3 Conclusions

The authors of the solution presented managed to develop and implement a real time algorithm for the detection and tracking of artificial markers and for the determination of their distance from the camera set based on their sizes. It should be noted that this kind of approach requires only one camera active, and no stereovision system is necessary to determine the distance. Ensuring the continuity of the processing speed and the limit of the speed of the objects being tracked on the scene to about 15 km/h (which is a contractual value at which objects are blurred in the image when acquired) and the visual contact with the tracked marker, the algorithm is capable of working continuously and faultlessly. Figs. 6.16–6.19 present the frames from a sequence of images analyzed by the algorithm:

Fig. 6.16 and 6.17 Frames taken from a sequence of images analyzed by the final algorithm. Faultless indication with relatively high speed of the marker (marker throwing).

Fig. 6.18 and 6.19 Frames taken from a sequence of images analyzed by the final algorithm. Faultless indication even with strong spatial deformations.

## References

Chen, X. and Yang, Y. H. (2011). Background Estimation and Its Applications. *Emerging Topics in Computer Vision and its Applications*, Vol. 1, Chapter 1.5, World Scientific Publishing Co. Pte. Ltd.

Fazl-Ersi, E., and Tsotsos, J.K. (2012). A performance evaluation of robot localization methods in outdoor terrains, *Emerging Topics in Computer Vision and its Applications*, Vol. 1, Chapter 3.7, World Scientific Publishing Co. Pte. Ltd.

Gomez, F., Althoefer, K. and Seneviratne, L.D. (2003). Modeling of ultrasound sensor for pipe inspection, *Robotics and Automation, Proceedings. ICRA '03.*

He, L., Wang, H. and Zhang, H. (2011). Object detection by parts using appearance, structural and shape features, *Mechatronics and Automation (ICMA)*, Beijing.

Hu, Z., Lamosa, F. and Uchimura, K. (2005). A complete U-V-disparity study for stereovision based 3D driving environment analysis, *3-D Digital Imaging and Modeling.*

Iida, K., Tang, Y., Mukai, T. and Nishimori, Y. (2004). Measurement of fish school volume by multi-beam sonar. Directional resolution and estimation error, *OCEANS '04. MTTS/IEEE TECHNO-OCEAN '04*.

Isenburg, M. and Shewchuk, J. (2009). Visualizing Lidar in Google Earth, *Geoinformatics*, Fairfax, VA.

Karakaya, S., Kucukyildiz, G., Ocak, H. and Bingul, Z. (2012). Obstacle and optimal heading direction detection algorithm on a mobile robot platform, *Signal Processing and Communications Applications Conference (SIU)*, Mugla.

Neto, A.M., Rittner, L., Leite, N., Zampieri, D.E., Lotufo, R. and Mendeleck, A. (2007). Pearson's Correlation Coefficient for Discarding Redundant Information in Real Time Autonomous Navigation System, *Control Applications*, Singapore.

Wang, B., Daly, C., Ryan, M. and Scott, M. (1993). Fuzzy logic methods for image motion detection and analysis, *Systems, Man and Cybernetics*, Le Touquet.

Wang, J., Huang, P., Chen, C., Gu, W. and Chu, J. (2011). Stereovision aided navigation of an Autonomous Surface Vehicle, *Advanced Computer Control (ICACC)*, Harbin.

Xu, C. and Velastin, S.A. (1993). A weighted Mahalanobis distance Hough transform and its application for the detection of circular segments, *Hough Transforms*, London.

Yan, G., Pan, Q. and Kang Y. (2005). Research on a new Gaussian self-adaptive smoothing algorithm in image processing, *VLSI Design and Video Technology*.

Zhang, H., Gao, W., Chen, X. and Zhao, D. (2005). Learning informative features for spatial histogram-based object detection, *Neural Networks, IJCNN '05*.

# CHAPTER 7

# PARTIAL FRACTIONAL-ORDER DIFFERENCE IN THE EDGE DETECTION

Piotr Duch, Rafał Jachowicz, Sylwester Błaszczyk, Maciej Łaski, Adam Wulkiewicz, Piotr Ostalczyk, and Dominik Sankowski

*Institute of Applied Computer Science*
*Lodz University of Technology*
*90-924 Lodz, ul. Stefanowskiego 18/22*
*{pduch, rjachowicz, sblaszc, mlaski, awulkie, dsan}@kis.p.lodz.pl,*
*piotr.ostalczyk@p.lodz.pl*

Edge detection is a fundamental tool in image processing algorithms. Edges are useful in determination of optical flow, stereovision algorithms, feature detection, feature extraction, object recognition and tracking. There are many methods for edge detection, one can easily distinguish those based on the first-order derivative expression computed from the image (Sobel operator), the second-order derivative (Laplacian of Gaussian) or the fractional order derivative (Mathieu *et al.*, 2003). A different group of edge detection algorithms are search-based methods, e.g. Canny (Canny, 1986). In this case, first the edge strength is computed and then the local directional maxima of the gradient magnitude are selected.

Fractional calculus is used in image processing for many different tasks, e.g. image texture enhancement (Gao *et al.*, 2011b,a), image segmentation (Ghamisi *et al.*, 2012), motion estimation (Chen *et al.*, 2010). A pioneer work for applying fractional calculus to edge detection is (Oustaloup *et al.*, 1991a,b; Mathieu *et al.*, 2003). The authors present 2D fractional-order backward difference (FOBD) edge detection operator.

The Chapter is organized as follows. First, a very short mathematical background concerning the fractional order on the backward difference/sum is given. Then, the partial fractional order backward difference definition is introduced and some numerical examples are presented. In the next section the main properties of the 2D-FOBD are shown. The definition of directional of 2D-FOBD is introduced in section 6 and the designed mask is presented in section 7. Finally, the experimental results of applying the edge detection method proposed and a short summary is given.

## 7.1    Fractional-order backward difference/sum (FOBD/S)

A FOBD/S is defined as a finite sum

$$
{}_0\Delta_k^{(\nu)} y_k = \sum_{i=0}^{k} a_i^{(\nu)} y_{k-i} \tag{7.1}
$$

where

$$
a_i^{(\nu)} = \begin{cases} 1 & \text{for} \quad i = 0 \\ (-1)^i \dfrac{\nu(\nu-1)\cdots(\nu-i+1)}{i!} & \text{for} \quad i = 1, 2, 3, \cdots \end{cases} \tag{7.2}
$$

Now we define a two discrete-variable function $y(k_1, k_2)$ where $k_1, k_2 \in \mathbf{Z}_+$ and $f(k_1, k_2) \in \mathbf{R}$. Further the function $y(k_1, k_2)$ will be denoted as $y_{k_1,k_2}$. It is clear that assuming $k_2 = const$ we can evaluate the FOBD with respect to $\nu_1 \in \mathbf{R}$. Hence we get a new two discrete-variable function, denoted as $g_{k_1,k_2}$

$$
{}_0\Delta_{k_1}^{(\nu_1)} y_{k_1-1,k_2} = \sum_{i=0}^{k_1} a_i^{(\nu)} y_{k_1-i,k_2} = g_{k_1,k_2} \tag{7.3}
$$

Next we can perform similar steps evaluating the FOBD with respect to the second variable $\nu_2 \in \mathbf{R}$

$$
{}_0\Delta_{k_2}^{(\nu_2)} g_{k_1,k_2} = \sum_{j=0}^{k_2} a_j^{(\nu_2)} g_{k_1,k_2-j} = \sum_{j=0}^{k_2} a_j^{(\nu_2)} \left( \sum_{i=0}^{k_1} a_i^{(\nu_1)} y_{k_1-i,k_2-j} \right) \tag{7.4}
$$

$$
{}_{0,0}\Delta_{k_1,k_2}^{(\nu_1\nu_2)} y_{k_1,k_2} = \sum_{j=0}^{k_2} \sum_{i=0}^{k_1} a_j^{(\nu_2)} a_i^{(\nu_1)} y_{k_1-i,k_2-j} \tag{7.5}
$$

## 7.2    Partial fractional-order backward difference definition

The sequence of two operations performed above, which defines a two dimensional backward difference, is denoted as ${}_{0,0}\Delta_{k_1,k_2}^{(\nu_1\nu_2)} y_{k_1,k_2}$.

The sum in equation (7.5) can be expressed as

$$
{}_{0,0}\Delta_{k_1,k_2}^{(\nu_1\nu_2)} y_{k_1,k_2} =
$$

$$
\begin{bmatrix} a_0^{\nu_2} & a_1^{\nu_2} & \cdots & a_{k_2}^{\nu_2} \end{bmatrix} \begin{bmatrix} y_{k_1,k_2} & y_{k_1-1,k_2} & \cdots & y_{0,k_2} \\ y_{k_1,k_2-1} & y_{k_1-1,k_2-1} & \cdots & y_{0,k_2-1} \\ \vdots & \vdots & \ddots & \vdots \\ y_{k_1,0} & y_{k_1-1,0} & \cdots & y_{0,0} \end{bmatrix} \begin{bmatrix} a_0^{\nu_1} \\ a_1^{\nu_1} \\ \vdots \\ a_{k_1}^{\nu_1} \end{bmatrix} \tag{7.6}
$$

Further, for simplicity, we introduce the following notation

$$\mathbf{a}_{k_1}^{(\nu_1)} = \begin{bmatrix} a_0^{(\nu_1)} \\ a_1^{(\nu_1)} \\ \vdots \\ a_{k_1}^{(\nu_1)} \end{bmatrix}, \quad \mathbf{a}_{k_2}^{(\nu_2)} = \begin{bmatrix} a_0^{(\nu_2)} \\ a_1^{(\nu_2)} \\ \vdots \\ a_{k_2}^{(\nu_2)} \end{bmatrix}, \quad {}_{0,0}\mathbf{Y}_{k_1,k_2} = \begin{bmatrix} y_{k_1,k_2} & y_{k_1-1,k_2} & \cdots & y_{0,k_2} \\ y_{k_1,k_2-1} & y_{k_1-1,k_2-1} & \cdots & y_{0,k_2-1} \\ \vdots & \vdots & \ddots & \vdots \\ y_{k_1,0} & y_{k_1-1,0} & \cdots & y_{0,0} \end{bmatrix}$$
(7.7)

Then

$$ {}_{0,0}\Delta_{k_1,k_2}^{(\nu_1,\nu_2)} y_{k_1,k_2} = \left[\mathbf{a}_{k_2}(\nu_2)\right]^T {}_{0,0}\mathbf{Y}_{k_1,k_2}\mathbf{a}_{k_1}(\nu_1) $$
(7.8)

Now one defines $(k+1) \times (k+1)$ matrix

$$\mathbf{N}_{k,k} = \begin{bmatrix} 0 & 0 & \cdots & 0 & 1 \\ 0 & 0 & \cdots & 1 & 0 \\ \vdots & \vdots & \ddots & \vdots & \vdots \\ 0 & 1 & \cdots & 0 & 0 \\ 1 & 0 & \cdots & 0 & 0 \end{bmatrix}$$
(7.9)

One immediately proves that

$$\mathbf{N}_{k,k} \cdot \mathbf{N}_{k,k} = \mathbf{N}_{k,k}^2 = \mathbf{I}_{k,k}$$
(7.10)

where $\mathbf{I}_{k,k}$ denotes $(k+1) \times (k+1)$ identity matrix. Then equation (7.8) may be expressed in the form

$$\begin{aligned} {}_{0,0}\Delta_{k_1,k_2}^{(\nu_1,\nu_2)} y_{k_1,k_2} &= \left[\mathbf{a}_{k_2}(\nu_2)\right]^T \mathbf{I}_{k_2,k_2}{}_{0,0}\mathbf{Y}_{k_1,k_2}\mathbf{I}_{k_1,k_1}\mathbf{a}_{k_1}(\nu_1) \\ &= \left[\mathbf{a}_{k_2}(\nu_2)\right]^T \mathbf{N}_{k_2,k_2}\mathbf{N}_{k_2,k_2}{}_{0,0}\mathbf{Y}_{k_1,k_2}\mathbf{N}_{k_1,k_1}\mathbf{N}_{k_1,k_1}\mathbf{a}_{k_1}(\nu_1) \end{aligned}$$
(7.11)

Simple calculations show

$$\begin{aligned} {}_{k_1,k_2}\widetilde{\mathbf{Y}}_{0,0} = \mathbf{N}_{k_2,k_2}{}_{0,0}\mathbf{Y}_{k_1,k_2}\mathbf{N}_{k_1,k_1} &= \mathbf{N}_{k_2,k_2} \begin{bmatrix} y_{k_1,k_2} & y_{k_1-1,k_2} & \cdots & y_{0,k_2} \\ y_{k_1,k_2-1} & y_{k_1-1,k_2-1} & \cdots & y_{0,k_2-1} \\ \vdots & \vdots & \ddots & \vdots \\ y_{k_1,0} & y_{k_1-1,0} & \cdots & y_{0,0} \end{bmatrix} \mathbf{N}_{k_1,k_1} \\ &= \begin{bmatrix} y_{0,0} & y_{1,0} & \cdots & y_{k_1,0} \\ y_{0,1} & y_{1,1} & \cdots & y_{k_1,1} \\ \vdots & \vdots & \ddots & \vdots \\ y_{0,k_2} & y_{1,k_2} & \cdots & y_{k_1,k_2} \end{bmatrix} \end{aligned}$$
(7.12)

$$\widetilde{\mathbf{a}}_{k_1}(\nu_1) = \mathbf{N}_{k_1,k_1} \begin{bmatrix} a_0^{(\nu_1)} \\ a_1^{(\nu_1)} \\ \vdots \\ a_{k_1}^{(\nu_1)} \end{bmatrix} = \begin{bmatrix} a_{k_1}^{(\nu_1)} \\ a_{k_1-1}^{(\nu_1)} \\ \vdots \\ a_0^{(\nu_1)} \end{bmatrix} \tag{7.13}$$

$$\widetilde{\mathbf{a}}_{k_2}(\nu_2) = \mathbf{N}_{k_2,k_2} \begin{bmatrix} a_0^{(\nu_2)} \\ a_1^{(\nu_2)} \\ \vdots \\ a_{k_2}^{(\nu_2)} \end{bmatrix} = \begin{bmatrix} a_{k_2}^{(\nu_2)} \\ a_{k_2-1}^{(\nu_2)} \\ \vdots \\ a_0^{(\nu_2)} \end{bmatrix} \tag{7.14}$$

Finally

$$_{0,0}\Delta_{k_1,k_2}^{(\nu_1,\nu_2)} y_{k_1,k_2} = [\widetilde{\mathbf{a}}_{k_2}(\nu_2)]^T {}_{k_1,k_2}\widetilde{\mathbf{Y}}_{0,0}\widetilde{\mathbf{a}}_{k_1}(\nu_1) \tag{7.15}$$

## 7.3 Numerical examples

In Fig. 7.1 the picture created by randomly created numbers from a range is presented (left) and then transformed according to equation (7.15) with $\nu_1 = -0.5$, $\nu_2 = 0.5$ and then $\nu_1 = 0.1$, $\nu_2 = 0.9$

## 7.4 2D - FOBD properties

$$_{0,0}\Delta_{k_1,k_2}^{(n_1,n_2)} y_{k_1,k_2} = {}_{0,0}\Delta_{k_2,k_1}^{(n_2,n_1)} y_{k_1,k_2} \tag{7.16}$$

For $n_1, n_2, n_3, n_4 \geq 0$ one can prove that

$$_{0,0}\Delta_{k_1,k_2}^{(n_3,n_4)} y_{k_1,k_2} {}_{0,0}\Delta_{k_1,k_2}^{(n_1,n_2)} y_{k_1,k_2} = {}_{0,0}\Delta_{k_2,k_1}^{(n_1+n_3,n_2+n_4)} y_{k_1,k_2} \tag{7.17}$$

## 7.5 2D - FOBD definition

The fractional-order backward difference (left-sided fractional-order difference)

$$_{k_1-L,k_2}\Delta_{k_1,k_2}^{(\nu)} y_{k_1,k_2} = \sum_{i=0}^{L} a_i^{(\nu)} y_{k_1-i,k_2} \tag{7.18}$$

The fractional-order upper difference (right-sided fractional-order difference)

$$_{k_1+R,k_2}\Delta_{k_1,k_2}^{(\nu)} y_{k_1,k_2} = \sum_{i=0}^{R} a_i^{(\nu)} y_{k_1+i,k_2} \tag{7.19}$$

Fig. 7.1 Images created by randomly numbers (a, c) transformed according to equation (7.15) with b) $\nu_1 = -0.5$, $\nu_2 = 0.5$ and d) $\nu_1 = 0.1$, $\nu_2 = 0.9$.

The fractional-order upper difference (upper-sided fractional-order difference)

$$_{k_1,k_2+U}\Delta^{(\nu)}_{k_1,k_2}y_{k_1,k_2} = \sum_{i=0}^{U} a_i^{(\nu)} y_{k_1,k_2+i} \qquad (7.20)$$

The fractional-order down difference (down-sided fractional-order difference)

$$_{k_1,k_2-D}\Delta^{(\nu)}_{k_1,k_2}y_{k_1,k_2} = \sum_{i=0}^{D} a_i^{(\nu)} y_{k_1,k_2-i} \qquad (7.21)$$

The fractional-order inclined upper left difference

$$_{k_1-U_L,k_2+U_L}\Delta_{k_1,k_2}^{(\nu)}y_{k_1,k_2} = \sum_{i=0}^{U_L} a_i^{(\nu)} y_{k_1-i,k_2+i} \qquad (7.22)$$

The fractional-order inclined upper right difference

$$_{k_1+U_R,k_2+U_R}\Delta_{k_1,k_2}^{(\nu)}y_{k_1,k_2} = \sum_{i=0}^{U_R} a_i^{(\nu)} y_{k_1+i,k_2+i} \qquad (7.23)$$

The fractional-order inclined down left difference

$$_{k_1-D_L,k_2-D_L}\Delta_{k_1,k_2}^{(\nu)}y_{k_1,k_2} = \sum_{i=0}^{D_L} a_i^{(\nu)} y_{k_1-i,k_2-i} \qquad (7.24)$$

The fractional-order inclined down right difference

$$_{k_1+D_R,k_2-D_R}\Delta_{k_1,k_2}^{(\nu)}y_{k_1,k_2} = \sum_{i=0}^{D_R} a_i^{(\nu)} y_{k_1+i,k_2-i} \qquad (7.25)$$

## 7.6   Simplified form of the PFOBD

One can prove that for $0 < \nu < 1$

$$\lim_{i\to\infty} a_i^{(\nu)} = 0 \qquad (7.26)$$

so for sufficiently large $L$ one can put

$$a_i^{(\nu)} = 0 \quad \text{for} \quad i = L, L+1 \qquad (7.27)$$

This assumption leads to a simplified form of the PFOBD. For $L_1$, $L_2$

$$_{0,0}\Delta_{k_1,k_2}^{(\nu_1,\nu_2)}y_{k_1,k_2} = \begin{cases} \sum_{j=0}^{k_2} \sum_{i=0}^{k_1} a_j^{(\nu_2)} a_i^{(\nu_1)} y_{k_1-1,k_2-j} & \text{for} \quad i < L_1, j < L_2 \\ \sum_{j=0}^{L_2} \sum_{i=0}^{L_1} a_j^{(\nu_2)} a_i^{(\nu_1)} y_{k_1-1,k_2-j} & \text{for} \quad i \geq L_1, j \geq L_2 \end{cases} \qquad (7.28)$$

Then an equivalent form is as follows

$$_{0,0}\Delta_{k_1,k_2}^{(n_1,n_2)}y_{k_1,k_2} = \begin{cases} \left[\mathbf{a}_{k_2}^{(\nu_2)}\right]^T {}_{0,0}\mathbf{Y}_{k_1,k_2} \mathbf{a}_{k_1}^{(\nu_1)} & \text{for} \quad i < L_1, j < L_2 \\ \left[\mathbf{a}_{k_2}^{(\nu_2)}\right]^T {}_{k_1-L_1,k_2-L_2}\mathbf{Y}_{k_1,k_2} \mathbf{a}_{L_1}^{(\nu_1)} & \text{for} \quad i \geq L_1, j \geq L_2 \end{cases} \qquad (7.29)$$

where

$$k_1-L_1,k_2-L_2\mathbf{Y}_{k_1,k_2} = \begin{bmatrix} y_{k_1,k_2} & y_{k_1-1,k_2} & \cdots & y_{k_1-L_1,k_2} \\ y_{k_1,k_2-1} & y_{k_1-1,k_2-1} & \cdots & y_{k_1-L_1,k_2-1} \\ \vdots & \vdots & \ddots & \vdots \\ y_{k_1,k_2-L_2} & y_{k_1-1,k_2-L_2} & \cdots & y_{k_1-L_1,k_2-L_2} \end{bmatrix} \qquad (7.30)$$

A range of elements of matrix (7.30) taken into account in PFOBD (7.29) is plotted in Fig. 7.2.

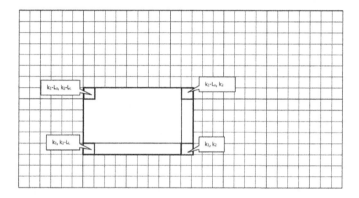

Fig. 7.2   A range of elements of matrix (7.30).

## 7.7   PFOBD edge detector

As the partial, fractional-order edge detector one considers an image transformation described by the equation

$$y_{k_1,k_2} = \frac{1}{4}\left\{ w_1\delta_{l,u} + w_2\delta_{r,u} + w_3\delta_{l,d} + w_4\delta_{r,d} \right\} \qquad (7.31)$$

where

$$\delta_{l,u} = \left[\mathbf{a}_{L_2}(\nu_u)\right]^T {}_{k_1-L_1,k_2-L_2}\mathbf{Y}_{k_1,k_2}\mathbf{a}_{L_1}(\nu_l) \qquad (7.32)$$

$$\delta_{l,d} = \left[\mathbf{a}_{L_2}(\nu_d)\right]^T (\mathbf{N}_{L_2,L_2})_{k_1-L_1,k_2-L_2}\mathbf{Y}_{k_1,k_2}\mathbf{a}_{L_1}(\nu_l) \qquad (7.33)$$

$$\delta_{r,u} = \left[\mathbf{a}_{L_2}(\nu_u)\right]^T {}_{k_1-L_1,k_2-L_2}\mathbf{Y}_{k_1,k_2}(\mathbf{N}_{L_1,L_1})\mathbf{a}_{L_1}(\nu_r) \qquad (7.34)$$

$$\delta_{r,d} = \left[\mathbf{a}_{L_2}(\nu_d)\right]^T (\mathbf{N}_{L_2,L_2})_{k_1-L_1,k_2-L_2}\mathbf{Y}_{k_1,k_2}(\mathbf{N}_{L_1,L_1})\mathbf{a}_{L_1}(\nu_r) \qquad (7.35)$$

and $w_1$, $w_2$, $w_3$ and $w_4$ mean weights (positive or negative). The pixel $(k_1, k_2)$ denoted by a red square with the ranges defined by matrix (7.30) and matrices in (7.32)–(7.35) is depicted in Fig. 7.3.

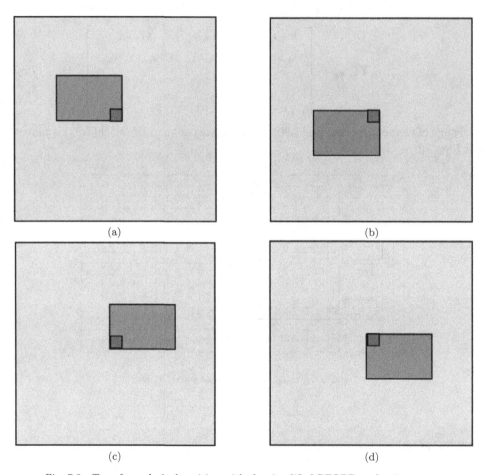

Fig. 7.3 Transformed pixel position with the simplified PFOBD evaluation range.

## 7.8 PFOBD edge detector application example

Grey-scale image

Fig. 7.4 Image taken with a simple night vision device.

The application of the PFOBD in a robot vision system is presented in the following example. An image, presented in Fig. 7.4 was taken with a simple night vision device and was converted to grayscale.

Fig. 7.5 Edges detected on Fig. 7.4 using PFOBD left-upper edge detector.

Fig. 7.6 Edges detected on Fig. 7.4 using PFOBD right-down edge detector.

Fig. 7.7   Edges detected on Fig. 7.4 using PFOBD left-down edge detector.

Fig. 7.8   Edges detected on Fig. 7.4 using PFOBD right-upper edge detector.

In the PFOBD edge detectors described by equations (7.32)–(7.35) equal lengths $L_1 = L_2 = 10$ and orders $\nu_u = \nu_d = \nu_l = \nu_r = 0.7$ are assumed. The edge detection results are presented in Fig. 7.5–7.8.

Fig. 7.9   Sum of images presented in Fig. 7.5–7.8.

In Fig. 7.9 the image obtained by the transformation described by equation (7.31) with $w_1 = w_2 = w_3 = w_4 = 1$ is presented.

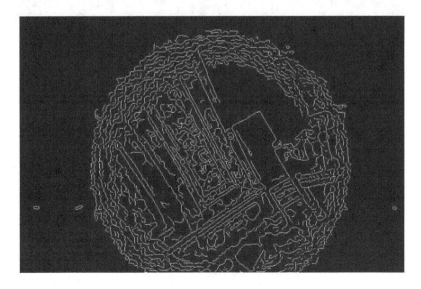

Fig. 7.10   Edges detected on Fig. 7.4 using Canny operator.

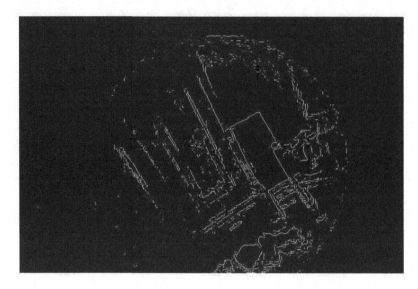

Fig. 7.11   Edges detected on Fig. 7.4 using Prewitt operator.

Fig. 7.12   Edges detected on Fig. 7.4 using Laplacian of Gaussian.

To compare the results obtained, classical methods: Canny, Prewitt, Laplasian and zero-cross filters are applied to the image presented in Fig. 7.4. The results are presented in Figs. 7.10–7.13, respectively.

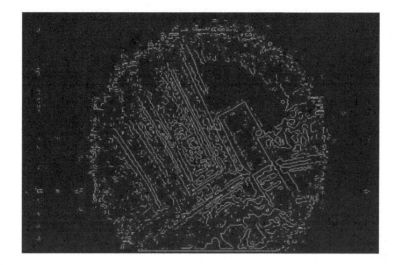

Fig. 7.13    Edges detected on Fig. 7.4 using zero-cross method.

## 7.9    2D - FOBD'S linear combination

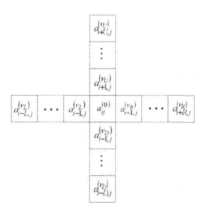

Fig. 7.14    The two-dimensional FOBD edge detection operator.

In practical applications the 2D-FOBD can be presented as a two-dimensional matrix. The edge detection process is simplified to the convolution of 2D-masks presented in Fig. 7.14 and Fig. 7.15 with image. The lengths of each of the star arms can be set up separately.

| $a_{i-L,j+U}^{(v_{U_L})}$ | | | $a_{i,j+U}^{(v_U)}$ | | | $a_{i+U_R,j+U_R}^{(v_{U_R})}$ |
|---|---|---|---|---|---|---|
| | | | $\vdots$ | | $\cdot\cdot$ | |
| | | $a_{i-1,j+1}^{(v_{U_R})}$ | $a_{i,j+1}^{(v_D)}$ | $a_{i+1,j+1}^{(v_{U_R})}$ | | |
| $a_{i-L,j}^{(v_L)}$ | $\cdots$ | $a_{i-1,j}^{(v_L)}$ | $a_{ij}^{(0)}$ | $a_{i+1,j}^{(v_R)}$ | $\cdots$ | $a_{i+R,j}^{(v_R)}$ |
| | | $a_{i-1,j-1}^{(v_{D_L})}$ | $a_{i,j-1}^{(v_D)}$ | $a_{i+1,j-1}^{(v_{D_R})}$ | | |
| | $\cdot\cdot$ | | $\vdots$ | | | |
| $a_{i-D_L,j-D_L}^{(v_{D_L})}$ | | | $a_{i,j-D}^{(v_D)}$ | | | $a_{i+D_R,j-D_R}^{(v_{D_R})}$ |

Fig. 7.15 The two-dimensional FOBD edge detection operator.

## 7.10   Results of applying the FOLBD/S edge detector

The FOLBD/S edge detector with its mathematical background was described in Chap. 4. For $w_i = 1$, $\nu_i = const$ and lengths $L_i = L_d = const$ for $i = 1, 2, \cdots, 8$ the FOLBD/S edge detector takes the form

$$p(k_1, k_2) = \frac{1}{8} \sum_{i=1}^{8} {}_{0}^{L} \Delta_{L_d}^{(\nu_i)} y[\mathbf{c}_i(k)] \tag{7.36}$$

In this Section its effectiveness is demonstrated as a numerical example.

A noisy artificial image is created. It is presented in Fig. 7.16.

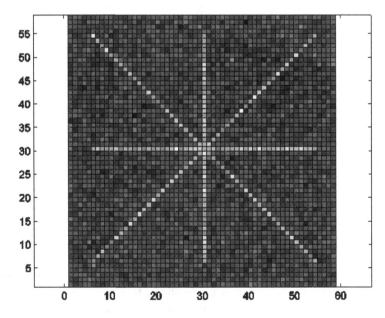

Fig. 7.16 The noised original image.

For $\nu = -0.75$ and constant length $L_d = 5$ the FOLBD/S edge detector (7.36) action result is presented in Fig. 7.17.

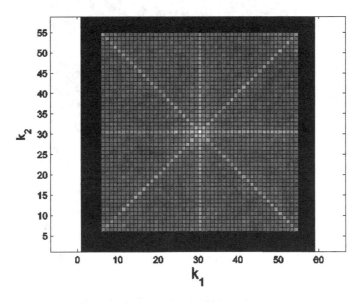

Fig. 7.17   The FOLBS detector processed noised image.

Applying a threshold $h = 0.65$ to the image presented in Fig. 7.17 one gets an image presented in Fig. 7.18.

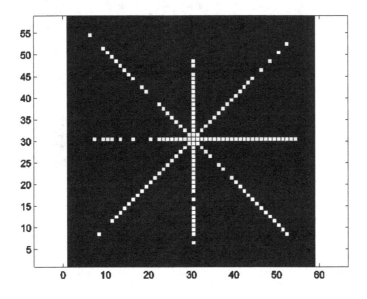

Fig. 7.18   The FOLBS detector processed noised image with a threshold.

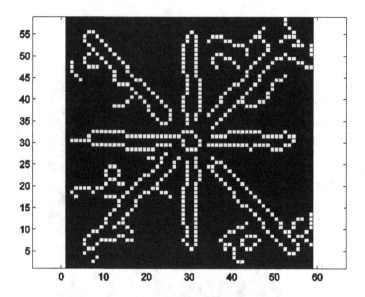

Fig. 7.19   Image edges obtained by the Canny mask.

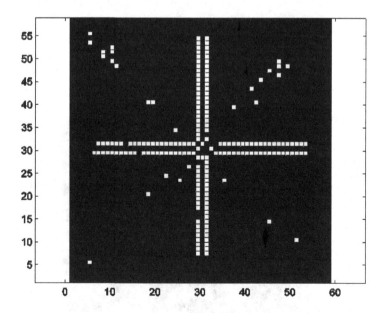

Fig. 7.20   Image edges obtained by the Prewitt mask.

To compare the edge detection results, commonly used masks: Canny, Prewitt, Roberts, zero-cross method and Laplacian were also applied to Fig. 7.16. The results are presented in Figs. 7.19–7.23, respectively.

Finally the original image (without noise) is presented in Fig. 7.24.

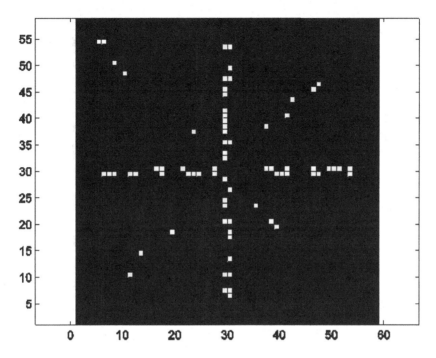

Fig. 7.21   Image edges obtained by the Roberts mask.

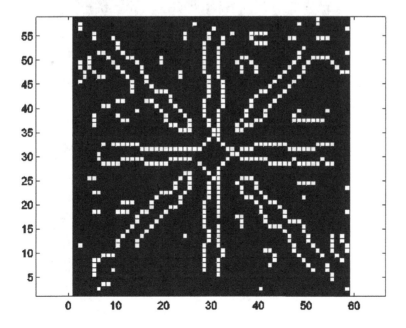

Fig. 7.22   Image edges obtained by the Zero-cross method.

*P. Duch et al.*

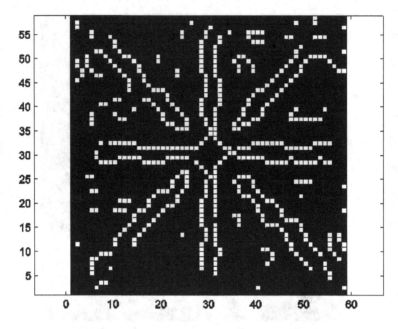

Fig. 7.23   Image edges obtained by Laplasian mask.

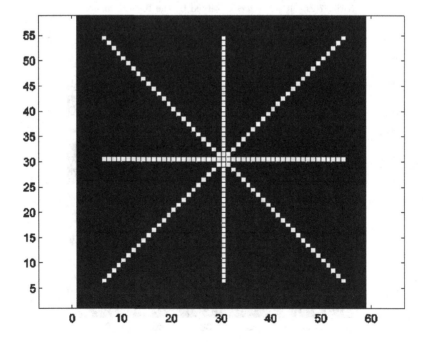

Fig. 7.24   Original image.

## 7.11 Conclusions

In this Chapter two original edge detection methods are presented. The algorithms proposed allow one to detect edges on the images regardless of the level of its noise. The construction of the masks offers possibility to detect edges in different directions with the constant accuracy. The main defect of the algorithm presented is its slow computational time. Our further work will aim at the reduction of this time.

The results presented may be applied to natural images, which proves its vast possibilities in edge detection. A wide selection of possible parameters enables allows the algorithm to be used for different tasks.

## References

Canny, J. (1986). A computational approach to edge detection, *IEEE Transactions on Pattern Analysis and Machine Intelligence* **6**, 8, pp. 679–698.

Chen, D., Chen, Y. and Sheng, H. (2010). Fractional variational optical flow model for motion estimation, in *Proceedings of the 4$^{th}$ IFAC Workshop on Fractional Differentiation and its Applications*. (Badajoz, Spain).

Gao, C., Zhou, J., Hu, J. R. and Lang, F., Lang (2011a). Edge detection of colour image based on quaternion fractional differential, *IETImageProcessing* **5**, 3, pp. 261–272.

Gao, C., Zhou, J., Zheng, X. and Lang, F. (2011b). Image enhancement based on improved fractional differentiation, *Journal of Computational Information Systems* **7**, 1, pp. 257–264.

Ghamisi, P., Couceiro, M. S., Benediktsson, J. A. and Ferreira, N. M. F. (2012). An efficient method for segmentation of images based on fractional calculus and natural selection, *Expert Systems with Applications* **39**, pp. 12407–12417.

Mathieu, B., Melchior, P., Oustaloup, A. and Ceyral, C. (2003). Fractional differentiation for edge detection, *Signal Processing* **83**, 3, pp. 2421–2432.

Oustaloup, A., Mathieu, B. and Melchior, P. (1991a). Edge detection using non integer derivation, in *Presented at the IEEE European Conference on Circuit Theory and Design (ECCTD91)*. (Copenhagen, Denmark).

Oustaloup, A., Mathieu, B. and Melchior, P. (1991b). Robust edge detector of non integer order: the CRONE detector, in *Presented at the 8$^{th}$ Congres de Cytometrie en Flux et dAnalyse dImage (ACF91)*. (Mons, Belgium).

# CHAPTER 8

# APPLICATION OF FRACTIONAL-ORDER DERIVATIVE FOR EDGE DETECTION IN MOBILE ROBOT SYSTEM

Sylwester Błaszczyk, Rafał Jachowicz, Piotr Duch, Maciej Łaski, Adam Wulkiewicz, Piotr Ostalczyk, and Dominik Sankowski

*Institute of Applied Computer Science*
*Lodz University of Technology*
*90-924 Lodz, ul. Stefanowskiego 18/22*
*{sblaszc, rjachowicz, pduch, mlaski, awulkie, dsan}@kis.p.lodz.pl,*
*piotr.ostalczyk@p.lodz.pl*

In this Chapter some chosen methods of edge detection based on the Fractional Calculus are discussed. Two original edge detection methods based on fractional-order derivatives are compared with the well-known classical edge detectors such as Sobel, Prewitt operator, Robert cross, zero-cross method and with a method based on Laplasian. The investigations are illustrated by numerical examples.

## 8.1 Introduction

Edge detection is a fundamental tool in image processing algorithms. Edges are useful in determination of optical flow, stereovision algorithms, feature detection, feature extraction, object recognition and tracking. There are many methods for edge detection; one can easily distinguish those based on the first-order derivative expression computed from the image (Sobel operator), the second-order derivative (Laplacian of Gaussian) or the fractional order derivative (Mathieu *et al.*, 2003). The different group of edge detection algorithms are search-based methods, e.g. Canny (Canny, 1986). In this case, first the edge strength is computed and next the local directional maxima of the gradient magnitude are selected.

Fractional calculus is used in image processing for many different tasks, e.g. image texture enhancement mage texture enhancement (Gao *et al.*, 2011b,a), image segmentation (Ghamisi *et al.*, 2012), motion estimation (Chen *et al.*, 2010). A pioneer work for applying fractional calculus to edge detection is (Oustaloup *et al.*, 1991a,b; Mathieu *et al.*, 2003). The authors present a CRONE edge detection operator, and a CRONE mask, as an extension to 2D image signal. The idea of using fractional differentiation for edge detection was further investigated by many researchers and new methods based on fractional calculus were developed (Yang et al., 2010).

## 8.2  Chosen methods for edge detection

The main task in image analysis and processing is information extraction from the image. Important information is included at the image edges. It is often possible to identify an object exclusively of the basis of information on its edge. In this Chapter selected methods of edge detection based on the use of the masks defined by the difference or sum of fractional orders are presented.

### 8.2.1  *Operator CRONE*

The CRONE operator is similar to the Prewitt operator (Mathieu *et al.*, 2003), except that classical integer order differentiation is generalized to the fractional order. To detect the edges in the image one should expand the CRONE one-dimensional operator to two dimensions. For this purpose, the operator must be considered as a vector of two independent components. Each of these components is formed on the basis of a one-dimensional CRONE operator. The one dimensional CRONE operator is defined as follows:

$$\underset{\leftrightarrow}{D}{}^{n} f(x) = \frac{1}{h^n} \sum_{k=0}^{\infty} a_k [f(x - kh) - f(x + kh)] \tag{8.1}$$

where

$$a_k = (-1)^k \binom{n}{r} = (-1)^k \frac{n(n-1)\cdots(n-k+1)}{k!} \tag{8.2}$$

and $f(x)$ is the function representing an image, a product $kh$ defines the pixel position, $k$ is an integer and $h$ is the calculation step. The order $n$ may be integer or fractional. A mask corresponding to a horizontal component is defined as

$$\begin{bmatrix} +a_m \\ \vdots \\ +a_1 \\ 0 \\ -a_1 \\ \vdots \\ -a_m \end{bmatrix} \tag{8.3}$$

where the above vector elements $a_m$ are defined by (8.1) and $m$ indicates the length of a mask. The order $n$ may be chosen from the interval [1,2] to improve detection selectivity or from the interval [-1,0] to improve noise rejection. It was proved (Mathieu *et al.*, 2003) that the CRONE operator is characterized by better effectiveness than the Prewitt operator.

## 8.3 The edge detection based on a fractional-order difference — Yang algorithm

In (Yang *et al.*, 2011) a method of edge detection operator construction was proposed which bases on the fractional-order difference gradients. In this approach a mask is created by a difference filter described by the discrete transfer function

$$D^\nu(z) = \left(\frac{1 - z^{-1}}{T}\right)^\nu \approx \frac{1}{T^n} \sum_{i=0}^{N} (-1)^i \frac{\Gamma(\nu + 1)}{\Gamma(i + 1)\Gamma(\nu - i + 1)} z^{-1} \qquad (8.4)$$

For digital images a formula defining the fractional-order difference gradients may be defined for different directions. For the horizontal direction one has

$$D^\nu_{XL \leftrightarrow XR} = D^\nu_{XL} - D^\nu_{XR} = a_1 I(x-1, y) - a_1 I(x+1, y) + \ldots + a_n I(x-1, y) - a_n I(x+1, y) \qquad (8.5)$$

where coefficients $a_i$ are defined by (8.1).

For $n = 3$ and any fractional order $\nu$, four $5 \times 5$ masks related to different directions are presented in Fig. 8.1.

### 8.3.1 *The edge detection based on a fractional-order difference — Gan algorithm*

The Gan algorithm (Gan and Yang, 2010) may serve as an example of a successful application of the Fractional Calculus in edge detection. Here a mask is created on the basis of Grünwald - Letnikov fractional-order backward difference definition. Image processing consists of the discrete image convolution with a $w \times h$ mask described by the formula

$$g(x, y) = \sum_{i=\frac{-w}{2}}^{\frac{w}{2}} \sum_{i=\frac{-h}{2}}^{\frac{h}{2}} t(i, j) g_o(x + i, y + j) \qquad (8.6)$$

where:

| | | |
|---|---|---|
| $t(i, j)$ | - | the mask value at $i, j$ position |
| $g_o(x + i, y + j)$ | - | the pixel value at coordinates $x + i$, $y + j$ in an original image |
| $i, j$ | - | the mask coordinates $T$ |

| 0 | 0 | 0 | 0 | 0 |
|---|---|---|---|---|
| 0 | 0 | 0 | 0 | 0 |
| (v²-v)/2 | -v | 0 | v | (v-v²)/2 |
| 0 | 0 | 0 | 0 | 0 |
| 0 | 0 | 0 | 0 | 0 |

(a)

| 0 | 0 | (v²-v)/2 | 0 | 0 |
|---|---|---|---|---|
| 0 | 0 | -v | 0 | 0 |
| 0 | 0 | 0 | 0 | 0 |
| 0 | 0 | v | 0 | 0 |
| 0 | 0 | (v-v²)/2 | 0 | 0 |

(b)

| (v²-v)/2 | 0 | 0 | 0 | 0 |
|---|---|---|---|---|
| 0 | -v | 0 | 0 | 0 |
| 0 | 0 | 0 | 0 | 0 |
| 0 | 0 | 0 | v | 0 |
| 0 | 0 | 0 | 0 | (v-v²)/2 |

(c)

| 0 | 0 | 0 | 0 | (v²-v)/2 |
|---|---|---|---|---|
| 0 | 0 | 0 | -v | 0 |
| 0 | 0 | 0 | 0 | 0 |
| 0 | v | 0 | 0 | 0 |
| (v-v²)/2 | 0 | 0 | 0 | 0 |

(d)

Fig. 8.1  Directional 5 × 5 masks. (a) Horizontal direction mask. (b) Vertical direction mask. (c) Direction 135° mask. (d) Direction 45° mask.

The mask used in convolution operations is a $n \times n$ matrix $T$ where $n$ is an odd integer. Below an example of the mask is given

$$T = \begin{bmatrix} T_1 & T_1 & T_1 \\ T_1 & T_0 & T_1 \\ T_1 & T_1 & T_1 \end{bmatrix} \tag{8.7}$$

In the mask one can distinguish 8 directions characterized by the angles: $0$, $\frac{\pi}{8}$, $\frac{\pi}{4}$, $\frac{3\pi}{8}$, $\frac{\pi}{2}$, $\frac{5\pi}{8}$, $\frac{3\pi}{4}$ and $\frac{7\pi}{8}$. For fractional orders $\nu \in (0,1)$ $i^{th}$ $(T_i)$ layer value $T$ of the mask $T$ is evaluated according to the formula:

$$T_i = \frac{1}{\Gamma(-\nu)} \left[ \frac{\Gamma(i - \nu + 1)}{(i+1)!} \left( \frac{\nu}{4} + \frac{\nu^2}{8} \right) + \frac{\Gamma(i-\nu)}{i!} \left( 1 - \frac{\nu^2}{4} \right) \right.$$
$$\left. + \frac{\Gamma(i-\nu-1)}{(i-1)!} \left( -\frac{\nu}{4} + \frac{\nu^2}{8} \right) \right] \tag{8.8}$$

To get a zero sum of all the mask elements as the element $T_0$ one should take

$$T_0 = -1 \sum_{i=1}^{n} (8 \cdot i \cdot T_i) \tag{8.9}$$

The resulting image contains the extracted edges.

### 8.3.2 *The likewise-radar fractional-order edge detector*

In (Duch *et al.*, 2013) a new approach to edge detectors was proposed. The algorithm is based on the mask created on the basis of the Sobel operator and fractional differentiation

$$G = \begin{bmatrix} T_2 & 0 & T_2 \\ T_1 & 0 & T_1 \\ T_2 & 0 & T_2 \end{bmatrix} \tag{8.10}$$

Values $T_1$ and $T_2$ are calculated according to formula (8.8). In this approach the mask is rotating through 180° with a desired step. For each pixel of the input image an convolution operation of the image with rotated masks is applied .

$$G_\Theta = rot(G, \Theta) * f(x, y) \tag{8.11}$$

where:

$G_\Theta$      -    the result of convolution of the mask with the image

$rot(G, \Theta)$    -    the rotation function of mask $G$ with angle $\Theta$

$f(x, y)$     -    the pixel value of input image

The results of the particular convolution operation are summed and the following function evaluates a new pixel value

$$g(x, y) = \left( \sum_{\Theta=0}^{\frac{180}{\Delta\Theta}} |G_{\Theta \cdot \Delta\Theta}|^\alpha \right)^{\frac{1}{\beta}} \tag{8.12}$$

where:

$g(x, y)$    -    the result of convolution of the image with the mask

$G_\Theta$       -    the result of convolution of the image with the mask rotated with $\Theta$

$\Delta\Theta$      -    the rotation step

$\Theta$        -    the angle of rotation

$\alpha$        -    the power of $G_{\Theta \cdot \Delta\Theta}$

$\beta$        -    the root order

After this operation a new image with detected edges is created. In the experiments masks rotated with 0°, 45°, 90°, 135° were used. Examples of the rotated masks are presented below.

$$G_{45°} = \begin{bmatrix} T_1 & T_2 & 0 \\ T_2 & 0 & T_2 \\ 0 & T_2 & T_1 \end{bmatrix} \qquad G_{135°} = \begin{bmatrix} 0 & T_2 & T_1 \\ T_2 & 0 & T_2 \\ T_1 & T_2 & 0 \end{bmatrix} \tag{8.13}$$

## 8.4 Evaluation of edge detection methods

To ensure the quality of experiments optimal parameters for all the selected edge detection algorithms were chosen. For each tested image the same procedure was applied. Next, some methods (Sobel operator, Roberts cross, Gan algorithm  presented in Sec. 8.3.1 and Likewise radar edge detector  presented in section Sec. 8.3.2) were tested against noise sensitivity and the quality of each method was determined.

### 8.4.1  *The test images*

Fig. 8.2 shows four images which were used in the experiment.

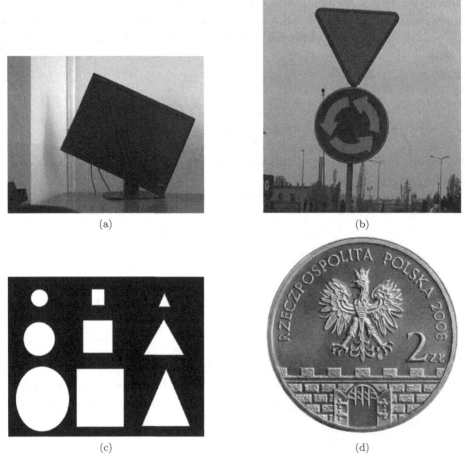

(a)                                        (b)

(c)                                        (d)

Fig. 8.2   Images used in the experiments.

For the experiment one artificial image Fig. 8.2(c) and three natural images Fig. 8.2(a,b,d) were chosen. The artificial image server as the basis for the comparison and evaluation of edge detection methods in an easy and objective way. In the case of natural images, each of them contains a single complete object. It helped in the visual evaluation of the result of the edge detection method used in the experiment.

Table 8.1   Range of parameters.

| Method | Threshold | Order |
|---|---|---|
| Sobel | 10 - 250 (step 15) | - |
| Roberts | 10 - 250 (step 15) | - |
| Gan algorithm | 75 - 150 (step 10) | 0.3 - 0.9 (step 0.3) |
| Likewise radar | 75 - 150 (step 10) | 0.3 - 0.9 (step 0.3) |

Table 8.2   Optimal parameters.

| Method | Threshold | Order |
|---|---|---|
| Sobel | 100 | - |
| Roberts | 55 | - |
| Gan algorithm | 95 | 0.3 |
| Likewise radar | 115 | 0.7 |

### 8.4.2   *Optimal parameter selection*

For the purpose of the optimal parameter selection a method proposed by Yitzhaky (Yitzhaky and Peli, 2003) was applied. In this method automatic statistical analysis of the correspondence of detected edges using different algorithm parameters is applied. Statistical measures are used to determine the ground truth image. In the next step the edges generated using the given algorithm with different parameters are compared to the previously created ground truth image and the best parameters are selected.

The unknown parameter in Roberts cross and Sobel operator methods is threshold. For methods based on fractional differentiation there are two parameters to set: a fractional order and a threshold. First, for each edge detection method a range of parameters was chosen, Table 8.1.

Next, for those edge detection methods the algorithm described in this section was applied independently for each of the images tested. Table 8.2 presents examples of optimal parameters calculated for Fig. 8.2(a).

Fig. 8.3 shows the results of applying edge detection methods with parameters presented in Table 8.2 on Fig. 8.2(a).

To all the images test the white noise with mean 0 and variance between 0.001 and 0.01 with step 0.001 was added. Next, each method was applied to detect edges in images with noise. To compare the quality of the tested methods the PSNR (peak signal to noise ratio) (8.14) index was used.

$$PSNR(f_0, f) = 10log_{10}\frac{255^2}{\frac{1}{N \cdot M}\sum_{x=0}^{M}\sum_{y=0}^{N}\left[f(x,y) - f_0(x,y)\right]^2} \tag{8.14}$$

where

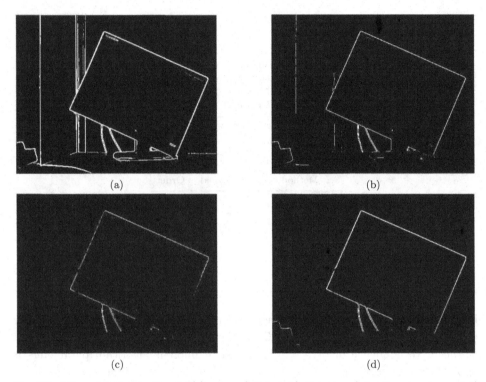

Fig. 8.3   Edges detected in Fig. 8.2(a) using a) Sobel, b) Roberts, c) fractional order and d) likewise - radar method.

| $f_0$ | - | the edges of the image with noise |
| $f$ | - | the edges of the image without noise |
| $M, N$ | - | the image dimension |
| $x, y$ | - | the pixel coordinates |

A greater value of PSNR index indicates that the extraction of edge information is better.

### 8.4.3   Results of experiments

Each of the methods examined was tested against noise sensitivity on the images presented in Fig. 8.2 with added noise. In Fig. 8.4 examples of images with added noise (Fig. 8.4(a) - mean 0, variance 0.005, Fig. 8.4(b) - mean 0, variance 0.009) are presented.

Fig. 8.5 and Fig. 8.6 present the result of applying different edge detection methods on the images from Fig. 8.4.

Fig. 8.7 and Fig. 8.8 present the value of PSNR index calculated for the selected edge detection method applied on the images with added noise with variance between 0.001  0.01. A greater value of this index indicates that the method is more insensitive to noise.

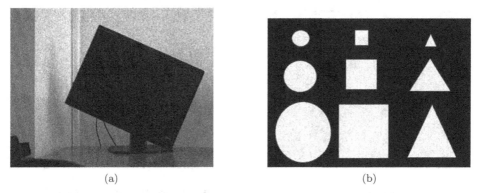

(a)                                            (b)

Fig. 8.4   Images with added noise.

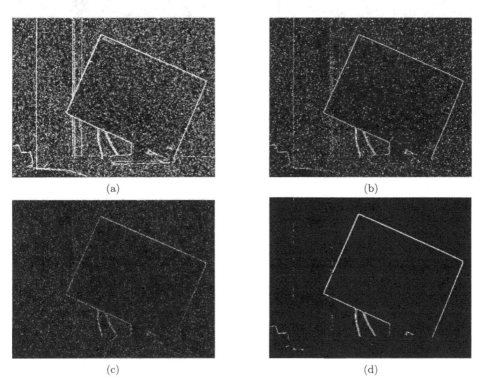

Fig. 8.5   Edges detected in Fig. 8.4(a) using a) Sobel, b) Roberts, c) fractional order and d) likewise-radar method.

The results presented in the Fig. 8.7 and Fig. 8.8 show that the algorithm proposed by the authors is more insensitive to noise, regardless of it intensity. The PSNR index is smaller for natural images, which is caused by additional noise generated by a camera.

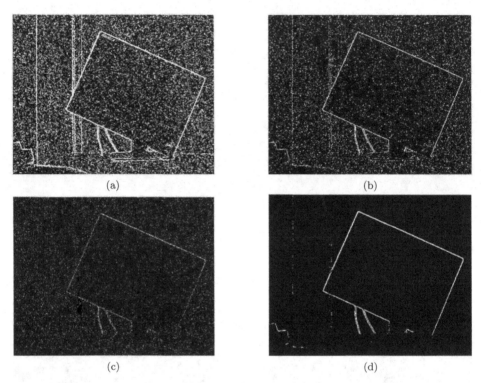

Fig. 8.6   Edges detected in Fig. 8.4(b) using a) Sobel, b) Roberts, c) fractional order and d)
likewise-radar method.

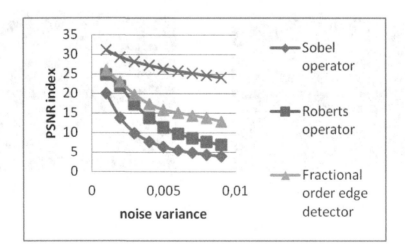

Fig. 8.7   The value of PSNR index calculated for selected edge detection methods applied in
Fig. 8.2(a).

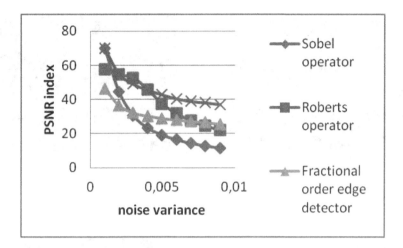

Fig. 8.8 The value of PSNR index calculated for selected edge detection methods applied in Fig. 8.2(c).

## 8.5 Results of applying likewise-radar edge detection method

Fig. 8.9 shows the results of applying the proposed likewise-radar method with different values of the threshold and the order on Fig. 8.2(d).

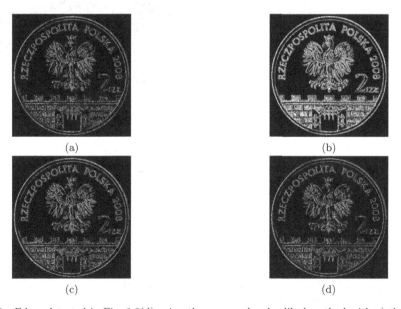

Fig. 8.9 Edges detected in Fig. 8.2(d) using the proposed radar liked method with a) threshold = 85 and order = 0.3, b) threshold = 85 and order = 0.9, c) threshold = 135 and order = 0.9, d) threshold = 145 and order = 0.7.

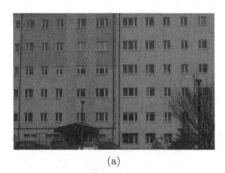 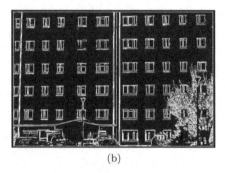

(a)                                                    (b)

Fig. 8.10   Edges detected using the proposed radar liked method: (a) Original image (b) edges detected using proposed likewise - radar method with threshold = 30 and order = 0.3.

 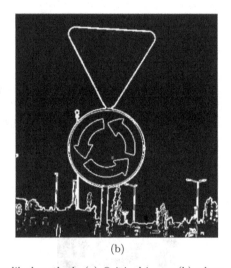

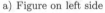
a) Figure on left side                                 (b)

Fig. 8.11   Edges detected using the proposed radar liked method: (a) Original image (b) edges detected using the proposed likewise - radar method with threshold = 37 and order = 0.3.

Fig. 8.10 and Fig. 8.11 present examples of the edges detected using the method proposed.

## References

Canny, J. (1986). A computational approach to edge detection, *IEEE Transactions on Pattern Analysis and Machine Intelligence* **6**, 8, pp. 679–698.

Chen, D., Chen, Y. and Sheng, H. (2010). Fractional variational optical flow model for motion estimation, in *Proceedings of the 4th IFAC Workshop on Fractional Differentiation and its Applications.* (Badajoz, Spain).

Duch, P., Blaszczyk, S., Laski, M., Krzeszewski, R. and Ostalczyk, P. (2013). Texture enhancement through multiscale mask based on rl fractional differential, in *Fractional Differentiation and its Applications*, Vol. 6 (Grenoble, France), pp. 647–652.

Gan, Z. and Yang, H. (2010). Texture enhancement through multiscale mask based on rl fractional differential, in *International Conference of Information Networking and Automation*, Vol. 1 (Kunming, China), pp. 333–337.

Gao, C., Zhou, J., Hu, J. R. and Lang, F., Lang (2011a). Edge detection of colour image based on quaternion fractional differential, *IET Image Processing* **5**, 3, pp. 261–272.

Gao, C., Zhou, J., Zheng, X. and Lang, F. (2011b). Image enhancement based on improved fractional differentiation, *Journal of Computational Information Systems* **7**, 1, pp. 257–264.

Ghamisi, P., Couceiro, M. S., Benediktsson, J. A. and Ferreira, N. M. F. (2012). An efficient method for segmentation of images based on fractional calculus and natural selection, *Expert Systems with Applications* **39**, pp. 12407–12417.

Mathieu, B., Melchior, P., Oustaloup, A. and Ceyral, C. (2003). Fractional differentiation for edge detection, *Signal Processing* **83**, 3, pp. 2421–2432.

Oustaloup, A., Mathieu, B. and Melchior, P. (1991a). Edge detection using non integer derivation, in *Presented at the IEEE European Conference on Circuit Theory and Design (ECCTD91)*. (Copenhagen, Denmark).

Oustaloup, A., Mathieu, B. and Melchior, P. (1991b). Robust edge detector of non integer order: the CRONE detector, in *Presented at the 8th Congres de Cytometrie en Flux et dAnalyse dImage (ACF91)*. (Mons, Belgium).

Yang, H., Ye, Y., Wang, D. and Jiang, B. (2010). A novel fractional - order signal processing based edge detection method, in *Control Automation Robotics & Vision (ICARCV), 2010 11th International Conference on* (Singapore).

Yang, Z., Lang, F., Yu, X. and Zhang, Y. (2011). The construction of fractional differential gradient operator, *Journal of Computational Information Systems* **7**, 12, pp. 4328–4342.

Yitzhaky, Y. and Peli, E. (2003). A method for objective edge detection evaluation and detector parameter selection,, *IEEE Transactions on Pattern Analysis and Machine Intelligence* **25**, 8, pp. 1027–1033.

# CHAPTER 9

# VISION BASED HUMAN-MACHINE INTERFACES: VISEM RECOGNITION

Krzysztof Slot, Agnieszka Owczarek, and Maria Janczyk

*Institute of Applied Computer Science*
*Lodz University of Technology*
*90-924 Lodz, ul. Stefanowskiego 18/22*
*kslot@kis.p.lodz.pl*

Human-computer interfacing is recently becoming one of the major scientific aspects in designing systems and machines that are capable of performing complex, possibly, autonomous tasks. The objective of the research is to provide means for natural, easy and seamless communication with computers, which will accelerate the proliferation of ubiquitous computing and reshape a way of experiencing technology advancements. For example, a use of speech or gestures as sources of control commands can significantly simplify operation of machines or systems, increase efficiency of task execution and minimize human distraction. Introduction of auditory and visual feedback will improve performance assessment and system-operation friendliness. An increase in human-system communication simplicity will reduce the exclusion of groups that have insufficient background for using modern technology or groups that, due to physical disabilities, cannot operate conventional computer interfaces. An important domain of exploration is also concerned with affective computing, with objectives to identify and properly respond to human emotional states, thus improving the quality of human-machine interaction.

Research on human-computer interfaces is very broad and covers all conceivable communication modalities, starting from speech and auditory events, through vision (gestures, gait, gaze or facial expressions) to bio-signals. Each of these fields have already been intensely studied and several remarkable achievements have been made, such as significant maturity in natural language processing, the development of operational gesture recognition systems or accomplishments in an emerging field of neuroprosthetics.

Despite enormous advances in the field, there exist several open problems in each of the aforementioned research tracks, such as speech recognition under presence of noise, gesture recognition against cluttered background, emotion recognition or robust gaze estimation. One of the ideas considered for improving speech recognition performance in realistic machine operation contexts, featuring high-levels of noise, is speech-reading. This domain has become a subject of our research and the following part of the chapter summarizes the ideas proposed, their implementation and the results obtained. A particular objective of the research was to develop a method for accurate identification of phones, uttered by a speaker, using face images as the only source of input information. Uttered phones are represented by a particular appearance of the mouth, which, by analogy to speech-signal analysis, is referred to as a *visual phonem* or a *visem*. The accurate recognition of visems is a core task for speech reading, as the remaining part of speech reading methodology exploits one of the two standard approaches: modeling visem sequences with Hidden Markov Models [Rabiner (1989)] or matching these sequences with prototypes using Dynamic Time Warping [Deller et al. (1987)].

The particular goal of the research, reported in the remaining part of this chapter, was to develop a robust method for visem recognition, based purely on outer lip contour modeling. The main ideas that have been proposed and discussed in detail, are utilization of previously unconsidered methods for contour extraction and parametrization, the introduction of novel dimensionality reduction strategies and the development of techniques for region of interest identification.

The structure of the Chapter is the following. After a concise presentation of research context, a general outline of the proposed algorithm is provided. Next, face image preprocessing methodology that has been used for extraction of region of interest for subsequent analyses is explained. The following section describes a strategy proposed for creating parametric outer lip-contour representation. As this representation is redundant, a novel feature vector dimensionality reduction technique is introduced. Finally, presentation of adopted feature vector classification methods and experimental evaluation of the algorithm's performance are provided.

## 9.1 Domain overview

Visem recognition is a difficult, open pattern recognition problem due to a few fundamental reasons. Firstly, lips and mouth play a secondary role in phone articulation, so their appearance provides insufficient amount of information on speech that is being uttered. Secondly, an individually learned way of speaking combined with elasticity of lips and mouth is highly diverse and variable, resulting in significant appearance differences for a given phone repeated by the same speaker, even larger deviations among different speakers and, finally, possible appearance similarities for uttering different phones by different speakers (large within-class and low between-class variability is the worst context in pattern recognition).

The challenging nature of the problem results in several different approaches that have been proposed for the task realization. The first group of visem recognition strategies, termed *low-level* or *pixel-based* methods, exploits textural information extracted from the mouth region to identify a visem. The main focus of reported research is derivation of compact and discriminative descriptors for a two-dimensional image region that contains the mouth. This can be accomplished in a variety of ways, for example, through principal component analysis applied directly to raw intensity images [Bregler and Konig (1994)] or by considering discrete cosine transform (DCT) coefficients [Potamianos et al. (2003)]. The second direction in research on visem recognition is based on lip geometry modeling (*high-level* or lip contour based approaches). The methods involve lip contour extraction combined with model fitting techniques followed by computation of a variety of features such as contour perimeter, lip area or horizontal to vertical opening ratios. Deformable templates [Foo et al. (2004); Yuille et al. (1992)], active shape models (ASM) [Wang et al. (2004); Perez et al. (2005); Matthews et al. (2002)], and snakes [Kass et al. (1987)] are sample representatives of contour extraction methodologies used. Finally, several methods that combine low-level and high-level approaches have also been proposed. For example, visem representation can be formed by concatenating the Karhunen Loeve transform (KLT) coefficients of the inner-outer lip contour points with texture information embedded in eigenlip images (an analog to eigenfaces) [Dupont and Luettin (2000)]. Similarly, mouth shape descriptors are combined with KLT derived for color images in [Chiou and Hwang (1997)]. The hybridization of the techniques may lead to solutions that better deal with drawbacks of individual techniques, however, no reliable rules for combining the two different pools of features have been proposed.

## 9.2 Outline of the proposed algorithm

Visem identification is a pattern recognition task, so the adopted data processing methodology conforms to a general pattern recognition framework. The proposed procedure, depicted schematically in Fig. 9.1, consists of several steps that assign class labels – visems – to input images (consecutive frames of input video sequence).

The first phase of the proposed algorithm: image preprocessing, is concerned with face detection and extraction of a mouth region that will be examined by subsequent data analysis procedures. Having the region of interest identified, a procedure for outer lip contour extraction is executed. The contour obtained is then modeled by a set of appropriately derived features. Finally, feature vector classification procedure is applied to establish class membership of a phone that is being uttered in some currently analyzed image.

Of the four main procedure blocks depicted in Fig. 9.1, the first one (image preprocessing) and the last one (recognition) are performed using standard, well-known approaches. The other two blocks, concerned with the the derivation of quantitative lip representation, provide a set of novel ideas that are discussed in detail in the remaining part of the chapter (these blocks are emphasized in the figure by using a solid gray background).

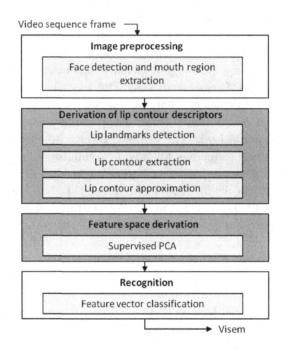

Fig. 9.1 Block diagram of the proposed algorithm. Modules involving novel concepts: lip descriptor derivation and feature space derivation, are shown grayed.

## 9.3 Frame preprocessing: mouth region extraction

The first step of the visem recognition procedure is extraction of a region of interest from each frame of an input video sequence. This region will be the domain for lip contour modeling, so its correct identification is of the uttermost importance from the point of view of recognition performance. The task is difficult, as it implicitly involves face detection and localization, if a fully automated recognition procedure is to be developed.

Face detection is one of the basic problems in computer vision, so a multitude of different approaches have been proposed to perform the task. One of the most commonly used techniques, due to its computational efficiency and reliability, is the algorithm proposed by Viola and Jones [[Viola and Jones (2004)]] and its extensions. The algorithm exploits Haar-like wavelets and searches for image regions that match some predefined face model, given in terms of spatial binary patterns. The main drawback of Viola-Jones approach is its sensitivity to face rotations and pose variations, so appropriate measures need to be taken to provide rotation and pose invariance, such as the prior execution of a set of appropriate affine transformations.

Having a face region extracted, one can easily determine an approximate location of the mouth region, using basic knowledge on face proportions (see Fig. 9.2). Such an approximate location is sufficient for executing some lip-shape finding algorithms, such as active contours. However, the procedure proposed requires detection of four lip landmarks, including the two corners, and two extreme lip points located along a vertical lip symmetry axis (see Fig. 9.3).

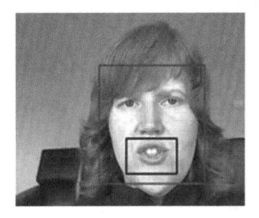

Fig. 9.2    Haar-like wavelet based face detection result with extracted mouth region.

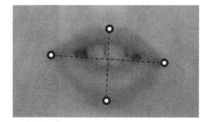

Fig. 9.3    Lip landmarks to be used in lip contour derivation.

Two different approaches have been examined for lip landmark extraction. The first, simpler procedure, depicted schematically in Fig. 9.4, is aimed at the extraction of lip corners and begins with the enhancement of an input image. This is accomplished by changing image representation from RGB to *Lab* color space followed by the subtraction of complement of the component 'b' from 'a' (as suggested in [Yao et al. (2010)]). The resulting image is converted to a binary form by thresholding, with the threshold determined using Otsu's method [Otsu (1979)]. Finally, a search for lip left and right corners is made in two phases. First, a silhouette of mouth region objects is formed from row-wise projections and a contiguous region corresponding to the mouth region is determined. A range of the region determines the height of a rectangle, where a column-wise search for the left-most and for the right-most boundary points is performed. The determined corner locations are then used to determine the two remaining landmarks: the 'top' and the 'bottom' lip endpoints.

The second method for lip landmark detection that has been examined uses an Active Appearance Model (AAM) [Cootes et al. (2001)] of a mouth. Active Appearance Models are considered to be one of the most robust techniques for modeling of objects that can undergo nonlinear deformations, and several AAM applications for handling the most challenging tasks, such as pose and rotation invariant face detection, gesture recognition or face image warping have been already considered [Xinbo et al. (2010)].

Active Appearance Models for lips were constructed based on a set of training images, with manually marked landmark points. Landmark points are points that are considered to be distinctive, repetitive and representative for an object. A total

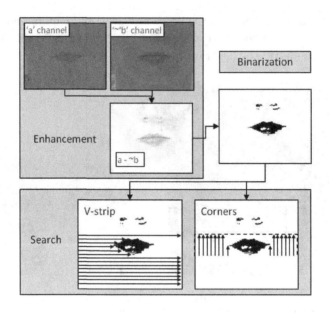

Fig. 9.4   Basic steps of lip corner detection procedure.

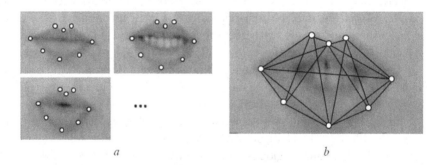

Fig. 9.5   Active Appearance Model of lips: sample training images with landmark locations represented by dots (a) and landmark detection result in previously unknown face, drawn with mesh lines (b).

of eight landmarks were used for constructing AAM lip models for the considered case, including the previously defined four points of interest. Sample training images and the landmark detection result produced by AAM algorithm, are shown in Fig. 9.5.

## 9.4   Lip-shape descriptor derivation

Instead of using active-contour models, adopted as the most common approach to lip boundary modeling, we propose a different approach that is optimal, computationally inexpensive and produces closed contours. The procedure consists of two steps. The first one is concerned with the derivation of an outer-lip contour and it is implemented with dynamic programming. As the resulting boundary is non-smooth and cannot be parametrized, which is crucial for forming feature vectors for subsequent classification procedures, the obtained contour is approximated during the second step using either a piecewise-cubic curve or a set of B-splines. The details of both steps are presented in the following subsections.

### 9.4.1 *Coarse lip contour derivation with dynamic programming*

Dynamic Programming is a well-known strategy for solving optimization problems, introduced by Bellman [Bellman and Dreyfus (1971)]. The idea of the approach is to decompose a given problem into a collection of subproblems that could be solved separately, providing individual components of an overall task solution. A problem to be solved is often expressed in terms of finding an optimal path through an oriented graph, composed of nodes, where nodes are arranged into a regular, locally-connected grid and inter-node transitions are assigned with some non-negative costs.

A central point to dynamic programming is a principle, referred to as *Bellman's optimality principle*. It states that if some optimal (i.e. minimum-cost) path between two nodes of a graph: $n_b$ and $n_e$ includes a node $n_i$ then any path's fragments: e.g. from $n_b$ to $n_i$ and from $n_i$ to $n_e$, are also optimal (there are no better ways to get from $n_b$ to $n_i$ and from $n_i$ to $n_e$). Applying this principle recursively, one can conclude that the optimal-path can be considered as a collection of optimal, local transitions between subsequent nodes.

Let's consider two consecutive nodes: $n_{k-1}$ and $n_k$ of some node sequence. The cost of transferring from a node $n_{k-1}$ to a node $n_k$ consists of two components: an inter-node transition cost $c^{k-1 \to k}$ and a node-visiting cost $c^k$. According to Bellman's principle, if a node $n_k$ is to be an optimal-path element, an accumulated cost at this node is given by:

$$C(n_k) = \min_i \left\{ C(n_{k-1}^i) + c_i^{k-1 \to k} \right\} + c^k \tag{9.1}$$

where all possible predecessors $n_{k-1}^i$ of the node $n_k$, with their minimum-accumulated costs $C(n_{k-1}^i)$, are considered.

Between-node transitions are typically restricted to a narrow subset of available possibilities. Commonly used transition patterns, depicted in Fig. 9.6a and Fig. 9.6b, assume that only a few neighboring nodes can be feasible predecessors of the given one. This reflects some basic constraints imposed on the problem to be solved, such as unidirectional path construction.

Dynamic programming is a two-phase algorithm: the first phase computes accumulated costs along the way from some origin to some destination, whereas the second phase uses backtracking to restore an optimal path, i.e. such a sequence of nodes between the destination and the origin, which ensures the minimum accumulated cost.

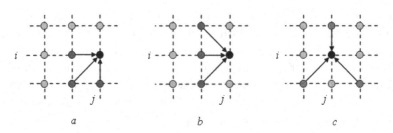

Fig. 9.6 Sample inter-node transition patterns used in dynamic programming (a), (b) and a sample illegal transition pattern (c) .

Taking into account numerous advantages of dynamic programming, including optimality and computational efficiency we decided to employ this method for solving the outer-lip contour extraction task (a similar approach was presented in [Shi et al. (2009)]). Lip-contour extraction with dynamic programming required an appropriate adaptation of the basic algorithm to the considered, a specific image processing task, involving an appropriate task redefinition, the specification of a cost function and legal node-transition patterns as well as the determination of endpoints of the search procedure.

A region that contains a mouth, which has been extracted during face-image preprocessing step of the procedure, becomes a domain for further analyses. Each pixel of this region becomes a node of the grid, so that an optimal path through a grid corresponds to a curve in the image domain. The region of interest has been divided into four rectangular parts, defined by previously detected landmarks: two lip corners and lip top and bottom points (see Fig. 9.8). This way, an outer lip contour will be composed of four quadrants determined by separate dynamic programming procedures.

The detection of object's contours is typically made based on image gradients, low-pass filtered gradients or image second-derivatives, as large local gray-level variability is one of the basic characteristics of boundary points. Therefore, a spatial distribution of mouth-region gradient magnitudes becomes input to dynamic programming (Fig. 9.9). Each node of the search graph is assigned a fitness (instead of a cost, as the procedure's objective is dual, i.e. to maximize an overall fitness, rather than to minimize the total cost), which is a gradient magnitude calculated at the corresponding image pixel. No fitness is assigned to between-node transitions, as all of the admitted transitions (specified by the adopted node interconnection patterns) are considered equally important. Hence, the fitness calculation rule of forward-phase of dynamic programming (9.1) assumes the form:

$$C(n_{p,q}) = \max_{i,j}\{C(n_{i,j})\} + c^{p,q} = \max_{i,j}\{C(n_{i,j})\} + |\nabla I(p,q)|, \quad (i,j) \in N(p,q)$$

where $C(..)$ denotes the accumulated cost, $|\nabla I(p,q)|$ is image gradient magnitude at a pixel located in $p$-th row and $q$-th column, and $N(p,q)$ denotes the set of legal predecessors of a node $(p,q)$.

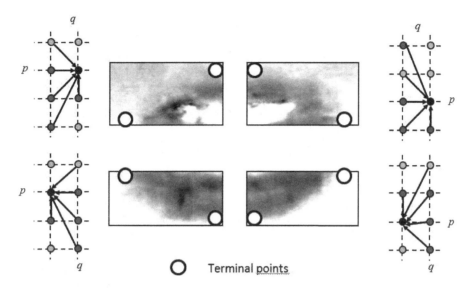

Fig. 9.7   Four images used as input to four separate dynamic programming-based contour derivation procedures along with admissible between-node transitions for each of the quarters. The
circles indicate locations of path endpoints.

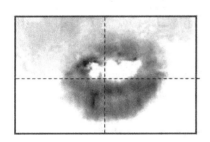

Fig. 9.8   Gradient magnitude image used as
input for contour extraction.

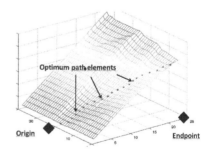

Fig. 9.9   Cost-map corresponding to the
upper-left quadrant of a mouth region.

In order to eliminate possible errors caused by strong gradients originating from
the inner lip-contour and teeth, an additional weighting of the gradient map, used as
dynamic programming input, has been introduced. The weighting scheme reduces
gradient magnitudes in an inversely-proportional manner as one moves out of the
main mouth axis (the lip corner-connecting line):

$$w_{i,k} = 1 - e^{\frac{d_i}{\sigma_k}}$$

where $d_i$ is the distance of a pixel from the main axis and $\sigma_k$ is the standard deviation
at an ordinate $k$ that varies along the main mouth axis in a piecewise-linear manner
(see Fig. 9.10).

A sample result of the presented procedure application for outer lip boundary
extraction has been shown in Fig. 9.11. Four procedures that separately handle

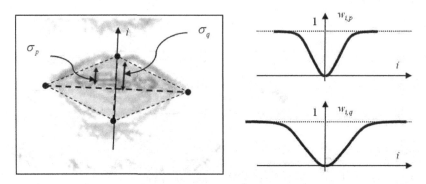

Fig. 9.10 Mouth region gradient magnitude image with boundaries of a region where weights are applied (left) and weighting functions at sample ordinates $p$ and $q$, corresponding to differing parameters $\sigma_p$ and $\sigma_q$ at these locations (right).

each of the four subregions, and which differ only in admissible node-transition patterns (as presented in Fig.9.7) and endpoint locations, are executed. To ensure the continuity of the produced curve, neighboring image blocks overlap by one row or one column and share endpoints of consecutive contour segments.

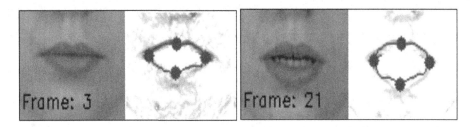

Fig. 9.11   Sample lip contours extracted by dynamic programming.

### 9.4.2 *Lip contour approximation*

As can be seen, contours extracted using dynamic programming are non-analytical and non-smooth. For the purpose of lip shape recognition, a contour should be represented by a possibly small set of features. Therefore, it is necessary to derive some parametric representation of the obtained closed curve, henceforth referred to as a *coarse lip contour*.

The simplest way of the coarse lip contour approximation would involve polynomials. However, as the lip shape varies significantly along their circumference, one needs to use polynomials of some unknown, high degree. Alternatively, one can apply piecewise approximation (for instance, for each of the available contour quadrants), using some low-order curves, such as cubic polynomials, bearing in mind that only zero-order smoothness will be provided at segment merger points. Yet another possibility is to make contour approximation, using composite curves such as Bezier-curves or B-splines, providing contour smoothness up to some required order.

As the exact relation between accurate modeling of lip-shape details and image-based phone representation is unclear, we decided to verify the following two methods for approximation of the coarse lip contour. The first one is piecewise-cubic modeling, where each quarter of the coarse contour is separately approximated using a third-order polynomial. The second approach assumes B-spline modeling of two separate segments of the coarse contour: the fragments corresponding to the upper and to the lower lip.

The piecewise-cubic approximation of the coarse lip contour $\ell^C$ is a solution to four separate mean-square error minimization problems, performed for consecutive contour quarters $\ell_k^C = [\ell_0^k, \ell_1^k \ldots \ell_{n_k-1}^k]$, $k = 0..3$, of the form:

$$(A^k, B^k, C^k, D^k) : e = \sum_{i=0}^{n_k}((A^k i^3 + B^k i^2 + C^k i + D^k) - \ell_i^k)^2 \to min \qquad (9.2)$$

where $A^k, B^k, C^k, D^k$ are polynomial parameters to be determined and the contour $\ell_k^C$ is forced to be a function of the horizontal variable by replacing all occurrences of points defined for the same ordinate by their mean value.

Arranging the variables into vectors and matrices, such that:

$$\mathbf{a}^k = \begin{bmatrix} A^k \\ B^k \\ C^k \\ D^k \end{bmatrix} \qquad \mathbf{X} = \begin{bmatrix} \mathbf{x}_0 \ldots \mathbf{x}_j \ldots \mathbf{x}_{n_k} \end{bmatrix} \qquad \mathbf{x}_j = \begin{bmatrix} j^3 \\ j^2 \\ j \\ 1 \end{bmatrix}$$

the contour approximation criterion (9.2) can be compactly rewritten as:

$$\mathbf{a}^k : e = (\mathbf{X}^T \mathbf{a}^k - \ell_k^C)^T (\mathbf{X}^T \mathbf{a}^k - \ell_k^C) \to min$$

or, equivalently, as:

$$\mathbf{X}\mathbf{X}^T \mathbf{a}^k = \mathbf{X}\ell_k^C$$

yielding a well-known solution that involves pseudoinverse of $\mathbf{X}$:

$$\mathbf{a}^k = (\mathbf{X}\mathbf{X}^T)^{-1}\mathbf{X}\ell_k^C$$

Observe that by centering input samples $\ell_k^C$ around their mean, the last component of the vector $\mathbf{a}^k$ vanishes, so that each quarter of the coarse lip contour is modeled using three parameters, yielding a twelve-element representation for the whole contour. The produced vector $\mathbf{F}^a$:

$$\mathbf{F}^a = [(\mathbf{a}^0)^T, (\mathbf{a}^1)^T, (\mathbf{a}^2)^T, (\mathbf{a}^3)^T]^T \qquad (9.3)$$

becomes a feature vector that represents a visem portrayed in some input sequence frame.

The second method for approximating the coarse lip contour uses composite B-splines [Lee (1986)]. A spline of the order $d$, denoted by $s_{i,d}$, is a curve generated by the recursive formula that involves a convex combination of two splines of the order $d - 1$:

$$s_{i,d}(t/t_{i-d}...t_{i+d-1}) = \frac{t_\beta - t}{t_\beta - t_\alpha} s_{i-1,d-1} + \frac{t - t_\alpha}{t_\beta - t_\alpha} s_{i,d-1}, \quad t \in [t_\alpha, t_\beta]$$

where $t_{i-d}, t_\alpha, t_\beta, t_{i+d-1}$ are parameters, refered to as *knots* that need to be chosen in a way ensuring combination convexity and monotonicity, i.e. $t_{i-d} \leq t_\alpha \leq t_\beta \ldots t_{i+d-1}$, and $i$ is the spline index.

At the basis of the recurrence hierarchy, there is an initial convex combination of individual points $\mathbf{c}_i$, referred to as spline *control points* that produces line segments:

$$s_{i,1}d(t) = \frac{t_\beta - t}{t_\beta - t_\alpha} \mathbf{c}_{i-1} + \frac{t - t_\alpha}{t_\beta - t_\alpha}, \quad t \in [t_\alpha, t_\beta] \mathbf{c}_i$$

To approximate or interpolate a set of data points one can use splines of an arbitrary order, however, the best strategy is to combine a set of some low-order splines instead. This way dataset approximation can be expressed as a set of $n$ spline segments of some order $d$, defined over consecutive intervals of the underlying parameter $t$:

$$f(t) = \begin{cases} s_{d,d}(t) & t \in [t_{d-1}, t_d], \\ s_{d+1,d}(t) & t \in [t_d, t_{d+1}], \\ \vdots \\ s_{n-1,d}(t) & t \in [t_{n-1}, t_n], \end{cases} \tag{9.4}$$

The piecewise approximation given in (9.4) can be compactly expressed by introducing an auxiliary function $B_{i,0}$, defined as:

$$B_{i,0} = \begin{cases} 1 & t \in [t_{i-1}, t_i], \\ 0 & \text{otherwise} \end{cases} \tag{9.5}$$

Using definition (9.5) one can rewrite (9.4) in the following manner:

$$f(t) = \sum_{i=d}^{n-1} s_{i,d} B_{i,0} \tag{9.6}$$

As control points form a core for composite spline based approximation, expression (9.6) can be rewritten in the form where they appear explicitly:

$$f(t) = \sum_{i=1}^{n} \mathbf{c}_i B_{i,d}(t)$$

where the piecewise-constant function $B_{i,0}$ is replaced by a nonlinear basis function $B_{i,d}(t)$, referred to as a *B-spline* function, defined recursively as:

$$B_{i,d}(t) = \frac{t - t_i}{t_{i+d} - t_i} B_{i,d-1}(t) + \frac{t_{i+1+d} - t}{t_{i+1+d} - t_{i+1}} B_{i+1,d-1}(t)$$

Third-order composite B-splines have been chosen as a contour modeling tool, as the lip shape can be correctly approximated by third-order polynomials and the considered B-splines ensure smooth contours (at each joint, both a curve and its first and second derivatives are continuous). We decided to approximate outer contours of upper and lower lips separately. In both cases, a different number of spline segments were considered. For the upper lip, from two to eight-element B-splines were considered, whereas test configurations for the lower lip contour modeling included from one up to four segments. The best approximation performance over the set of training lip-images under consideration was observed for the four-segment B-spline, in case of upper-lip contour modeling and for single-segment B-spline, in case of lower-lip contour modeling, yielding the following equations for the upper $f^u(t)$ and the lower $f^l(t)$ contour segments:

$$f^u(t) = \begin{cases} s_{2,3}(t/c_0^u...c_3^u; t_{-1}^u...t_4^u) & t \in [t_1^u, t_2^u], \\ s_{3,3}(t/c_1^u...c_4^u; t_0^u...t_5^u) & t \in [t_2^u, t_3^u], \\ s_{4,3}(t/c_2^u...c_5^u; t_1^u...t_6^u) & t \in [t_3^u, t_4^u], \\ s_{5,3}(t/c_3^u...c_6^u; t_2^u...t_7^u) & t \in [t_4^u, t_5^u], \end{cases} = \sum_{i=2}^{5} s_{i,3} B_{i,0} = \sum_{j=0}^{6} c_j^u B_{j,3} \quad (9.7)$$

$$f^l(t) = \left\{ s_{2,3}(t/c_0^l...c_3^l; t_{-1}^l...t_4^l) \; t \in [t_1^l, t_2^l], \right. = \sum_{i=2}^{2} s_{i,3} B_{i,0} = \sum_{j=0}^{3} c_j^l B_{j,3} \quad (9.8)$$

As it was the case for the presented cubic approximation of the coarse lip contour, also for B-spline based approximation a mean-square error minimization criterion is used to choose the optimal locations of control-points. The resulting feature vector that represents a visem, which is to be recognized, comprises control points of B-splines given by (9.7) and (9.8), and has the form:

$$\mathbf{F}^s = [\mathbf{c}_0^u, \mathbf{c}_1^u, \mathbf{c}_2^u, \mathbf{c}_3^u, \mathbf{c}_4^u, \mathbf{c}_5^u, \mathbf{c}_6^u, \mathbf{c}_0^l, \mathbf{c}_1^l, \mathbf{c}_2^l, \mathbf{c}_3^l] \quad (9.9)$$

Since control points are defined over 2-dimensional image space, the feature vector $\mathbf{F}^s$ contains 22 components.

## 9.5 Feature space derivation with supervised PCA

The derived, twenty-two-element outer-lip contour descriptor in case of B-spline approximation (9.9) and the 12-element one, in case of piecewise-cubic approximation (9.3), need to be significantly reduced to provide a feature space that offers reasonable class-modeling. The one offered by vectors $\mathbf{F}^s$ and $\mathbf{F}^a$ is seriously impaired due to the curse of dimensionality and due to the noise originating from potentially irrelevant features. Commonly used dimensionality-reduction strategies belong to one of the two groups: supervised methods use class labels to find the most discriminative directions in feature space, whereas unsupervised ones produce the most expressive directions. The most commonly used technique of the latter group is PCA (principal component analysis) and its various extensions (such as weighted PCA or kernel-PCA). Supervised techniques are the natural choice for problem dimensionality reduction if class membership of data samples is known.

The simplest method for supervised feature extraction, LDA (linear discriminant analysis), provides a subspace that maximizes linear data separability. As object categories in real-world classification problems are typically not linearly-separable, LDA extensions that handle such cases have been developed (e.g. kernel PCA combined with LDA). However, these approaches are difficult to integrate within the probabilistic pattern recognition framework, which is essential for the analysis of processes that evolve in time, such as the considered case of speech. Therefore, the objective of our research was to develop a computationally efficient feature-space reduction method for generation of low-dimensional vector sequences that are to be analyzed under the HMM (Hidden Markov Model) framework.

The supervised PCA methods that are proposed in this section aim at deriving a feature space that emphasizes between-class variability. This is accomplished by adjusting scatter matrix eigenvalues, so that their magnitudes reflect also between-class scatter of the considered dataset. The methods can be considered as a supervised extension to both classical PCA analysis, as well as to its nonlinear version (kernel-PCA).

In case of the proposed, supervised extension of linear PCA (henceforth, referred to as a *linear SPCA*), the first step of the algorithm is to determine eigenvalues $\lambda_i$ and eigenvectors $\mathbf{e}_i$ of the co-variance matrix, computed for all the available samples from all the classes considered (referred to as *global*-PCA):

$$\mathbf{E} = [\mathbf{e}_0, \ldots \mathbf{e}_{d-1}], \qquad \Lambda = [\lambda_0, \ldots \lambda_{d-1}] \tag{9.10}$$

Next, PCA is executed separately for each class $c_k$ producing eigenvalues and eigenvectors representing within-class distributions for all $m$-classes (henceforth, referred to as *class*-PCA):

$$\mathbf{E}^k = [\mathbf{e}_0^k, \ldots \mathbf{e}_{d-1}^k], \qquad \Lambda^k = [\lambda_0^k, \ldots \lambda_{d-1}^k] \tag{9.11}$$

The objective of the proposed linear SPCA is to combine information on overall data scatter, given by parameters (9.10) with information on within-class scatters (9.11), to determine directions of a feature space that focuses on between-class data variability. The adopted methodology aims to decrease the importance of these directions of global-PCA (9.10), where data variability is caused by within-class scatters rather than by between-class scatters. To implement this concept, we define weighting coefficients $w_i$, which will quantify contributions of within-class scatters in data variability along consecutive principal components. This contribution is calculated as a total projection of all class-PCA eigenvectors, weighted by their corresponding eigenvalues onto a global-PCA eigenvector $\mathbf{e}_j$, accumulated over all the considered classes:

$$w_j = \sum_{k=0}^{K-1} \sum_{i=0}^{d-1} \lambda_i^k \cdot | < \mathbf{e}_j, \mathbf{e}_i^k > | \tag{9.12}$$

The expression (9.12) measures an amount of data variability induced by within-class scatter in total data variability. The larger it gets, the less discriminative becomes the global-PCA derived direction $\mathbf{e}_j$. A measure of actual discriminativeness of global-PCA directions is thus given by adjusting original eigenvalues by the derived weights (9.12):

$$\tilde{\lambda}_j = \frac{\lambda_j}{w_j} \tag{9.13}$$

The proposed approach can easily be extended to the case of nonlinear, kernel-PCA analysis. Recall that kernel-PCA is concerned with the implicit derivation of principal components in some auxiliary, very high-dimensional feature spaces. A central concept of the method is to analyze data scatter in a new, possibly very high-dimensional feature space, without making explicit feature-space transformations.

Assume, we are given a set of $N$-input samples $\mathbf{X} = \{\mathbf{x}_0, \ldots \mathbf{x}_{N-1}\}$ defined over $\Re^d$ space and some nonlinear mapping $\Phi(\mathbf{x})$ that transforms original samples to a $D$-dimensional space, i.e. $\Phi(\mathbf{x}) : \Re^d \to \Re^D, D > d$. Eigen-analysis in a target $D$-dimensional space explores scatter of projections of original samples, by solving the equation:

$$(\sum_{i=0}^{N-1} \Phi(\mathbf{x}_i)\Phi^T(\mathbf{x}_i))\mathbf{V} = \lambda\mathbf{V} \equiv \Psi\Psi^T\mathbf{V} = \lambda\mathbf{V} \equiv \mathbf{C}\mathbf{V} = \lambda\mathbf{V} \tag{9.14}$$

where $\Psi = [\Phi(\mathbf{x}_0) \ldots \Phi(\mathbf{x}_{N-1})]$ is the matrix of original data projections, $\mathbf{C}$ is the co-variance matrix of size $D \times D$ computed for these projections, $\mathbf{V}$ and $\lambda$ denote eigenvectors and eigenvalues of $\mathbf{C}$, and points in the target space are assumed to be centered around their mean. The given eigen-problem (9.14) is computationally intractable due to large values of $D$, so it is appropriately transformed. First, one can observe that premuliplying the equation (9.14) by a transpose of the matrix $\Psi$ gives the following result:

$$\Psi^T\Psi\Psi^T\mathbf{V} = \lambda\Psi^T\mathbf{V} \equiv (\Psi^T\Psi)(\Psi^T)\mathbf{V} = \lambda(\Psi^T\mathbf{V}) \equiv \mathbf{K}\mathbf{W} = \lambda\mathbf{W} \tag{9.15}$$

where $\mathbf{W} = \Psi^T\mathbf{V}$ is an eigenvector of the *kernel* matrix $\mathbf{K}$, composed of dot products of sample projections:

$$\mathbf{K} = \begin{bmatrix} <\Phi(\mathbf{x}_0), \Phi(\mathbf{x}_0)> \ldots <\Phi(\mathbf{x}_0), \Phi(\mathbf{x}_j)> \ldots \\ \vdots \quad\quad \vdots \quad\quad \vdots \quad\quad \vdots \\ <\Phi(\mathbf{x}_i), \Phi(\mathbf{x}_0)> \ldots <\Phi(\mathbf{x}_i), \Phi(\mathbf{x}_j)> \ldots \\ \vdots \quad\quad \vdots \quad\quad \vdots \quad\quad \vdots \end{bmatrix}$$

Instead of computing dot products of projections $\Phi(..)$ (this would occur in very high-dimensional spaces), one can introduce (this can be done under certain assumptions) appropriate, nonlinear kernel functions $K(..)$ that operate on the original data :

$$<\Phi(\mathbf{x}_i), \Phi(\mathbf{x}_j)> = K(\mathbf{x}_i, \mathbf{x}_j)$$

Kernel-PCA analysis determines projections $f(..)$ of unknown input data samples $\mathbf{u}$ onto some eigenvector $\mathbf{e}^\alpha$ of the co-variance matrix $\mathbf{C}$ of (9.15) by computing the following expression:

$$f(\mathbf{u}) = < \Phi(\mathbf{u}), \mathbf{e}^\alpha > = \sum_{i=0}^{N-1} \mathbf{W}_i^\alpha K(\mathbf{u}, \mathbf{x}_i) \qquad (9.16)$$

where $\mathbf{W}_i^\alpha$ are eigenvectors of $\mathbf{K}$ (9.15), $K(..)$ is the selected kernel (e.g. radial basis or sigmoid function) and $\mathbf{x}_i$ are the original data points.

The selection of eigenvectors $\mathbf{W}^\alpha$ that form a new space is based on the analysis of eigenvalues $\lambda$ (9.15): one leaves only the dominant components, dropping all vectors that correspond to the negligible data scatter. It is evident that supervised analysis presented in case of linear PCA can also be applied to the case considered, using analogous methodology. The proposed supervised kernel-PCA procedure is the following. First, kernel-PCA is performed for the whole dataset, yielding the corresponding eigenvector and eigenvalue sets, as in the case of eq. (9.10). Similarly, the kernel-PCA analyses are then executed separately for all classes, yielding within-class eigenvector and eigenvalue sets (as in eq. (9.11)). Finally, eigenvalues determined for the whole dataset are adjusted using eigenvectors and eigenvalues derived for within-class case, analogously as in eq. (9.12 and 9.13)). The resulting, reordered eigenvalue set is a basis for constructing a novel, discriminative feature space.

## 9.6   Feature vector classification

The derived feature space provides a domain for phone classification. The first issue that needs to be addressed at the classifier design stage is determination of categories that will be considered in recognition. Speech-reading and visem classification, unlike speech-signal based recognition, has its own specificity, resulting in the severe reduction in a pool of identifiable phone classes. Firstly, no reliable visual representation can be constructed for several phones, such as plosives ('p', 'b' etc.) as their duration is insufficient for being reliably captured using typical video frame rates (for instance, 25 frames per second rate implies sampling interval of 40 milliseconds, whereas plosive phone duration can be of the order of milliseconds). Secondly, articulation differences for some phones are caused by internal fragments of the vocal tract and have no visible representation in mouth or lip appearance. Therefore, a pool of visems that can be used in speech reading is relatively narrow and need to be tailored to a particular speaker. Seven different visems, corresponding to stable fragments of phones 'a', 'o', 'm', 'u', 's', 'sh' and 'v', together with neutral mouth appearance (Fig. 9.12), which were found to be relatively easy for manual identification and labeling, were selected for identification.

Each of these visems, represented by feature vectors derived using the previously described procedure, were subject to classification in the last step of the algorithm. Two different, well-established classification strategies were considered:

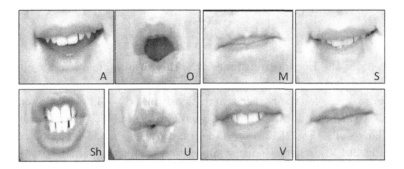

Fig. 9.12   Visem categories considered in recognition.

Ada-Boosting [Shapiro (2003)] and Gaussian Mixture Modeling (GMM) [Reynolds et al. (2000)]. The former approach resolves class membership by combining decisions made by several appropriately trained weak, linear classifiers, whereas probabilistic, parametric class models are built and utilized for recognition in the latter case.

## 9.7   Experimental evaluation of the procedure

The experimental evaluation of the presented procedure was made using a database of recordings that included a total of approximately one thousand images of the considered visems and one thousand other visems, all manually labeled, extracted from video sequences of different resolutions, taken under different recording conditions. In addition to frame labeling, every image containing a visem of interest was subject to content-labeling. This included the manual determination of mouth region and lip landmarks. The dataset used for algorithm evaluation involved recordings taken from five speakers. Speaker-dependent recognition scenario was considered. The proposed algorithm was subject to several evaluations, aimed at determining properties of the proposed, individual procedures as well as at determining an overall visem classification performance.

The first part of the experiments was concerned with the evaluation of accuracy of lip landmark extraction. The correct localization of the four landmark points determines the accuracy of the subsequent processing procedures, so we were interested in confronting the performance of the two considered, alternative landmark detection strategies. Landmark detection accuracy was measured by computing a mean-square displacement between manually-established locations of landmarks ($m_i$) and locations given by the corner-search algorithm ($c_i$) and the AAM algorithm ($a_i$), averaged over all test images:

$$e_c^l = \frac{1}{n} \sum_{k=0}^{n-1} \sum_{i=0}^{3} d(m_i^k - c_i^k) \tag{9.17}$$

$$e_a^l = \frac{1}{n} \sum_{k=0}^{n-1} \sum_{i=0}^{3} d(m_i^k - a_i^k) \tag{9.18}$$

where $d$ denotes the Euclidean distance, $k$ is the training image index and $i$ is the index of a landmark.

An average error for the corner-search algorithm $(e_c^l)$ – 3.2% – is lower than the one for the AAM-based localization $(e_a^l = 5.6\%)$. Although typically, both methods yield similar results (Fig. 9.13a), in some frames AAM-based detection completely fails (see Fig. 9.13b), whereas such significant errors for the corner-based search do not appear.

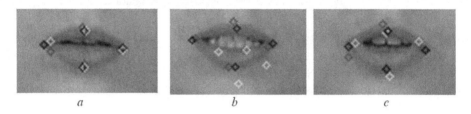

$a$ $\qquad\qquad\qquad\qquad\qquad$ $b$ $\qquad\qquad\qquad\qquad\qquad$ $c$

Fig. 9.13 Landmark localization accuracy evaluation: manually-selected landmark locations are shown in black, locations determined by the corner extraction algorithm - in gray and locations determined by the AAM method are given in white.

The second part of the experiments was concerned with the evaluation of mouth-region extraction accuracy. This was performed by computing lip segmentation error $e^s$, defined as a relative difference between a manually extracted region of the mouth, and the interior of a closed contour, returned by a dynamic-programming procedure:

$$e^s = \frac{n_B + n_F}{n_m} \cdot 100\%$$

where $n_B$ is the number of background pixels that were erroneously enclosed within the produced contour, $n_F$ is the number of mouth pixels that remained outside of the contour an $n_m$ is the total number of pixels that belong to the manually extracted mouth region.

For the considered set of 35 images (five images per visem), dynamic programming based contour extraction resulted in 12.3 % error. The error jumped to 18.3% after B-spline approximation has been generated and to 21.2% for the piecewise-cubic approximation. Sample binary images that show segmentation results, based on automatically extracted lip contours approximated with B-splines, have been presented in Fig. 9.14.

Visem recognition results for different configurations of the proposed procedure, involving two alternative lip landmark detection methods (corner-detection based and AAM-based), two alternative contour approximation strategies (piecewise-cubic and B-splines) and two alternative classification methodologies (Ada-Boost and

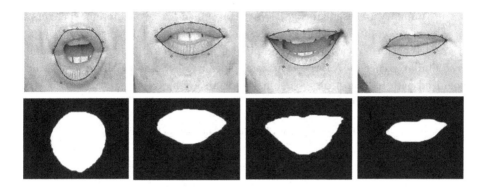

Fig. 9.14   Mouth segmentation results based on contours extracted with dynamic programming.

GMMs) are summarized in Table 9.1. For the used training data set, alternative dimensionality reduction methods (linear supervised PCA and supervised kernel PCA) performed similarly, so the simpler one – linear SPCA – was applied throughout the remaining experiments.

Table 9.1   Visem recognition results: FAR rates (at FRR set to 10%) averaged over all speakers and over all considered visems

|  | Ada-Boost | | GMM | |
|---|---|---|---|---|
|  | Piecewise Cubic | B-splines | Piecewise Cubic | B-Splines |
| Landmarks: Corner detection | 8.8% | 4.96% | 8.6% | 5.4% |
| Landmarks: AAM | 9.6% | 6.9% | 7.9 | 6.8% |

For the best performing visem recognition configuration: the corner-detection procedure used for lip landmark extraction, B-spline approximation of contours produced by dynamic programming, linear SPCA-based feature space derivation and Ada-Boost classification, relative operating characteristics (ROC) have been plotted for all visems in Fig. 9.15.

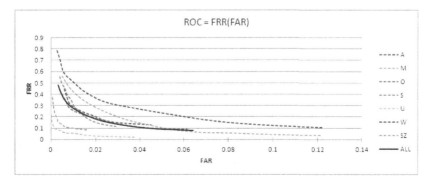

Fig. 9.15   Relative operating characteristic (ROC) curves for the phones considered in recognition.

The summary of visem recognition performance for two different speakers are represented by confusion matrices, shown in tables 9.2–9.3. As it can be seen, recognition performance for different phones is, in general, speaker-dependent (it has a similar structure for the remaining speakers). Some of the phones can be recognized with high accuracy (such as 'u' or 'm'), whereas the rates for other vary significantly among the speakers (such as 'a' or 'v').

Table 9.2   Confusion matrix for the Speaker A

| Visem | A | M | O | S | U | V | Sh |
|-------|------|------|------|------|------|------|------|
| A | **0.36** | 0.00 | 0.09 | 0.00 | 0.00 | 0.00 | 0.02 |
| M | 0.01 | **0.88** | 0.00 | 0.09 | 0.00 | 0.22 | 0.01 |
| O | 0.06 | 0.00 | **0.38** | 0.00 | 0.01 | 0.00 | 0.04 |
| S | 0.04 | 0.00 | 0.00 | **0.65** | 0.00 | 0.01 | 0.00 |
| U | 0.02 | 0.05 | 0.50 | 0.00 | **0.99** | 0.05 | 0.19 |
| V | 0.01 | 0.06 | 0.00 | 0.24 | 0.00 | **0.69** | 0.00 |
| Sh | 0.49 | 0.01 | 0.03 | 0.03 | 0.00 | 0.03 | **0.74** |

Table 9.3   Confusion matrix for the Speaker B

| Visem | A | M | O | S | U | V | Sh |
|-------|------|------|------|------|------|------|------|
| A | **0.74** | 0.02 | 0.19 | 0.02 | 0.01 | 0.01 | 0.02 |
| M | 0.00 | **0.83** | 0.00 | 0.00 | 0.00 | 0.00 | 0.00 |
| O | 0.10 | 0.00 | **0.78** | 0.00 | 0.14 | 0.00 | 0.05 |
| S | 0.03 | 0.04 | 0.00 | **0.67** | 0.00 | 0.15 | 0.00 |
| U | 0.01 | 0.06 | 0.02 | 0.17 | **0.82** | 0.00 | 0.00 |
| V | 0.02 | 0.05 | 0.00 | 0.13 | 0.03 | **0.82** | 0.01 |
| Sh | 0.10 | 0.00 | 0.02 | 0.01 | 0.00 | 0.00 | **0.92** |

The results obtained are difficult to compare with those of other reports on visem recognition due to different databases used as well as the emphasis on isolated word recognition performance which is the focus of the majority of recent publications in the field. However, the produced visem recognition rates are sufficient for the successful extension of the proposed procedure to the isolated word recognition case.

## 9.8   Conclusion

The main objective of the chapter was to evaluate the possibility of obtaining robust recognition of the selected visems by focusing purely on outer lip-shape modeling. The outer lip contour is probably the main source of information on phones being uttered (it can be supplemented only with information extracted from the inner lip contour shape and the overall mouth texture). Therefore, it is important to estimate limits of correct visem recognition by experimenting with a variety of lip-contour modeling and data classification approaches.

The presented ideas focus on both of the aforementioned tasks. Dynamic programming-based contour extraction provides a high level of tolerance against variability in image acquisition conditions. B-spline modeling better reflects the actual lip shape than currently used approaches, such as active-contours that in fact provide only piecewise-linear contour approximation. Also, the presented novel supervised extensions to well-known PCA analysis enable the determination of features that combine the expressiveness and class distinctiveness of the underlying data set. Both of the presented novel ideas are quite general, and can be considered for other image analysis and pattern recognition problems.

The presented outer lip modeling and visem recognition methodology can easily be applied to inner lip contour modeling to generate additional phone-related visual descriptors. The resulting feature vectors could be eventually combined with appropriately derived textural descriptors to provide the comprehensive representation of visems, which could provide a good basis for robust speech-reading.

# References

Bair, E., Hastie, T., Paul, D. and Tibshirani, R. (2006). Prediction by supervised principal components. *J. Amer. Statist. Assoc.*, pp. 119-137.

Bellman R.E., Dreyfus S.E.(1971). *Applied dynamic programming* (Princeton University Press).

Bregler C., Konig Y.(1994) Eigenlips for robust speech recognition, Proc. IEEE Conf. Acoustics, Speech and Signal Processing, pp. 669672.

Bestbury, B. W. (2003). *R*-matrices and the magic square, *J. Phys. A* **36**, 7, pp. 1947–1959.

Chiou G.I., Hwang J.-N. (1997), Lipreading from color video, Trans. Image Processing, vol. 6, pp. 1192–1195.

Cootes, T.F., Edwards, G.J., Taylor, C.J. (2001). Active appearance models, *IEEE Transactions on Pattern Analysis and Machine Intelligence*, **23**, 6, pp.681–685.

Deller J.R., Proakis J.G., Hansen J.H. (1987). *Discrete-Time Processing of Speech Signals* (Prentice Hall, NJ).

Dupont S,. Luettin J. (2000), Audio-visual speech modeling for continuous speech recognition, IEEE Trans. Multimedia, vol. 2, no. 3, pp. 141-151.

Foo S. W., Lian Y., Dong L. (2004), Recognition of visual speech elements using adaptively boosted hidden Markov models, *IEEE Trans. on Circuits Systems and Video Technology*, **14**, 5, pp. 693-705.

Kass M., Witkin A., Terzopoulos D.(1987), Snakes: Active contour models, *Internaltional Journal of Computer Vision*, pp. 321–331.

Lee E. T. Y. (1986). Comments on some B-spline algorithms, *Computing* (Springer-Verlag), **36**, 3, pp. 229–238.

Matthews I., Cootes T. F., Bangham J. A., Cox S., Harvey, R. (2002), Extraction of visual features for lipreading, *IEEE Trans. on Pattern Analysis and Machine Intelligence*, **24**, 2, pp. 198-213.

Otsu N. (1979). A threshold selection method from gray-level histograms, *IEEE Trans. on Systems Man and Cybernetics* **9**, 1, pp. 62-66.

Perez J. F. G. , Frangi A. F., Solano E. L., Lukas K. (2005), Lip reading for robust speech recognition on embedded devices, Proc. Int. Conf. Acoustics, Speech and Signal Processing, vol. I, pp. 473476,

Potamianos G., Neti C., Gravier G., Garg A., Senior A. W. (2003). Recent advances in
    the automatic recognition of audio-visual speech, *Proceedings of the IEEE*, **91**, 9,
    pp. 1306-1326.
Rabiner L. (1989). A Tutorial on Hidden Markov Models and Selected Applications in
    Speech Recognition, *Proc. of the IEEE*, **77**, 2, pp. 257–286.
Reynolds D.A., Quatieri T.F., Dunn R.B. (2000). Speaker Verification Using Adapted
    Gaussian Mixture Models, *Digital Signal Processing*, **10**, pp. 19-41.
Schapire R.E. (2003). *The boosting approach to machine learning: An overview Nonlinear
    Estimation and Classification* (Springer).
Shi W., Kim H., Chen H. (2009), Contour Tracking Using Centroid Distance Signature
    and Dynamic Programming Method, Proc. of IEEE Conf. Computational Sciences
    and Optimization (1), pp. 274–278.
Viola P., Jones M. J. (2004), Robust Real-Time Face Detection, *Information Journal of
    Computer Vision*, **57**, 2, pp. 137–154.
Wang S. L., Lau W. H., Leung S. H., Yan H. (2004), A real-time automatic lipreading
    system, Proc. 2004 Int. Symp. Circuits and Systems, **2**, pp. 101104.
Xinbo G., Ya S., Xuelong L., Dacheng T. (2010). A Review of Active Appearance Models,
    *IEEE Transactions on Systems, Man and Cybernetics, Part C: Applications and
    Reviews*, **40**, 2, pp.145–158.
Yao W., Liang Y., Du M. (2010), A real-time lip localization and tacking for lip reading,
    Proc. of Int. Conf. on Advanced Computer Theory and Engineering (ICACTE), **6**,
    pp.363–366.
Yuille A., Hallinan P., Cohen D. (1992), Feature extraction from faces using deformable
    templates, *Int. J. Computer Vision*, **8**, 2, pp. 99–111.

Part 3

# Industrial Applications of Computer Vision in Process Tomography, Material Science and Temperature Control

# CHAPTER 10

# HYBRID BOUNDARY ELEMENT METHOD APPLIED FOR DIFFUSION TOMOGRAPHY PROBLEMS

Jan Sikora[1], Maciej Pańczyk[2], and Paweł Wieleba[1]

*[1]Electrotechnical Institute*
*04-703 Warsaw, ul. Pożaryskiego 28*
*j.sikora@iel.waw.pl*
*[2]Institute of Computer Science*
*Lublin University of Technology*
*ul.Nadbystrzycka 36b*
*20-618 Lublin*
*m.panczyk@pollub.pl*

## 10.1 Boundary Element Application for Tomography

### 10.1.1 *Introduction*

The identification of unknown internal shapes $\Gamma_2$ and $\Gamma_3$ with the aid of the boundary element method and the level set method is discussed in this chapter. In the proposed mathematical model, the conductivity of an immersed object or objects is constant and known. The main aim is to find the shape of internal structures (objects) based on the measurements of voltages on the external boundary $\Gamma_1$ for different current excitations (so called projection angles). The velocity $v$ on the interface is determined by solving the Laplace equation to obtain boundary values of the potential $\Phi$ and the normal derivative $\dfrac{\partial \Phi}{\partial n}$ (Jabłoński, 2009; Mitchell, 2007). Next, the adjoint state is defined by the Poisson equation. As a source function of the Poisson equation the difference between the solved $\Phi$ and the observed data $u_0$ is applied (Jabłoński, 2009). The calculated velocity on the interface should be extended (for example, see Osher and Fedkiw, 2003) and the extended velocity is used to update the level set function by solving the Hamilton-Jacobi equation. A numerical example is presented at the end of the chapter in order to prove the efficiency of the proposed approach.

## 10.1.2 *Boundary Element Method for spatially inhomogeneous subregions*

The field studies may be split into two main branches. The first one relies on finding the field distribution in the area under consideration. In that case the following are defined: the topology of the structure (interface boundary – outside and/or inside), boundary conditions, material coefficients (e.g. conductivity), the internal source or sources etc. The second case concerns the inverse problem solution. In this case, the unknown parameters are searched when the field distribution is known. The unknown shape of the interface could play the role of the unknown parameters and the inverse problem could, for example, be called, Electrical Impedance Tomography (EIT). Normally, we only know the field distribution on the most external boundary of the object.

An example of the EIT problem with two internal objects is presented in Fig. 10.1. The conductivities of both objects are fixed and known.

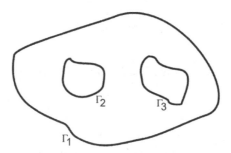

Fig. 10.1 An example of the structure: region $\Omega_1$ limited by border $\Gamma_1$ with two internal objects: $\Omega_2$ limited by border $\Gamma_2$ and $\Omega_3$ limited by border $\Gamma_3$ .

The aim is to find the interface (inside boundary) using voltage measurements on the periphery of the region. The boundary $\Gamma_1$ is known, but the starting shape of the interface in request has been chosen randomly (at the beginning of the iteration process). For so called "current electrodes" (nodes of the boundary elements) the Dirichlet boundary conditions are imposed (emulate the voltage source), for the rest (the "voltage electrodes") the homogeneous Neumann boundary condition $\dfrac{\partial \Phi}{\partial n} = 0$ is imposed.

The partial differential equation in its integral form was defined in order to compute the field distribution for the inhomogeneous regions presented in Fig 10.1. (Sikora, 2007a):

$$c_i \Phi_1(r_i) + \oint_{\Gamma_1+\Gamma_2+\Gamma_3} \Phi_1(r) \frac{G_1(r_i,r)}{\partial n} d\Gamma = \oint_{\Gamma_1+\Gamma_2+\Gamma_3} G_1(r_i,r) \frac{\partial \Phi_1(r)}{\partial n} d\Gamma,$$

$$c_i \Phi_2(r_i) + \oint_{\Gamma_2} \Phi_2(r) \frac{G_2(r_i,r)}{\partial n} d\Gamma = \oint_{\Gamma_2} G_2(r_i,r) \frac{\partial \Phi_2(r)}{\partial n} d\Gamma, \quad (10.1)$$

$$c_i \Phi_3(r_i) + \oint_{\Gamma_3} \Phi_3(r) \frac{G_3(r_i,r)}{\partial n} d\Gamma = \oint_{\Gamma_3} G_3(r_i,r) \frac{\partial \Phi_3(r)}{\partial n} d\Gamma,$$

where the value of $c_i$ coefficient is defined by the location of the point indicated by the position vector $r_i$ (Sikora, 2007a):

$$c_i = \begin{cases} 1, & r_i \in \Omega \\ \dfrac{1}{2}, & r_i \in \partial\Omega \\ 0, & r_i \notin \Omega \end{cases} \quad (10.2)$$

The state function (electric potential) and its normal derivative on the particular boundary $\Gamma_1, \Gamma_2, \Gamma_3$ of each substructure $\Omega_1, \Omega_2$ and $\Omega_3$ are denoted as:

$$\Phi_1(\Gamma_1) = \Phi_1^1, \quad \Phi_1(\Gamma_2) = \Phi_2^1, \quad \Phi_1(\Gamma_3) = \Phi_3^1, \quad \Phi_2(\Gamma_2) = \Phi_2^2, \quad \Phi_3(\Gamma_3) = \Phi_3^3$$

$$\frac{\partial \Phi_1(\Gamma_1)}{\partial n} = \left(\frac{\partial \Phi}{\partial n}\right)_1^1, \quad \frac{\partial \Phi_1(\Gamma_2)}{\partial n} = \left(\frac{\partial \Phi}{\partial n}\right)_2^1, \quad \frac{\partial \Phi_1(\Gamma_3)}{\partial n} = \left(\frac{\partial \Phi}{\partial n}\right)_3^1, \quad (10.3)$$

$$\frac{\partial \Phi_2(\Gamma_2)}{\partial n} = \left(\frac{\partial \Phi}{\partial n}\right)_2^2, \quad \frac{\partial \Phi_3(\Gamma_3)}{\partial n} = \left(\frac{\partial \Phi}{\partial n}\right)_3^3.$$

Substitute:

$$\widehat{A_{ij}} = \oint_{\Gamma_j} \frac{\partial G(r_i,r)}{\partial n} d\Gamma \quad (10.4)$$

Where

$$A_{ij} = \begin{cases} \widehat{A_{ij}}, & i \neq j \\ \widehat{A_{ij}} + c_i, & i = j \end{cases}$$

and

$$B_{ij} = \oint_{\Gamma_j} G(r_i, r) \, d\Gamma \tag{10.5}$$

we will rewrite Eq.(10.1) in matrix form. The same manner was used to mark the matrices $\mathbf{A}$ and $\mathbf{B}$ as the state function and its normal derivative (see equation (10.3)):

$$
\begin{aligned}
&A^1(\Gamma_1) = A_1^1, \quad A^1(\Gamma_2) = A_2^1, \quad A^1(\Gamma_3) = A_3^1, \\
&A^2(\Gamma_2) = A_2^2, \quad A^3(\Gamma_3) = A_3^3, \\
&B^1(\Gamma_1) = B_1^1, \quad B^1(\Gamma_2) = B_2^1, \quad B^1(\Gamma_3) = B_3^1, \\
&B^2(\Gamma_2) = B_2^2, \quad B^3(\Gamma_3) = B_3^3.
\end{aligned}
\tag{10.6}
$$

The voltage on the internal boundary (interface) $\Gamma_2$ or $\Gamma_3$ fulfills the continuous conditions:

$$
\begin{aligned}
\Phi_1(\Gamma_2) &= \Phi_2(\Gamma_2) \\
\Phi_1(\Gamma_3) &= \Phi_3(\Gamma_3)
\end{aligned}
\tag{10.7}
$$

If the conductivity $\Omega_1$ is equal to $\sigma_1$, $\Omega_2$ is equal to $\sigma_2$ and the conductivity of $\Omega_3$ is $\sigma_3$ (as previously noted: the conductivity in objects is constant), then:

$$
\begin{aligned}
\sigma_1 \frac{\partial \Phi_1(\Gamma_2)}{\partial n} &= -\sigma_2 \frac{\partial \Phi_2(\Gamma_2)}{\partial n} \\
\sigma_1 \frac{\partial \Phi_1(\Gamma_3)}{\partial n} &= -\sigma_3 \frac{\partial \Phi_3(\Gamma_3)}{\partial n}
\end{aligned}
\tag{10.8}
$$

where: the minus means the opposite direction of the normal unit vector to the border of $\Omega_1$ - $\Omega_2$ and $\Omega_1$ - $\Omega_3$.

After rewriting the equations for inhomogeneous regions in matrix form we have:

$$
\begin{bmatrix}
A_2^1 & -B_2^1 & A_3^1 & -B_3^1 \\
A_2^2 & -\left(-\dfrac{\sigma_1}{\sigma_2}\right)B_2^2 & 0 & 0 \\
0 & 0 & A_3^3 & -\left(-\dfrac{\sigma_1}{\sigma_3}\right)B_3^3
\end{bmatrix}
\begin{bmatrix}
\Phi_2^1 \\
\left(\dfrac{\partial \Phi}{\partial n}\right)_2^1 \\
\Phi_3^1 \\
\left(\dfrac{\partial \Phi}{\partial n}\right)_3^1
\end{bmatrix}
=
\begin{bmatrix}
-A_1^1 \Phi_1^1 + B_1^1 \left(\dfrac{\partial \Phi}{\partial n}\right)_1^1 \\
0 \\
0
\end{bmatrix}
\tag{10.9}
$$

where the following values $\Phi_2^1$, $\left(\dfrac{\partial \Phi}{\partial n}\right)_2^1$, $\Phi_3^1$, $\left(\dfrac{\partial \Phi}{\partial n}\right)_3^1$ are unknown. It is necessary to solve equation (10.9) for all projection angles.

To solve the inverse problem with the aid of SLM we need the adjoint equation defined as below:

$$-\nabla^2 \Lambda = \Phi - u_0 \tag{10.10}$$

Solving the state and the adjoint equation, the steepest decent direction $v_{i,j}$ is computed (Chen *et al.*, 2009):

$$v_{i,j} = \left(\frac{\partial \Phi}{\partial n}\right)_{2,i,j}^1 \left(\frac{\partial \Lambda}{\partial n}\right)_{2,i,j}^1 \tag{10.11}$$

Finally, the velocity of the $i$ - $th$ boundary element is calculated:

$$V_i = \frac{\displaystyle\sum_{j=1}^{p} v_{i,j}}{p} \tag{10.12}$$

where $p$ is the total number of projection angles.

### 10.1.3 *Level Set Method coupled with BEM*

The level set method is known to be a powerful and versatile tool to model the evolution of interfaces. The idea is merely to define a smooth function which represents the boundary. The interface information, such as tangential and normal, the derivatives and the curvature at projections are obtained from the values of the level set function at the grid point plus a bilinear interpolation.

In the application presented in this chapter velocity is defined only on the interface (border $\Gamma_2$ and $\Gamma_3$ see Fig. 10.1). Velocity will be used to update the level set function $\varphi$. It is necessary to extend the velocity in the neighborhood of the interface, by defining velocity along the normal direction, which is done by solving equation (Osher and Fedkiw, 2003):

$$V_t + S(\varphi)\frac{\nabla \varphi}{|\nabla \varphi|} \cdot \nabla V = 0 \tag{10.13}$$

a)                                                    b)

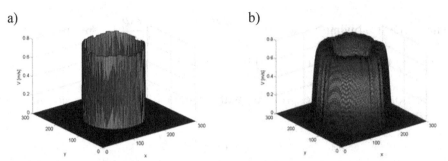

Fig.10.2 The velocity counted for the first iteration step, (a) before extension, (b) extended in 60 steps.

After the velocity is calculated and extended, it is time to update the $\varphi$ function by solving the Hamilton-Jacobi equation:

$$\varphi^{n+1} = \varphi^n - \Delta t\ V \cdot \nabla \varphi^n \tag{10.14}$$

The Hamilton-Jacobi equation was repeatedly updated causing the deformation of the function $\varphi$. Reinitialization is changing the function $\varphi$ to a different one with the same zero level. This is done by solving the reinitalization equation (Osher and Fedkiw, 2003):

$$V_t + S(\varphi)(|\nabla \varphi| - 1) = 0 \tag{10.15}$$

In each case $\Delta t$ should be chosen as Courant-Friedreichs-Lewy (CFL) conditions – the speed of the numerical wave $\frac{\Delta x}{\Delta t}$ must be at least the same as the speed of $|\overrightarrow{V^n}|$, which can be achieved by the value $\Delta t$limitation using the CFL condition (Osher and Fedkiw, 2003):

$$\Delta t < \frac{\Delta x}{\max(|V|)} \tag{10.16}$$

a)                                                    b)

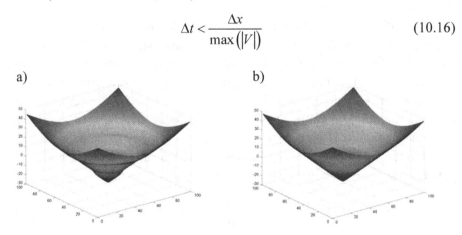

Fig.10.3 The function $\varphi$ in 20 step of iteration (the reinitialization proceed every 5 iterations), (a) before reinitialization process, (b) after reinitialization process.

### 10.1.4 *Numerical example*

Fig. 10.5a shows an example of the ETI problem. An existing object with unknown inside topology was reconstructed. In this case, the level set function $\varphi$ is used (Fig. 10.4a) with four zero level objects (Fig 10.4b). For each simulated object velocity is calculated, assuming that the conductivities of all objects are known. Two unknown objects are marked by a blue line.

The following algorithm is given to find unknown internal objects:

a)  make discretization of the outside boundary $\Gamma_1$ (on 32 elements),

b)  make measurements on the electrodes for all unique projections (for 32 elements of discretization it will be 16*32=512 values), marked as $u_0$,

c)  build simulation, the outside boundary $\Gamma_1$ is known, random inside objects (e.g. four – shown in Fig. 10.4b),

d)  use the Laplace equation and compute measurements $\Phi$ for all the electrodes with the aid of BEM,

e)  find difference between the measured $u_0$ and the computed $\Phi$ data to use it for the adjoint equation as in (10.11),

f)  compute velocity using equation (10.12),

g)  extend the velocity,

h)  use the extended velocity to update the Hamilton-Jacobi equation,

i)  check whether it is necessary to reinitialize the $\varphi$ function,

j)  get a zero level contour of the updated $\varphi$ function, choose discretization points,

k)  the cost function calculations according to equation

$$F_c = \frac{1}{2}\sum_{j=1}^{p} F_j = \frac{1}{2}\sum_{j=1}^{p}\left\{ \left(\Phi_{1j}^1 - \mathbf{u}_{0,j}\right)^T \left(\Phi_{1j}^1 - \mathbf{u}_{0,j}\right)\right\}$$

where: $\Phi$ – is the value computed by the program, by solving the Laplace equation for the object under consideration; $\mathbf{u}_0$ – is the value measured for the real object, if the result is satisfying, then "end", or else go to the point (d) and repeat the process from the beginning.

Below, in Fig. 10.4a is the Level Set Function $\varphi$ with four zero level contours presented in Fig. 10.4b.

a)                                              b)

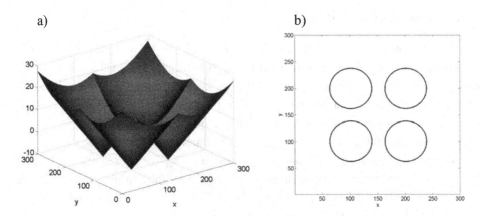

Fig.10.4 The level set function φ with 4 zero level contours.

a)                                              b)

c)                                              d)

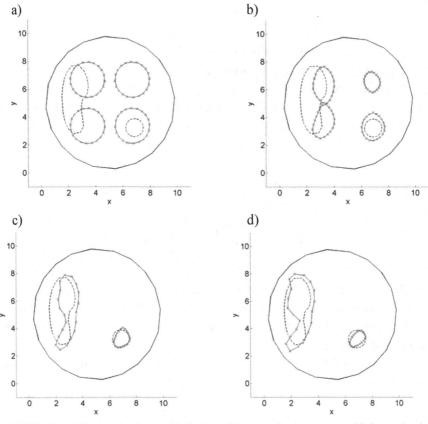

Fig. 10.5 The black line marks the outside border of the examined structure with internal unknown objects (marked by a blue line). The red line marks simulated objects of zero level contours (a) 1st iteration step, (b) for 98 iterations, one step before merging two objects (c) for 225 iterations with the lowest value of the cost function: 0,96 (d) for last 300 iteration steps.

Results of the iteration process as described above are shown in Fig. 10.5. Unknown structures are marked by the blue line; simulated objects are marked by the red line. For last 300 iterations step the unknown structure was found. The result is presented in Fig. 10.5d.

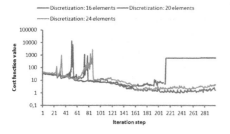

Fig. 10.6 The value of the cost function depends on the number of discretization elements.

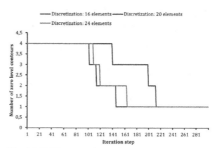

Fig.10.7 The number of zero level contours which represent simulated objects depends of discretization elements.

### 10.1.5 *Conclusion*

The problem of identification of unknown objects from boundary measurements of the potential was presented in this paper. The inverse problem was solved using the combination of the level set method and the boundary element method (solving the Laplace equation and the Poisson equation with the adjoint variable). Our method shows a way to compute the velocity on the inside border by updating the Hamilton-Jacobi equation.

Finally information about the reinitialization is given to fix the set level function and the signed distance function. Additionally there is given the step-by-step algorithm to solve the presented problem. Experimental results show that our models are able to identify (reconstruct) unknown objects.

## 10.2  Infinite Boundary Element Method and its Application for Mammography

### 10.2.1 *Introduction*

Optical Tomography and Electrical Impedance Tomography can use Boundary Element Method (BEM) for forward problem solution (Sikora, 2007b). In the area of human female breast cancer investigations it is only possible to make measurements on the skin. The boundary surface between the breast and the chest remains unavailable for detectors. It is also difficult to set a precise boundary condition on that surface.

A typically simple BEM solution is to extend the boundary element mesh outside the zone of interest and to truncate it at a large distance away so that the new boundary does not influence the results. Such a solution generates a large number of additional boundary elements and if that truncation occurs too near, it can introduce an unknown error. A more effective method is to incorporate infinite boundary elements (Beer and Watson, 1989; Bettess, 1992) into the conventional BEM analysis. The use of infinite elements was first adopted in wave propagation and geotechnical problems (Beer *et al.*, 1987). In our case, the aim is to estimate the differences between models and to consider whether the model with the infinite elements implemented offers a similar accuracy to those with an extended area.

### 10.2.2 *Theory*

There are two main lines of infinite element development: mapped infinite elements, where the element is transformed from the finite to the infinite domain, and the decay function infinite elements, which uses special decay functions in conjunction with ordinary boundary element interpolation functions (Sikora, 2007b)).

Both types offer a similar accuracy. For the present investigations the decay function infinite elements based on standard eight nodes quadrilateral isoparametric boundary elements was chosen (Fig.10.1). Obviously, the final mesh will consists of both ordinary and infinite boundary elements.

The basis interpolation functions $N_i$ (index $i$ corresponds to element nodes $i=0...7$) for ordinary quadrilateral boundary elements are given by the following equations (10.17):

$$N_0(\xi,\eta) = -(1-\xi)(1-\eta)(1+\xi+\eta)/4,$$
$$N_1(\xi,\eta) = (1-\xi^2)(1-\eta)/2,$$
$$N_2(\xi,\eta) = -(1+\xi)(1-\eta)(1-\xi+\eta)/4,$$
$$N_3(\xi,\eta) = (1+\xi)(1-\eta^2)/2,$$
$$N_4(\xi,\eta) = -(1+\xi)(1+\eta)(1-\xi-\eta)/4, \qquad (10.17)$$
$$N_5(\xi,\eta) = (1-\xi^2)(1+\eta)/2,$$
$$N_6(\xi,\eta) = -(1-\xi)(1+\eta)(1+\xi-\eta)/4,$$
$$N_7(\xi,\eta) = (1-\xi)(1-\eta^2)/2.$$

The basic idea of the decay infinite elements is that the ordinary basis interpolation functions $N_i$ (10.17) are multiplied by the decay functions $D_i$ (10.19). This combination ensures that the behavior of the element causes the field variable to tend to its value at infinity. Basis interpolation functions for the decay infinite elements are then constructed as:

$$M_i(\xi,\eta) = N_i(\xi,\eta)D_i(\xi,\eta). \tag{10.18}$$

The infinite basis interpolation functions $M_i$ (10.18) must be unity at its own node. This means that the decay functions $D_i$ must also be equal to one at its own node and tend to the far field value at infinity. The transformation of 8 node quadrilateral boundary element into a relevant decay infinite element which extends into infinity in one (positive) $\xi$ direction is presented in Fig. 10.8a.

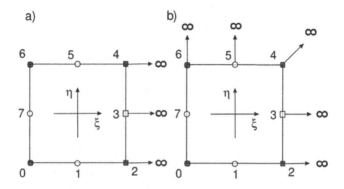

Fig. 10.8 Transformation of standard 8 node quadrilateral boundary element into decay type infinite element: a) in one $\xi$ direction, b) in two $\xi$ and $\eta$ directions.

Decay infinite basis interpolation functions can be also developed for both $\xi$ and $\eta$ directions. The infinite element, which extends into infinity in both (positive) $\xi$ and $\eta$ directions, is presented in Fig. 10.10b.

Further considerations will be related to reciprocal decay functions of the form:

$$D_i(\xi,\eta) = \left(\frac{\xi_i - \xi_0}{\xi - \xi_0}\right)^n \tag{10.19}$$

where: the exponent $n$ has to be greater than the highest power of $\xi$ encountered in $M$. $\xi_0$ is some origin point located outside the element on the opposite side to that which extends to infinity. Its role is to avoid singularity within the element.

In the case of transforming the element into infinity in both directions $\xi$ and $\eta$ the reciprocal decay function is written as:

$$D_i(\xi,\eta) = \left(\frac{\xi_i - \xi_0}{\xi - \xi_0}\right)^n \left(\frac{\eta_i - \eta_0}{\eta - \eta_0}\right)^m, \tag{10.20}$$

where: the exponent $m$ has to be greater than the highest power of $\eta$ encountered in $M$.

For the quadratic basis interpolation function, exponent $n=3$ infinite functions $M_i$ are given by the formulas:

$$M_0(\xi,\eta) = -(1-\xi)(1-\eta)(1+\xi+\eta)/4\left[(1-\xi_0)/(\xi-\xi_0)\right]^3,$$

$$M_1(\xi,\eta) = (1-\xi^2)(1-\eta)/2\left[(-\xi_0)/(\xi-\xi_0)\right]^3,$$

$$M_2(\xi,\eta) = -(1+\xi)(1-\eta)(1-\xi+\eta)/4\left[(1-\xi_0)/(\xi-\xi_0)\right]^3,$$

$$M_3(\xi,\eta) = (1+\xi)(1-\eta^2)/2\left[(1-\xi_0)/(\xi-\xi_0)\right]^3,$$

$$M_4(\xi,\eta) = -(1+\xi)(1+\eta)(1-\xi-\eta)/4\left[(1-\xi_0)/(\xi-\xi_0)\right]^3,$$ (10.21)

$$M_5(\xi,\eta) = (1-\xi^2)(1+\eta)/2\left[(-\xi_0)/(\xi-\xi_0)\right]^3,$$

$$M_6(\xi,\eta) = -(1-\xi)(1+\eta)(1+\xi-\eta)/4\left[(-1-\xi_0)/(\xi-\xi_0)\right]^3,$$

$$M_7(\xi,\eta) = (1-\xi)(1-\eta^2)/2\left[(-1-\xi_0)/(\xi-\xi_0)\right]^3.$$

For debugging purposes in the case of the decay type infinite basis interpolation function, the same test as for ordinary basis interpolation functions can be used. Its aim is to check whether all the basis interpolation functions sum to unity and all the derivatives to zero. The simple test is to check whether each function has a unit value on their own node and zero on the others. Infinite decay type basis interpolation function graphs are presented in Fig. 10.9.

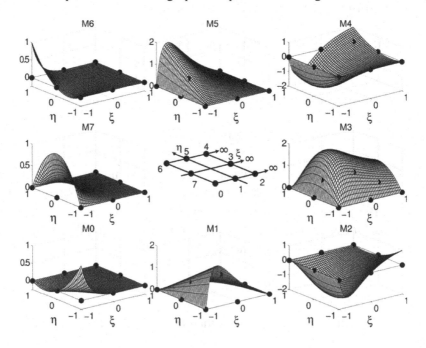

Fig. 10.9 Basis interpolation functions M0, M1, M2, M3, M4, M5, M6 and M7, equation (10.18), for decay type infinite boundary element.

To study boundary elements which are two-dimensional structures placed in the 3D space, first we need to define the way in which we can pass from the $x,y,z$ global Cartesian system to the $\xi,\eta,\zeta$ system defined over the element, where $\xi$ and $\eta$ are oblique coordinates and $\zeta$ is in the direction of the normal. The transformation for a given function $\Phi$ is related through the following:

$$
\begin{bmatrix} \dfrac{\partial \Phi}{\partial \xi} \\[2mm] \dfrac{\partial \Phi}{\partial \eta} \\[2mm] \dfrac{\partial \Phi}{\partial \zeta} \end{bmatrix} = \begin{bmatrix} \dfrac{\partial x}{\partial \xi} & \dfrac{\partial y}{\partial \xi} & \dfrac{\partial y}{\partial \xi} \\[2mm] \dfrac{\partial x}{\partial \eta} & \dfrac{\partial y}{\partial \eta} & \dfrac{\partial z}{\partial \eta} \\[2mm] \dfrac{\partial x}{\partial \zeta} & \dfrac{\partial y}{\partial \zeta} & \dfrac{\partial z}{\partial \zeta} \end{bmatrix} \begin{bmatrix} \dfrac{\partial \Phi}{\partial x} \\[2mm] \dfrac{\partial \Phi}{\partial y} \\[2mm] \dfrac{\partial \Phi}{\partial z} \end{bmatrix} \tag{10.22}
$$

where the square matrix is the Jacoby matrix.

The transformation of this type allows us to describe differentials of a surface in the Cartesian system in terms of the curvilinear coordinates. A differential of the area will be given by:

$$
\partial \Gamma = |\mathbf{n}| \partial \xi \partial \eta = \left| \frac{\partial \mathbf{r}}{\partial \xi} \times \frac{\partial \mathbf{r}}{\partial \eta} \right| \partial \xi \partial \eta = \sqrt{n_x^2 + n_y^2 + n_z^2} \, \partial \xi \partial \eta, \tag{10.23}
$$

where:

$$
n_x = \frac{\partial y}{\partial \xi} \frac{\partial z}{\partial \eta} - \frac{\partial y}{\partial \eta} \frac{\partial z}{\partial \xi} \;,\quad n_y = \frac{\partial x}{\partial \xi} \frac{\partial z}{\partial \eta} - \frac{\partial x}{\partial \eta} \frac{\partial z}{\partial \xi} \;,\quad n_z = \frac{\partial x}{\partial \xi} \frac{\partial y}{\partial \eta} - \frac{\partial x}{\partial \eta} \frac{\partial y}{\partial \xi}
$$

and $\mathbf{n}$ is the normal vector directed outward to the surface element $\left( \mathbf{n} = \dfrac{\partial \mathbf{r}}{\partial \xi} \times \dfrac{\partial \mathbf{r}}{\partial \eta} \right)$.

This mapping introduces the Jacobian $J$ (determinant of the Jacoby matrix – see Eq. (10.22) proportional to the magnitude of the area of the mapped boundary element. The first derivatives of the mapped interpolation functions with respect to the $\xi$ for infinite element are given by:

$$\frac{\partial M_0(\xi,\eta)}{\partial \xi} = -\frac{(\xi_0+1)^3(\eta-1)\left(2\xi\eta+\xi_0\eta-3\eta+\xi^2+2\xi_0\xi-3\right)}{4(\xi-\xi_0)^4},$$

$$\frac{\partial M_1(\xi,\eta)}{\partial \xi} = \frac{\xi_0^3(\eta-1)\left(\xi^2+2\xi_0\xi-3\right)}{2(\xi-\xi_0)^4},$$

$$\frac{\partial M_2(\xi,\eta)}{\partial \xi} = -\frac{(\xi_0-1)^3(\eta-1)\left(-2\xi\eta-\xi_0\eta-3\eta+\xi^2+2\xi_0\xi-3\right)}{4(\xi-\xi_0)^4},$$

$$\frac{\partial M_3(\xi,\eta)}{\partial \xi} = -\frac{(\xi_0-1)^3(\eta^2-1)(2\xi+\xi_0+3)}{2(\xi-\xi_0)^4},$$

$$\frac{\partial M_4(\xi,\eta)}{\partial \xi} = \frac{(\xi_0-1)^3(\eta+1)\left(2\xi\eta+\xi_0\eta+3\eta+\xi^2+2\xi_0\xi-3\right)}{4(\xi-\xi_0)^4},$$

$$\frac{\partial M_5(\xi,\eta)}{\partial \xi} = -\frac{\xi_0^3(\eta+1)\left(\xi^2+2\xi_0\xi-3\right)}{2(\xi-\xi_0)^4},$$

$$\frac{\partial M_6(\xi,\eta)}{\partial \xi} = \frac{(\xi_0+1)^3(\eta+1)\left(-2\xi\eta-\xi_0\eta+3\eta+\xi^2+2\xi_0\xi-3\right)}{4(\xi-\xi_0)^4},$$

$$\frac{\partial M_7(\xi,\eta)}{\partial \xi} = \frac{(\xi_0+1)^3(\eta^2-1)(2\xi+\xi_0-3)}{2(\xi-\xi_0)^4},$$

(10.24)

$$\frac{\partial M_0(\xi,\eta)}{\partial \eta} = \frac{(\xi_0+1)^3(\xi-1)(2\eta+\xi)}{4(\xi-\xi_0)^3},$$

$$\frac{\partial M_1(\xi,\eta)}{\partial \eta} = -\frac{\xi_0^3(\xi^2-1)}{2(\xi-\xi_0)^3},$$

$$\frac{\partial M_2(\xi,\eta)}{\partial \eta} = \frac{(\xi_0-1)^3(\xi+1)(\xi-2\eta)}{4(\xi-\xi_0)^3},$$

$$\frac{\partial M_3(\xi,\eta)}{\partial \eta} = \frac{(\xi_0-1)^3(\xi+1)\eta}{(\xi-\xi_0)^3},$$

$$\frac{\partial M_4(\xi,\eta)}{\partial \eta} = -\frac{(\xi_0-1)^3(\xi+1)(2\eta+\xi)}{4(\xi-\xi_0)^3},$$

$$\frac{\partial M_5(\xi,\eta)}{\partial \eta} = \frac{\xi_0^3(\xi^2-1)}{2(\xi-\xi_0)^3},$$

$$\frac{\partial M_6(\xi,\eta)}{\partial \eta} = -\frac{(\xi_0+1)^3(\xi-1)(\xi-2\eta)}{4(\xi-\xi_0)^3},$$

$$\frac{\partial M_7(\xi,\eta)}{\partial \eta} = \frac{(\xi_0+1)^3(1-\xi)\eta}{(\xi-\xi_0)^3}.$$

(10.25)

The boundary integral equation containing both finite (surface covered by ordinary boundary elements) and infinite boundary elements (surface covered by infinite decay function boundary elements) after discretization will take the form:

$$
\begin{aligned}
c(\mathbf{r})\Phi_i(\mathbf{r}) + \sum_{i=0}^{std-1}\sum_{k=0}^{7}\int_{-1}^{+1}\int_{-1}^{+1}\Phi(\mathbf{r}')N_k(\xi,\eta)\frac{\partial G(|\mathbf{r}-\mathbf{r}'|)}{\partial n}J^N(\xi,\eta)\partial\xi\partial\eta + \\
+\sum_{i=0}^{inf-1}\sum_{k=0}^{7}\int_{-1}^{+1}\int_{-1}^{\infty}\Phi(\mathbf{r}')M_k(\xi,\eta)\frac{\partial G(|\mathbf{r}-\mathbf{r}'|)}{\partial n}J^M(\xi,\eta)\partial\xi\partial\eta = \\
=\sum_{i=0}^{std-1}\sum_{k=0}^{7}\int_{-1}^{+1}\int_{-1}^{+1}\frac{\partial\Phi(\mathbf{r}')}{\partial n}N_k(\xi,\eta)G(|\mathbf{r}-\mathbf{r}'|)J^N(\xi,\eta)\partial\xi\partial\eta + \\
+\sum_{i=0}^{inf-1}\sum_{k=0}^{7}\int_{-1}^{+1}\int_{-1}^{\infty}\frac{\partial\Phi(\mathbf{r}')}{\partial n}M_k(\xi,\eta)G(|\mathbf{r}-\mathbf{r}'|)J^M(\xi,\eta)\partial\xi\partial\eta
\end{aligned}
\tag{10.26}
$$

For singularity treatment as one of the most effective, the regularization method (Abramowitz and Stegun, 1973) was used. The advantage of using that method is that the calculation schema remains unchanged for infinite elements.

### 10.2.3 *Models*

Four simple theoretical models of the human breast were investigated. For all the models one placement of the light source was presented – located near the bottom of the hemisphere model. The first model presented in Fig. 10.19 corresponds to the pure hemisphere. The second model was extended by adding a cylinder in the bottom (Fig. 10.20). The intention was to avoid possible errors at the bottom of the hemisphere by adding a cylinder with the identical height but a greater diameter (Fig. 10.21). Its aim was to eliminate the errors near the basis circumference. All the models were constructed from 1536 second order eight nodes quadrilateral boundary elements and 4610 nodes. Half of the elements covers the hemisphere.

The governing equation for the problem is the diffusion approximation of the transport equation (Sikora, 2007b) (Helmholtz – assuming that scattering and absorption are homogeneous):

$$
\nabla^2\Phi(\mathbf{r},\omega) - k^2\Phi(\mathbf{r},\omega) = -\frac{q_0(\mathbf{r},\omega)}{D}, \quad \forall \mathbf{r}\in\Omega/\Gamma,
\tag{10.27}
$$

where $\Phi$ stands for the photon density, $k = \sqrt{\dfrac{\mu_a}{D} - j\dfrac{\omega}{cD}}$ the complex wave number, $D = \left[3(\mu_a + \mu_S')\right]^{-1} [mm^{-1}]$ the diffusion coefficient, $\mu_S'$ is the reduced scattering coefficient, $\mu_a$ is the absorbing coefficient, $c$ the speed of light in the medium, $q_0$ is the source of light (number of photons per volume unit emitted by a concentrated light source located in the position **r** with a modulation frequency $\omega$).

Generally, in the Diffusive Optical Tomography, the distribution of $\mu_a$ and $\mu_s'$ is investigated. There are Robin boundary conditions on surfaces (Arridge, 1999; Tarvainen, 2006; Zacharopoulos *et al.*, 2006):

$$\Phi(\mathbf{r},\omega) + 2\alpha D \frac{\partial \Phi(\mathbf{r},\omega)}{\partial n} = 0, \quad \forall \mathbf{r} \in \Gamma. \tag{10.28}$$

with different coefficients for a breast tissue and for skeletal muscles on the basis (Arridge, 1999; Tarvainen, 2006; Zacharopoulos *et al.*, 2006) imposed. In the analyzed example the following breast tissue properties were taken (Arridge, 1999; Zacharopoulos *et al.*, 2006):

$$\mu_a = 0.025[mm^{-1}], \mu_s' = 2[mm^{-1}], \alpha = 1, f = 100 MHz$$

The last open boundary model consists of 768 standard boundary elements and 64 infinite decay type elements based on eight nodes second order quadrilateral boundary elements (Bettess, 1992; Watson, 1979). The number of nodes is reduced to 2561 nodes in that case (see Fig.10.13).

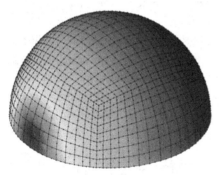 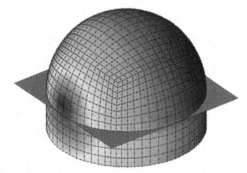

Fig. 10.10 Base model of the breast – hemisphere    Fig. 10.11 Extended model with additional part of chest, hemisphere with cylinder at the bottom

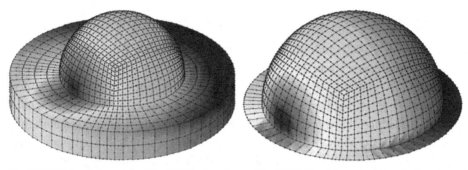

Fig. 10.12 Extended model with additional part of the chest, hemisphere with a wider cylinder at the bottom

Fig. 10.13 Open boundary hemisphere breast model with infinite boundary elements at the bottom

Decay function infinite basis approximation functions (Bettess, 1992) were used for element transformation, as presented in Fig. 10.8. The relevant boundary integral equation for surfaces covered by standard and infinite elements can be written as:

$$C(\mathbf{r})\Phi(\mathbf{r}) + \int_{\Omega} \frac{\partial G(|\mathbf{r}-\mathbf{r'}|,\omega)}{\partial n}\Phi(\mathbf{r'})d\Omega +$$

$$+ \int_{\Omega_{\infty}} \frac{\partial G(|\mathbf{r}-\mathbf{r'}|,\omega)}{\partial n}\Phi(\mathbf{r'})d\Omega_{\infty} = \int_{\Omega} G(|\mathbf{r}-\mathbf{r'}|,\omega)\frac{\partial \Phi(\mathbf{r'})}{\partial n}d\Omega + \qquad (10.29)$$

$$+ \int_{\Omega_{\infty}} G(|\mathbf{r}-\mathbf{r'}|,\omega)\frac{\partial \Phi(\mathbf{r'})}{\partial n}d\Omega_{\infty} - \sum_{S=0}^{n_S-1} Q_S G(|\mathbf{r_S}-\mathbf{r}|,\omega)$$

where $Q_S$ is the magnitude of the concentrated source ($q_0 = Q_S\delta(r_S)$) and $n_S$ is the number of these sources, $\Phi$ stands for the photon density and $G$ is the fundamental solution for the diffusion equation (Arridge, 1999; Tarvainen, 2006; Zacharopoulos *et al.*, 2006). In a 3D space for the diffusion equation, the fundamental solution is (Sikora, 2007b):

$$G(|\mathbf{r}-\mathbf{r'}|,\omega) = \frac{1}{4\pi|\mathbf{r}-\mathbf{r'}|}e^{-k|\mathbf{r}-\mathbf{r'}|}. \qquad (10.30)$$

The normal derivative of the Green function in the direction n can be written:

$$\mathbf{n}\cdot\nabla G = \mathbf{n}\cdot\frac{\mathbf{r}-\mathbf{r'}}{|\mathbf{r}-\mathbf{r'}|}\left(\frac{-1}{4\pi|\mathbf{r}-\mathbf{r'}|^2} - \frac{k}{4\pi|\mathbf{r}-\mathbf{r'}|}\right)e^{-k|\mathbf{r}-\mathbf{r'}|}. \qquad (10.31)$$

### 10.2.4 *Results*

The values of $\partial\Phi/\partial n$ module and phase of the light at the hemisphere circumference cross-section for $y=0$ are presented in Fig. 10.14 and Fig. 10.15, respectively. To estimate the solution differences, models with an extended bottom part – Fig. 10.11, 10.12 and 10.13 were compared to the basic hemisphere – Fig. 10.10. Generally, three extended models offer similar results, remarkably different from those achieved while using the simplest hemisphere. It should be noted that there is a logarithmic scale on $\partial\Phi/\partial n$ module, see Fig. 10.14. The medium approximation differences for the module reach 50% and for the phase – oscillate largely around 3%.

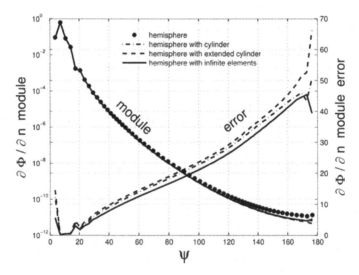

Fig. 10.14 Comparison of results for $\partial\Phi/\partial n(\Psi)$ module and solution differences compared to the hemisphere model.

### 10.2.5 *Conclusion*

Generally, three extended breast models (those with the added additional cylindrical part on the hemisphere basis or with infinite elements incorporated) offer a similar accuracy – figures 10.14 and 10.15. Significantly worst results were achieved from the pure hemisphere model. Medium differences between the hemisphere and the extended models for the module reach 50% and for the phase about 3%. Fifty percent differences are sufficiently significant to extend the mesh and to include in calculations not only the breast tissue represented by a hemisphere but also a part of the chest with muscles and bones relevant to the additional cylindrical part or ring consisting of infinite elements.

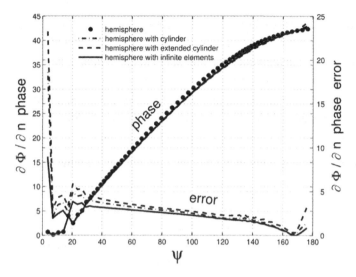

Fig. 10.15 Comparison of results for $\partial\Phi/\partial n(\Psi)$ phase and solution differences compared to the hemisphere model.

The advantage of using infinite elements is to shorten the calculation time and keep the accuracy similar to that of extended models. Reducing the number of mesh elements almost to 50% is fundamental for inverse problem solution when the forward problem has to be calculated many times. All the extended models were built from the same number of standard 8-node second order quadrilateral boundary elements. Mesh density on the additional surface related to the cylindrical part of the model was lower than that on the hemisphere surface. This is a typical practical solution as the additional part represents the region outside the zone of interest, and is included only to improve the accuracy. Except the models mentioned above, which consist of 1536 elements and 4610 nodes, calculations were also carried out for models covered by 384 elements and 1154 nodes as well as 6144 elements and 18434 nodes.

The results calculated from the simplest model built from 384 elements had little oscillations instead of a smooth solution. The model with the highest mesh density, 6144 elements, required too much memory and took too long calculation time without significant accuracy improvement. Calculations take 18 seconds for 384 node models and 4 minutes and 47 seconds in the case of 4160 node models.

The model with infinite elements built from 832 elements, which is relevant to the standard one built from 1536 elements and 4610 nodes, required 1 minute and 24 seconds for calculations. Dedicated to this particular problem a mesh generator was used to build the presented meshes containing only quadrilateral elements and to create open boundary models with infinite boundary elements.

The application of infinite boundary elements in the Boundary Element Method improves computational efficiency in comparison to mesh truncation.

The process of incorporating infinite elements into the BEM calculation scheme is rather logical. The presented application Optical or Electrical Impedance mammography can be used as a screening examination in breast cancer detection, but infinite elements can also be used for other purposes.

## 10.3 Open Source Software for Boundary Element Calculation

### 10.3.1 *Introduction*

For a number of decades a rapid development of boundary element method (BEM) has been observed (Banerjee, 1994; Becker, 1992; Beer, 2001; Beer and Watson, 1992; Tanaka and Du, 1987; Bonnet, 1995; Chen and Zhou, 1992). During that time the range of BEM's application has increased (Amaratunga, 2000; Frijns *et al.*, 2000; Guven and Madenci, 2003; Sikora *et al.*, 2006). Among other things, it is used in electromagnetic, thermal or optical analysis. However, it is not easy to find ready-to-use BEM implementations. The situation becomes more difficult if one tries to find the free opensource software. It is even worse if specialized BEM software applicable, for example, to diffusion optical tomography is needed.

One of the reasons why this state is maintained, might be the complexity of integration (in particular singular integrals) which has to be done in BEM calculations. However, this is not a problem which cannot be overcome (Aliabadi and Hall, 1989). The need for the BEM calculations software exists and is unquestionable, but it has been only insignificantly implemented (Table A.1). It appears that industrial and scientific groups would like to have a well-designed platform for BEM calculations. It should be universal but at the same time its modularity should easily enable application in very specific tasks.

If necessary, writing one's own plugins/modules, extending its functionality of solving other boundary integral equations (BIE), the use of other integration algorithms, for example, or a set of linear equations solvers, must be simple. Moreover, the possibility of using only selected, single modules in external projects is expected. The needs and expectations mentioned above caused the platform for BEM calculations to be realized. Particular elements of the project are introduced in the following sections.

### 10.3.2 *BEM and diffusion optical tomography*

Before beginning work on a model of the project, the method and its example application has to be introduced. Here, the boundary element method in diffusion optical tomography (DOT) will be used. Why DOT?

There are plenty of excellent papers in which the implementation of the finite element method is presented (see, for example the works of Prof. Arridge (Arridge *et al.*, 1993; Arridge *et al.*, 2000)). However, only the BEM offered a very simple and effective of good quality surface meshes of the layers of the patient's head – also taking into account the cerebrospinal fluid (CSF) layer (Riley *et al.*, 2000). Especially in three dimensions, the generation of good quality surface meshes is much easier and more effective in BEM than good quality volume mesh generation in FEM (Arridge, 1999). The description here will be limited only to general concepts important to the model and its implementation.

The main purpose of the library is the capability of doing all essential calculations to solve the specific problem. The visualization of the solution, using the input mesh and the output data, can be done using MATLAB.

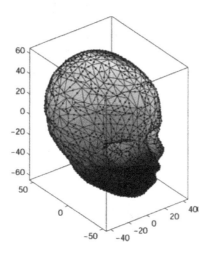

Fig. 10.16 Head with the distribution of photon density.

Fig. 10.16 presents a baby's head with the distribution of photon density calculated by this library. The main interest will be devoted to the diffusion equation, as it is the approximation used in DOT (Sikora *et al.*, 2006; Muller *et al.*, 1993; Riley *et al.*, 2001; Ripoll, 2000). The following equation is presented in its differential form:

$$\nabla^2 \Phi(\mathbf{r}) - k^2 \Phi(\mathbf{r}) = b, \tag{10.32}$$

where the potential function $\Phi$ represents the photon density, $k$ is the wave number and $b$ represents internal sources.

As BEM uses Boundary Integral Equations (BIE) the integral form of Eq. (10.32) is:

$$C(\mathbf{r})\Phi(\mathbf{r}) + D\int_{\Gamma} \frac{\partial G(R)}{\partial n} \Phi(\mathbf{r}')d\Gamma(\mathbf{r}') = D\int_{\Gamma} G(R)\frac{\partial \Phi(\mathbf{r}')}{\partial n} d\Gamma(\mathbf{r}') + \int_{\Omega} G(R)b \, d\Omega(\mathbf{r}') \tag{10.33}$$

where $C(\mathbf{r})$ for the smooth surfaces takes the value 1/2, $D$ is the diffusion parameter (to see details consult Zacharopoulos et al., 2006) and $R = |\mathbf{r} - \mathbf{r}'|$ is the distance between the fixed interior or here a boundary point $\mathbf{r}$ and any point $\mathbf{r}'$ on the boundary $\Gamma$ of the region $\Omega$.

For the simplicity of further equations, $b=0$ will be assumed. After applying the discretization of the boundary $\Gamma$ into J elements, the following equation will be obtained:

$$C(\mathbf{r})\Phi(\mathbf{r}) + D\sum_{j=1}^{J} \int_{\Gamma_j} \frac{\partial G(R_j)}{\partial n} \Phi(\mathbf{r}_j')d\Gamma = D\sum_{j=1}^{J} \int_{\Gamma_j} G(R_j)\frac{\partial \Phi(\mathbf{r}_j')}{\partial n} d\Gamma \tag{10.34}$$

Special attention must be drawn to the integrals in Eq. (10.34). If $\mathbf{r}$ belongs to the same boundary element as $\mathbf{r}_j'$ then singularity occurs and singular integration has to be used instead of the non-singular. Finally, the set of $J$ (for the constant boundary element) linear boundary equations will be obtained by choosing point $\mathbf{r}$ from a different boundary element for each equation, which will result in Eq. (10.35) in matrix form. Generally, the number of equations is represented by the number of nodes, which is applicable for boundary elements of the higher order. After applying Neumann, Dirichlet and/or Robin boundary conditions, Eq. (10.36) will be obtained.

$$\mathbf{A}\Phi = \mathbf{B}\frac{d\Phi}{dn}, \tag{10.35}$$

$$\mathbf{a}\begin{bmatrix} \Phi \\ \dfrac{d\Phi}{dn} \end{bmatrix} = \mathbf{b}, \tag{10.36}$$

$\Phi$ and/or $\dfrac{d\Phi}{dn}$, in the Eq. (10.36), represent only unknown boundary values as known ones (boundary conditions) are integrated into a and b BEM matrices. Now, the system of linear Eq. (10.36) can be solved with an appropriate solver, for example, the iterative GMRES algorithm or LU decomposition (Saad and Schultz, 1986).

When the calculations are finished, all the values of $\Phi$ and $\dfrac{d\Phi}{dn}$ on the boundary are known. And now, the value of the function $\Phi$ in any point $\mathbf{r}$ from the inside of the considered region can be calculated using only simple operations as shown in Eq. (10.37). Eq. (10.37) is valid in the case of discretization with constant boundary elements:

$$\Phi(\mathbf{r}) = \sum_{j=1}^{J} \left( \frac{\partial \Phi(R_j)}{\partial n} \int_{\Gamma_j} G(R_j) d\Gamma - \Phi(R_j) \int_{\Gamma_j} \frac{\partial G(R_j)}{\partial n} d\Gamma \right). \qquad (10.37)$$

$C(\mathbf{r})$ was skipped since for any internal point it takes the value of 1 (Tanaka and Du, 1987).

### 10.3.3 *BEM implementation scheme*

After the theoretical and mathematical description of the algorithm, the subsequent stage of the project is to create its activity diagrams, which will help to divide the project into components. The general BEM algorithm description from one of the previous sections has to be 'translated' into the programming language.

Finally, a large amount of source code needs to be generated where obviously assimilability is not its essence. Clear drawings or diagrams are comprehended much better when their elements are appropriately grouped, while confusion in drawings can make them worthless. To avoid such a situation, it is better to use a formal notation of drawn components, which will prevent inaccuracy and misunderstanding.

Therefore, such a 'language' of making diagrams was created. Now, the standard is the unified modeling language (UML), which will be used for making diagrams throughout the project process (Ambler, 2004; Fowler, 2003; Sommerville, 2004). BEM algorithm elements are presented one by one in Fig. 10.17.

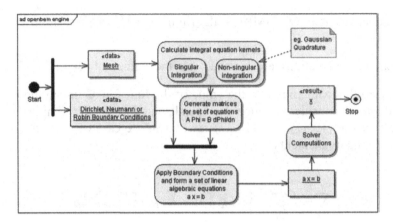

Fig. 10.17 BEM engine activity diagram.

There are two input data types for BEM library computations: the generated mesh and the boundary conditions. There is also the third one: the parameters for internal sources defined on the right hand side of Eq. (10.32).

The problem solved using the library will be encapsulated into the program that will comply with the chosen library components. Mesh generation is out of the scope of this chapter, therefore external programs have to be used. One of them is NETGEN. Together with the mesh, boundary conditions have to be supplied.

When all input data are available, calculations can be started. First, all integral equation kernels from Eq. (10.34) have to be calculated $\left(\int_{\Gamma_j} \frac{\partial G(R_j)}{\partial n} \Phi(\mathbf{r}_j) d\Gamma\ \text{and}\right.$

$\left.\int_{\Gamma_j} G(R_j) \frac{\partial \Phi(\mathbf{r'}_j)}{\partial n} d\Gamma\right)$. The number of singular and non-singular kernels is evaluated in the following way: every point times every boundary element times every point in the boundary element. In most cases, integration is made numerically, where, for example, Gaussian Quadrature can be used (Aliabadi and Hall, 1989). The singular integration occurs when a kernel is calculated for two points belonging to the same element. Green's function $G$ is the fundamental solution specific for the specific BIE. For the two dimensional diffusion equation, it would be $G = \frac{1}{2\pi} K_0$ (Abramowitz and Stegun, 1973), where $K_0$ is the Bessel function of the second kind and zero order $\left(K_\alpha(x) = \frac{1}{\pi}\int_0^\pi \sin(x\sin\Theta - \alpha\Theta)d\Theta - \frac{1}{\pi}\int_0^\infty \left[e^{\alpha t} + (-1)^\alpha e^{-\alpha t}\right]e^{-x\sinh t}dt\right)$.

After all the kernels are calculated, a set of linear equations can be generated as shown in Eq. (10.35). This equation has twice as many unknowns ($2J$) as equations ($J$). Additional equations will be provided by the Boundary Conditions (BC). There must be $J$ BCs and they can be of the Dirichlet $\left(\Phi_j\right)$, Neumann $\left(\dfrac{\partial \Phi_j}{\partial n}\right)$ and/or Robin $\left(\dfrac{\partial \Phi_j}{\partial n} = -\dfrac{1}{2D}\Phi_j\right)$ type – one per boundary element.

When BCs are applied, a set of linear equations from Eq. (10.36) can be generated. Now, very time-consuming calculations are required since for homogeneous regions, matrix **A** is asymmetric and fully populated. Any of the solvers included in the library can be used as they all have a common interface. They are grouped in the 'Solvers' component.

After solving Eq. (10.36) all the boundary values are known. Therefore, now the function $\Phi$ can be calculated as well as $\dfrac{d\Phi}{dn}$ values in the points from the inside of the region according to Eq. (10.37). The unquestionable advantage of BEM is that calculations can be made only in necessary internal points – not in all of those defined by the volume mesh, as in the case of the FEM.

### 10.3.4 *Structure*

The structure of the library can be introduced on the basis of the already created activity diagram describing the BEM algorithm presented in Fig. 10.18.

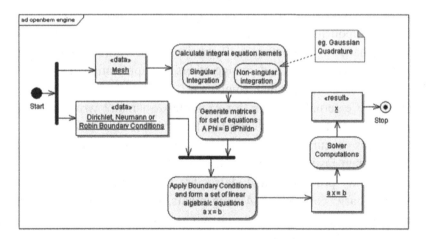

Fig. 10.18 BEM engine activity diagram

The main activities without feedback, isolated in the above-mentioned diagram, help in creating proper components of the library. The library assumption of the component built and independence of particular elements led to the component diagram presented in Fig. 10.19.

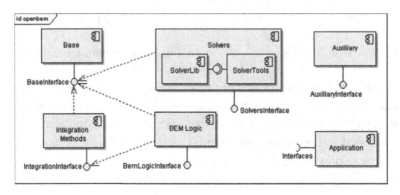

Fig. 10.19 BEM library component diagram.

In such a project, structures for storing and exchanging data between the components, the user and the external programs are required. They are combined in the 'Base' component, which is used by all the other components. This component controls importing and exporting data from/to MATLAB matrix format. It will cover both complex and real number representation.

According to the activity diagram (Fig. 10.18), the next component required should cover integration algorithms. It exports the common interface which can be implemented by the internal library or self-implemented algorithms and by wrappers to the external libraries.

Another component is the 'BEM Logic'. It is composed of subcomponents which are in charge of generating **A**, **B** matrices from Eq. (10.35) and **a**, **b** from Eq. (10.36) for enabled BIEs and, among others, the application of BCs.

The 'Solvers' is still another component. At this stage, it is composed of wrappers to the external libraries (for example CSparskit2 wrapper to Sparskit library) as solver computation is complex. However, if a demand for own solver implementations occurs, they can be developed by realizing the 'Solver Interface' and then easily included in the library.

Some auxiliary procedures and classes are also needed. They are combined into the 'Auxiliary' component. Among others, they are used for logging and statistics. The 'Application' component represents the specific user's program, which solves the specified problem.

It uses selected classes and procedures implementing other components interfaces. Each component exports and imports an interface. It is also made up of two subcomponents as shown for the 'Solver'. One is the library, and the other one represents the program, which allows one to make calculations the component externally. The input and export data are in MATLAB format.

### 10.3.5 *Technology*

There were two main factors which affected the choice of technology. The library is an open source project primarily to be run in the Linux/Unix environment. The second one is the selected programming language.

#### *Licensing*

The aim of the license is to give users a right for software running, analyzing, copying, modifying and distributing it with improvements. The license chosen is widely used in an open source environment: the GNU public license (GPL). Among others, it allows one to redistribute modified binary software only if the source code is open. Particular license conditions in GPL are grouped in 4 liberties (0–3). There is also a possibility that, in the future, some parts of the code will be distributed with the Berkeley software distribution (BSD) compliant license.

#### *Programming language*

There are two programming languages which can be considered for this kind of project: C and C++. Other programming languages such as Java or C# have many advantages. They have an open source implementation (Mono for C#). They are easy to use, objective and have good documentation (especially Java). They are portable because they use virtual machines, available on most operating systems, to run programs compiled to byte code. But they have one main disadvantage. They are slow for numerical computations. As it is easier to design and implement software using objective programming, the project is being implemented in C++ (Eckel, 2000; Eckel, 2003). The GNU C++ compiler is used as it is very popular and available on nearly all platforms.

#### *Source Code Version Control*

Source Code Version Control allows one to keep the source code under control and follow the changes. It is also the only possible way of working with the code by many people. There are many source code version control systems but as the project is going to be open source, one of the following was considered as the project choice: Concurrent Versions System (CVS) and Subversion (commonly referred to as SVN). Now they are both popular but as SVN is the successor of CVS and lacks the predecessor's disadvantages, this project will use

SVN as its code version control system. In addition, there is supposed to be a web interface for SVN, which sometimes facilitates work on the code.

*Source code manager*

One of the main requirements is that the user of any software is able to use it. The open source community mainly uses GNU 'autotools' (Tromey *et al.*, 2000) to make distributions from the C/C++ source code and afterwards configure software properly on the client's machine, build and finally install in the client's operating system. Autotools is a common referrer for naming GNU automake, autoconf and libtool. They are used in this project despite their disadvantages (complicated configuration, unclear source code or differences of usage between platforms, which has to be manually patched). Another possibility considered was the use of CMake, which is used in the inter alia K Desktop Environment (KDE) project, but it is not a standard yet and may not become one.

*Documentation*

Documentation is essential for the majority of people as it makes the use of software easier. It should follow the user's needs, and it seems that there should be three major types of documentation: installation instructions, tutorial and code documentation. These types help users at different stages of the library usage. It appears that the tutorial or, in the case of the Unix/Linux environment, the HOWTO is very important from the user's point of view. It enables the user to start using the software in a short time.

Sometimes it may be the only encouragement for choosing the exact software. There are tutorials covering the major functionalities of the software. They practically illustrate simple and advanced features through examples. They should not be verbose, but have to show only the library's essence. Tutorials will be supported by UNIX style manuals, especially for enclosed programs. As users are becoming skilled in using software or they would like to improve software, they will need source code documentation. Unlike Java, C++ does not have a native built-in documentation tool like 'javadoc'.

All programming languages, which are part of .net (including C++) have Microsoft xml-based documentation format. However, this format is not interpreted by any appropriate multiplatform documentation generators when applied in C++ code. For example, mono, mentioned in Subsection entitled *Programming language* does not support C++. As C/C++ is widely used in computer operating systems and applications, some documentation syntaxes and generators came into being. This library is using Doxygen, which is also used by

MySQL database developers. It is not perfect but substantially better than other available open source generators.

The good source code documentation can be generated only if the source code itself is properly documented. Javadoc is to be used as the code documentation syntax. The other considered here was Qt style, but it seems to be less legible.

### Portal

There must exist a 'good open source project' on the Internet, the most popular and accessible place for distributing software and documentation, submitting bug reports and improvements, exchanging ideas. The project portal should have at least downloads, documentation and news sections. Mailing lists and a forum are advisable. Library sources, binary software packages in DEB and RPM file format for Linux distributions will be available as well as ports for FreeBSD and Gentoo Linux.

The portal will be managed by a group of people. Therefore, the Content Management System (CMS) such as Drupal or Joomla will be used. Documentation will be maintained with MoinMoin wiki. The next step will be preparing a mailing list. It will be managed by Mailman. If necessary, a forum will be created. The considered possibilities, also open source, are phpBB (inter alia used by Gentoo) or less probably, Java based jforum.

### 10.3.6 *Conclusions*

There are not any open source universal BEM projects available for the community. There are only a small number of implementations, mostly in Fortran, and only applicable for restricted problems. In this situation a properly designed, developed and maintained open source project may popularise BEM and its application. As the library and enclosed programs are compatible with the Linux/Unix products style, its use is intuitive and does not require any special technological knowledge. A modular and component architecture will enable one to use only selected parts in one's own programs and implementations. If the project is sufficiently good, it will gather people creating the community, who will improve and expand it.

### References

Abem. <http://www.boundary-element-method.com/>.
Abramowitz, M. and Stegun, I. A. (1973). *Handbook of mathematical functions with formulas, graphs and mathematical tables*. John Wiley, New York.

Aliabadi, M. H. (2002). The boundary element method, *Applications in solids and structures*, Vol. 2. John Wiley & Sons LTD.

Aliabadi, M. H. and Hall, W. S. (1989). The regularizing transformation integration method for boundary element kernels. *Comparison with series expansion and weighted Gaussian integration methods. Eng. Anal. Bound. Elements*, Vol. 6, No. 2, pp. 66-70.

Amaratunga, K. A. (2000). Wavelet-based approach for compressing kernel data in large-scale simulations of 3D integral problems. *Comput. Sci. Eng.*, pp. 34-45.

Ambler, S. W. (2004). The object primer agile model-driven development with UML 2.0. *Cambridge University Press*.

Arridge, S. R. (1999). Optical tomography in medical imaging, *Inverse Problems*, Vol. 15, No. 2, pp. 41-93.

Arridge, S. R. (1999). Optical tomography in medical imaging. *Inverse Problems*, Vol. 15, No. 2, pp. 41-93.

Arridge, S. R., Dehghani, H., Schweiger, M. and Okada, E. (2000). The finite element model for the propagation of light in scattering media: a direct method for domains with nonscattering regions. *Med. Phys.*, Vol. 27, No. 1, pp. 52-64.

Arridge, S. R., Schweiger, M., Hiraoka, M. and Delpy, D.T. (1993). A finite element approach for modeling photon transport in tissue. *Med. Phys.*, Vol. 20, No. 2, pp. 299-309.

Banerjee, P. K. (1994). The boundary element methods in engineering. *McGraw-Hill Book Company*.

Beasy. <http://www.beasy.com/>.

Becker, A. A. (1992). The boundary element method in engineering. A complete course. *McGraw-Hill Book Company*.

Beer, G. (2001). Programming the boundary element method, *An introduction for engineers*, John Wiley & Sons.

Beer, G. and Watson, J. O. (1989) Infinite Boundary Elements, *International Journal for Numerical Methods in Engineering*, Vol. 28, pp. 1233-1247.

Beer, G. and Watson, J. O. (1992). Introduction to finite and boundary element methods for engineers. John Wiley & Sons.

Beer, G., Watson, J. O. and Swoboda, G. (1987). Three-dimensional analysis of tunnels using infinite boundary elements, *Computers and Geotechnics*, Vol. 3, pp. 37-58.

Bemlib. <http://dehesa.freeshell.org/BEMLIB/bemlib.shtml>.

Bettess, P. (1992). Infinite Elements, *Penshaw Press*.

Biepack. <http://www.math.uiowa.edu/atkinson/bie.html>.

Bonnet, M. (1995). Boundary integral equation methods for solid and fluids, John Wiley & Sons.

BSD license. <http://www.opensource.org/licenses/bsd-license.html>.

Chen W., Cheng J., Lin J., W. (2009). A level set method to reconstruct the discontinuity of the conductivity in EIT, *Science in China Series A: Mathematics*, Vol. 52.

Chen, G. and Zhou, J. (1992). Boundary element methods. *Academic Press*.

CMake (cross-platform make). <http://www.cmake.org/>.

Concurrent versions system (CVS). <http://www.nongnu.org/cvs/>.

CSparskit2. <http://www.iem.pw.edu.pl/wielebap/csparskit2/>.

Doxygen. <http://www.stack.nl/dimitri/doxygen/>.

Drupal cms. <http://www.drupal.org/>.

Eckel, B. (2000). Thinking in C++, *Introduction to standard C++, 2nd ed.*, Vol. 1, Prentice-Hall.

Eckel, B. (2003). Thinking in C++, *Practical programming, 2nd ed.*, Vol. 2. Prentice-Hall.

Fowler, M. (2003). UML distilled: a brief guide to the standard object modeling language, *3rd ed. Addison-Wesley.*

FreeBSD. <http://www.freebsd.org/>.

Frijns, J. H. M., de Snoo, S. L. and Schoonhoven, R. (2000). Improving the accuracy of the boundary element method by the use of second-order interpolation functions. *IEEE Trans. Biomed. Eng.*, Vol. 47, No. 10, pp. 1336-46.

Gentoo Linux. <http://www.gentoo.org/>.

GNU C++ compiler. <http://gcc.gnu.org/>.

GNU public license. <http://www.gnu.org/copyleft/gpl.html>.

GNU. <http://www.gnu.org/>.

Gratkowski, S. (2009). Asymptotyczne warunki brzegowe dla stacjonarnych zagadnień elektromagnetycznych w obszarach nieorganicznych - algorytmy metody elementów skończonych, *Wydawnictwo Uczelniane Zachodniopomorskiego Uniwersytetu Technologicznego w Szczecinie,* (in Polish).

Guven, I. and Madenci, E. (2003). Transient heat conducting analysis in a piecewise homogeneous domain by a coupled boundary and finite element method. *Int. J. Numer. Meth. Eng.*, Vol. 56, pp. 351-80.

Integrated engineering software. <http://www.integratedsoft.com/bem.asp>.

Ito K., Kunisch K. and Li Z. (2001) Level-set function approach to an inverse interface problem, *Inverse Problems*, Vol. 17, No. 5, pp. 1225-1242.

Jabłoński, P. (2009). Engineering Physics – Electromagnetism Handbook (EFE, sem. 2), *Częstochowa University of Technology.*

Jforum. <http://www.jforum.net/>.

Joomla cms. <http://www.joomla.org/>.

K desktop environment (kde). <http://www.kde.org/>.

Libbem. <http://www.mis.mpg.de/scicomp/software/libbem/libbem.html>.

Mailman. <http://www.gnu.org/software/mailman/>.

MATLAB. <http://www.mathworks.com/products/matlab/>.

Mitchell I. M. (2007). A Toolbox of Level Set Methods.

Moinmoin wiki engine. <http://moinmoin.wikiwikiweb.de/>.

Mono. <http://www.mono-project.com/>.

Muller, G., Chance, B., Alfano, R., Arridge, S. R., Beuthan, J., Gratton, E., Kaschke, M., Masters, B., Svanberg, S. and van der Zee, P., Editors. (1993). The forward and inverse problems in time-resolved infrared imaging, *Proceedings of the SPIE, Medical Optical Tomography: Functional Imaging and Monitoring.*

MySQL application programming interface (API). <http://dev.mysql.com/ sources/doxygen/mysql-5.1/>.

NETGEN. <http://www.hpfem.jku.at/netgen/>.

Osher, S. and Fedkiw R. (2003a). Level Set Methods and Dynamic Implicit Surfaces, *Springer.*

phpBB. <http://www.phpbb.com/>.

Riley, J. D., Arridge, S. R., Chrysanthou, Y., Dehghani, H., Hillman, E. M. C. and Schweiger, M. (2001). Radiosity diffusion model in 3D, *Stefan Andersson-Engels, Michael F. Kaschke, Editors. Photon migration, optical coherence tomography, and microscopy. Proceedings of SPIE*, Vol. 18-21, pp. 53-64.

Riley, J. D., Dehghani, H., Schweiger, M., Arridge, S. R., Ripoll, J. and Nieto-Vesperinas, M. (2000). 3D optical tomography in the presence of void regions. *Opt. Express*, Vol. 7, No. 13.

Ripoll, J. (2000). Light diffusion in turbid media with biomedical application. PhD Thesis, University of Madrid.

Saad, Y. and Schultz, M. H. (1986). GMRES: a generalized minimal residual algorithm for solving nonsymmetric linear systems, *SIAM J. Sci. Statist.Comput.*, Vol. 7, No. 3, pp. 56-69.

Sikora J. (2007). Boundary Element Method for Impedance and Optical Tomography.

Sikora, J. (2007b). Boundary Element Method for Impedance and Optical Tomography, *Oficyna Wydawnicza Politechniki Warszawskiej*, Warsaw.

Sikora, J., Zacharopoulos, A. D., Douiri, A., Schweiger, M., Horesh, L. and Arridge, S. R. (2006). Diffuse photon propagation in multilayered geometries. *Phys. Med. Biol.*, Vol. 51, No. 3, pp. 97-516.

Sommerville, I. (2004). Software engineering, *7th ed. Addison-Wesley.*

Sparskit. <http://www-users.cs.umn.edu/saad/software/SPARSKIT/sparskit.html>.

Subversion (SVN). <http://subversion.tigris.org/>.

Tanaka, M. and Du, Q., Editors. (1987) Theory and applications of boundary elements methods, *Proceedings of 1st Japan–China symposium on boundary element methods*, Pergamon.

Tarvainen, T. (2006). Computational Methods for Light Transport in Optical Tomography, PhD Thesis, Department of Physics, University of Kuopio.

Tromey, T., Lance, I., Gary, T., Vaughan, V. and Elliston, B. (2000). GNU autoconf, automake, and libtool, *Sams.*

Watson, J. O. (1979). Advanced implementation of the boundary element method for two- and three-dimensional elastostatics, *Developments in Boundary Element Methods - 1 (Editors P.K. Banerjee and R. Butterfield)*, Elsevier Applied Science Publishers, Vol. 61, pp. 31-63.

Wu, T. W. (2000). Boundary element acoustics: fundamentals and computer.

Zacharopoulos, A., Arridge, S.R., Dorn, O., Kolehmainen, V. and Sikora, J. (2006). Three-dimensional reconstruction of shape and piecewise constant region values for optical tomography using spherical harmonic parameterization and a boundary element method, *Inverse Problems*, Vol. 22, pp.1-24.

# Appendix A

## BEM software

There is no wide choice of any software destined for BEM calculations. Extensive searches resulted only in the products presented in Table A.1.

Table A.1. BEM library review.

| Library | Environment (language) | Distribution conditions | Purpose |
|---|---|---|---|
| ABEM | Fortran | Commercial open source | Acoustics |
| LibBem | C++ | Semi commercial | Laplace eq. |
| BEMLIB | Fortran | GPL | Laplace, Helmholtz, Stokes flow |
| BIEPACK | Fortran | Free open source | Laplace eq. |
| BEA (Wu, 2000) | Fortran | Distributed with the book | Acoustics |
| BEASY | Windows or Unix binaries | Commercial | Construction engineering, acoustics |
| Integrated engineering software | Windows binaries | Commercial | Fields, wave, thermal analysis |

CHAPTER 11

# TWO-PHASE GAS-LIQUID FLOW STRUCTURES AND PHASE DISTRIBUTION DETERMINATION BASED ON 3D ELECTRICAL CAPACITANCE TOMOGRAPHY VISUALIZATION

Robert Banasiak[1], Radosław Wajman[1], Tomasz Jaworski[1], Paweł Fiderek[1], Jacek Nowakowski[1], and Henryk Fidos[2]

[1]*Institute of Applied Computer Science*
[2]*Faculty of Process and Environmental Engineering*
*Lodz University of Technology*
*90-924 Łódź, ul. Stefanowskiego 18/22*
*{rbanasi, rwajman, t.jaworski, p.fiderek, jacnow}@kis.p.lodz.pl, fidos@p.lodz.pl*

## 11.1 Introduction

The application of electrical capacitance tomography for the two-phase gas-liquid mixtures visualization and the phase distribution calculation is becoming popular. This non-invasive imaging technique is very useful, especially when key flow parameters are necessary for industrial applications. These particularly include: efficient non-invasive automatic phase fraction calculation and flow structure identification in the vertical and horizontal pipelines. These key flow parameters can be effectively determined by using three-dimensional electrical capacitance tomography combined with non-deterministic evaluation techniques for the analysis of volumetric images. In this chapter a solution to automated two-phase gas-liquid flow pattern identification based on a fuzzy inference of a series of reconstructed 3D images is discussed. Measurement data that include the spatial information of the flow is obtained by using an electrical capacitance tomography system and non-linear-type reconstruction algorithms. Finally, a set of fuzzy-based features is computed for flow classification. As a result of the fuzzy logic analysis performed, the classification of a given volumetric image into one of known flow regime structures is carried out.

## 11.2  Key problems of the two-phase gas-liquid flows diagnosis

Two-phase gas-liquid flows are a very important component of many industrial processes (Plaskowski et al., 1995). There are several possibilities for a potential industrial application of an electrical tomography based diagnosis and monitoring system. It can be applied for the investigation of aeration processes in chemical reactors, flotation processes, water and sewage aeration systems. The main task of aeration systems is to produce a proper fraction of aerated liquid and oxygen. The oxygen injection process is important due to the fact that liquid circulation must be achieved. It helps intensifying a mass transfer process. One of the fundamental problems is the proper evaluation of the inter-phase interface – there are really important parameters from the mass transfer point of view. Biological sewage treatment plants can serve as an example of water and sewage aeration. An important role in these processes is played by aerobic bacteria (aerobic), which grow only in the presence of free oxygen from the atmosphere or dissolved in water (Plaskowski et al., 1995). The level of aeration must be within a specified range, which depends on water temperature. One way of aeration is to inject compressed air into the system, causing the gas bubbles to move up freely in the liquid column. The two-phase flow processes also occur in bubble columns. Their purpose is to implement various physical and chemical processes. Controlling the size of the interface often determines the intensity of progress of these processes. For example, in air-lift columns and ejectors (Plaskowski et al., 1995) the movement of the liquid stream is forced by the gas stream. Such devices are commonly used in the extractive industries (e.g. flotation processes) or for the precipitation of a fraction of the liquid in the sedimentation processes, such as degreasing, where the size of bubbles is significant. There is also a separate group of industrial processes in which gas bubbles may be formed in the liquid as a result of chemical reactions. It can occur, for example, in chemical reactors or in the process of electrolysis, where the gas phase is a product (often a by-product) of a chemical reaction. Then, the appearance of bubbles indicates the quality of ongoing changes, and the measurement of bubble size provides information about the process. There are industrial processes in which the occurrence of bubbles is undesirable, such as those taking place in heat exchangers or heating devices. In these systems the appearance of unwanted gas is the evidence of a boiling liquid phenomenon, and necessitates the signaling of the state of emergency. This is similar to the cavitation phenomenon in rotational pumps, caused by a rapid decrease in the pressure below the pressure of the boiling liquid. Such a phenomenon is undesirable because it leads to the erosion of the pump blades and, at the same time, may also be the evidence of the system leakage. The most fundamental issues the knowledge of which is necessary to describe the hydrodynamics of two-phase flow mixtures include: the determination of mixtures of two-phase flow patterns, the determination of the void fraction in the flowing mixture and the flow resistance of mixtures. In addition to the description of mass transfer in two-phase flow systems, we need to know the interfacial surface area, the coalescence of gas bubbles and the mass transfer coefficient. Up to date, none of these issues have been satisfactory presented in the literature (Chhabra and Richardson, 1986). This is due to the complicated mechanism of flow dynamics, often connected with difficulties in its description from the mathematical point of view, and is connected with complicated measurement methods typically used in the research of two-phase flows. One of the main challenges in two-phase flows is a possibility of a priori prediction of the characteristics and the type of mixture flow on the basis of known values of the apparent velocity and the properties of individual phases, as well as the flow geometry (the diameter and angle of wire, the placement and shape of structures). There are different observation techniques of the resulting mixture flow structures developed by various research teams. They vary from the simplest visual observations by different techniques for photographing and filming to the opto-electronic detectors to the newest methods of using two- and three-dimensional computed tomography (Xie et al., 1992; Plaskowski et al., 1995; Xie et al., 2004; Marashdeh et al., 2007; Warsito et al., 2007). The most commonly observed structures of a gas-liquid mixture flow in the pipes of the vertical ascending movement are consistent with those presented in the past (Nicklin et al., 1962). These structures generally classify this type of flow. There are five basic flow patterns known: bubble flow, slug flow, foam flow, ring flow and dispersion flow (as can be seen in Figs 11.1A and 11.1B).

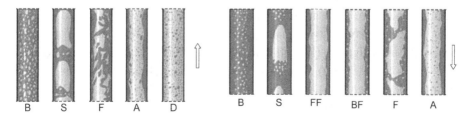

Fig. 11.1A  Two-phase flow structures in bottom-up vertical direction (B – bubble flow, S – slug low, F – foam flow, A – annular flow, D – dispersed flow).

Fig. 11.1B  Two-phase flow structures in top-down vertical direction (B – bubble flow, S – slug flow, FF – film flow, BF – bubble-film flow, F – foam flow, A – annular flow).

These flows can succinctly be characterized as follows:

- The bubble flow occurs when gas bubbles are dispersed in the liquid stream that fills the entire cross section and their speed is close to the speed of the liquid;
- The slug flow occurs when large bubbles of gas, typically with a diameter similar to the diameter of the pipeline flow alternating with portions of the liquid which additionally may be tiny bubbles of gas;
- The foam flow is usually defined as oscillatory and unoriented – this is the chaotic flow of two phases creating a rare form of the resembling foam;
- The annular flow occurs when the gas flows at a high speed in the central part of a pipe while the liquid forms a thin layer adhering to the walls of the pipe;

Dispersion occurs when the gas stream flows around the wire carrying the cross-tiny liquid droplets.

In the case of a flow in vertical pipelines, with a two-phase mixture flowing from top to bottom (so-called counter-flow), flow structures were already defined in the literature in the past (Oschinowo and Charles, 1974). It is worth noting that the flow structures described in this work relate to two-phase mixtures of liquids with very high viscosity – not exceeding 100 mPa·s. For liquids with higher viscosity, the flow structure should be defined differently, as indicated in the literature (Loebker and Empie, 1998). The ranges of occurrence of various flow patterns are typically presented on graphs and called flow maps. The area of such plots, which are made in different coordinate systems, is divided on the lines demarcating the subareas where a given structure of the flow exists. The concept of boundary lines means a certain narrow band of the flow map area. In the real world of flows there are no sharp boundaries between various flow structures. The transition from one structure to another proceeds in conditions similar to those precisely defined by the boundary line map, with approximately 10% deviation. In the world literature of the subject many of two-phase gas-liquid mixture flow maps for vertical pipes were investigated (Hewitt and Roberts, 1969; Oshinowo and Charles, 1974; Taitel et al., 1980). The proposers of this application present only one flow map described by Dziubinski and Fidos (Dziubinski et al., 2004) – as presented in Figure 11.2.

The flow map presented typically yields good results for viscous liquid-air mixtures. Mixtures of this type are used as a medium of a flow during the experimental part of this study.

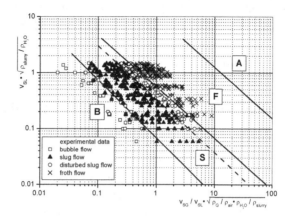

Fig. 11.2  Example of the flow map of two-phase gas-liquid flow.

## 11.3  Two-phase flows characterization using electrical process tomography

The characterization of the gas-liquid flow is very important for the design and implementation of an industrial-scale research facility and for the process of the research aimed to verify the results obtained. The continuous monitoring and diagnosing of any abnormalities can provide valuable information about their dynamic state and allow for continuous and automatic monitoring and control. Two-phase gas-liquid flows diagnosis belong to the most rapidly growing trend in fluid mechanics research. In recent years there has been significant, but still insufficient progress in the development of knowledge about these industrial processes. Two-phase flows arouse a growing interest because of their great practical significance. They are closely related to the rapidly developing field of research in bioprocess engineering, biotechnology, environmental engineering, energy, and many other related areas. Nowadays electrostatic field sensing based imaging techniques are sufficiently mature to be used for the non-invasive visualization of two-phase flows (Xie et al., 1992) in both: a cross-sectional (2D) (Plaskowski et al., 1995; Yang and Peng, 2003; Xie et al., 2004), a volumetric-oriented (3D) mode (Wajman et al., 2006; Soleimani et al., 2007; Fei et al., 2010) and even a real-time volumetric (4D) mode (Soleimani et al., 2009). Formerly conducted research has shown that three-dimensional electrical tomography imaging is to become an important tool in industrial processes imaging where the cross-sectional electric-field based systems provide inaccurate and incomplete information about a process (Yang and Peng, 2003). Much work in the field of volumetric tomography imaging was devoted to the development of sensing and inverse and forward problem solving techniques (Wajman et al., 2006; Marashdeh et al., 2007; Soleimani et al., 2007; Banasiak et al., 2010). The 3D ECT inspection of two-phase flows is a relatively new approach applied to gas-liquid mixtures (Marashdeh et al., 2007; Wang et al., 2009); however those studies focused only on the observation of 3D structures of a flow by volumetric images reconstruction.

## 11.4 Two-phase flows characterization using non-deterministic methods

While a diagnosis process of the flow is conducted important is to identify the flow regime, its structures and determining a void fraction. In this study, to classify the two-phase vertical and horizontal flow regimes, the information obtained from the spatial analysis of a set of objects from 3D reconstructed images is used. The fuzzy logic inference will be applied for the first time to three-dimensional reconstructed images as an efficient evaluator. Classical logic applied for two-phase flow images is based on two values, usually represented by 0 and 1, or terms true and false. By using them, we can classify a flow as liquid or air regions (e.g. air bubbles, slugs). The boundary between them is more or less clearly defined and constant. A fuzzy logic approach is an extension of classical reasoning the intention of which is to be 'more human' (Dubois et al., 1980). It introduces standard values between 0 and 1; it fuzzifies (blurs) the crisp boundaries, giving them the possibility of values within this range (i.e. almost false, half true). In this book, the uncertainty (values between true and false) can be transposed to gas/liquid 3D reconstruction in which every point can be considered as only a liquid, only a gas and a state between them. There have already been a number of studies in this field, with the use of fuzzy logic (Ji et al., 2004; Shi, 2008) and without the use of fuzzy logic (Zhou et al., 2008; Wang et al., 2009); however this research work only partially considers the spatial nature of a sensor (Li et al., 1992).

## 11.5 Fundamentals of ECT

The principle of an ECT technique is based on measuring the changes in the electric capacitance between the capacitor planes as a result of changes in the dielectric object located between these planes (Olszewski et al., 2006). The dielectric object, in the case of ECT flow imaging, is typically a liquid with gas bubbles. The general capacitance tomography measurement concept is as follows: the positive potential on one of the electrodes (excited electrode) is set, while others are grounded. The measured values of capacitances are collected and then the other electrodes are excited. A limited set of measurements can then be used in image building.

To extend this theory to make it more detailed, we can claim that the three-dimensional image reconstruction of two-phase flows using tomographic data is implemented herein as two separate cyclic stages: the forward problem stage and the inverse problem stage (Banasiak et al., 2010). This first one is related to the 3D electric field modeling using the ECT mathematical model. The forward problem for 3D ECT tomography is the simulation of measurement data for a given value of excitation and gas-liquid spatial distribution. The inverse problem is the imaging result for a given set of simulated or experimental data. To solve the inverse problem, we need to solve the forward problem first. At this stage, for a certain computational simplification, we assume that there is no wave effect and we use a low frequency approximation to Maxwell's equations. In a simplified ECT mathematical model, the electrostatic field approximation:

$$\nabla \times E = 0 \tag{11.1}$$

is taken, effectively ignoring the effect of wave propagation. Now let us define $E$ as:

$$E = -\nabla \varphi \tag{11.2}$$

and assume the lack of internal charges (Banasiak et al., 2010). Then, the following equation holds:

$$\nabla \cdot \varepsilon \nabla \varphi = 0 \quad in \quad \Omega \tag{11.3}$$

Where: $\varphi$ is the electric potential distribution, $\varepsilon$ is the dielectric permittivity distribution of a gas-liquid mixture and $\Omega$ represents the region containing the distribution of the electric field. The potential on each electrode is defined as:

$$u = v_k \quad at \quad e_k \tag{11.4}$$

Where: $e_k$ is the k-th electrode held at the potential $v_k$. Using finite modeling we can compute the RHS of this equation:

$$Y(\varepsilon)\varphi = F \tag{11.5}$$

Where: the matrix Y is the discrete representation of the operator $\nabla \cdot \varepsilon \nabla$, the matrix F includes both: Neumann-type and Dirichlet-type boundary conditions and $\varphi$ represents an electric potential solution. The very important task of this step is to determine the Jacobian matrix J that contains sensitivity maps for the sensor. The value of sensitivity in a specific area of the sensor (which is typically a voxel in 3-dimensional region) is defined as a change in the inter-electrode capacitance as a response to the permittivity distribution change in this voxel.

This nature of the electric field induces the 3D ECT inverse problem to be ill-posed. The calculated Jacobian J is the most important element in the deterministic inverse processing. Its task is the linearization of the electric field and it provides a mapping of the approximated permittivity distribution image $\varepsilon$ with the measured or simulated capacitance data $\chi$.

$$\hat{\varepsilon} = J^{+} \cdot \chi \qquad (11.6)$$

In the ECT imaging the Jacobian is a rectangular-type matrix. Its dimensions depend on the number of independent measurements and on a much greater number of pixels. Hence, it is impossible to directly calculate $J^{-1}$. Therefore, various known image reconstruction methods use its pseudo-inverse form $J^{+}$. For the purpose of two-phase flow imaging, various image reconstruction methods were studied (Yang and Peng, 2003; Warsito et al., 2007; Li and Yang, 2008; Banasiak et al., 2010; Fei et al., 2010). There are a number of linearly-schemed algorithms commonly used for the purpose of flow imaging. The most popular, the Linear Back-Projection technique, is simple and fast but inaccurate. In this case, the pseudo-inverse form of the Jacobian matrix is computed by its simple transposition. This method has also several iterative modalities and one of them is known as Landweber's iterations technique (Li and Yang, 2008). This algorithm is still fast but, having some internal optimization features, is typically more accurate, which can be sufficient if a high image resolution and the visibility of small objects in a gas-liquid mixture are not important. This method is defined as follows:

$$\hat{\varepsilon}_{k+1} = \hat{\varepsilon}_k - \alpha J^{T} \left( J\hat{\varepsilon}_k - \chi \right) \qquad (11.7)$$

where $k$ is the number of the current iteration, $\alpha$ is the relaxation parameter for convergence rate control that can be selected empirically or by calculation. For the task of flow pattern classification, we must know more about the internal structure of the mixture investigated. To ensure the above, a more sophisticated reconstruction algorithm must be used. In this paper the non-linear inversion is considered as an efficient solution for the proper revealing of gas-liquid phases distribution inside the sensor volume using experimental data. The inverse solution was computed using a slightly modified 3D ECT complete forward model reported in (Banasiak et al., 2010) according to the new 3D ECT sensor design with internally mounted electrodes. To obtain the proper capacitance data and volumetric image mapping, the traditional voxel-based sensitivity matrix approach was applied during the nonlinear three-dimensional image reconstruction process. The three-dimensional reconstruction of the two-phase flow internal structure can be briefly described in several main steps. The initial step of the whole non-linear reconstruction process is to find a starting approximation of the flow interior $\varepsilon_{3D}(0)$ by using experimental data $\chi_m$ and a truncated linear back-projection method with an initial Jacobian $J_0$. In this study, for the initial calculation of the starting Jacobian $J_0$, the simulated homogeneous permittivity of the gas-liquid mixture ($\varepsilon_r=37$) was applied. The value of the initial dielectric permittivity value was chosen according to the known maximum ($\varepsilon_r=37$ - glycol polypropylene) and the minimum ($\varepsilon_r=1$ – the air) values of permittivity of a true gas-liquid mixture. The next step is to find a nonlinear forward solution $\chi_c(k)$ and $\varphi_0$ by using 3D ECT forward modeling for the initial flow image $\varepsilon_{3D}(0)$, where $\chi_c(k)$ is the calculated measurement capacitance data and $\varphi_0$ is the electrical field distribution. At this stage a new updated Jacobian matrix must be calculated. The k-th image $\varepsilon_{3D}(k)$ can be found using conventional iterative back-projection optimization. Each k-th step of the whole 3D image reconstruction process includes a new nonlinear forward problem solution for $\varphi_k$ and $\chi_c(k)$ and the Jacobian $J_k$, which is the updated basis of the current k-th reconstructed image. The convergence of image reconstruction can be adjusted by the relaxation factor $\alpha$. The stopping criterion is aided by a dynamic change in the relaxation parameter and performed by tracing the progress in capacitance residual decrementing. The process is stopped if the progress is not sufficient or is controlled by a fixed value. To give a brief overview of the efficiency of this algorithm there are two examples of 3D images presented in Figure 3 for both a horizontal and a vertical flow.

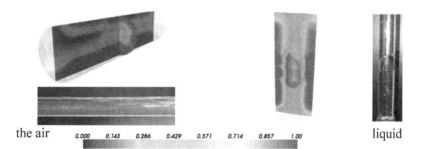

the air  0.000  0.143  0.286  0.429  0.571  0.714  0.857  1.00  liquid

Fig. 11.3   Example 3D images of flow interior reconstruction for both: horizontal and vertical subsections of test facility pipes (40mm in diameter) with equivalent CCD camcorder view.

## 11.6 Electrical capacitance tomography for two-phase flow imaging

Knowing the relationship between the spatial concentration of $\rho$ and the spatial dielectric permittivity distribution $\varepsilon$, it is possible to determine the liquid phase spatial distribution of $\rho_{liquid}(x, y, z)$ and gas $\rho_{gas}(x, y, z)$. This, in turn, allows the identification of the flow structure and the measurement of the void fraction. This type of measurement technique seems, in fact, to be the most effective in applications for flows, mainly because of its speed and non-destructive nature. In the research concerning the ECT application for two-phase gas-liquid flows monitoring (Xie et al., 2004; Wajman et al., 2009) only the two-dimensional images were produced with the flow structures and the phase distribution. The two-dimensional image cannot fully reveal the information about the flow in the measuring area. It is built by averaging the signal across the electrodes ring creating a cross-sectional image of the pipeline. In this case, the spatial phenomena of the electromagnetic field distribution are neglected and additionally, in the case of long electrodes this averaging prevents accurate measurement. It was, therefore, necessary to extend this technique to the three-dimensional mode (Wang and Zhang, 2009). The information on the entire structure of the flow is obtained by the 3D visualization of the material distribution. This type of imaging requires a specific deployment of the measuring electrodes as well as an appropriate measurement procedure (Soleimani et al., 2007).

## 11.7 Fuzzy logic techniques for flow characterization

For the purpose of automatic flow structure identification a new algorithm was developed using a fuzzy logic technique. The concept of the proposed solution is illustrated in Figure 11.4.

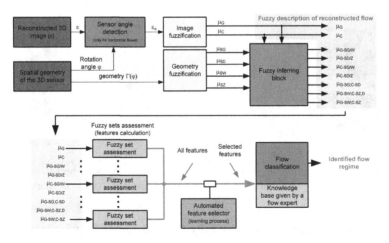

Fig. 11.4 Flow recognition algorithm diagram for both horizontal and vertical flows; G – gas; C – liquid; SG – upper zone; SD – lower zone; SW – inner zone; SZ – outer zone; SG/W - upper or inner zone; SD/Z – lower or outer zone; SZ,D – outer and lower zones;

There are two types of data that must be provided as the input. The first of them is the reconstructed 3D ECT image including its mesh geometry. The other is the flow regime knowledge base (typically it can be developed by process engineering expert observations). In this study there are two assumptions that must be done:

For the purpose of fuzzy logic evaluation, the input 3D ECT image was obtained by using a nonlinear reconstruction method to provide high spatial resolution input data;

In the case of a horizontal flow, the geometry of the sensor is rotated around the horizontal axis, which is parallel to the flow direction. The angle is determined using a sensor calibration routine.

The idea of 3D ECT sensor angle determination is to eliminate a potential error of the sensor placement on a horizontal pipeline. Ultimately, it facilitates the use of fuzzy linguistic values, such as "upper / lower zone sensor" in the fuzzy inference block. The angle detection algorithm assumes that the input image is a three-dimensional reconstruction of the pipeline section which, during the calibration, is filled with liquid in its lower half and with gas in its upper half. In this study a polyethylene glycol was chosen as the test liquid. This situation is depicted in Figure 5.

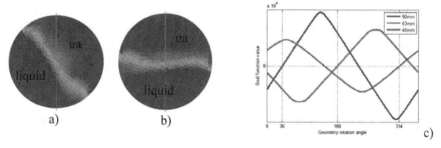

Fig. 11.5 3D ECT Sensor's XY plane projection view (a cross-section) for a horizontal flow) before the geometry rotation procedure, b) after the procedure (when angle is determined), c) a plot of the goal function (8).

In order to determine the rotation angle for the geometry, the proposed algorithm performs the analysis of the objective function defined as:

$$Jrot(\varphi) = \sum_{p \in \Gamma(\varphi)} sign(p_x)V_p I_p \qquad (11.8)$$

where p is a voxel of the three-dimensional flow approximation $\varepsilon$, $\Gamma(\varphi)$ is the sensor geometry rotated by angle $\varphi$, px is the x coordinate component of the center of gravity of the voxel p, $V_p$ is the volume of the voxel p and Ip is the value of the reconstructed voxel p in the image $\varepsilon$ and can be denoted as Ip = $\varepsilon$(p).

In this study a set of three angular characteristics were obtained for sensors with diameter $d_z$ = 90, 63 and 40mm. They are presented in Figure 11.5c. Equation (11.8) generates two solutions, while the appropriate one can be found only on the rising edge of the objective function. After the angle of rotation has been determined, such a geometry can be used for further processing. An example of visualization of the reconstructed image cross-section is shown in Figure 11.5b. Obviously, in the case of vertical flow identification, the sensor angle correction is not performed.

After the correction (if any) of a given geometry, a reconstructed image is transferred to the fuzzy inference block. The use of fuzzy logic, as part of the identification process, the authors' opinion, is dictated by the natural ambiguity of the results. An arbitrary point of the reconstructed image (in this case a voxel) on the canvas of classical logic can be interpreted as a "liquid", "gas" or "intermediate state". A quantitative boundary between these terms is debatable and values, if fixed, can dramatically affect the outcome of the identification process. Fuzzy logic, by its nature, is capable of using uncertainty intrinsic to every ECT reconstruction, allowing one to determine a membership of a voxel to both: the concept of a "gas" and a "liquid". The intermediate state is here "built-in" to the concept of the membership value. It is worth mentioning that although the probability and membership use the same unit interval [0;1], the physical interpretation is quite different. For example, if the membership value of a voxel to term "gas" / "liquid" is 0.5, it can be said that it contains both a gas and a liquid in a similar amount while the probability of 0.5 indicates that the point can be a gas or a liquid (the same probability).

The fuzzy inference block begins its work from its first phase – the fuzzification of a reconstructed image, i.e. the calculation of two fuzzy sets that describe the input image in terms of "it is a gas" and "it is a liquid". These sets, identified by the gas (G) and the liquid (C) (illustrated in Figure 11.6) describe the extent to which each of the points in the input image can be attributed to the concepts of gas / liquid.

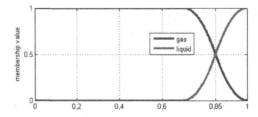

Fig. 11.6 Two membership functions used for fuzzification of an input image.

The second phase is fuzzification process of the spatial geometry of the sensor, defined in this article as a setup of areas inside the sensor, which can be identified by linguistic values, such as the upper zone (SG), the lower zone (SD), the inner zone (SW) and the outer zone (SZ). In the case of a horizontal sensor, membership values are determined by the percentage of the Y component (height). Its values are varying from 0 (not upper zone) to 1 (upper zone). The characteristic functions of fuzzy sets are shown in Figure 11.7a along with the visualization of the lower zone (SD) set in Figure 11.7b.

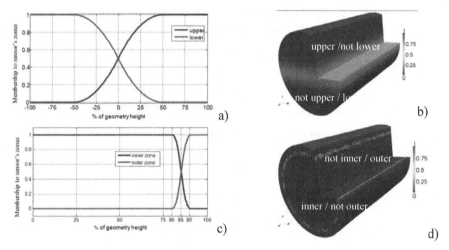

Fig. 11.7  Geometry fuzzification: a characteristic function for zones: a) upper and lower, and b) lower zone visualization, c) inner and outer, and d) outer zone visualization.

It shows that the values of membership vary smoothly from full membership (Y max) to no membership (Y minimum). Moreover, it should be noted that the idea behind two fuzzy sets describing a zone is not to discriminate between inner and outer but between inner and not inner since inner/outer and the upper/lower boundary is not always easy to be precisely distinguished.

The fuzzification of vertical geometries is quite similar. An example of the inner and the outer zone for vertical flows is shown in Figure 11.7c. In this case, the value of membership to the inner (SW) and the outer (SZ) zone depends on the percentage distance of the point of interest from the center of gravity of geometry in the XY plane. An example of visualization for this case is shown in Figure 11.7d. After both the image and the geometry are fuzzified using the previously generated fuzzy sets, the flow identification algorithm uses fuzzy rules in the inferring procedure. The designed fuzzy rule base module is presented in Table 11.1.

Table 11.1 Fuzzy rule base for the inference process along with accumulating expressions.

| Group 1 (inference – with fuzzy implication) |
|---|
| If zone is *upper* and image is *gas* then **gas above (G-SG)**; |
| If zone is *lower* and image is *gas* then **gas below (G-SD)**; |
| If zone is *upper* and image is *liquid* then **liquid above (C-SG)**; |
| If zone is *lower* and image is *liquid* then **liquid below (C-SD)**; |
| **Group 2 (inference – with fuzzy implication)** |
| If zone is *inner* and image is *gas* then **gas inside (G-SW)**; |
| If zone is *outer* and image is *gas* then **gas outside (G-SZ)**; |
| If zone is *inner* and image is *liquid* then **liquid inside (C-SW)**; |
| If zone is *outer* and image is *liquid* then **liquid outside (C-SZ)**; |
| **Group 3 (accumulation – with fuzzy product)** |
| *gas above* (*G-SG*) and *liquid below* (*C-SD*) →**gas above liquid (G-SG;C-SD)** |
| *gas inside* (*G-SW*) and *liquid outside* (*C-SZ*) →**gas inside liquid (G-SW;C-SZ)** |
| *gas inside* (*G-SW*) and *liquid outside* (*C-SZ*) and zone jest *lower* (*SD*)→**gas inside; liquid below and outside (G-SW;C-SZ,D)** |

The fuzzy sets marked in italics are the input sets while fuzzy sets marked in bold are the resulting sets describing the conclusion of the rule.

For a collection of fuzzy sets obtained in steps described previously, each fuzzy set from the collection is assessed by a set of features. It is important to mention that for every set from the collection the same set of features is calculated. Along with commonly accepted methods of assessment for sets of values: the mean, standard deviation or fuzzy set descriptions (Alsharhan et al., 2001), energy, the length of support, the kernel or its entropy, the authors have proposed their own evaluation features corresponding to the specificity of the flow structure identification problem. Below is the list of the most important features of the acquired fuzzy sets along with proposed ones:

The cardinal number of the fuzzy set A:

$$card(A) = \sum_{x} \mu_A(x) \tag{11.9}$$

where A is a fuzzy set, $\mu A$ is a membership function of the fuzzy set A (also called the characteristic function of the set).

The Local Average entropy of the fuzzy set A (proposed):

$$E_{SL}(A) = \frac{1}{|A|} \sum_{p \in A} \left( \frac{1}{|NB(p)|} \sum_{p \in NB(p)} -SE(\mu_A(p)) \right) \tag{11.10}$$

where A is a fuzzy set, NB ($\bullet$) is a list of neighbors of $\bullet$ and SE ($\bullet$) is an entropy measure for a particular value of membership, expressed as:

$$SE(\mu) = \mu \log \mu + (1 - \mu) \log(1 - \mu) \qquad (11.11)$$

The entropy of the fuzzy set A:

$$E(A) = \frac{1}{|A|} \sum_{p \in A} SE(\mu_A(p)) \qquad (11.12)$$

The fuzzy entropy "AnotA":

$$F_{A\bar{A}}(A) = \frac{card(A \wedge \bar{A})}{card(A \vee \bar{A})} \qquad (11.13)$$

where $\bar{A}$ is a complement of the fuzzy set, usually expressed as $\mu_{\bar{A}} = 1 - \mu_A$.

The maximum local uniformity of the spatial fuzzy set A (proposed):

$$H_{LM}(A) = \frac{1}{\sum |NB(p)|} std\left(\bigcup_{p \in A} \max\left(\{|\mu_A(p) - \mu_A(q)| : q \in NB(p)\}\right)\right) \qquad (11.14)$$

where std ($\bullet$) is the standard deviation, expressed by the formula:

$$std(X) = \sqrt{\left(\sum_{x \in X} (x - \bar{X})^2\right) / \bar{X}} \qquad (11.15)$$

The local uniformity of the spatial fuzzy set A (proposed):

$$H_L(A) = \sum_{p \in A} \frac{std(\{\mu_A(q) : q \in NB(p)\})}{\{\mu_A(q) : q \in NB(p)\}} \qquad (11.16)$$

Gas against the liquid separation over the reconstruction area (proposed):

$$S^{r,s}(G,C) = \frac{\sum_{p \in G} p_Y^r p_Z^s \mu_G(p)}{card(G)} - \frac{\sum_{q \in C} q_Y^r q_Z^s \mu_C(q)}{card(C)} \qquad (11.17)$$

where PY and PZ are Y and Z components of the center of gravity of the voxel p. G and C are gas and liquid fuzzy sets. The implementation uses S1, 0 for horizontal flows and S0, 1 for vertical flows.

With the feature values of experimental flows already calculated, the authors began to prepare the learning and test sets for the final classifier (see Figure 11.5). For each of the selected and reconstructed flow spatial image, the process engineering expert (one of the authors) carried out a description. During the experiments, the spatial image analysis highlighted a number of classes corresponding to the observed flow structures. A set of classes is presented in Table 11.2.

Table 11.2  List of classes of recognizable structures, divided by diameters and types of flows.

| Type | Diameter $d_z$ | Class of flow |
|---|---|---|
| horizontal flow | 90mm, 63mm, 40mm | only liquid (1), laminar (2), slug (3), bubble (4) |
| vertical flow | 90mm, 63mm, 40mm | slug (1), bubble (2), annular (3), only liquid (4) |

Each of the flow images was assigned to one of the classes by the expert. As for the classifier, in this paper the SVM (Cortes and Vapnik, 1995), Support Vector Machine was used. The idea of SVM algorithm is to separate a set of points into two different classes (labeled as 1 and -1) by a hyperplane. These points in this study are images described by feature vectors of equal length.

Since the SVM classifier is inherently two-class only (class labels are 1 and -1) and the expert identified more than two classes, the ONE-versus-ALL method was used (Rifkin and Klautau, 2004). Its idea is to use a K number of classifiers for K different classes, where each classifier develops a separating hyperplane between its class and all the other points. The final decision is to select the class for which the classified point is farthest from the separating hyperplane in the direction normal to this plane.

Due to a large number of features describing each of the images (302 elements in the vector), the authors decided to use the method for automatic selection of features, namely SVM-RFE (Guyon et al., 2002). Its aim is to eliminate the insignificant features and improve the classification stage performance. Limiting the number of features, in terms of the classification process, also allowed the computational effort to be reduced, which is necessary to calculate a feature set for every image, and thus to increase the efficiency of the whole system.

In addition, the authors compared the results of SVM classifier against the FCM (Fuzzy C-Means), which is one of the most widely used fuzzy clustering algorithms (Dunn, 1974). The main difference between the FCM and SVM algorithm is the ability to assign one point to different classes with their appropriate degrees of membership. When comparing the results, the authors assumed that the image belongs to the class for which the value of membership is the highest.

## 11.8 The experimental setup

The experimental part of this study was entirely performed in the Process Tomography Research Laboratory at the Institute of Applied Computer Science at the Lodz University of Technology. The Laboratory is equipped with a semi-industrial scale two-phase gas-liquid flow research test facility. This flow facility allows one to conduct experiments in both: the horizontal and vertical parts of a pipeline. Each of the sections (horizontal and vertical) is built using PVC (polyvinyl chloride) pipes (with PVC valves) with three different internal diameters $d_z$ – 34mm, 53.6mm and 81.4mm, respectively. This enables one to test two-phase flows at different scales. The length of the horizontal section under study was approximately 7.5m, while the length of the vertical one was 4m. All the sections studied contain transparent subsections of the pipes that allow one to observe the flow structure movement and record it by using a video camera. The flow facility can be filled with a fluid by a pump system which consists of two storage tanks and a multi-stage centrifugal pump. In one of the tanks there is a propeller stirrer, which allows one to prepare research liquids, such as aqueous solutions of various substances. There are also Kobold oval gear wheel flow meters that make it possible to precisely measure the volumetric flow rate of a liquid with a relatively high viscosity value. CRN 45 pumps produced by the Grundfos company provide the maximum flow rate of a liquid (with a viscosity close to the viscosity of water) pumped into the pipeline at a rate of 50m3/h. In addition, the screw compressor DMD 100CR applied allows a volume stream of air at 1000 l/min with pressure about 7bar to be obtained. For the experiments, a two-phase mixture of pure glycol polypropylene (99,9% pure component) and air was used. The viscosity value of the liquid was equal to 0,055 Pa•s and was measured using the Anton Paar rheometer at a temperature of 22°C. The conditions mentioned let all types of two-phase flows occurring in industrial environments to be obtained – from a bubble flow to an annular flow.

In the research, to perform the visualization of the two-phase flow structures in the vertical and horizontal sections of pipelines, two independent 3D ECT hardware systems (Olszewski et al., 2006) were applied.

A crucial part of the ECT 3D system is a capacitance sensor. For the purpose of this research a new three-dimensional capacitance sensor structure was proposed. The new sensor includes a set of electrodes that are located inside a pipe with the use of non-conductive isolation from the gas-liquid mixture – see Figure 9. This sensor is specially designed to make the non-intrusive inspection of liquids with high values of the relative dielectric constant.

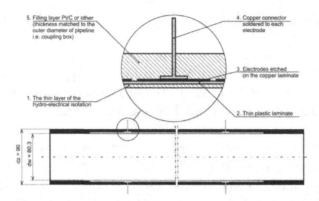

Fig. 11.9  The new type 3D ECT sensor construction with the layout of inner electrodes.


The design of the sensor structure improves the effect of deeper electric field penetration inside the process volume even in the case of a mixture of conductive and dielectric media. This phenomenon decreased the heterogeneity of the sensor sensitivity distribution. In order to design a proper 3D sensor structure the high-precision impedance meter Agilent E4980A was used.


### 11.9  Two-phase flow characterization — experimental results

With implementations of the previously described components of the system along with learning and testing data collections, the authors proceeded to the stage of teaching the classifiers of flow patterns. Table 3 shows the results of teaching and the classification process of selected teaching patterns for different experiments. The process was verified by the minus one element method (called Leave-One-Out) (Evgeniou et al., 2004).

The results of the classification for the stage of learning, for the both SVM and FCM classifiers are presented in the form of the confusion matrix (Table 11.3) for the learning sets for each of the flow classes, which the flow engineering expert was able to see in the experiments described. The rows of the matrix represent the expected outcome classification, while the columns the obtained one (classified). Each row has a label and an ID assigned by the expert, equal to those in Table 11.3.

Table 11.3   Result of the classifier learning effectiveness, learning data was used as test points (decisions closer to expert marked with gray color; gray color shows also selected equal results for SVM and FCM).

| Sensor | Diameter | Number of images | Class name (class number) / Expert evaluation | SVM classification 1 | 2 | 3 | 4 | Sensitivity | Specificity | Accuracy | FCM classification 1 | 2 | 3 | 4 | Sensitivity | Specificity | Accuracy | Patterns/im ages |
|---|---|---|---|---|---|---|---|---|---|---|---|---|---|---|---|---|---|---|
| Horizontal | d=90mm | 200 | only liquid (1) | 48 | 1 | 1 | 0 | 96.00% | 99.33% | 98.50% | 34 | 1 | 0 | 15 | 68.00% | 98.00% | 90.50% | 50 |
| | | | laminar (2) | 0 | 48 | 2 | 0 | 96.00% | 98.00% | 97.50% | 0 | 49 | 1 | 0 | 98.00% | 88.67% | 91.00% | 50 |
| | | | slug (3) | 1 | 2 | 47 | 0 | 94.00% | 98.00% | 97.00% | 3 | 16 | 31 | 0 | 62.00% | 99.33% | 90.00% | 50 |
| | | | bubble (4) | 0 | 0 | 0 | 50 | 100.00% | 100.00% | 100.00% | 0 | 0 | 0 | 50 | 100.00% | 90.00% | 92.50% | 50 |
| | d=63mm | 191 | only liquid (1) | 49 | 0 | 1 | 0 | 98.00% | 97.87% | 97.91% | 49 | 0 | 1 | 0 | 98.00% | 94.33% | 95.29% | 50 |
| | | | laminar (2) | 0 | 48 | 2 | 0 | 96.00% | 95.04% | 95.29% | 1 | 49 | 0 | 0 | 98.00% | 92.20% | 93.72% | 50 |
| | | | slug (3) | 3 | 7 | 39 | 1 | 78.00% | 96.45% | 91.62% | 7 | 11 | 27 | 5 | 54.00% | 99.29% | 87.43% | 50 |
| | | | bubble (4) | 0 | 0 | 2 | 39 | 95.12% | 99.33% | 98.43% | 0 | 0 | 0 | 41 | 100.00% | 96.67% | 97.38% | 41 |
| | d=40mm | 193 | only liquid (1) | 43 | 0 | 0 | 0 | 100.00% | 99.33% | 99.48% | 37 | 1 | 5 | 0 | 86.05% | 99.33% | 96.37% | 43 |
| | | | laminar (2) | 1 | 49 | 0 | 0 | 98.00% | 100.00% | 99.48% | 1 | 48 | 1 | 0 | 96.00% | 99.30% | 98.45% | 50 |
| | | | slug (3) | 0 | 0 | 50 | 0 | 100.00% | 100.00% | 100.00% | 0 | 0 | 50 | 0 | 100.00% | 95.80% | 96.89% | 50 |
| | | | bubble (4) | 0 | 0 | 0 | 50 | 100.00% | 100.00% | 100.00% | 0 | 0 | 0 | 50 | 100.00% | 100.00% | 100.00% | 50 |
| Vertical | d=90mm | 200 | slug (1) | 32 | 13 | 3 | 2 | 64.00% | 89.33% | 83.00% | 23 | 22 | 4 | 1 | 46.00% | 84.00% | 74.50% | 50 |
| | | | bubble (2) | 11 | 39 | 0 | 0 | 78.00% | 88.67% | 86.00% | 20 | 30 | 0 | 0 | 60.00% | 83.33% | 77.50% | 50 |
| | | | annular (3) | 3 | 4 | 17 | 26 | 34.00% | 79.33% | 68.00% | 4 | 3 | 20 | 23 | 40.00% | 88.67% | 76.50% | 50 |
| | | | only liquid (4) | 2 | 0 | 28 | 20 | 40.00% | 81.33% | 71.00% | 0 | 0 | 13 | 37 | 74.00% | 84.00% | 81.50% | 50 |
| | d=63mm | 200 | slug (1) | 37 | 1 | 12 | 0 | 74.00% | 93.33% | 88.50% | 21 | 12 | 17 | 0 | 42.00% | 76.00% | 67.50% | 50 |
| | | | bubble (2) | 2 | 43 | 5 | 0 | 86.00% | 95.33% | 93.00% | 2 | 48 | 0 | 0 | 96.00% | 89.33% | 91.00% | 50 |
| | | | annular (3) | 8 | 6 | 35 | 1 | 70.00% | 86.67% | 82.50% | 33 | 4 | 13 | 0 | 26.00% | 75.33% | 63.00% | 50 |
| | | | only liquid (4) | 0 | 0 | 3 | 47 | 94.00% | 99.33% | 98.00% | 1 | 0 | 20 | 29 | 58.00% | 100.00% | 89.50% | 50 |
| | d=40mm | 186 | slug (1) | 20 | 9 | 13 | 8 | 40.00% | 74.26% | 65.05% | 32 | 7 | 2 | 9 | 64.00% | 66.91% | 66.13% | 50 |
| | | | bubble (2) | 14 | 20 | 4 | 12 | 40.00% | 82.35% | 70.97% | 5 | 20 | 7 | 18 | 40.00% | 83.09% | 71.51% | 50 |
| | | | annular (3) | 13 | 3 | 34 | 0 | 68.00% | 87.50% | 82.26% | 40 | 0 | 10 | 0 | 20.00% | 93.38% | 73.66% | 50 |
| | | | only liquid (4) | 8 | 12 | 0 | 16 | 44.44% | 86.67% | 78.49% | 0 | 16 | 0 | 20 | 55.56% | 82.00% | 76.88% | 36 |

In order to improve readability, the column names are just identifiers. The quality of the classification learning stage (same as for the test data classification) for the two classifiers was determined by three measures of evaluation: sensitivity, specificity and total accuracy. Sensitivity, defined by (11.18), is the classification probability of an image in a class determined by the expert (the probability of correct classification).

$$M_{\text{Sensitivity}} = \frac{TP}{TP + FN} \tag{11.18}$$

Specificity, defined by (19), is the probability of a point not being classified in a class that it was not assigned to, according to the expert.

$$M_{\text{Specificity}} = \frac{TN}{TN + FP} \tag{11.19}$$

Accuracy (or total accuracy) given by (20), is a point correct classification probability.

$$M_{\text{Accuracy}} = \frac{TP + TN}{TP + FP + FN + TN} \tag{11.20}$$

where values of TP, TN, FP, FN in these expressions are determined for each confusion matrix. Their meanings are given below:

- FN (False Positive) - the number of classifications of points from class X to class X;
- TN (True Negative) - the number of classifications points from a class other than X to classes other than X;
- FP (False Positive) - the number of classifications of points from an outside class X to class X;
- FN (False Negative) - the number of classifications of points from class X to classes other than X.

In the case of experiments with the classification of horizontal flows, it is clearly visible that the SVM classifier is the right tool to be used (the classification accuracy is higher than FCM for all diameters). However, this does not disqualify the FCM classifier, as its accuracy (except the slug flow / $d_z$ = 63mm) remains above 90%. Such a low score of classification, at a level of 87%, can be explained by the similarity of tomographic flow images of slug to laminar. It is also worth mentioning that there is a very good distinction between a bubble flow and the pipeline full of liquid. During the preparation of the test data, the expert often could not decide which of these classes should be assigned to the flow analyzed.

Apparently, a visibly lower accuracy of classification in the learning stage was obtained for vertical flows. In this case, in the authors' opinion, the reason for a lower accuracy is the effect of a flow reverse (seen as "falling structures"). In addition, many images could not be classified by the expert due to the lack of visual belonging to any recognizable flow structure. In this case, the lowest classification accuracy was obtained for slug flows in the sensor of diameter $d_z$ = 40mm. Owing to the nature of the phenomenon in this diameter almost every image obtained using tomography was similar to the slug flow.

Table 11.4 Result of the classifier learning effectiveness, test data was used as test points; decisions closer to expert marked with gray color).

| Sensor | Diameter | Number of images | Expert evaluation | SVM classifier 1 | 2 | 3 | 4 | Sensitivity | Specificity | Accuracy | FCM classifier 1 | 2 | 3 | 4 | Sensitivity | Specificity | Accuracy | Patterns/ Images |
|---|---|---|---|---|---|---|---|---|---|---|---|---|---|---|---|---|---|---|
| Horizontal | $d_z$=90mm | 4590 | only liquid (1) | 1532 | 0 | 40 | 4 | 97.21% | 99.95% | 99.19% | 578 | 0 | 1 | 997 | 36.68% | 75.83% | 64.97% | 1576 |
| | | | laminar (2) | 0 | 2439 | 160 | 0 | 93.84% | 100.00% | 97.19% | 0 | 2089 | 510 | 0 | 80.38% | 98.31% | 90.11% | 2599 |
| | | | slug (3) | 2 | 0 | 413 | 0 | 99.52% | 96.20% | 96.45% | 0 | 52 | 349 | 14 | 84.10% | 90.30% | 89.85% | 415 |
| | | | bubble (4) | 0 | 0 | 0 | 1094 | 100.00% | 99.91% | 99.93% | 993 | 0 | 0 | 101 | 9.23% | 77.97% | 64.74% | 1094 |
| | $d_z$=63mm | 2311 | only liquid (1) | 760 | 0 | 3 | 0 | 99.61% | 99.87% | 99.78% | 763 | 0 | 0 | 0 | 100.00% | 99.87% | 99.91% | 763 |
| | | | laminar (2) | 0 | 1328 | 31 | 0 | 97.72% | 99.90% | 98.62% | 0 | 993 | 366 | 0 | 73.07% | 99.16% | 83.83% | 1359 |
| | | | slug (3) | 2 | 1 | 143 | 3 | 95.97% | 98.34% | 98.18% | 2 | 8 | 60 | 79 | 40.27% | 81.28% | 78.64% | 149 |
| | | | bubble (4) | 0 | 0 | 2 | 40 | 95.24% | 99.87% | 99.78% | 0 | 0 | 39 | 3 | 7.14% | 96.52% | 94.90% | 42 |
| | $d_z$=40mm | 1021 | only liquid (1) | 0 | 0 | 43 | 0 | 0.00% | 100.00% | 95.79% | 0 | 0 | 28 | 15 | 0.00% | 100.00% | 95.79% | 43 |
| | | | laminar (2) | 0 | 567 | 0 | 0 | 100.00% | 100.00% | 100.00% | 0 | 567 | 0 | 0 | 100.00% | 100.00% | 100.00% | 567 |
| | | | slug (3) | 0 | 0 | 307 | 1 | 99.68% | 79.41% | 85.52% | 0 | 0 | 7 | 301 | 2.27% | 81.51% | 57.63% | 308 |
| | | | bubble (4) | 0 | 0 | 104 | 0 | 0.00% | 99.89% | 89.73% | 0 | 0 | 104 | 0 | 0.00% | 65.58% | 58.90% | 104 |
| Vertical | $d_z$=90mm | 894 | slug (1) | 98 | 47 | 14 | 27 | 52.69% | 65.11% | 62.53% | 9 | 134 | 15 | 28 | 4.84% | 84.60% | 68.01% | 186 |
| | | | bubble (2) | 123 | 103 | 10 | 1 | 43.46% | 90.87% | 78.30% | 100 | 137 | 0 | 0 | 57.81% | 71.99% | 68.23% | 237 |
| | | | annular (3) | 75 | 11 | 77 | 102 | 29.06% | 85.06% | 68.46% | 9 | 41 | 48 | 167 | 18.11% | 92.37% | 70.36% | 265 |
| | | | only liquid (4) | 49 | 2 | 70 | 85 | 41.26% | 81.10% | 71.92% | 0 | 9 | 33 | 164 | 79.61% | 71.66% | 73.49% | 206 |
| | $d_z$=63mm | 748 | slug (1) | 131 | 14 | 73 | 0 | 60.09% | 93.21% | 83.56% | 97 | 52 | 35 | 34 | 44.50% | 71.70% | 63.77% | 218 |
| | | | bubble (2) | 3 | 75 | 7 | 0 | 88.24% | 96.08% | 95.19% | 2 | 83 | 0 | 0 | 97.65% | 90.50% | 91.31% | 85 |
| | | | annular (3) | 27 | 12 | 100 | 5 | 69.44% | 60.43% | 62.17% | 82 | 11 | 31 | 20 | 21.53% | 94.04% | 80.08% | 144 |
| | | | only liquid (4) | 6 | 0 | 159 | 136 | 45.18% | 98.88% | 77.27% | 66 | 0 | 1 | 234 | 77.74% | 87.92% | 83.82% | 301 |
| | $d_z$=40mm | 240 | slug (1) | 1 | 68 | 2 | 1 | 1.39% | 97.02% | 68.33% | 5 | 43 | 24 | 0 | 6.94% | 91.67% | 66.25% | 72 |
| | | | bubble (2) | 5 | 58 | 15 | 3 | 71.60% | 6.92% | 28.75% | 10 | 8 | 63 | 0 | 9.88% | 42.14% | 31.25% | 81 |
| | | | annular (3) | 0 | 44 | 7 | 0 | 13.73% | 91.01% | 74.58% | 0 | 49 | 2 | 0 | 3.92% | 37.04% | 30.00% | 51 |
| | | | only liquid (4) | 0 | 36 | 0 | 0 | 0.00% | 98.04% | 83.33% | 4 | 0 | 32 | 0 | 0.00% | 100.00% | 85.00% | 36 |

Table 11.4 presents the example classification results of the test images obtained in the experiments and classified by the experts using observation. The images and flows used for test purposes were not used in the learning stage due to the fact that every flow type has its own regime and can simultaneously generate similar images, i.e. a bubbly flow pattern. Therefore, the images of flow structures that were tested, were previously excluded from the learning stage. The identification experiment demonstrated that the results obtained were similar to the classification of the training data. Therefore, the SVM classifier proved to be a correctly chosen instrument for the identification of flows in both types of

*R. Banasiak et al.*

sensors. In the case of horizontal flows, the SVM classifier shows a higher overall accuracy than the FCM. In the case of vertical flows, the FCM is slightly better, which can be explained by the effect of the glycol "aeration" due to a high viscosity of the medium used. It was observed that long-term flows create an effect of imprisonment of gas in a liquid in the form of suspension. This effect also interfered with the process of calibration and normalization of measurements due to the always-existing residual gas suspended in the liquid. Figure 11.10 shows several samples of the input reconstructed volumetric images of the flow patterns analyzed by the chemical engineering expert during the preparation of data for the learning and testing stages.

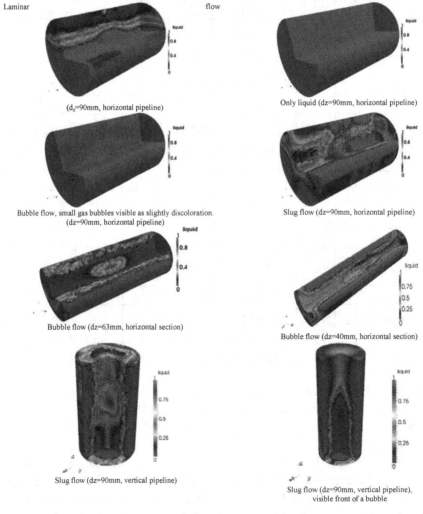

Laminar                                                flow

(d_z=90mm, horizontal pipeline)                        Only liquid (dz=90mm, horizontal pipeline)

Bubble flow, small gas bubbles visible as slightly discoloration.    Slug flow (dz=90mm, horizontal pipeline)
(dz=90mm, horizontal pipeline)

Bubble flow (dz=63mm, horizontal section)              Bubble flow (dz=40mm, horizontal section)

Slug flow (dz=90mm, vertical pipeline)                 Slug flow (dz=90mm, vertical pipeline),
                                                       visible front of a bubble

Fig. 11.10  Samples from reconstructed flow images used by the expert to create learning and testing data sets.

## 11.10 Summary

The present study has demonstrated that the identification of the two-phase vertical and horizontal flow patterns is possible using information provided by the spatial analysis of a large set of 3-dimensional images generated by a noninvasive 3D electrical capacitance tomography system. The flow pattern identification can be successfully performed by using fuzzy logic inference with a very high accuracy of classification (in this study the results varied mainly from 85% to 99% in the horizontal case or 65% and above in the vertical one). The classification process of the two-phase flow which is investigated by means of a 3D ECT system can provide valuable information for the construction of more accurate two-phase flow maps as well as reliable industrial process control and diagnosis. Owing to the efficiency of the proposed flow evaluation technique and fast 3D image reconstruction using a CUDA approach, this process could be performed in an on-line mode if the 3D ECT hardware were capable of to providing sufficient input measurement data with a high frame-rate.

## Acknowledgements

This work is supported by the Polish Committee of Science and Research (Project number 4664/B/T02/2010/38). The work is also partially co-funded by The Rector of Lodz University of Technology of (TUL) in Poland and The Dean of The Faculty of Electrical, Electronic, Computer and Control Engineering at the TUL, within the framework of the young researchers support program.

## References

Alsharhan S. , Karray F. , Gueaieb W. , Basir O., Fuzzy entropy: a brief survey Fuzzy Systems. in the Proceedings of the 10th IEEE International Conference on Fuzzy Systems (FuZZ-IEEE 01), pp. 1135-1139, December 2001, Melbourne, Australia

Banasiak R., Wajman R., Sankowski D., Soleimani M., Three-dimensional nonlinear inversion of electrical capacitance tomography data using a complete sensor model, Progress In Electromagnetics Research PIER, (2010), 100, pp. 219-234;

Chabra R.P., Richardson J.F., Encyclopedia of Fluid Mechanics, Vol. 3, Section II, p. 563, Gulf Publishing Company, Houston (1986)

Cortes C. , Vapnik V., Support-Vector Networks Mach. Learn., Kluwer Academic Publishers, (1995), 20, 273-297;

Dubois, D. J. &Prade, H. Fuzzy Sets and Systems: Theory and Applications, Academic Press, New York, (1980)

Dunn, J., A Fuzzy Relative of the ISODATA Process and its Use in Detecting Compact, Well Separated Clusters. J. Cyber, (1974), 3, 32-57

Dziubiński M., Fidos H., Sosno M., The flow pattern map of two-phase non-Newtonian liquid-gas flow in the vertical pipe, Int. J. Multiphase Flow (2004), 29, 132

Evgeniou, T.; Pontil, M. &Elisseeff, A., Leave One Out Error, Stability, and Generalization of Voting Combinations of Classifiers Mach. Learn., Kluwer Academic Publishers, (2004), 55, 71-97

Fei W., Marashdeh Q., Fan L.S., Warsito W., Electrical Capacitance Volume Tomography: Design and Applications, Sensors (2010), 10, 1890-1917

Guyon I., Weston J., Barnhill S., Vapnik V., Gene Selection for Cancer Classification using Support Vector Machines Mach. Learn., Kluwer Academic Publishers, (2002), 46, 389-422

Hewitt, G.F., and Roberts, D.N., Studies of Two-Phase Flow Patterns by Simultaneous X-ray and Flash Photography, AERE-M 2159, HMSO, (1969)

Ji H., Huang Z., Wang, B., Li H., Demidenko S., Ottoboni R., P.D.P.V.W.D. (Ed.) Monitoring system of gas-liquid two-phase flow Conference Record - IEEE Instrumentation and Measurement Technology Conference, (2004), 3, 2298-2301

Li H., Zhou Z., Hu C., Measurement and evaluation of two-phase flow parameters IEEE Transactions on Instrumentation and Measurement, (1992), 41, 298-303

Li Y, Yang WQ, Image reconstruction by nonlinear Landweber iteration for complicated distributions, Meas. Sci. Technol., (2008), 19 094014

Loebker, D.W.,Empie H.L. "Effervescent spraying: a new approach to spraying high solids black liquor"; TAPPI Engineering Conference Miami Beach, FL September 13-17, 1998

Marashdeh Q., Wang F, Fan L.S., Warsito, W., Velocity Measurement of Multi-Phase flows Based on Electrical Capacitance Volume Tomography, Sensors, (2007) IEEE, pp.1017-1019

Nicklin, D. J., Wilkes, J. O., Davidson, J. F. Two phase flow in vertical tubes, Trans. Inst. Chem. Engs., (1962), 40, 61-68

Olszewski T., Brzeski P., Mirkowski J., Plaskowski A., Smolik W., Szabatin R., Modular Capacitance Tomograph, proc 4th International Symposium on Process Tomography in Warsaw, (2006)

Oshinowo T., Charles M.E., Vertical two-phase flow. Part I. Flow pattern correlations, The Canadian. J. Chem. Engng, (1974), 52, 25-35,

Pląskowski A., Beck M.S., Thorn R. and Dyakowski T., Imaging Industrial Flows – Applications of Electrical Process Tomography, Institute of Physics Publishing, (1995)

Rifkin, R. &Klautau, A., In Defense of One-Vs-All Classification J. Mach. Learn. Res., JMLR.org, (2004), 5, 101-141

Shi L., Fuzzy recognition for gas-liquid two-phase flow pattern based on image processing, IEEE International Conference on Control and Automation, ICCA, (2008), 1424-1427

Soleimani M., Wang H., Li Y, Yang W., A Comparative Study Of Three Dimensional Electrical Capacitance Tomography, International Journal For Information Systems Sciences, (2007), Vol.3, No.2.

Soleimani M., Mitchell C. N., Banasiak R., Wajman R., Adler A. (2009) Four-dimensional electrical capacitance tomography imaging using experimental data; Progress In Electromagnetics Research PIER, 90, str. 171-186

Taitel, Y., Barnea, D., Dukler A. E., Modelling flow pattern transitions for steady upward gas-liquid flow in vertical tubes, AIChE J, (1980), 26, 345-354

Wajman R., Niedostatkiewicz M., Banasiak R., Grudzień K., Chaniecki Z., Romanowski A., Sankowski D., 3D Visualization of the Bulk Solid Flow in Slender Silo using ECT Method; Proc. 3nd International Workshop on Process Tomography, (2009), Tokyo, Japan

Wang, H.X. and Zhang L. F., Identification of two-phase flow regimes based on support vector machine and electrical capacitance tomography, Meas. Sci. Technol. 20 (2009) 114007

Warsito W., Marashdeh Q., Fan L-S. (2007) Electrical Capacitance Volume Tomography, IEEE Sensors Journal, Vol. 7, No. 4, pp. 525-535

Xie CG., Huang SM., Hoyle BS., Thorn R., Lenn C., Snowden D., Beck MS., Electrical capacitance tomography for flow imaging: system model for development of image reconstruction algorithms and design of primary sensors; IEE Proceedings - G, Vol. 139, No. 1, February 1992, pp. 89-98

Xie D., Ji H., Huang Z., Li H., An online flow pattern identification system for gas-oil two-phase flow using electrical capacitance tomography, Instrumentation and Measurement Technology Conference, 2004. IMTC 04. Proceedings of the 21st IEEE , vol.3, no., pp. 2320- 2325 Vol.3, 18-20 May 2004

Yang W.Q., Peng L.H. (2003) Image reconstruction algorithms for electrical capacitance tomography; Meas. Sci. Technol. 14 R1-R13

Zaman, A.; Fricke, A. "Newtonian viscosity of high solids kraft black liquors: effects of temperature and solids concentrations" Ind. Eng. Chem. Res. (1994), 33,428-435

Zhou Y., Chen F., Sun B., Identification Method of Gas-Liquid Two-phase Flow Regime Based on Image Multi-feature Fusion and Support Vector Machine Chinese Journal of Chemical Engineering, (2008), 16, 832-840

# CHAPTER 12

## TOMOGRAPHIC VISUALIZATION OF DYNAMIC INDUSTRIAL SOLID TRANSPORTING AND STORAGE SYSTEMS

Zbigniew Chaniecki, Krzysztof Grudzień and Andrzej Romanowski

*Institute of Applied Computer Science*
*Lodz University of Technology*
*90-924 Łódź, ul. Stefanowskiego 18/22*
*{z.chaniecki, k.grudzien, a.romanowski}@kis.p.lodz.pl*

### 12.1 Introduction

Tomographic visualization is a collection of means for representing diverse information derived from tomography measurements in the form of images and diagrams. Visualization for industrial purposes aims to translate the acquired measured signals values into graphical messages that are more meaningful to field experts. Industrial tomographic visualization can be considered from two aspects. Images can be the representation of the actual state of the process being scanned with tomography equipment as well as can constitute the representation of abstract process-related data and parameters. In most visualization methods it is regarded as tomographic imaging, e.g. representing the internal structures of the scanned process in the form of a reconstructed image. On the other hand, it can be treated as information visualization, e.g. visualizing important values or parameters associated with the quantities measured by means of tomographic devices. The first sense may further be considered as 2D images or 3D models of the process reactors or pipelines measured, revealing both spatial relations of the internal structures of a process and its individual fractions. The second connotation means not only single images but also time sequences of images, raw measurement data visualization or topogram images built out of sections of reconstructed cross-section images taken out of a time series. Each of the approaches provides significant process-related information. However, the choice of proper visualization methods depends on the purposes of a given study. First, the aim of measurements should be taken into account. In-process measurement is often carried out for one or more of the following purposes:

- Gaining a fundamental understanding of the process phenomena (in space and time);
- Devising (or validating) a new (or existing) model for the process phenomena;
- Inspecting the performance of the state of the process (e.g. in/out of specification) for fault detection or advice;
- Quantitative measurement of the process (e.g. component flow rate or flux);
- On-line control of a process using the feedback from the information derived.

In each case the design of measurement and data visualization methods needs to be matched to the intended purpose, since the elements of sensor design and location, measurement protocol and data analysis are rarely three independent stages. These stages are inherently linked to the objective of the measurement and the explicit nature of the process (geometry, flow regimes, process dynamics). As the number of combinations of process types, measurement purposes and tomography modalities result in multiple ways for visualization, this chapter will focus on the electrical tomography application to granular solids flow, i.e. pneumatic conveying transportation and silo storage systems.

## 12.2 Granular flow

Numerous industrial processes consist of successive phases of conveying, storage and discharging bulk solids. About 60% of all industrial products are produced as powders or grains (Seville et al., 1997). Many of them are stored in containers, a majority in silos, and are transported between points of reception in an industrial installation by means of a pneumatic conveying system. The storage of the material and the transport system are crucial elements of the production process (Pląskowski et al., 1995; Sideman and Hijikata, 1993; Klinzing et al., 1997; Scott and McCann, 2005).

The development of production technologies demands increasingly accurate information about the processes conducted. Standard methods for measuring the physical quantities of material are no longer sufficient. Other existing non-standard measurement techniques such as ultrasounds, stereo-photogrametric, termovision, Particle Image Velocimetry and nucleonic, are point-wise in their nature, provide the data limited to the vicinity of the pipe wall or the material free surface, are invasive or introduce an unacceptable level of radiation (Chaniecki,

2007). On the other hand, process tomographic techniques as non-invasive methods both to the monitored process as well as to the process reactor are to address these issues even in real time.

The efficacy of electrical capacitance process tomography in the diagnostics and control of bulk solid flow industrial processes is based on its capability of measuring a granular material concentration change during the flow. The material concentration as well as the flow velocity are crucial parameters for the proper monitoring of the industrial process described. The knowledge about the values of change in the concentration level and velocity during the flow allow one to track the state of the process. Concentration, for bulk solids, can be determined as the ratio of volume fraction $V_{mixture}$ occupied by material $V_{granular\ material}$:

$$k = \frac{V_{granular\ material}}{V_{mixture}} \tag{12.1}$$

Bulk solid relative concentration defined as a relation of the bulk solid amount to the total mixture volume can be mapped to <0, 1> range. Material density can be defined by the formula:

$$\rho = (1 - k)\rho_g \tag{12.2}$$

where $\rho_g$ is the real (physical) value of the granular material density. The phase velocity is the mean velocity of all granular elements in the direction, usually, parallel to the pipeline wall (Mills, 2004; Jones et al., 1999; Klinzing et al., 1997).

The knowledge of these two main flow parameters, concentration and velocity, is analyzed in most flow monitoring systems. Additionally, a very important aspect of flow monitoring is the understanding of flow phenomena, taking into account changes in the concentration and the velocity profile. Such a priori knowledge is very useful in the control module.

### 12.2.1 Pneumatic conveying of granular materials

The pneumatic conveying of solid materials is a well-known method of transportation systems in some industries. It is common to find such installations in the food (large scale wheat containers loading), chemical, construction and other type of branches. Pneumatic conveying usually aims at transporting bulk solid or powder using pressurized air along pipeline installations. Figure 12.1 shows a scheme of research facilities and a photograph of the measurements section in The Tom Dyakowski Process Tomography Laboratory (Institute of Applied Computer Science, Lodz University of Technology, Poland).

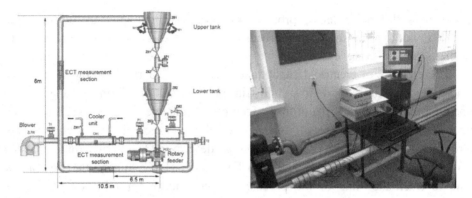

Fig. 12.1. Pneumatic flow rig at the Process Tomography Laboratory in LUT: Photography of ECT, a) scheme of installation, b) photography of ECT measurements section.

This type of conveying is cheaper and easier than others, unlike transportation by means of belt conveyors (complex and expensive construction, high cost of power supply, noise, possibility of external conditions affecting the transported material, dropping of the material from the conveyor, high cost of maintenance) (Dyakowski 1996, McKee et al. 1996, Plaskowski et al. 1995). However, pneumatic systems are not perfect. In most cases, the airflow velocity controls the material flow. The relationship between the material density and the air velocity causes a number of flow patterns to occur. Figure 12.2 presents an example of the flow pattern characteristic, dependent on the air velocity and the conveying material mass flow rate.

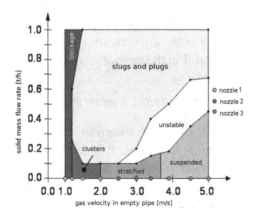

Fig. 12.2. Flow regimes pattern for pneumatic conveying horizontal section with fixed gas flow rate and mass flow rate based on flow patterns recorded on UMIST flow rig (Jaworski and Dyakowski et al., 2002).

The flow pattern describes behavior of flowing material during conveying in an installation and the changes of material distribution in spatial and temporal domain. Insufficient air velocity causes stationary slugs to form in the horizontal sections of pipelines, whereas in vertical sections material cannot overcome the gravity force and falls downward causing hold-ups and pipeline blockages. Unblocking may require demounting of the pipe sections and cause many hours' production breaks, resulting in substantial economic losses. Too high airflow velocity may, in turn, cause material degradation and escalate disadvantageous electrostatic phenomena (Neuffer et al., 1999; Bridle et al., 1995). Moreover, the shortage of air pressure, or velocity, with respect to the overload of material introduced into the pipeline may result in pipeline blockage. The careful adjustment of transport medium parameters can be performed on the basis of parameters measurement records such as an amount/weight of the material allowed into the installation and collected at the end of the pipeline segment in time unit with respect to such parameters as air pressure and bulk rotary valve operation speed. Visual inspection could also serve as the ultimate criterion for determining the material flow regime. However, this inspection is not possible at all times, since in most cases pipelines are opaque.

The monitoring of the gas-solids flow phenomena can be beneficial for the design and operation of pneumatic transport installations. Examples include the design of transportation systems with an optimized flow regime when solid material is transported at a required velocity with minimized energy consumption. The monitoring and control strategies that could be derived on the basis of the knowledge about the behavior of the multiphase flow were reported (Jaworski and Dyakowski 2002, Arko et al. 1999, Deloughry et al. 1999). This can prevent the incidence of unwanted events (blockage monitoring and prevention) and hence the improvement in the overall performance of pneumatic installations in terms of efficiency, minimizing energy consumption and avoiding transported component degradation. In addition to the flow regime identification itself and the detection of the component fraction, it is also necessary to determine the distribution of the distinct phase fraction within the flow composition. This is particularly important when the conveying installation is part of a bigger system and its contribution to this system performance must be considered as the continuous feeding of a certain amount of a particular medium with the smallest possible variation.

## 12.2.2  Silo flow of granular materials

Silos or hoppers are receptacles for protecting, storing and loading solid materials into process machinery. Detailed theoretical studies on the monitoring and control of particulates or a powder flow were conducted by Seville (1997), Bridgwater and Scott (1983). Materials stored in hoppers are mostly particulate materials or powders and receptacles may have one of a variety of bin shapes and sizes. There may occur different types of granular flow in the silo and a given type depends on the material and the silo geometrical parameters (see Fig. 12.3). The silo flow discharging process starts after opening the silo outlet, located at the silo bottom, automatically or forced mechanically, which depends on the material cohesion.

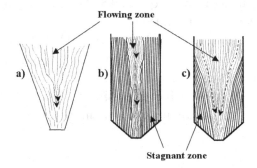

Fig. 12.3. Principal discharge hopper powder flow (a- mass flow, b – funnel flow, c – semi-mass flow) – (Bates, 1999).

Figure 12.4 presents the silo discharging process, taking place in a rectangular silo model and the scheme of the flow pattern changes from the funnel to the mass flow. The mass flow is characterized by all the material being discharged uniformly at the same downward velocity across the entire cross section area, hence at a constant mass discharge rate. In contrast to the 'funnel' or the 'core' flow, as the name suggests, the material is flowing only in the center of the hopper giving rise to stagnant zones of material at the walls of the container.

The operation of silos for storing friable materials is a complex problem due to safety issues. This is caused by a possibility of uneven material distribution, resulting in the unsymmetrical stress of the walls and bottom both during loading and discharging (Horabik et al. 1992, Munch-Andersen and Nielsen 1990, Zhong et al. 2001). The most recurrent phenomena causing problems with the use of silos are: shear zone localizations, increasing pressure in the points of the silo geometry transformation (when changing from the bin to the cone part of the container), and dynamic effects (Niedostatkiewicz and Tejchman, 2003). Dynamic effects are practically inherent during the discharging process. For cohesionless materials, it appears as pulsation, and for cohesion materials as pulsations and strokes (Tejchman 1987, Nielsen and Ruckenbrod 1988, Tejchman 1998, Wensrich 2002, Muite et al. 2004, Dhoriyani et al. 2006, Buick et al. 2005, Grudzien et al. 2010).

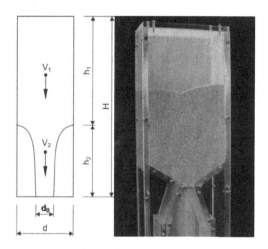

Fig. 12.4. Diagram of bulk solid flow phases during silo unloading (d – silo diameter, H – silo height, $d_0$ – outlet diameter, $h_1$ – mass flow, $h_2$ – core flow, $v_1$ i $v_2$ – solid material velocity respectively in mass and core flow phases) – (Niedostatkiewicz, 2003).

Strong dynamic effects are the result of the dynamic interaction between the vibrating material while unloading it from the silo and the silo structure (Wilde et al. 2007, Niedostatkiewicz and Tejchman 2008). The more slender the silo is the greater the probability of an occurrence of strong dynamic effects. Resonance vibration raises the pressure on the walls and the bottom, which in turn may cause malfunction or even a construction catastrophe. One of the most effective means of dynamic effect reduction is increasing the silo wall roughness. On the industrial scale it consists in mounting special corrugated sheets on the inner side of the silo walls (Tejchman, 1998). Increasing the silo wall roughness results in the shear at the walls, hence causing additional transversal strain waves, which raise the frequency of vibrations of the flowing particles. This in turn shifts the frequency from the construction vibrations, and results in omitting the resonance (Niedostatkiewicz and Tejchman, 2003). On the other hand, increasing the wall roughness does not change the bulk solid flow nature – in the upper part of the container the mass flow regime is maintained. On the experimental basis, several sources of extensive dynamic effects in silos may cause a number of unwanted phenomena (Buick et al., 2005; Wensrich, 2002; Baxter and Behringer, 1990; Moriyama, 1988):

- Slip-stick behaviour between the stored solids and the silo walls,
- Grain collisions and the frozen disorder of the bulk solid,
- Insufficient flow ability of the silo fill,
- Propagating longitudinal stress waves due to a resonant interaction between the granular material and the silo structure, which were induced at the outlet,
- Alternating flow patterns during the flow,
- Non-linear change in the wall friction with flow velocity,
- Acceleration and deceleration of the granular material during the transition between the bin and the hopper,
- Internal slip-stick,
- Solid dilation during the flow.

## 12.3 Imaging — process visualization

As the word tomography suggests, it is "writing the slice" of the inside of the object even if it is a nontransparent object. Hence, it is usually identified with visualization. Over the recent years there has been a development of increasingly precise image reconstruction algorithms. Therefore, reconstruction methods aiming to produce 2D and 3D images of the processes examined have widely been expanded. The review of image reconstruction algorithms for soft-field tomography was presented in chapter 1. Visualization is usually the first step in the majority of industrial process tomography applications; yet in some cases it can be the final step as well (York et al., 2003; Mazurkiewicz et al., 2004; Mann et al., 1997). The 2D/3D imaging is the primary method for material distribution visualization in non-transparent systems. The next stage on the way of tomography data analysis usually is the application of chosen post-processing algorithms. Often the information retrieved from tomography images gives knowledge about the process states described by characteristic parameters of the process in real time.

The reconstructed image gives the information about material distribution inside the sensor. The pixel values represent the dielectric permittivity estimated during the reconstruction process. The value of capacitance between each pair of ECT sensor electrodes is changing if the permittivity inside the sensitive area of the measurement electrode pair is changing. During the granular flow, the capacitances increase or decrease depending on the material packing density changes. Such a situation is the cause of the existing relationship between the dielectric permittivity value and the packing density of the material. This chapter is not devoted to the discussion on the relationships between the permittivity value and the material packing density. Depending on the model used, which describes this relationship, a change in the level of the measurement caused by changes in packing density can be different. However, as was shown in (Grudzien et al., 2009), the application of a different models causes only a change in the range of the material packing density, i.e. it is a question of relative values. In the context of tomography data visualization swapping the model of the capacitance/permittivity relationship is not crucial and implies only adjustment of the scale of data visualization. The methods of flow visualization based on tomography data presented in this chapter can be applied for all known models (Yang and Byars, 1999; Xie and Huang, 1992; Zhiheng, 2011). It is the user who decides which model is the best for monitoring the industrial process. The selected results presented in this chapter show that the dielectric permittivity distribution can be directly related to the material concentration distribution inside the measurement ECT sensor. The local solids concentration is associated with the measured permittivity of the mixed solids with air. The denserthe packing, the greater the value of permittivity in a given region. Therefore, a change in the packing density during the flow conveys basic information describing the nature of the process. The tomography system works on the basis of the normalized measurements in range of <0,1>, which implies that the estimated dielectric permittivity and hence the packing density is considered in the same normalized scope of values (Williams and Beck, 1995; Pląskowski et al., 1995).

## 12.4 Process visualization for diagnosis purposes

Depending on the aim of data visualization, specifics of investigated process as well as the absolute level of the process dynamics, appropriate data processing and analysis methods have to be chosen. In terms of the diagnostic system for dynamic granular flows data visualization algorithms are often applied. They give different data presentation forms from single 2D tomograms (Fig. 12.5a) to sequences of those to single and sequences of 3D tomograms to topograms (Fig. 12.5d). Whereas classical tomograms are 2D visualization of the sensor cross-section; topograms are formed as a transverse-section sequence of tomography images of the process. Another approach is to show material concentration changes over time for the chosen localization, a pixel or a group of pixels, in the cross-section of the sensor (see Fig. 12.5d).

The main problem in process tomography data visualization is how large sets of images can be valuably presented, especially when it is necessary to conduct comparative analysis of a large number of frames. This aspect is related to high temporal resolution of process tomography measurements. In contrast to the application of tomography in medicine, the temporal changes in the monitoring process force to acquire a large number of measurements per second. As a result of this requirement, sequences of tomograms are often obtained regardless the particular dynamics of the given process that result in maladjustment of measurement frequency and parameters. Then conclusions are drawn based on such a set, which unfortunately often results in the loss of important phenomena and the true nature of the analyzed process. Because of this, assuming that the results should be presented clearly for the user, it is proposed that data be visualized in a continuous manner in the form of characteristics. The tracking of changes of the crucial parameter value in the determined characteristic allows one to monitor the process for the entire duration of the measurement.

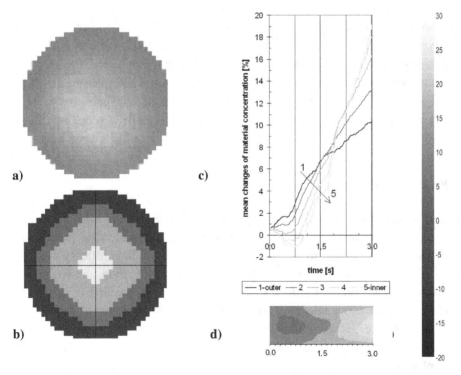

Fig. 12.5. Different type of tomography data visualization, a) tomogram – horizontal cross section in 3.0 s; b) shape of the concentric areas on a sample tomogram for appointing the characteristic (c), c) diagrams of mean changes in material concentration, d) topogram – longitudinal sequence of the flow, e) scale of relative changes in material concentration for (a) and (d).

## 12.4.1 Raw data visualizations

The basic, and the most easily accessible, information about the process state is available in the form of raw (unprocessed) measurement data. In general, on the basis of the analysis of the raw measurement data, it is possible to determine the feasibility of the process state visualization in successive time points and hence, the feasibility of the process dynamics studies. In the case of the gravitational and pneumatic granular flows, changes in the measured capacitance in time correspond to the dynamic material concentration changes. Fig. 12.6 shows two plots of raw data changes during the unloading of a silo, in the case of the funnel flow, for adjacent and opposite pairs of electrodes (Fig. 12.6a – adjacent and Fig. 12.6b – opposite electrodes). Such plots are the simplest form of raw data graphical visualization. With the application of such a visualization method, it is possible to distinguish 4 main characteristic process stages (in relation to the ECT sensor position) of silo unloading. These stages are indicated as follows: the phase of funnel propagation directly in the sensing zone (A), funnel stabilization (B), the upper material surface reaching the sensor zone (C) and the material finally leaving the sensor zone (D). The corresponding schematic diagram of the silo discharging process is presented in Fig. 12.8, while a detailed description can be found in Romanowski (2006). The visible difference in the shape between the plots of adjacent (Fig. 12.6a) and opposite (Fig. 12.6b) electrodes is significant from the point of view of material behavior. The analysis of those graphs allows one to study the material packing density in the central cross-section region and near-wall regions. The presented results concern the gravitational flow of rice grains of an ellipsoidal shape of ~5mm long and 3mm wide (Fig. 12.7a). The measurement data are obtained from a sensor located 150 mm above the silo outlet, so the capacitance measurements between the adjacent electrodes cover an area ofso called stagnant zone (see Figs 12.7b and Fig. 12.8). The increasing packing density at the walls results in the rise of the measured inter-electrode capacitance for the adjacent electrodes as shown in segments (A, B, C) in (Fig. 12.5a). The material inside this area, after reaching the maximum concentration in the stagnant zone (see Fig. 12.8), does not move during the flow and the granular concentration is almost constant. In the case of the opposite electrodes, changes in the measured capacitance are associated with the funnel area appearing in the central part of the receptacle cross section area and these correspond to the decreased material concentration in this part, where the material moves during the silo discharging process.

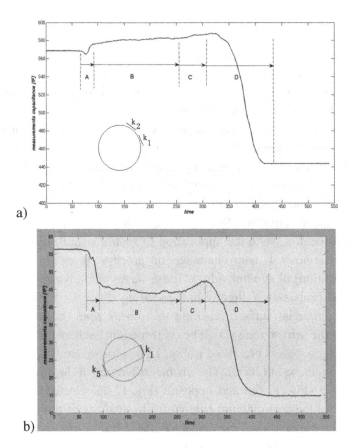

Fig. 12.6. Measurements capacitance changes for adjacent (a) and opposite (b) electrodes, for sensor presented in Fig. 12.7.

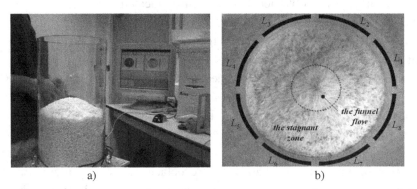

Fig. 12.7. Photos of the hopper – ECT experimental setup (University of Leeds, Institute of Particle Science and Engineering). Photos ofthe hopper filled with rice grains: a) side view of the hopper and the computer system in the background; b) top view of the container with schematically depicted sensor electrodes around the silo (8 electrodes sensor).

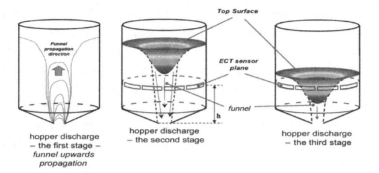

Fig. 12.8. Schematically shown hopper discharge stages (funnel propagation, funnel flow, top surface passing through the measurement space).

Such visualization can be prepared based on the normalized value of measurements. The results are in range <0, 1>. Percentage changes of the material concentration relative to the initial packing density, before opening the silo outlet, can be determined on the basis of these values. Then this information can be used to diagnose the unit of flow in control systems in order to track the level of material concentration changes at the silo wall for the mass flow. An example of such characteristic, but for the normalized capacitance data, is shown in Fig. 12.9. The interpretation of concentration changes presented in Fig. 12.9 allows measuring the material vibrations during container unloading in the upper part of the silo model used. The results present the gravitational flow in the smooth wall silo and initial dense packing of material fill-up. Material fill-up achieved with the aid of sieve forced the initial dense packing of sand particles (before the discharging process). Such initial conditions cause pulsations to occur during the flow itself– this phenomenon is visible in the form of rapid changes in measurement values.

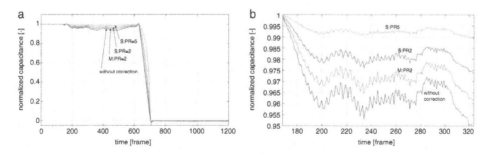

Fig. 12.9. Concentration changes (taken from the measurement between the adjacent electrodes) at the silo wall during the silo discharging for initially dense packing with the smooth silo wall and height of sensor position $h = 0.75$ m. (a) Plot of a complete discharging process, (b) magnification of the fragment associated with sand pulsations.

To sum up, depending on the silo construction and the position of the ECT sensor, changes in the packing density in the stagnant zone and the funnel area (for the funnel flow) can be determined based on this kind of tomography measurement visualization (the raw and normalized). As far as the mass flow is concerned, changes in measurements during the silo discharging process provide information about the level of interaction between the silo construction and the material structure in the form of the oscillation and vibration of the flowing material. For the second type of gravitational flow, the mass flow, such information is important especially for a higher position of the silo construction. Additionally, information to be extracted from the raw data is the size of the stagnant zone and the shape of the funnel area in order to diagnose the abnormal behavior of the granular flow (see Chapter 13).

The main advantage of raw data visualization and processing is a possibility of extracting the knowledge about the process state in real time, which can be exploited on the control field of industrial processes. However, the basic type of information conveyed by raw data, as presented in this section, is not always sufficient for process diagnosis. The quality of information about the process states, obtained from the raw data, depends on the kind of process that is being visualized. In the case of pneumatic conveying raw data give more significant knowledge than for the silo flow, where the interpretation of values recorded between the opposite electrodes is difficult and can be confusing.

In the pneumatic conveying flow it is possible to visualize the material concentration changes in the form of continuous characteristic of the raw data changes in time. A sequence of the raw measurement data can be visualized in the form of a two dimensional image. The visualization results of the measurement data sequence, produced for a chosen range of the measurement vector, give a two-dimensional image of the longitudinal-section of the flow (see Fig. 12.10). In Fig. 12.11 an example of a frame sequence of the normalized raw data of pneumatic conveying in the vertical part of the installation is presented. On the basis of characteristic diagrams (see Fig. 12.11) one can say that the flow is not continuous –it has a character of plug transport. There can be long gaps or small regions of the transported substance between the consecutive plugs (even as many as 500 frame gaps can occur). The brightness of each point is directly proportional to the dielectric permittivity. The analysis of the whole sequence derived shows the necessity of the development of an automatic method for the evaluation of the flow time parameters for the computational subsystem. Such visualization allows monitoring the frequency of plug sand their length, and thus helps preventing a blockage of the pipeline.

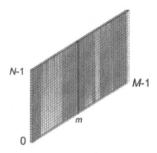

Fig. 12.10. Sequence of $M$ measurement vectors forming a two-dimensional image (presented in the gray scale) of the longitudinal section of the flow; m data vector is marked, $m = 0. . .M-1$.

Fig. 12.11. Raw data sequence for: a) 1000 frames of raw data, b) 250 frames magnified four times.

In contrast to the silo flow, raw data visualization in the case of pneumatic conveying seems to be more fruitful in diagnose information. For the silo flow such visualization is only preliminary knowledge about the process state. In order to enhance information about the process, the visualization of the spatial distribution of material during the process should be conducted. The reconstruction process solves the inverse problem and as a solution it delivers an image of the material distribution inside the measurement space. The application of image reconstruction practice to the silo flow is presented in the next section.

### 12.4.2 2D imaging of the granular flow

As described in Chapter 1, a reconstructed image is built with the aid of a limited number of measurements. When a sensor used for measurements consists of, for example, 8 electrodes, the number of measurements is equal to 28 for each time point. The tomogram usually consists of 1024 pixels (32x32) with 812 corresponding to the internal sensor space for the circular cross-section of the installation. The choice of a particular image reconstruction method is a trade-off between the amount of time devoted to the reconstruction procedure and the anticipated image quality. The choice of reconstruction methods, in the context of a diagnosis system, must be related to the time that is sufficient to detect anomalies in the monitored process, and hence, the dynamics of the process. It is obvious that the use of a computer system with higher performance computing allows applying more sophisticated and thus more computationally demanding methods for inverse problem solving.

In the case of silo-flow visualization, example images are produced with the aid of iterative methods - SIRT (Su et al., 1999). Fig. 12.12 reveals a single frame with the material concentration distribution divided into 3 main regions. Three distinct regions differing in material concentration can be distinguished: the funnel flow region in the center, by-the-wall region, and the transient area between these two. A detailed decomposition model for a tomographic hopper flow image is presented in Romanowski (2006).

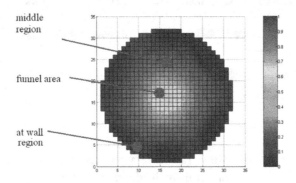

Fig. 12.12. Tomographic image reconstructed for hopper flow with characteristic funnel flow areas indicated.

As one can see, the obtained image is of moderate quality. Nevertheless, such image sequences can demonstrate the dynamic changes of the industrial process state. For process studies it is appropriate to consider more than one frame at a time. Fig. 12.13 presents an example of a sequence of reconstructed images placed on a silo flow characteristic plot. The arrows indicate their association with the corresponding time points during hopper unloading.

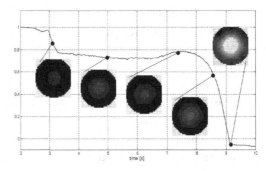

Fig. 12.13. Sequence of reconstructed images for the silo flow with indication of their association with the corresponding points on a silo flow characteristic plot.

Figure 12.14 shows another method for silo flow visualization. This kind of visualization exposes full information about the material distribution inside the sensor. It presents material distribution along a radial line passing through the cross-section of the sensor vs. time on the horizontal axis (namely – topogram as introduced in section #4 of this chapter). The examples show a relative symmetry of material distribution concentration changes during the flow for two different initial packing densities of the material. For the dense flow (Fig. 12.14a) the concentration in the center of a silo is increased in comparison to the initial stage, while for the silo wall area, the material concentration decreased. For the loose flow (Fig. 12.14b) the situation is the opposite.

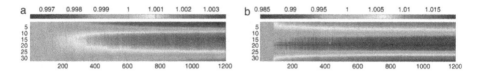

Fig. 12.14. Topogram of silo, a) topogram for sieve fill up, b) topogram for funnel fill up (Grudzien et al., 2010).

This kind of visualization can be used for the silo flow as well as for the pneumatic conveying of solids. In Figure 12.15a set of topograms for pneumatic conveying is presented. From the monitoring point of view it does not matter whether we operate on raw or reconstructed data. The quality of information is the same; hence, the visualization of raw data, without triggering the image reconstruction process, is a better option. Another important advantage of topogram visualization form is the possibility to show all the process data with historical imaging. Based on such information the present and the future state of the process can be concluded.

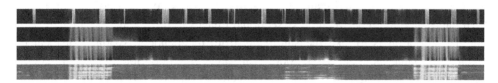

Fig. 12.15. Set of topogram vertical plugs transport material. a) sequence of 8.000 frames, b) and c) 500 magnified fragment shown in two perpendicular planes, d) raw topogram data for b/c) sequence.

An additional noteworthy visualization pattern, in the case of the silo flow, can be obtained by combining a sequence of 2D images in the form of a stack. Such a stack is presented in Fig. 12.16. All the distinct images are collected with the sensor placed at the same height during hopper unloading. The frames presented were taken in distinct time points. This type of image presentation allows better process dynamics visualization from a different point of view, i.e. it is possible to choose the type of stacked image (reconstructed or further processed) and adjusting the parameters on the stacked images.

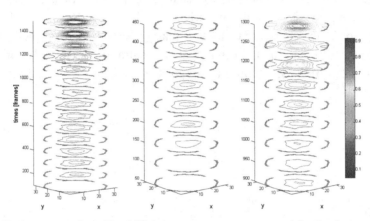

Fig. 12.16. Stack patterns of combined 2D image sequence reconstructed for fixed sensor location and different funnel size. One frame corresponds to 20ms.

### 12.4.3  3D imaging of the silo flow

The last type of visualization described in this chapter is often used in medical applications, i.e. three-dimensional reconstruction. A 3D image obtained with X-ray tomography system gives full information about the investigated object. In the case of process tomography, 3D capacitance imaging has been dynamically developed over the past few years. There are two approaches existing in the research community. First one is to use the capacitance measurements can obtained from electrodes spatially located using ring-electrode sensors (Warsito 2003, 2007) and then interpolate them between cross-sectional images to form a 3D-like images. An alternative option is to take measurements from independent, spatially dislocated electrodes (Banasiak et al., 2004, 2006; Wajman et al., 2006; Soleimani et al., 2006, 2009; Warsito and Fan, 2003) in order to directly reconstruct an image in 3D space.

If the process investigated is relatively slow, it is possible to collect measurements data for the same state of the process. The 3D visualization of pneumatic conveying seems to be unreasonable. The dynamic changes of material distribution inside the sensor sometimes causes the speed of the acquisition system to reach an order of 100 frames per second, which can insufficiently fast to monitor the process using 2D reconstructed images. One must however bear in mind that 3D visualization requires much more measurement data for the reconstruction process (based on inverse problem solving). In the case of the silo flow, a 3D image illustrates the material flow in space, not only in the cross-section of the sensor.

Due to the similarities to 2D image reconstruction methodology, described by (Wajman 2006), 3D capacitance tomography can make use of the same reconstruction methods as classic 2D tomography. The main difference in the reconstruction method between 2D and 3D reconstruction is time and computing resource consumption for data processing. For 3D imaging the number of voxels is usually much greater than that for the corresponding pixels in 2D visualization. Generally, to reduce computation time, GPU computing is applied (Matusiak et al, 2013). However, even with application of GPU computing, reconstruction time is not sufficiently short to use such visualization in an on-line control system, especially for iterative algorithms. Several stages of the funnel propagation during initial hopper discharging phases are presented in 3D in Fig 12.17.

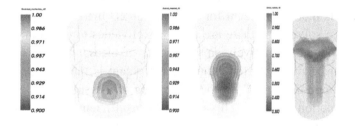

Fig. 12.17. 3D visualization of the funnel propagation phenomenon; a) and b) successive frames of funnel propagation occurring during the initial phase of hopper discharging; c) funnel flow.

Fig. 12.18 shows examples of different 3D images for the ongoing funnel flow occurring during the hopper unloading process.

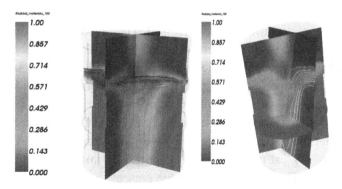

Fig. 12.18. 3D reconstructed image for reveals material concentration distribution inside measurement space; a) x-y and y-z x cross-sectional view; b) image with different x-y-z cross-section angle.

Significant advantage of 3D visualization lies in the possibility of an illustrative perspective of the presentation showing material concentration distribution. The material concentration distribution inside the measurement space can be indicated by iso-surfaces similar to spatial layers, where different packing density surfaces can be represented by hue and shades of colors. This type of information visualization graphically conveys material concentration variations mapped to 3 dimensional spaces. This can further be of use to instantly reveal volumetric dimensions of the phenomena monitored, e.g. size and geometry, which can be especially beneficial for field experts in order to detect abnormal occurrences and flow instabilities. Moreover, dynamic 3D animation is possible. In general, 3D visualization allows comprehensive observation of a 3D shape evolution in successive process phases. It is also possible to observe funnel propagation along the sensor axis. On the other hand, 3D imaging for electrical process tomography is unfortunately associated with some issues. The most significant is that spatial resolution (similarly to 2D images) is limited; however, this limitation is more significant for 3D than for 2D imaging. This is related to inverse problem solving, as there are more voxel values to determine than pixel values in 2D, having a relatively small amount of measurement data. In effect, 3D images can be more blurred. The other problem is the high computational resource consumption demand for the processing of measurement data using 3D image reconstruction methods. This issue with insufficient computational power to deal with a large complexity of the 3D reconstruction process is especially acute when considering on-line performance.

### *12.5 Conclusions*

As the visualization in the field of process tomography serves particular purposes varying by several factors discussed in this chapter, the form of graphical representation may significantly vary as well. The raw data analysis reveals some information about the process performance. It is possible to imagine a simple control mechanism (possibly indicating a process failure) using such a type of information. 2D image reconstruction makes the information wealthier, since it maps more information in space, although presented on a flat surface of the image frame. Distinct images or whole sequences can be analyzed or post-processed in order to determine characteristic process parameter values. The examples presented show a great potential of 3D capacitance imaging for the hopper flow study application. The application of image presentation algorithms such as multi-iso-surfaces and intersections cut along an arbitrary plane and angle allows, better than 2D, imaging of structure materializing during the process. On the other hand, the current stage of development, the quality and spatial resolution of 3D images is still unsatisfactory. Therefore, for monitoring and control purposes the raw data, 2D image based visualizations, and post-processing methods are usually employed – especially when tens or hundreds of frames per second are needed in real time. Wealthier visualization containing spatial features is included in volumetric representations; however, it is time consuming to produce a 3D image, so it justified only for some experimental applications such as feature extraction, the study of phenomena behavior, model development and validation.

## References

Arko, A., Waterfall, R.C., Beck, M.S., Dyakowski, T., Sutcliffe, P. and Byars, M. (1999). Development of Electrical Capacitance Tomography for solids mass flow measurement and control of pneumatic conveying systems, *Proc. 1st World Congress on Industrial Process Tomography*, pp. 140–146.

Banasiak, R., Wajman, R., Sankowski, D. and Soleimani, M. (2010). Three-dimensional nonlinear inversion of electrical capacitance tomography data using a complete sensor model, *Progress in Electromagnetics, Research PIER*, 100, pp. 219-234.

Bates, L. (1999). Problems In Particulate Flow, Part A -- Flow Regimes, *Ajax Equipment Ltd., Educational Resources for Particle Technology*, http://www.erpt.org.

Baxter, G.W. and Behringer, R.P. (1990). Pattern formation and time-dependence in flowingsand, in: Two-phase Flows and Waves, New York, Springer Verlag.

Beck, M.S. and Pląskowski, A. (1987). Cross Correlation Flowmeters: Their Application and Design, Adam Hilger, Bristol.

Bridle, I., Woodhead, S.R., Burnett, A.J., and Barnes, RN. (1995). A Review of Techniques for the Investigation of Particle Degradation in Pneumatic Conveying Systems, *Proc. 5th International Conference on Bulk Materials Storage, Handling and Transportation*, pp. 205-210.

Brzeski, P.J., Mirkowski, T., Olszewski, T., Pląskowski, A., Smolik, W. and Szabatin, R. (2003). Multichannel capacitance tomograph for dynamic process imaging, *Opto-Electronics Review*, 11, pp. 175-180.

Buick, J.M., Chavez-Sagarnaga, J., Zhing, Z., Ooi, J.Y., Pankai, Y. and Cambell, D.M. (2005). Investigation of silo-honking: Slip-stick excitation and wall vibration, *Journal of Engineering Mechanics ASCE,* 131, *pp.* 299-307.

Chaniecki, Z. and Sankowski, D. (2007). Monitoring and diagnosis of dynamics states using process tomography, Diagnostics of Processes and Systems, EXIT, pp. 387-394, (in polish).

Chaniecki, Z., Dyakowski, T., Niedostatkiewicz, M., Sankowski, D. (2006). Application of electrical capacitance tomography for bulk solids flow analysis in silos, *Particle & Particle Systems Characterization*, 23, pp. 306–312.

Deloughry, R., Young, M., Pickup, E. and Barratt, L. (2001). Variable density flow meter for loading road tankers using process tomography, *Proc. Process Imaging for Automatic Control*, 273.

Dhoriyani, M.L., Jonnalagadda K.K., Kandikatla R.K., Rao K.K. (2006). Silo music: Sound emission during the flow of granular materials through tubes, *Powder Technology*, 167, pp. 55–71.

Dyakowski, T., Edwards, R.B., Xie, C.G. and Williams, R.A, (1997). Application of capacitance tomography to gas- solid flows, *Chemical Engineering Science*, 52, pp. 2099-2110.

Dyakowski, T., Wang, S.J., Xie, C.G., Williams, R.A. and Beck M.S. (1993). Real time capacitance imaging of bubble formation at the distributor of a fluidised bed, *Chemical Engineering Journal*, 56, pp.95 -100.

Grudzień, K. and Sankowski, D. (2010). Tomography image post-processing (virtual channel concept), Electrical Capacitance Tomography. Theoretical Basis and Applications, Warszawa, ed. Sankowski D. and Sikora J., pp. 160-175.

Grudzień, K., Romanowski, A., Aykroyd, R.G. and Williams, R.A. (2006). Advanced Statistical Analysis as a Novel Tool to Pneumatic Conveying Monitoring and Control Strategy Development, *Particle & Particle System Characterisation.* 23, pp. 289-296.

Grudzień, K., Romanowski, A., Aykroyd, R.G., Williams, R.A., West, R.M. and Meng S. (2004). Application of the Bayesian Approach to Powder Flow Investigation, *3rd International Symposium on Process Tomography,* Poland, pp. 72-76.

Grudzień, K., Romanowski, A., Chaniecki, Z., Niedostatkiewicz, M. and Sankowski, D. (2010). Description of the silo flow and bulk solid pulsation detection using ECT, *Flow Measurement and Instrumentation,* 21, pp. 198-206.

Huang, X.M., Dyakowski, T., Pląskowski A. and Beck, M.S. (1988). A tomographic flow imaging system based on capacitance measuring techniques, *Proc. IX International Conference on Pattern Recognition,* Italy, pp. 570-572.

Jaworski, A.J. and Dyakowski, T. (2002). Investigations of flow instabilities within the dense pneumatic conveying system, *Powder Technology,* 125, pp. 279-291.

Klinzing, G.E., Marcus, R.D., Rizk, F. and Leung, LS. (2010). Pneumatic conveying of solids - theoretical and practical approach, Springer Science+Business Media B.V.

Li, Y. and Yang, WQ. (2008). Image reconstruction by nonlinear Landweber iteration for complicated distributions, *Meas. Sci. Technol.*, 19, pp. 094014.

Mann, R., Dicking, F.J., Wang, M., Dyakowski, T. and Williams, R.A., Edwards R.B., Forrest A.E., Holden P.J., (1997), Application of electrical resistance tomography to interrogate mixing processes at plant scale, Chem. Eng., Sci., 52, pp. 2087-2097.

Matusiak, B., Romanowski, A., Sankowski, D. and Grudzień K. (2013). Forward problem solution acceleration with graphical processing units (GPU) for electrical tomography imaging, *Proc. 7th World Congress on Industrial Process Tomography*, Poland, pp. 234-243.

Mazurkiewicz, Ł., Banasiak, R., Wajman, R., Dyakowski, T. and Sankowski, D. (2005). Introduction to Capacitance Horikography, *Proc. 4th World Congress on Industrial Process Tomography*, Japan, pp. 147-152.

Michalowski, R.L. (1984). Flow of granular material through a plane hopper, *Powder Technology*, 39, pp. 29–40.

Moriyama R, Jimbo G. Reduction of pulsating wall pressure near the transition point in a bin. Bulk Solid Handling 1988;8:4215.

Mute, B.K., Quinn, F.S., Sundaresan, S. and Rao, K.K. (2004). Silo music and silo quake: granular flow-induced vibration, *Powder Technology*, 145, pp. 190-202.

Neuffer, D., Alvarez, A., Owens, D.H., Ostrowski, K., Luke, S.P. and Williams R.A. (1999). Control of pneumatic conveying using ECT, *Proc 1st World Congress on Industrial Process Tomography*, UK, pp. 71–77.

Niedostatkiewicz M., Tejchman J., Reduction of dynamic effects during granular flow in silos, Powder Handling and Processing, 2008.

Niedostatkiewicz, M., Tejchman J. (2003). Experimental and theoretical studies on resonance dynamic effects during silo flow, *Powder Handling & Processing*, 15, pp. 36–42.

Niedostatkiewicz, M., Tejchman, J., Grudzień., K. and Chaniecki, Z. (2010). Application of ECT to solid concentration measurements during granular flow in a rectangular model silo, *Chemical Engineering Research & Design*, 88, pp. 1037-1048.

Ostrowski, K., Luke, S.P., Bennett, M.A. and Williams, R.A. (1999). Application of capacitance electrical tomography for on-line and off-line of flow pattern in horizontal pipeline of pneumatic conveyor, *Proc. the 1st World Congress on Industrial Process Tomography*, UK, pp.96-99.

Pląskowski, A., Beck, M.S., Thorn, R. and Dyakowski T. (1995). Imaging Industrial Flows, Institute of Physics Publishing, Bristol.

PTL, http://www.tomography.com/pdf/TFL%20R5000%20Flow%20Measu-rement.pdf

Romanowski, A., Chaniecki, Z., Grudzień, K., Matusiak, B., Sankowski, D., Betiuk, J. (2010). Investigation of Pneumatic Conveying Pseudo-Emergency Induced Operation with use Electrical Capacitance Tomography and raw data analysis, *Proc. 6th World Congress on Industrial Process*, China, pp. 196-203.

Romanowski, A., Grudzień, K. and Williams, R.A. (2006). Analysis and interpretation of hopper flow behavior using electrical capacitance tomography, *Particle & Particle System Characterisation*, 23, pp. 297–305.

Romanowski, A., Grudzień, K., Banasiak, R., Williams, R.A. and Sankowski D. (2007). Hopper Flow Measurement Data Visualization, Developments Towards 3D, *Proc. 5th World Congress on Industrial Process*, Norway, pp. 986-993.

Scott, D.M. and McCann H. (2005). Process Imaging for automatic control, Taylor and Francis Group.

Seville, J.P.K., Tuzun, U. and Clift, R. (1997). Processing of Particulate Solids, Blackie Academic, London.

Sideman, S. and Hijikata, K. (1993). Imaging in Transport Processes, Begell House.

Soleimani, M. (2006). Three-dimensional electrical capacitance tomography imaging, Insight, Non-Destructive Testing and Condition Monitoring, 48, pp. 613-617.

Soleimani, M., Mitchell, C.N., Banasiak, R., Wajman, R. and Adler, A. (2009). Four-dimensional electrical capacitance tomography imaging using experimental data, *Progress In Electromagnetics Research-PIER*, 90, pp. 171-186.

Su, B.L., Peng, L.H., Yao, D. and Zhang, B.F. (1999). An improved Simultaneous Iterative Reconstruction Technique for Electrical Capacitance Tomography, *Proc. 1$^{st}$ World Congress on Industrial Process Tomography*, pp. 418-422.

Tejchman, J. (1998). Silo-quake-measurements, a numerical polar approach and a way for its suppression, *Thin-Walled Structures*, 31, pp. 137–158.

Tüzün, U. and Nedderman R.M. (1985). Gravity flow of granular materials round obstacles. I: Investigation of the effects of inserts on flow patterns inside a silo, *Chem. Eng. Sci.*, 40, pp. 325–336.

Wajman, R., Banasiak, R., Mazurkiewicz, Ł., Dyakowski, T. and Sankowski, D. (2006).Spatial imaging with 3D capacitance measurements, *Measurement Science and Technology*, 17, pp. 2113-2118.

Warsito, W. and Fan, L-S. (2003). Development of 3-Dimensional Electrical Capacitance Tomography Based on Neural Network Multi-criterion Optimization Image Reconstruction, *Proc. of 3rd World Congress on Industrial Process Tomography*, Canada, pp. 942-947.

Wensrich, C. (2002). Experimental behavior of quaking in tall silos, *Powder Technology*, 127, pp. 87–94.

Wilde, K., Rucka, K. and Tejchman, J. (2008). Silo music–mechanism of dynamic flow and structure interaction, *Powder Technology*, 186, pp. 113–129

Williams, R.A. and Beck, M.S. (1995). Process Tomography — Principles, Techniques and Applications, Butterworth-Heinemann, Oxford.

Yang, WQ. and Liu, S. (2000) Role of tomography in gas/solids flow measurement, *Flow Meas. and Instrum.*, 11, pp. 237-244.

York, T.A., Mazurkiewicz, Ł., Polydorides, N., Mann, R. and Grieve, B. (2003). Image Reconstruction for Pressure Filtration Using EIDORS 3D, *Proc. 3rd World Congress on Industrial Process Tomography*, pp. 559–564.

# CHAPTER 13

## TOMOGRAPHY DATA PROCESSING FOR MULTIPHASE INDUSTRIAL PROCESS MONITORING

Krzysztof Grudzień, Zbigniew Chaniecki, Andrzej Romanowski,
Jacek Nowakowski and Dominik Sankowski

*Institute of Applied Computer Science*
*Lodz University of Technology*
*90-924 Łódź, ul. Stefanowskiego 18/22*
*{k.grudzien, z.chaniecki, a.romanowski, jacnow, dsan}@kis.p.lodz.pl*

### 13.1 Introduction

Tomography (from the Greek *tomos* – to slice and *graph* – to write) means producing an image of the cross-section of an object of interest based on measurements acquired from its outer boundaries (Williams and Beck, 1995; Scott and McCann, 2005). The common approach to tomographic measurement data analysis is their processing in order to obtain a reconstructed image of the distribution of a required property in the cross section area examined. The procedure of constructing a tomographic image is called image reconstruction, which results in an inverse problem solution (Yang and Peng, 2003; Lionheart, 2004). In the case of electrical capacitance tomography (ECT) the reconstructed image is the dielectric permittivity distribution $\varepsilon$, which is obtained from a vector of independent measurements of capacitances $C$. This permittivity distribution $\varepsilon$ could be, in the case of a solid flow, directly related to the transported material concentration or presented as a percentage phase fraction for liquid flow in sensor space. In the case of a mixture of two materials (solid-gas), a value associated with each pixel, represents the relative value of a higher electric permittivity material (friable material) with respect to a lower electric permittivity material (e.g. air). The reconstructed image reveals a phase (gas, liquid, solid) distribution in the form of a 2D/3D picture consisting of discrete elements – pixels (Yang and Liu, 2003; Lionheart, 2004).

The quality of an output image depends mainly on the ratio between the number of independent measurement records and the number of pixels with respect to a single reconstructed image and of course image reconstruction algorithms used. In the case of rectangular pixels a typical 2D image consists of 32x32 elements. In depends on the type of discrete elements which divided the sensor space; the shape and the size of pixels can be different. In some cases the shape of pixels is triangular. However, the interpretation of material distribution inside a measurement sensor is independent of the pixel geometrical parameters. The presented algorithms of data processing can work based on the rectangular and triangular shape of pixels.

The reconstructed image or image sequence obtained is further processed and analyzed in order to acquire important information about the industrial process (Williams and Beck, 1995; Scott and McCann, 2005; Dyakowski et al., 1997; Jaworski and Dyakowski, 2002; Pląskowski et al. 1995; Nuffer et al., 1999; Beck and Pląskowski, 1987). The extraction of industrial process characteristic parameters from reconstructed images gives a possibility of predicting unwanted incidents and the feasibility of in-depth exploration of dynamic spatio-temporal phenomena occurring in industrial processes. Moreover, information derived from tomograms can be used as an input signal for monitoring or control of a given process (Williams and Beck, 1995; Scott and McCann, 2005; Nuffer et al., 1999; Beck and Pląskowski; 1987). The aspects of tomography data processing presented in this chapter focus on the application of an Electrical Capacitance Tomography system, which is a common solution for the measurement of material distribution during industrial multiphase flow processes (Pląskowski et al. 1995, Williams and Beck 1995, Scott and McCann, 2005).

## 13.2 Multiphase flow

Multiphase flow is characterized by a simultaneous flow of materials with different or the same phases (gas/liquid, gas/solid, gas/liquid/solid). Despite intensive worldwide research and development over the past three decades, the problem of how to accurately measure the flowrate of multiphase mixtures in a pipeline still remains a notorious task (Plaskowski et al., 1995; Williams and Beck, 1995; Govier and Aziz, 1972; Hetsroni, 1982; Miller, 1983; Woodhead et al., 1990; Nieuwland et al., 1996; Someya and Takei, 2005). Control modules of multiphase flow phenomena were investigated for many years by means of different measurement techniques (Arko et al., 1999; Pląskowski et al., 1995; Klinzing et al., 1997; Sideman and Hijikata, 1993; Seville et al., 1997; Baker and Hemp, 1981; Hoyle et al., 2005). The understanding of the nature of gas/liquid or gas/solid interactions is very important since such phenomena are widely present in different types of industry, from pharmaceutical, food, environmental to civil, or chemical engineering. The investigations concentrate mainly on developing robust measurements tools, both hardware and software, especially elaborated agile data processing methods to better extract information about processes from measurements data.

### 13.2.1  *Pneumatic conveying of solid*

Pneumatic conveying, which is the transportation method of granular solids in a pipeline using a gas stream, is prone to design problems. Pneumatic conveying is based on the principle that under some conditions, the air is capable of transporting solid state materials (Ostrowski et al., 1999; Klinzing et al., 1995; Neidigh, 2005; Jones et al., 1999). In natural conditions the air transports various materials e.g. snow, sand, leaves or seeds. The flow is the result of the difference in pressure between two points – the inlet and outlet of a pipeline in the case of a flow rig. A pneumatic transport system works in one of the two modes – under pressure or a sectorial mode. The present work focuses on the pressurized air mode. The air velocity depends on transported material properties but, usually for most of them, remains in the range between 2m/s and 30m/s. The relationships between the bulk particle density and the transporting gas velocity enables several patterns to be distinguished, i.e. flow regimes, as shown in Fig. 12.1 in Chapter 12. The predominant and necessary mechanism of transport of solids in dense phase pneumatic conveying is due to flow instabilities referred to as "slug" and "plug" (Jaworski and Dyakowski, 2002; Ostrowski et al., 1999). In this figure a photograph of slug formation in the horizontal section of the installation is shown.

The main goals of the study on a pneumatically driven flow are: flow regime identification, the nature of the phenomena associated with different types and eventually resulting information about the transported components fraction, solids fraction profile (heterogeneous or homogeneous structures), velocity, mass rate. Other issues to be addressed are monitoring and control strategies that could be devised on the basis of the knowledge about the behavior of the multiphase flow. This can obviate the occurrence of unwanted events (blockage monitoring and prevention) and hence improve the overall performance of pneumatic installations in terms of efficiency, minimizing the energy consumption and avoiding transported component degradation. This is particularly important when the transport installation is a part of the bigger system and the contribution to its performance must be considered as continuous feeding a certain amount of a particular medium with a small variance.

## 13.2.2  Gravitational flow of solid

The second type of granular flow described is often applied in different stages of production, especially when the storage of granular material is an important part of the whole industrial system. Specialized containers for the protection and storing of solid materials called silos are usually used for this task. There are different types of silos varying by size, shape, material of construction and stored powder properties. Different types of flow regimes may occur during discharging the silo. The two basic types of flow behavior, during the silo discharging process, are a 'mass flow' and a 'core' (or 'funnel') flow. The first is characterized by the situation when all the material within any cross section is simultaneously discharged from the container across the whole cross section. In the second type the material tends to flow only in the center of the cross section. When the material tends to flow mainly in the core region of the container the rest of the solid situated close to the walls tends to form the so-called stagnant zones. The visualizations of different kinds of the gravitational flow are presented in Chapter 12. For different types of silos that usually induce a particular flow type for a given material, different methods of tomography data analysis are applied to monitor this stage of the process.

## 13.2.3  Three-phase flow

Other types of flows playing a significant role in industrial applications are multi-phase flows. One of the most important areas of application is the undersea exploration of oil in the petroleum industry. Atkinson I., Theuveny B., et al. (Atkinson I., Theuveny B., et al. 2004) presented a review of current and possible future technologies in multiphase flow measurements. There are two main principles of multi-phase measurements: the separation of the flow components and subsequent independent component flow measurements. There were approximately 1300 such industrial installations in 2004. Another type of three phase meter uses a Venturi equipped with an absolute and differential pressure sensor and dual energy gamma ray detectors. This device measures the total mass flow and fractions of gas, oil and water. Compact three phase flow meters have existed for more than 30 years but they only work for well-mixed flows in vertical pipes. There is a demand for multiphase flow meters down the holes in the oil production industry, where the pipes are inclined or horizontally positioned. Different aspects of the application of gamma radiation in process tomography are presented by (Holstad 2003, 2004), (Tjugum et al. 2001, 2003). In the research performed within the framework of the DENIDIA MC ToK project in

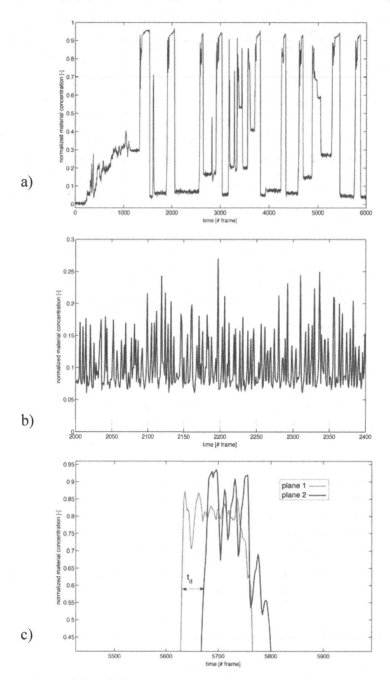

Fig 13.1. Characteristics of the average concentration of the material in the sensor space in the horizontal section of pipeline, setting inverters: a) blower 15Hz, rotator 80Hz, b) blower 19Hz, rotator 80Hz, c) the characteristic of the signal changes in the case of the two ECT sensor planes, with the marked time shift between the signals, associated with the distance between the planes of measurement.

cooperation of the Computer Engineering Department of the Lodz University of Technology and Department of Physics and Technology of the University of Bergen, a three phase flow meter based on combining three modalities – ECT, ERT and gamma ray radiation – was designed and tested. The system presented in Fig. 13.18 (at the end of this chapter) is built on a 90mm diameter PVC pipe and consists of 8 stainless steel pin shaped resistance electrodes and 8mm copper capacitance electrodes placed outside the pipe. The system uses ECT measurements from the copper electrodes when the mixture is oil or gas continuous and not conducting and the ERT mode when the mixture is water continuous and conductive. The ECT/ERT sensor is connected to a 16-channel electronic measurement unit, DENIDIA ECT, designed and manufactured in the Lodz University of Technology. In addition, a vertically and a horizontally directed narrow gamma beams from two pin sources are placed in the radial mid plane of the EIT system. The two beams are detected by two CTZ-sensors including all the necessary electronics for yielding a TTS shaped signal equal to the number of counts. The construction is placed in a protective steel housing. The measurement system presented was submitted to the European Patent Office.

## 13.3 Data processing

The main advantage of process tomography is the possibility to visualize, in the form of an image, the internal structures of objects. The information included in an image or images, or even in unprocessed measurement data (raw data), offers valuable knowledge about the process monitored contained in tomography data. The main task for data processing algorithms is to use the most appropriate data (an image, a sequence of images or raw data), in the context of type of process, to bring out significant information about the process state. The estimated parameters, which describe the states of a process, can be later employed by the monitoring unit in order to manage the process in the optimized way. In this chapter the authors will present a selection of methods with the focus on industrial processes, extensively relying on granular solid materials transportation and storage.

### 13.3.1 *Monitoring of pneumatic conveying*

In the pneumatic conveying system the control unit is typically based on pressure and air velocity measurements. The weight of the material transported is determined based on the reading of the beam load cells placed in the buffer tank.

A typical pneumatic conveying installation has two main elements that affect the operating parameters:

- a blower setting the parameters of the transporting medium – gas,
- a feeder rotary valve dosing the amount of solid material in the pipeline.

An appropriate choice of the solid material amounts and the velocity of the transporting gas is essential for optimum transport. Figs 13.1 and 13.2 present the characteristics of concentration changes during the flow for different values of gas velocity and solid material and different sections of the pipeline (horizontal – Fig. 13.1, vertical – Fig. 13.2). The concentration was estimated based on tomography data as an average of the capacitance measurements vector for each time measurement point.

The characteristics presented allow one to notice that the concentration changes demonstrate repeatability during the flow. Another important aspect visible in the characteristics shows that increasing the air flow speed, with the same solid volume loaded into the pipeline, results in a flow regime change, switching from the slug flow to the diluted flow. Such changes can damage the transported particles and lead to energy loss. To obtain the optimal material transport, which often depends on the material, it is necessary to ensure the proper functioning of the process of energy saving and avoid degradation of the material. The methods of tomography data analysis presented below used raw data and a sequence of images to determine the characteristic parameters of pneumatic conveying. Depending on the analysis purpose either raw data or images can be selected as the input data for processing algorithms. The algorithms of data analysis that generate diagnosis parameters for a chosen time point, typically for pneumatic conveying, run on a sequence of data for a user predefined time window. This can be defined as the temporal analysis of tomography data in contrast to the spatial analysis, which was earlier invoked for the silo flow.

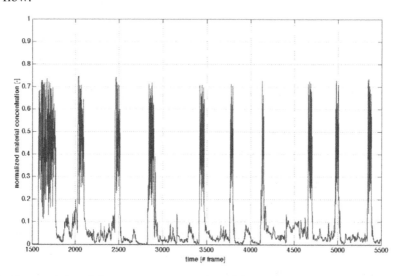

Fig. 13.2. Characteristics of the average concentration of the material in the sensor space in the vertical section of pipeline.

288          *K. Grudzień et al.*

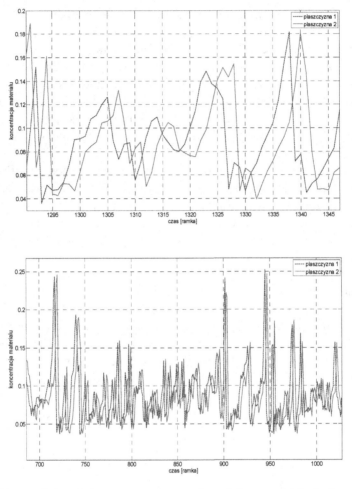

Fig. 13.3. Characteristics of the average concentration of the material in both cross-sectional planes. On the magnification characteristics was marked time shift td between signals from first and second sensors, associated with the distance between the planes of measurement. Setting inverters: 19,5Hz blower, rotator 100Hz.

### 13.3.1.1   *Tomography image analysis*

Since a tomography image provides information about material distribution inside the sensor space, the sequence of images allow monitoring the material concentration changes during the flow over time. A sequence of tomography images allows one to analyze the concentration changes in time for the chosen position in the pipe cross-section, based on the reconstructed pixels values. Statistical relationships between local changes of solids concentration in the sensor space can be used to split a cross-section into several regions in which the concentration changes are very similar. The analysis of behavior of homogenous pixel groups allows one to: firstly, understand flow phenomena and secondly, to prepare a signal for the control flow module (Sankowski et al., 2003; Sankowski et al., 2005; Dyakowski et al., 2003).

An individual reconstructed tomographic image is recorded in the form of a matrix $X$ consisting of $M$ elements (pixels): $M=32*32$, as given by Equation 1 below:

$$X_{i,j} = \begin{bmatrix} x_{1,1} & \cdots & x_{1,32} \\ \vdots & \ddots & \vdots \\ x_{32,1} & \cdots & x_{32,32} \end{bmatrix}_{32 \times 32} \tag{13.1}$$

where: $i, j$ – the image element indices, $i=1,...,32, j=1,...,32$.

Each matrix element $x_{i,j}$ denotes the value of the material concentration for the image pixel located in the point with coordinates $(i, j)$. The time sequence of tomograms, regarding its discrete character, can be denoted as $X(t_l)$ , where $l$ stands for the image (frame) number and is related to time $t$ in the form of $t_l=\Delta T * l$, where $\Delta T$ stands for the sampling time. A 2D tomogram sequence is considered as stochastic, $M$–dimensional image vector ($M$ – the number of stochastic processes associated with the number of tomogram pixels). The image vector $X(t)=[X_i(t_j)]_{N \times M}$ can be written in the following form:

$$X(t) = \left[ X_i(t_j) \right]_{\substack{i=1,...,M \\ j=1,...,N}} = \begin{bmatrix} x_{1,1} & x_{2,1} & \cdots & x_{M,1} \\ x_{1,2} & x_{2,2} & \cdots & x_{M,2} \\ \vdots & \vdots & & \vdots \\ x_{1,N} & x_{2,N} & \cdots & x_{M,N} \end{bmatrix} \begin{matrix} image\ 1 \\ image\ 2 \\ \vdots \\ imageN \end{matrix} \tag{13.2}$$

where: $i$ – the image pixel number, $i=1,...,M$, $j$ – the image number, $j=1,...,N$, $M$ – the number of image elements (pixels), $N$ – the number of processed images.

Each of vector $X_i$ elements is considered as an elementary stochastic process $EP_i$, which is defined with the aid of probabilistic values of the $i$-th pixel in the image sequence. The basic assumption is that each $EP$ process is a stationary process and its statistical parameters are constant throughout a plug body.

In order to identify homogeneous flow components, methods based on the spatio-temporal segmentation of the sensor zone are proposed. Separation is based on grouping of the elementary processes (*EPs*), defined by pixel values in a given time interval. A separated homogeneous flow component is a cross section region characterized by similar flow properties (concentration changes or velocity in the case of twin plane electrodes). In order to carry out the decomposition of a flow, based on criteria of the similar concentration changes, the principal component analysis and k-mean methods can be applied to assemble elementary processes in a homogenous group.

The cyclically conducted process identification of homogeneous components in a predetermined frame sequence allows one to track flow component location changes in time and to determine the level of turbulence during a pneumatic flow. The motion of each group in time, in the case of decomposition based on the similarity in concentration changes, is defined by the displacement of the center of a group, which has been determined as its center of gravity (see Fig. 13.4). Motions of each group are represented by a 2D image called a velocity map, in which each vector determines a group displacement direction and its length is proportional to the velocity magnitude.

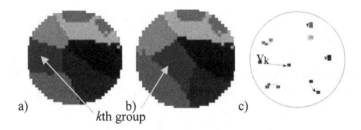

a)                          b)                          c)
                    kth group

Fig. 13.4. Results of consecutive groupings; (a, b) flow decomposition based on similar concentration changes for two successive time interval – group image, (c) respective velocity map of (a) and (b). This algorithm is based on the factor analysis of sequence images and the applied k-means procedure.

As can been seen in Fig. 13.4 a velocity map distinctly indicates motions of the groups within a plug body. A velocity magnitude for any group motion $v_k$, called a kth group velocity, is defined as:

$$v_k = \frac{l_k}{\Delta t} \tag{13.3}$$

where $k$ is the number of a group, $l_k$ is the length of a velocity vector of kth group in a velocity map, $\Delta t$ is the time delay between two consecutive groupings.

For example, in Fig. 13.4c the velocity magnitude of the chosen group $v_k = 54$ cm/s with the following parameters: $l_k = 1.62$ mm, $\Delta t = 0.03$s (with the pipe diameter equal of 57 mm). Figure 13.5 shows the level of the turbulence flow by characterized solids mixing processes within a plug body during pneumatic conveying of solid. The estimated groups of velocities are collected in Table 1.

Table 1. The group velocities (cm/s) for example in Fig. 13.8.

|  | $v_1$ | $v_2$ | $v_3$ | $v_4$ | $v_5$ | $v_6$ | $v_7$ |
|---|---|---|---|---|---|---|---|
| $t_1$ | 6,16 | 13,12 | 6,16 | 6,16 | 8,34 | 0,15 | 6,16 |
| $t_2$ | 20,41 | 57,32 | 18,30 | 76 | 13,12 | 20,41 | 32,53 |
| $t_3$ | 0,08 | 0,00 | 9,75 | 4,65 | 4,65 | 4,65 | 7,43 |
| $t_4$ | 0,13 | 48,34 | 6,16 | 50,45 | 6,16 | 13,12 | 13,12 |
| $t_5$ | 0,28 | 45,21 | 38,10 | 28,32 | 20,41 | 18,30 | 72,57 |
| $t_6$ | 0,08 | 27,43 | 8,53 | 13,20 | 30,25 | 17,30 | 54,73 |
| $t_7$ | 0,40 | 72,57 | 42,24 | 9,75 | 25,78 | 0,15 | 27,60 |

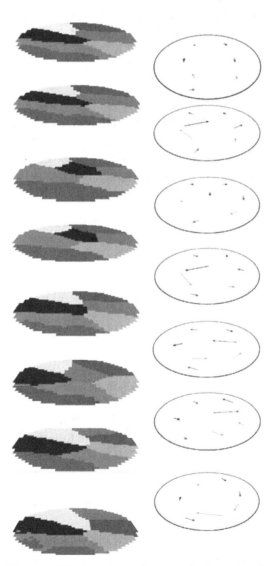

Fig. 13.5. Examples of images of the grouping and group velocity vectors during the plug-in of the vertical section of the installation.

The second homogeneity criterion of flow decomposition that can be used is based on a similarity in the flow velocity in each pixel position. It is based on the spatial and temporal variable segmentation of space taken between two sensor planes. To determine the flow velocity it is necessary to have two sensor planes located at a known distance. For each pixel the shift time, a signal delay between the first and the second sensor plane, is determined using the cross-correlation function and a sequence of tomographic images from two planes of sensors. The classical method for calculating solids velocity is based on the correlation of the corresponding pixels from the reconstructed images. For the reconstructed images the cross correlation can be calculated as in (Dyakowski 1996):

$$R_{x_n y_n}(p) = \frac{1}{M}\sum_{i=0}^{M-1} x_n(i) y_n(i+p), \qquad n = 0,1,...,N-1 \tag{13.4}$$

where p = ..., -1, 0, 1, ... is the shift time in the number of images; n is the pixel index, N is the total number of pixels; M is the number of images for which the cross-correlation is calculated; $x_n(i)$, $y_n(i+p)$ – are the numerical values associated with pixel n from i image obtained from the sensors of the plane A and from i+p image from the sensors of the plane B, respectively (see Fig. 13.6).

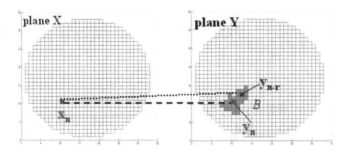

Fig. 13.6. Cross-correlation between pair of pixels of a plane X and plane Y. The region B in the plane Y represents a set of the best-correlated pixels with the pixel $x_n$ (Mosorov et al., 2002)

The knowledge of the transit time and distance d between the two sensors enables the solids velocity to be calculated from the following equation (13.5):

$$V = \frac{d}{p_\tau} * T_F \tag{13.5}$$

where TF is the time period of a single frame; $p_\tau$ is the shift for which correlation (1) reaches the maximum.

In total, N velocities, which have created the velocity profile, are calculated. The product of the concentration and the velocity profiles is integrated across the pipe cross-section, and then integrated over a period of time so that a mass flow rate can be obtained. However, in practice, the following factors have a profound effect on cross-correlation flowmeters: the fact that the turbulence is not always isotropic, a possibility of a change in the flow stream between the two sensors due to upstream conditions, and the nature of the flow. The result is that the flow velocity has components in three orthogonal directions; for example, the effect of gravity separation in a two-component flow. To overcome these difficulties the "best-correlated pixels" method can be applied. It is based on calculating the cross-correlation between a pixel from the first plane and some pixels from the second plane (Mosorov et al., 2002). The pixels from the second plane are chosen from the corresponding pixel and its neighborhood following the criteria $\max\{R_{Xn,Yn-j}\}j = B$, where B is the neighborhood of pixel n on the second plane.

The best-correlated pixels method has a disadvantage as well. In practice, one can often observe that some pixels from the first plane are best correlated with the single pixel from the second plane. Those pixels from the second plane are not used for calculating the velocity profile. The problem of the unambiguous determination of the pair of pixels is presented in Fig. 13.7. As shown in the figure, the particles have high and low velocities in different regions of the cross-section (see Fig. 13.7a). A similar case can be observed for an average concentration profile (see Fig. 13.7b). The investigation carried out allows one to conclude that the description of a flow as a set of independent variables changes dynamically with time. In order to solve these problems and, additionally, simplify the description of flow structures without any substantial loss of information about the total flow one can use a method to split the cross-section of a flow into a homogenous region of pixels, where the homogenous criterion is velocity (Eq. 13.6).

Thus, the flow structure detection allows one to analyze the behavior of variables of the separated structures as an alternative to the whole flow analysis. Then the *k*-means procedure, based on the estimated velocity in each pixel, makes the set of EPs (relative to the velocity) split into k homogeneous groups. Such homogeneous flow components are called virtual channels. The "virtual channel" is represented by a group of *EP* processes. Once the virtual channels have been established, a series of parameters such as the mean concentration, the velocity, the mass flow rate are determined. The detailed description of virtual channel determination can be found in (Sankowski et al., 2005; Grudzień et al., 2010).

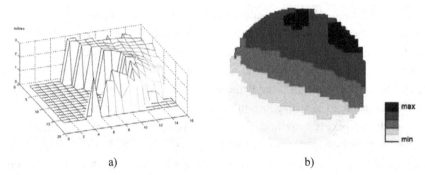

a)                                 b)

Fig. 13.7. Profile example in cross-section during a plug (frame numbers 885-930) a: velocity profile; b: average concentration profile (Dyakowski et al., 2003).

### 13.3.1.2 *Raw data analysis*

The pneumatic conveying flow is not continuous – it has a character of plug transport in the vertical section of pipelines (as shown in Figs. 13.1 and 13.2), or even a variety of flow structures occurring in horizontal sections. In a plug flow, short periods or long gaps between the consecutive transported material plugs can occur. The estimation of flow parameters, i.e. the determination of the beginning and the end of each plug sequence, is possible with raw data analysis algorithms, without taking reconstructed images into analysis. Such an approach is very convenient for the monitoring module, when the time factor is critical for control decisions. In order to have information about the frequency and size (time window) of plugs, the analysis of raw tomography data is sufficient. This knowledge can be employed directly in the control module to maintain flow stability.

Plug determination is based on the detected concentration value. The flow portion is assigned to the plug when the concentration of the material exceeds the defined level. Fig. 13.8 shows exemplary results of the evaluated parameters of the pneumatic conveying flow (Chaniecki 2007).

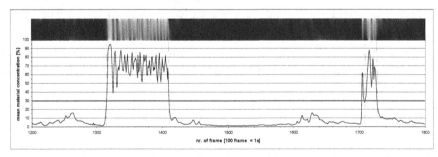

Fig. 13.8. Characteristic of mean concentration of material in cross-section in vertical section of pipe and topograms (longitudinal view). Pointed parameters of sequence plugs (beginning and end) for 30% level of classification.

The characteristic plot of concentration changes presented in Fig. 13.8 refers to the vertical section of a pipe. For this configuration the plugs moving upwards are separated by an air gap or a low material concentration, which causes some solid particles to gravitationally fall down. The allocation of the time window to the plug area uses two parameters – the threshold and the time window. The instantaneous values of the relative material concentration are not lower than the threshold value $P$ during $K$ time points (measurement frames). The selection of the threshold value $P$ depends on the characteristics of the material transported, and the installation parameters can be modified by the operator during the measurement. In addition, the length of the diagnosis time window $K$ can be regulated. In Fig. 13.9 a scheme of classification of the plug during the pneumatic conveying of bulk materials is presented.

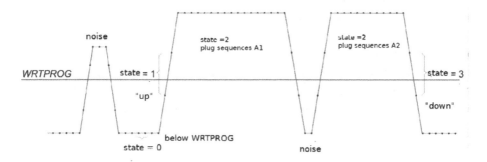

Fig. 13.9. The scheme of classification of the plug during the pneumatic conveying of bulk materials.

### 13.3.2 *Dynamic effect monitoring during the silo discharging process*

The silo discharging process can be associated with some undesirable dynamic effects. Strong dynamic effects are the result of the dynamic interaction between a vibrating material and the silo structure (Wilde et al. 2007, Niedostatkiewicz and Tejchman 2008). The resonance vibration increases the pressure on the walls and the bottom, which in turn may cause malfunction or even a collapse of the construction. One of the most effective means for the reduction of dynamic effects is increasing the silo wall roughness. On an industrial scale it consists in mounting special corrugated sheets on the inner side of the silo walls (Tejchman 1998). Increasing the silo wall roughness causes additional transversal strain waves, which raise the frequency of vibrations of flowing particles. In this situation, the so called shear zone is formed, which is visible along the silo wall. Such effects cause a decreased level of interaction between the material and the silo structure, which can be controlled by tracking the shear zone parameters.

Shear zones occur along the silo walls characterized by a lower material concentration than the surrounding areas. The accurate parametric description of shear localization zones in bulk solid is important since in practice they inherently influence the wall stresses. The complex knowledge about the shear zone material behavior needs information about the shear zone width and the level of material concentration in this zone. The tracking of the shear zone parameters during a gravitational flow is helpful from a monitoring point of view and allows one to react in time to unwanted phenomena described in chapter 12.

The example results shown in Figs 13.10 and 13.11 reveal the experiments with the initially loose sand packing density and rough silo walls (Grudzień et al., 2010; Grudzień et al., 2012). The diameter of the outlet, the slim silo with a flat bottom, was 0.07cm. The bulk solid changes during the silo discharging process illustrated in Figs 13.10 and 13.11 are shown in the form of characteristic concentration changes on a time plot (Fig. 13.6) and topograms (Fig. 13.6).

Such a configuration characterizes the tendency to contractancy in the cross-section center (pixel $p(15,15)$) for different heights of the sensor position (in the results presented h= 1 and 1.5m above silo outlet) for the mass flow regime. However, in the silo wall area (wall-adjacent zone (pixel $p(1,15)$) the material has a tendency to dilatancy. The shear zone can be observed at the silo wall area.

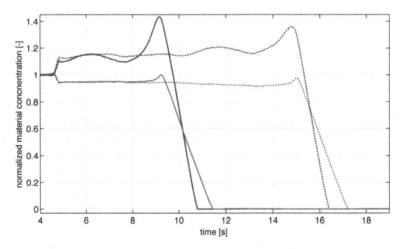

Fig. 13.10. Changes in loose sand concentration for rough silo walls. Continuous lines represent changes in wall-adjacent pixel $p(1,15)$ for the reconstruction based-plot. The center plot (for pixel $p(15,15)$) – dashed line. The pair of characteristics dropping at about 10s time is for h=1.5m sensor height, the pair of characteristics dropping at about 16s is for sensor location at h=1m.

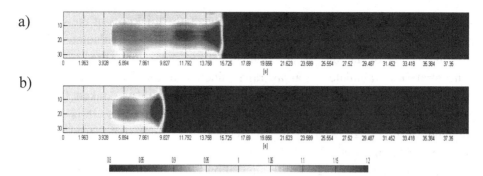

Fig. 13.11. Tomographic visualization (in form of topogram) of concentration changes — loose sand, rough silo walls, sensor location: a) h=1.0 m, b) h=1.5 m.

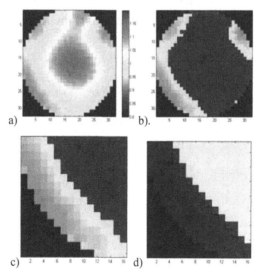

Fig. 13.12. Tomography image processing method stages: a) input image b) threshold image c) shear zone analysis image fragment d) boundary pixel determination.

The shear zone occurred in the wall-adjacent region, and dilatancy reached ~5-8% of the initial state. The thickness of the shear zone, determined with the simplified approximation of the pixel number, was about 20mm. To determine the shear zone parameters (e.g. concentration and size), the reconstructed image processing method can be applied. Distinct phases of the image-processing algorithm applied are presented in Fig. 13.12. The first step is to determine the indices of all the pixels belonging to the shear zone. Based on thresholding operation, pixels categorized as not belonging to the shear zone are found (Fig. 13.12b, zoom c). And after indicating the pixels lying on the boundary of the shear zone and the rest of the cross-section image, the thickness of the shear zone is obtained, as the difference between the silo radius and the calculated distance from the center of the silo to the boundary of the shear zone (Fig. 13.12d). The concentration is calculated as an average value for all the pixels belonging to the zone. In the example presented only part of the image was taken for further analysis because of the flow asymmetry of the rough walls. It is a circular fragment of the image, which correctly represents the material distribution in the wall-adjacent zone (Fig. 13.12c).

Such tomography image processing allows one to monitor the shear zone parameters during the entire silo discharging time. Fig. 13.13 presents plots of the shear zone parameter changes (including additional algorithm transient parameters) in a predefined time period 6.6-13.2s time. The starting point of the analysis was ~1.7s from the silo discharging process start (associated with flow stabilization) and the analysis was conducted until the material surface appeared at the level of the lower sensor location.

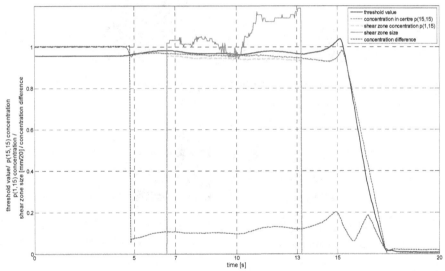

Fig. 13.13. Characteristic plot of shear zone parameters change during silo discharging bulk solid flow (Grudzień et al., 2012).

### 13.3.3 *Monitoring a silo funnel flow*

In the case of the funnel flow type, reconstructed images processing, enables one to examine the core flow area during the silo discharging process. Examples of reconstructed images for the core flow are presented in Chapter 12. To properly track a gravitational flow, the funnel monitoring parameters have to be determined. There is a set of parameters which characterize the dynamics of the silo flow: the solids concentration in the funnel $\xi_g$, the funnel size area $\eta_g$ and the funnel shape $\kappa_g$ (Grudzień, 2008).

For a symmetrical flow the material distribution takes the form of concentric zones (as shown in Fig. 13.14). Such knowledge about the process allows one to distinguish distinct regions in the reconstructed image as concentric areas: (1) wall, (2) transition and (3) central zones (the core of the flow). Additional two intermediate rings improve the legibility of transition between the areas (Chaniecki 2007).

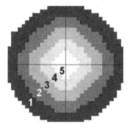

Fig. 13.14. Shape of the concentric areas on the sample tomogram.

In the following steps of processing the average concentration of the material for each distinct area at each frame of the measuring-sequence is determined. On this basis the characteristic plots of average changes in the porosity of the material are prepared. In Fig. 13.15 an example of changes in the material concentration for each of the concentric areas is presented (Chaniecki et al., 2006).

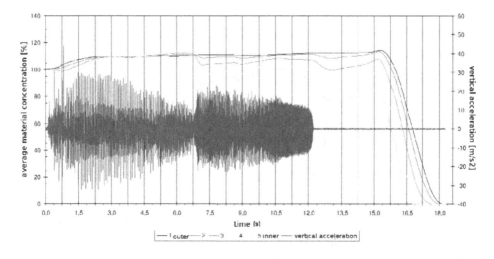

Fig. 13.15. Changes of material concentration for determined five regions of ECT sensor cross-section.

The method described above allows one to track material concentration changes in the core flow area. However, the assumed sizes of these zones are constant, hence it is not possible to monitor the changes of the core flow size. In Fig. 13.16 the changes in the funnel diameter size with respect to time for different heights of the sensor located above the outlet level – *H*, is shown, (Grudzień, 2008). In calculating these graphs, it is assumed that the particle packing density in the funnel flow zone is constant. The upper curve (red) represents a series for the normalized permittivity of the funnel for a range of up to a limit of 0.9 - $\rho_1$. The lower (blue) curve represents a range of up to a limit of 0.85 - $\rho_2$ . As can be seen the results at each height are consistent.

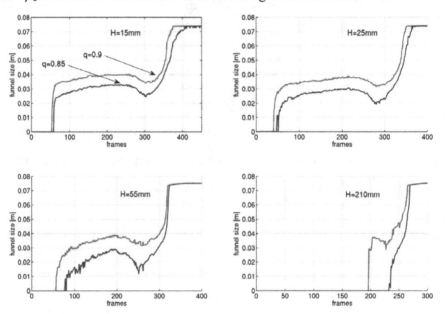

Fig. 13.16. Funnel size analysis for various sensor positions in experiments (with 50mm outlet orifice), time unit is 2 ms

In order to dynamically estimate the funnel flow parameters a geometrical model of the hopper flow based on the ellipse equation can be proposed:

$$\frac{(x-x_0)^2}{a_g^2} + \frac{(y-y_0)^2}{b_g^2} = 1 \tag{13.6}$$

where $(x_0, y_0)$ describe the coordinates of the gravity center of the funnel cross-section, $a_g$ and $b_g$ determine the half length of the major and minor ellipse axes.

The ellipse thus defined represents the approximated boundary between the area of the sensor cross-section occupied by the funnel zone – $a_1$ and the remaining part of the sensor area – $a_2$. The arrangement proposed makes it easy to monitor these boundary changes and hence allows one to diagnose unwanted phenomena during the hopper discharge process (e.g. flow blockage or partial blockage). As the first step of this procedure, the spatial segmentation of a tomography image is conducted. After the image reconstruction and thresholding (pixels are split into 2 groups – the funnel area and the rest of the sensor area) the ellipse axes are determined by applying the least-squares analysis and linear regression methods. The main steps of this algorithm are presented in Fig. 13.17.

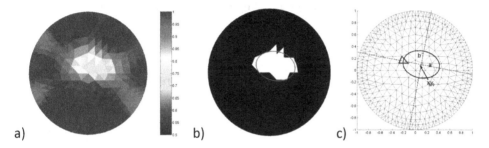

a)                              b)                              c)

Fig. 13.17. The main steps in defining the hopper flow parameters. a) reconstruction image, b) image thresholding, c) determined ellipse parameters.

The analysis of tomographic image presented can be carried out for a sequence of images obtained during the hopper discharge process. The shift of the funnel gravity center $(x_0, y_0)$ and the ratio of the major $(a_g)$ and minor ellipse $(b_g)$ axes gives the information about the correct/incorrect funnel flow.

### 13.3.4 *Data processing for gas/oil/water flow monitoring*

Before any measurements can be performed, the multimodality measurement system has to be calibrated for a given installation. The Electric Capacitance Tomography (ECT) system, equipped with capacitance-measurement electrodes, is calibrated in three steps: for pipe filled with gas, oil and distilled water

Pipe filled with gas: $C_i^{gas} = K_i^{gas} \varepsilon_0 \varepsilon_{gas}$ (13.7)

Pipe filled with oil: $C_i^{oil} = K_i^{oil} \varepsilon_0 \varepsilon_{oil}$ (13.8)

Pipe filled with distilled water: $C_i^{w_d} = K_i^{w_d} \varepsilon_0 \varepsilon_{w_d}$ (13.9)

where i = 1-2,1-3,…1-N, 2-3, 2-4…..2-N…… N= the total number of electrodes, $\varepsilon_0 \varepsilon_{gas}$, $\varepsilon_0 \varepsilon_{gas}$, $\varepsilon_{w_d}$ are the dielectric constant and relative permittivity of gas, oil and distilled water, respectively.

C is the capacitance measured between a particular pair of capacitance-measurement electrodes, Ki are the calibration constants,

$$K_i^{gas} = \frac{C_i^{gas}}{\varepsilon_0 \varepsilon_{gas}}$$ (13.10)

Thus, all the $K_i^{oil}$, $K_i^{w_d}$, $K_i^{gas}$ can be derived from equations corresponding to water and gas. The ERT-system can only be calibrated under salt water conditions. Since this tomograph is mainly intended for use in the North Sea, sea water or the water of the same conductivity (5 S/m) should be used.

Pipe filled with saline water: $R_i^{w_s} = K_i^{w_s} \dfrac{1}{\sigma_{w_s}}$ (13.11)

where the values for $i$ are the same for all the electrode pairs as for the ECT system, Thus

$$K_i^w = R_i^w \sigma_{w_s}$$ (13.12)

The system used consists of two perpendicular narrow beams. It is therefore not a full tomography system; nevertheless, out of necessity it is useful for a three-component flow, specially designed for horizontal pipes. The calibration procedure of the GRT-system is performed in a similar way, a sensor is filled with all the flow components: gas, oil and water.

Pipe filled with gas: $\qquad I_{gas} = I_0 e^{-\mu_{gas}D - 2\mu_d d}$ (13.13)

Pipe filled with oil: $\qquad I_{oil} = I_0 e^{-\mu_{oil}D - 2\mu_d d}$ (13.14)

Pipe filled with water: $\qquad I_w = I_0 e^{-\mu_w D - 2\mu_d d}$ (13.15)

Where $I_{gas}$, $I_{oil}$, $I_w$ are the number of counts elapsed for the sensor filled with gas, oil and water ,respectively, $I_0$ is the initial count elapsed with the empty gamma sensor.

When using the ECT, it is often difficult to determine the border between gas and oil. It is usually easier to determine the interphase between water and oil where the difference in electrical permittivities are large (71/2.3). This interphase border can be applied to increase the accuracy of determining the border between oil and gas using the GRT. If the sensor is filled with a mixture of the components: gas, oil and process water, the linear absorption coefficient along the gamma ray beam in an oil/gas/water- mixture can be presented in the following form:

$$\mu_{mix} = \mu_{gas}\alpha_{gas} + \mu_{oil}\alpha_{oil} + \mu_w\alpha_w \tag{13.16}$$

$$\alpha_{gas} + \alpha_{oil} + \alpha_w = 1 \text{, } \alpha_{oil} = 1 - \alpha_{gas} - \alpha_w \tag{13.17}$$

This gives:

$$\mu_{mix} = \alpha_{gas}(\mu_{gas} - \mu_{oil}) + \alpha_w(\mu_w - \mu_{oil}) + \mu_{oil} \tag{13.18}$$

where $\alpha_{gas}$, $\alpha_{oil}$ and $\alpha_w$ are the fractions of the components in the gamma beam and $\mu_{mix}$, $\mu_{mix}$, $\mu_{mix}$ are the linear absorption coefficients of gas, oil and water, respectively. The elapsed number of counts for the multi phase mixture can be described by the following formula.

$$I_{mix} = I_0 e^{-\mu_{mix}D - 2\mu_d d} = I_0 e^{-\alpha_{gas}(\mu_{gas} - \mu_{oil})D - \alpha_w(\mu_w - \mu_{oil})D - \mu_{oil}D - 2\mu_d d} \tag{13.19}$$

$$\frac{I_{mix}}{I_{gas}} = e^{-\alpha_{gas}(\mu_{gas} - \mu_{oil})D - \alpha_w(\mu_w - \mu_{oil})D - (\mu_{oil} - \mu_g)D}$$

Since $\alpha_w$ is known from the EIT image, we can find $\alpha_{gas}$ from gamma measurements. After calculations the fractions of different phases can be derived from the following formulae:

$$\alpha_{gas} = \frac{\ln\dfrac{I_{mix}}{I_{oil}} - \alpha_w \ln\dfrac{I_w}{I_{oil}}}{\ln\dfrac{I_{gas}}{I_{oil}}} \tag{13.20}$$

$$\alpha_{oil} = \frac{\ln\dfrac{I_{mix}}{I_w} - \alpha_{gas} \ln\dfrac{I_{gas}}{I_w}}{\ln\dfrac{I_{oil}}{I_w}} \tag{13.21}$$

Presented measurement principle was practically verified during experiments performed on a three phase flow rig in the University of Bergen. Fig. 13.19 shows partial view of the installation used for measurements.

*K. Grudzień et al.*

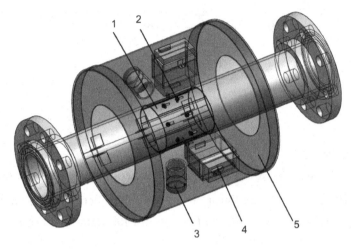

Fig. 13.18 Schematic view of the sensor: 1-ECT copper electrodes, 2- ERT pin electrodes, 3-gamma source holder, 4-evaluator 1000 gamma ray detector, 5-steel gamma ray protective housing.

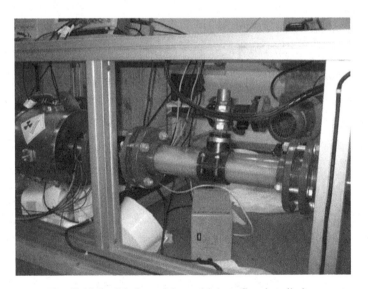

Fig. 13.19 Partial view of the multiphase flow installation.

Fig. 13.20a) presents example of flow obtained during pump start. Measured oscillations of the flow composition are presented on fig. 13.20b).

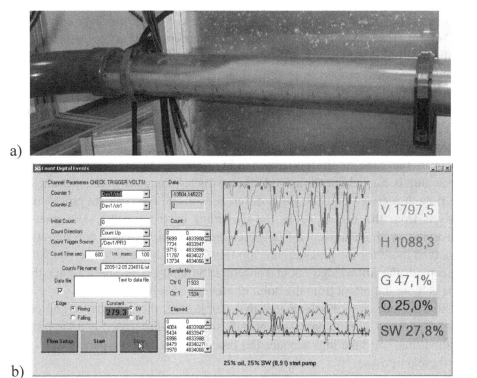

a)

b)

Fig. 13.20 a) example of flow obtained during pump start. b) Measured oscillations of the flow composition.

## 13.4 Conclusions

Multiphase industrial processes are a challenging field for imaging and visualization but the successful application of computer vision methods may be beneficial for process control in terms of economical and safety aspects. Process monitoring with the use of tomographic methods require the adjustment of data analysis methods in such a way as to allow one to retrieve significant process information with respect to its form, course, dynamics, parameters and other context related aspects. This makes it impossible to choose the all-embracing data analysis method to suit all cases.

In this chapter the authors presented several flow types coupled with a number of methods for the visualization and extraction of information from tomography measurement data. The processes of pneumatic conveying and storing as well as the unloading of silos examined herein vary in the progress speed, sizes of the reactors investigated and flow conditions. Data processing can be conducted based on single data (raw data vector or single image) or based on sequence of data. This depends on type of process monitoring and expected kind of information about process state. The spatio-temporal segmentation and analysis is proposed for the identification of homogeneous flow components in based on physical properties of flow, material concentration changes, or even be computed using measurement data from twin plane sensors (the flow velocity parameter). Such an analysis requires conducting calculation in time window to obtain control information about the flow states at a given moment. This approach is used during image and raw data processing. A different situation can occur in the case of the silo discharging process, where tomography image processing is carried out for a single reconstructed image to obtain parameters of the flow at a given moment. Such differences in the type of data processing are derived from the difference in the dynamics of the process. The pneumatic conveying is characterized by larger changes in the material concentration in time than those for the gravitational silo flow. As far as the second type of flow is concerned, the knowledge about concentration changes in space rather than in time is more important from the point of view of control.

To conclude, it is worth saying, that monitoring, the diagnostics and control of transport and storage processes of bulk solids require selecting an appropriate method that will take into account the process itself and the parameters crucial for these purposes.

## References

Arko A., Waterfall R.C., Beck M.S., Dyakowski T., Sutcliffe P., Byars M, (1999). Development of Electrical Capacitance Tomography for solids mass flow measurement and control of pneumatic conveying systems, *Proc 1st WCIPT*, Buxton, pp. 140-146.

Atkinson I., Theuveny B., et al. (2004). A New Horizon in Multiphase Flow Measurements Oilfield Review Winter 2004/2005 52-63

Baker, R.C.. Hemp, J. (1981). Slurry concentration meters: state of the art; review and bibliography, Cranfield, *BHRA Fluid Engineering*, Series, vol. 8., p.28.

Beck, MS., Plaskowski, A. (1987). *Cross Correlation Flowmeters: Their Application and Design*, Adam Hilger, Bristol, 1987.

Chaniecki, Z., (2007). Algorithms of data processing and analysis in electrical capacitance tomography for diagnosis of selected industrial processes, Zeszyty Naukowe. Elektryka / Politechnika Łódzka (in polish), Issue 111, p. 5-12.

Chaniecki, Z., Dyakowski, T., Niedostatkiewicz, M., Sankowski, D. (2006). Application of Electrical Capacitance Tomography for bulk solids flow analysis in silos, *Particle & Particle Systems Characterization*, v. 23, 3-4, pp. 306-312.

Dyakowski T., Mosorow W., Sankowski D., M. Seaton, Mazurkiewicz Ł., Grudzień K., Jaworski A., 2003, Virtual Channel Concept for Measuring Solids Mass Flow Rate. 3rd World Congress on Industrial Process Tomography, The Rockies Alberta Canada, pp. 535-540.

Dyakowski, T., Edwards, RB., Xie, CG., Williams, RA. (1997). Application of capacitance tomography to gas-solid flows, *Chemical Engineering Science*, Vol.52, pp. 2099-2110, 1997.

Govier G.W., Aziz K. (1972). The flow of complex mixtures in pipes. Van Nostrand Reinhold, New York.

Grudzień, K., Chaniecki, Z., Romanowski, A., Niedostatkiewicz, M. and Sankowski, D. (2012). ECT Image Analysis Methods for Shear Zone Measurements during Silo Discharging Process, Chinese Journal of Chemical Engineering, 20, pp. 337-345.

Grudzien K., Romanowski A., Chaniecki Z., Niedostatkiewicz M., Sankowski D.:, Description of the silo flow and bulk solid pulsation detection using ECT. Flow Measurements and Instrumentation, Vol. 21, 2010, p. 198-206.

Grudzien K., Romanowski A., Williams R.A., Application of a Bayesian approach to the tomographic analysis of hopper flow, Particle & Particle Systems Characterization, Volume 22, Issue 4, p.246-253.

Grudzien, K., (2008). Algorithms of data processing and analysis in electrical capacitance tomography for diagnosis of selected industrial processes, Zeszyty Naukowe. Elektryka / Politechnika Łódzka (in polish), Issue 111, p. 5-12.

Grudzien, K., Romanowski, A. ; Sankowski, D. ; Aykroyd, R.G. ; Williams, R.A, (2005). Advanced statistical computing for capacitance tomography as a monitoring and control tool, Intelligent Systems Design and Applications, 2005. ISDA '05. Proceedings. 5th International Conference, p. 49-54.

Hetsroni G., 1982: Handbook of Multiphase Systems, New York, McGraw-Hill.

Holstad, M.B. (2004). Gamma-Ray Scatter Methods Applied to Industrial Measurement Systems, PhD Thesis, University of Bergen, ISBN 82-497-0240-9.

Holstad, M.B., Johansen G.A., Jackson P., Eidsens K.S. (2003). Scattered gamma radiation Utilized for Level Measurements in Gravitational Separators, Sensors Journal, IEEE, 5, pp. 175-182.

Hoyle B., McCann H., Scott DM. (2005). Process Tomography, edit by Scott & McCann, Process Imaging for automatic control, Taylor and Francis Group, pp.85-126.

Jain A.K., Murty M.N., Flynn P.J., 1999, Data Clustering: A Review, ACM Computing Surveys; 31, 3, p. 264-323.

Jaworski, AJ., Dyakowski, T. (2002). Investigations of flow instabilities within the dense pneumatic conveying system, *Powder Technology*, Vol. 125, pp. 279– 291.

Jones, M., Mason, D., Buchan, L. (1999). Approaches to the Design of Pneumatic Conveyors in Large Scale Applications, *PORT TECHNOLOGY INTERNATIONAL*, pp. 169-172.

Klinzing GE., Marcus RD., Rizk F., Leung LS., 1997: Pneumatic conveying of solids - theoretical and practical approach, Chapman & Hall.

Lionheart W.R.B., 2004, Review: Developments in EIT reconstruction algorithms: pitfalls, challenges and recent developments, Physiol. Meas., Vol. 25, pp. 125-142.

Lionheart, WRB, (2004). Review: Developments in EIT reconstruction algorithms: pitfalls, challenges and recent development. *Physiol. Meas.*, Vol. 25, 125-142.

Miller RW. (1983). Flow Measurement Engineering Handbook, New York, McGraw-Hill, p.184.

Mosorov V., Sankowski D., Mazurkiewicz L., Dyakowski T., 2002, The application of electrical capacitance tomography for solids mass flow measurements. Measurement Science and Technology, 13(12), 2002, 1810-1814.

Neidighs S. (2005). Introduction to the Theoretical and Practical Principles of Pneumatic Conveying, Neuero Corporation, West Chicago, IL, USA.

Neuffer, D., Alvarez, A., Owens, DH., Ostrowski, KL., Luke, SP., Williams, RA. (1999). Control of pneumatic conveying using ECT, *Proc. 1st World Congress on Industrial Process Tomography*, Buxton, pp. 71 – 77.

Niedostatkiewicz M., Tejchman J., 2003, Experimental and theoretical studies on resonance dynamic effects during silo flow, Powder Handling & Processing, 15, 1, pp. 36-42.

Niedostatkiewicz M., Tejchman J., Chaniecki Z., Grudzień K.: Determination of bulk solid concentration changes during granular flow in a silo with ECT sensors. Chemical Engineering Science. 64, 2008, pp. 20-30.

Nieuwland J.J., Meijer R., Kuipers JAM., van Swaaij WPM., (1996). Measurement of solids concentration and axial solids velocity in gas-solid two-phase flows, Powder Technology 87, pp.127-139.

Ostrowski K., Luke SP., Bennett MA. and Williams RA., (1999). Real time visualisation and analysis of dense phase powder conveying, *Powder Technology*, Vol. 102, pp. 1–13.

Ostrowski, K., Luke, SP., Bennett MA., Williams RA. (1999). Application of capacitance electrical tomography for on-line and off-line analysis of flow pattern in horizontal pipeline of pneumatic conveyor, *Chemical Engineering Journal*, 3524, pp.1-8.

Plaskowski A., Beck M.S., Thorn R., Dyakowski T., 1995, Imaging Industrial Flows, Institute of Physics Publishing Ltd.

Romanowski A., Grudzien K., Williams R.A., (2006): Analysis and Interpretation of Hopper Flow Behaviour Using Electrical Capacitance Tomography, Particle & Particle Systems Characterization, Volume 23, Issue 3-4, p. 297-305.

S.L. McKeea, T. Dyakowski, R.A. Williamsa, T.A. Bell, T. Allen. (1995). Solids flow imaging and attrition studies in a pneumatic conveyor. Powder Technology, Volume 2, pp. 105+113.

Sankowski D., Mosorow W., Grudzień K., 2003, Statistical Method for Analysis of Multi-dimensional Tomographic Data. In Proc. IEEE VII-th CADSM, Lviv-Febr, pp.110-112.

Sankowski, D., Mosorov, V., Grudzien, K., Mass flow measurements based on a virtual channel concept, , Intelligent Systems Design and Applications, 2005. ISDA '05. Proceedings. 5th International Conference, p. 274-279.

Scott, DM., McCann, H. (2005). *Process Imaging for automatic control*, Taylor and Francis Group, p.439.

Seville JPK., Tuzun U., Clift R., 1997: Processing of Particulate Solids, Blackie Academic, London, p. 377.

Sideman, S., Hijikata, K. (1993). Imaging in Transport Processes, Begell House, p.621.

Someya S, Takei M. (2005). Direct Imaging Technology, edit by Scott & McCann, Process Imaging for automatic control, Taylor and Francis Group, pp.35-84.

Wilde K., Rucka M., Tejchman J.: Silo music-mechanism of dynamic flow and structure interaction. Powder Technology. 186, 2008, pp. 113-129.

Williams, R.A., Beck, M.S., (1995). *Process Tomography — Principles, Techniques and Applications*, Butterworth-Heinemann, Oxford, p.507.

Woodhead SR., Barnes RN., Reed AR. (1990). On-line mass flow measurement in pulverized coal injection system, Powder Handling and Processing, 2, 123-7.

Yang WQ, Peng L. (2003). Image reconstruction algorithms for electrical capacitance tomography, Meas. Sci. Technol. 14, R1-R13.

Yang, WQ., Liu S. (2000). Role of tomography in gas/solids flow measurement, *Flow Meas. and Instrum.*, Vol. 11, pp. 237-244.

Yang, WQ., Liu, S., Role of tomography in gas/solids flow measurement, *Flow Measurement and Instrumentation*, Volume 11, p. 237-244.

# CHAPTER 14

## DEDICATED 3D IMAGE PROCESSING METHODS FOR THE ANALYSIS OF X-RAY TOMOGRAPHY DATA: CASE STUDY OF MATERIALS SCIENCE

Laurent Babout [1] and Marcin Janaszewski

*Institute of Applied Computer Science*
*Lodz University of Technology*
*90-924 Łódź, ul. Stefanowskiego 18/22*
*{l.babout, janasz}@kis.p.lodz.pl*

### 14.1 Introduction

Image processing is one of the main branches of computer science and an important aspect of machine vision. Many books have been written in the last three decades, mainly about two-dimensional (2D) images, while rather few have been dedicated to three-dimensional (3D) images. Additionally, most of the applications, which have revealed the need of 3D image processing, have concerned biomedical science. This is until recently that materials science applications have come into play, thanks to the emergence of X-ray microtomography. Indeed, 3D microstructure of materials, revealed by this non-destructive technique, may reveal challenges for the image processing community. This chapter illustrates this aspect. Firstly, different 3D image processing procedures are presented and their application to 3D images of different material science scenarios are presented. These methods concern the use of edge-preserving smoothing, topological methods and image processing dealing with texture images.

It is worth mentioning that the authors have used different software such as Amira® and Matlab®, as well as the image processing library PINK (Couprie 2013) to generate the processed images.

Last but not least, the examples of microstructural images concern past and current studies carried out by the authors in collaboration with different

---

[1] l.babout@kis.p.lodz.pl

researchers from the School of Materials of the University of Manchester (i.e. Prof. T.J. Marrow (now at the University of Oxford), Prof. P. Prangnell, Prof. M. Preuss and Prof. P.J. Withers). Each example presented in this section has revealed challenging problems, as far as image processing is concerned. Some have been solved, while others are still under investigation. What all the studies have in common is to deal with key engineering applications. Indeed, the examples are closely related to cracking in nuclear or aero engine materials, metal welding or the fabrication process of cellular materials.

## 14.2  Non linear filtering methods

An important aspect in the different steps of image processing is image denoising or enhancement. A great number of methods have been thought to improve the quality of images, from linear to nonlinear methods. Most of them are associated with filtering, which means using information from the neighborhood of pixels (voxels) to modify their intensity and, ultimately, modifying the output image. If we had to pick the most well-known methods for filtering using linear and non-linear methods, that will be Gaussian filtering and median filtering, respectively. These two smoothing techniques are effective at removing noise in smooth patches or smooth regions of a signal, but adversely affect edges. This drawback is crucial:  whilst it is important to reduce the noise in a signal for better perception and future processing purposes, it is important to preserve the edges in images since they are of critical importance to their visual appearance, for example. To some extent, when the level of noise is moderate and of speckle or salt and pepper type, median filter does perform better than Gaussian filter to remove noise whilst preserving edges for a given, fixed window size, as verified by (Arias-Castro and Donoho 2009). However, for higher level of (Gaussian) noise, the method lacks in contour/surface preservation. Therefore, the current chapter aims at presenting two state-of-the-art methods, which have been originally designed for 2D images, but can easily be extended to N-D. These are the well-known mean shift and nonlinear diffusion filtering methods.

### 14.2.1  *Mean shift*

Firstly, how can we assess that the mean shift method proposed by (Fukunaga and Hostetler 1975), rediscovered by (Cheng 1995) and popularized by Comaniciu and Meer (Comaniciu and Meer 2002) is among the most well-known methods developed in the last decade? The first indicator is the number of times the last mentioned article has been cited: more than 2000 works have been referring to this approach[2], spanning a wide range of scientific domains such as automation control systems, telecommunication, medical imaging, geosciences, materials science and, of course, computer science. The wide range of applications also comes from the diversity of scenarios the method can deal with to analyze feature space: clusterization (Lee and Lee 2005, Wu and Yang 2007, Zhou et al. 2009), discontinuity preserving smoothing (Jarabo-Amores et al. 2011, Paris 2008), image segmentation (Liu et al. 2008, Mayer and Greenspan 2009, Tao et al. 2007), pattern detection (Huang and Zhang 2008, Kim et al. 2003) or object tracking (Comaniciu et al. 2003, Debeir et al. 2005, Yilmaz et al. 2006), to cite a few.

The main stream of the method relies in finding the location of the maxima of the density function from a discrete set of points or sampled data using kernel(s) of given bandwidth(s). This detection step is done by determining the weighted mean of neighbor points delimited by a kernel bandwidth. The value is used in the next iteration step until convergence. By translating the kernel position to the new position defined by the weighted mean of the previous iteration, the method defines a path to a stationary point. Since this point may be the *final destination* of different source points in the data set, clustering will take place in the data set. Alternatively, smoothing will also occur if the weighted mean value obtained at the last iteration step is re-allocated at the source point.

In the case of imaging, the *feature* space is defined as the combination of the *spatial* domain, which corresponds to the image lattice, and the *range* domain, which corresponds to the intensity (gray level or color) of the pixels/voxels (in 3D). Therefore, the weighted mean calculation is done by using the product of two kernels: a *spatial* kernel with bandwidth $h_s$ and a *range* kernel with bandwidth $h_r$. For a given joint vector $\mathbf{x}$, composed of the spatial location $\mathbf{x}^s$ and a range $\mathbf{x}^r$ (i.e. color, intensity), the *joint* kernel is expressed as:

$$K\left(\mathbf{x}\right) = k_{h_s}\left(\mathbf{x}^s\right) k_{h_r}\left(\mathbf{x}^r\right)$$ (14.1)

It has been shown that the mean shift vector $\mathbf{m}_j$, which corresponds to the translation of the joint kernel between 2 successive iterations $j$ and $j+1$, is given by

---

[2] *Source: Web of Science (April 2013).*

$$m_j = \frac{\sum_{i=1}^{n} x_i g_{h_s}\left(\left\|\frac{\left\|x_i^s - x_j^s\right\|^2}{h_s}\right\|\right) g_{h_r}\left(\left\|\frac{\left\|x_i^r - x_j^r\right\|^2}{h_r}\right\|\right)}{\sum_{i=1}^{n} g_{h_s}\left(\left\|\frac{\left\|x_i^s - x_j^s\right\|^2}{h_s}\right\|\right) g_{h_r}\left(\left\|\frac{\left\|x_i^r - x_j^r\right\|^2}{h_r}\right\|\right)} - x_j \qquad (14.2)$$

where the Kernel G, associated with the kernel profile g, is the derivative of the kernel K. For example, in the classical case of the kernel defined by the Epanechnikov profile, i.e.

$$k_E(x) \propto \begin{cases} 1 - \|x\|^2 & \|x\| \le 1 \\ 0 & \text{otherwise,} \end{cases} \qquad (14.3)$$

G is equivalent to a radially symmetric kernel of 1 for radius lower than the corresponding bandwidth. From equation (14.2), one can see that only the bandwidth parameter $h=(h_s, h_r)$ has to be set by the user.

Based on the previous equations, the mean shift filtering corresponds to the following steps. Let $x_i$ and $z_i$, $i=1,\ldots, n$ be the $d$-dimensional input and filtered output image in the joint spatial range domain (in the case of a grey level 3D image, $d=4$). Let assume that the Epanechnikov profile has been used. For each pixel/voxel:

- *Initialize $j=1$ and $y_{i,1}=x_i$*
- *Select the set of points $S_i$, which have Euclidean distance to $x_i$ lower than* **h**
- *Calculate the mean joint vector $m_{i,j}$ of $S_i$, and compute the new position $y_{i,j+1}=m_{i,j}+y_{i,j}$ based on equation (14.2) until convergence, $y=y_{i,c}$*
- *Assign $z_i=(x_i^s, y_{i,c}^r)$*

The method is illustrated in Fig. 14.1 for a 2D image of random values and for the first iteration. One can see that the kernel $k_{hs}$ (represented in grey color) moves towards the region of pixels with values contained in the range defined by the bandwidth $h_r$. The procedure will stop when the mean shift magnitude in the feature space is lower than a given threshold (e.g. lower than 1). This implies that both the kernel translation and the change in the range value are minimized.

| 130 | 205 | 231 | 54 | 254 | 75 | 81 | 115 | 206 | 147 | 83 |
|---|---|---|---|---|---|---|---|---|---|---|
| 58 | 93 | 95 | 136 | 179 | 174 | 227 | 145 | 63 | 125 | 167 |
| 140 | 186 | 65 | 162 | 133 | 108 | 119 | 114 | 67 | 59 | 223 |
| 150 | 145 | 223 | 10 | 105 | 186 | 235 | 40 | 26 | 39 | 194 |
| 207 | 80 | 190 | 239 | 53 | 161 | 171 | 148 | 73 | 183 | 207 |
| 208 | 83 | 245 | 243 | 205 | 233 | 31 | 113 | 41 | 153 | 121 |
| 54 | 101 | 92 | 100 | 102 | 67 | 116 | 216 | 187 | 131 | 72 |
| 179 | 25 | 245 | 247 | 26 | 2 | 85 | 21 | 112 | 214 | 115 |
| 224 | 69 | 216 | 143 | 230 | 185 | 205 | 218 | 189 | 72 | 73 |
| 67 | 7 | 170 | 240 | 245 | 230 | 134 | 140 | 254 | 127 | 91 |
| 214 | 168 | 70 | 54 | 159 | 96 | 24 | 252 | 210 | 27 | 156 |

(a)

| 130 | 205 | 231 | 54 | 254 | 75 | 81 | 115 | 206 | 147 | 83 |
|---|---|---|---|---|---|---|---|---|---|---|
| 58 | 93 | 95 | 136 | 179 | 174 | 227 | 145 | 63 | 125 | 167 |
| 140 | 186 | 65 | 162 | 133 | 108 | 119 | 114 | 67 | 59 | 223 |
| 150 | 145 | 223 | 10 | 105 | 186 | 235 | 40 | 26 | 39 | 194 |
| 207 | 80 | 190 | 239 | 53 | 161 | 171 | 148 | 73 | 183 | 207 |
| 208 | 83 | 245 | 243 | 205 | 233 | 31 | 113 | 41 | 153 | 121 |
| 54 | 101 | 92 | 100 | 102 | 67 | 116 | 216 | 187 | 131 | 72 |
| 179 | 25 | 245 | 247 | 26 | 2 | 85 | 21 | 112 | 214 | 115 |
| 224 | 69 | 216 | 143 | 230 | 185 | 205 | 218 | 189 | 72 | 73 |
| 67 | 7 | 170 | 240 | 245 | 230 | 134 | 140 | 254 | 127 | 91 |
| 214 | 168 | 70 | 54 | 159 | 96 | 24 | 252 | 210 | 27 | 156 |

(b)

Fig. 14.1 Example of mean shift procedure in 2D image with random intensity values. Initial joint spatial-range point: (6, 6, 233), **h**=(4, 15), Epanechnikov profile. (a) Pixels satisfying the range kernel of bandwidth $h_r$ and located in the spatial kernel of bandwidth $h_s$ (shown in grey) are shown in black boxes. (b) After the first iteration, the new joint point (4.81, 6.81, 235.3), marked by the vector, is used. One can clearly see the shift of the kernel mask, as well as new pixels used for the mean calculation. After this step, the new position in the joint spatial-range domain is: (4.31, 6.98, 237.9).

### 14.2.2 Nonlinear diffusion filtering

Another interesting approach, which is very popular in the image processing community (almost 4000 citations[3]) and has been considered to smooth images whilst preserving edges, is based on nonlinear diffusion filtering. This is also referred in the literature as anisotropic diffusion filtering. The concept has been proposed by Perona and Malik (Perona and Malik 1990) in the beginning of the 90s with the introduction of the physical concept of diffusion to image filtering. Based on the Fick's law of diffusion, the equation relates a filtered image $u(x,t)$ of an original image $f(x)$ in the $m$-dimensional space to a diffusivity parameter $g$ as:

$$\frac{\partial u}{\partial t} = \text{div}\left[ g\left(\left|\nabla u_\sigma\right|^2\right)\nabla u \right]$$

(14.4)

where $\nabla u$ and $\nabla u_\sigma$ correspond to the gradients of the original and smoothed version of u at "time" t. The smoothed version is obtained by convolving the original image with a (truncated) Gaussian mask of standard deviation $\sigma$ and width w

$$w = 4\sigma\sqrt{2\ln 2}$$

(14.5)

---

[3] *Source: Web of Science (April 2013).*

The "time" t is associated with the simplification (or degree of smoothing) of the image: the larger t, the simpler the image. However, the parameter that will affect the degree of edge preservation the most during smoothing is the choice of the diffusivity function g. The basic trend is that the diffusivity should decrease steeply as the gradient in the image increases, which reflects the fact that edges should not be filtered as strongly as on both their sides. A typical diffusivity function used in the literature is presented in Table 14.1 and illustrated in Fig. 14.2. In all these expressions, $\lambda$ plays the role of a contrast parameter: features with $\left|\nabla u_\sigma\right| > \lambda$ are regarded as edges (since the value is close to zero), while features with $\left|\left|\nabla u_\sigma\right|\right| < \lambda$ are considered interior regions.

Table 14.1 Popular diffusivity functions.

| Method | Diffusivity $g(s)$ |
|---|---|
| *(Charbonnier et al. 1994)* | $1\big/\sqrt{1+s^2\big/\lambda^2}$ |
| *(Perona and Malik 1990)* | $1\big/\left(1+s^2\big/\lambda^2\right)$ |
| *(Weickert et al. 1998)* | $\begin{cases} 1 & (s \le 0), \\ 1-\exp\left(-3.315\big/(s/\lambda)^4\right) & (s > 0). \end{cases}$ |

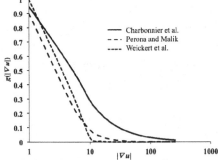

Fig. 14.2 Examples of diffusivity function vs. gradient for $\lambda=3$ and grey scale range (i.e. [0 255]). The diffusivity from (Weickert et al. 1998) presents the steepest decrease with diffusivity close to zero when gradient around 10.

In the case of images, it is common to express the diffusion equation in the discrete space. In the case of an image in the m-dimensional space, it has been shown in (Weickert et al. 1998) that the m-D case of the diffusion process (equation 14.4) can be decomposed in m 1-D cases, as follows:

$$\frac{\partial u}{\partial t} = \sum_{i=1}^{m} \partial_{x_i} \left[ g\left(\left|\nabla u_\sigma\right|^2\right) \partial_{x_i} u \right] \tag{14.6}$$

Therefore, the discretization step is performed sequentially *m*-times, using the following simplest approximation in the vector-matrix notation:

$$\frac{u^{k+1} - u^k}{\tau} = \left( \sum_{l=1}^{m} \mathbf{A}_l\left(u^k\right) \right) u^k \tag{14.7}$$

where $u^k$ denotes the approximation to u at discrete time $t_k = k\tau$ ($\tau$ being the time step size). Moreover, $\mathbf{A}_l(u^k) = [\mathbf{a}_{ijl}(u^k)]$ is a tridiagonal tensor of the diffusivity approximation along the coordinate axis l, where the matrix elements $\mathbf{a}_{ijl}(u^k)$ are defined as:

$$a_{ijl}\left(u^k\right) = \begin{cases} \dfrac{g_{il}^k + g_{jl}^k}{2} & \left[j \in \Omega(i)\right], \\[2ex] -\sum_{n \in \Omega(i)} \dfrac{g_{il}^k + g_{nl}^k}{2} & (j = i), \\[2ex] 0 & (else). \end{cases} \tag{14.8}$$

The term $g_{il}^k$ refers to the local diffusivity based on the grey values of the set $\Omega(i)$ of the two neighbors of pixel i along axis l and is expressed as:

$$g_{il}^k := g\left[ \frac{\left(u_{(i+1)l}^k - u_{(i-1)l}^k\right)^2}{4} \right] \tag{14.9}$$

It has been shown by (Weickert et al. 1998) that the explicit (forward) scheme to solve equation (14.7) guarantees discrete scale-space, but only for very a low time step value (i.e. $\tau < 1/2$), which is practically restrictive. However, the semi-implicit scheme that they propose assures a discrete nonlinear diffusion scale-space for arbitrary large time steps. Moreover, by using the Thomas algorithm to solve tridiagonal linear systems, they achieve a linear running time, which is only 2 times slower than the explicit scheme.

## 14.3 Methods based on digital topology

Digital spaces are increasingly important in everyday life as we use more and more digital devices, which generate an enormous amount of digital data. Usually the data has to be processed or analyzed to extract or enhance important information. Therefore we have more and more problems to solve in digital spaces. A significant number of such problems might be solved with methods based on digital topology.

Topology has been developed on the basis of geometry and set theory, and is the study of both the fine structure and global structure of space. The general topology considers continuous spaces in which any ball (even of a very small size) has an infinite number of points. The fundamental assumption is not fulfilled in discrete spaces where there are an infinite number of balls, which contain only one point. Therefore topology cannot be directly applied to digital spaces.

This is why in the last 40 years the dynamic development of digital topology, which is equivalent to general topology but applied to digital spaces, has been observed. In the following subsections we present selected digital topology notions and algorithms, which have found applications in image analysis and processing.

### 14.3.1 *Basic notions of digital topology*

In this section, we recall some topological notions for binary images, which are of importance in understanding the following part of the section. A more extensive review is provided in (Kong and Rosenfeld 1989). We will present definition for three-dimensional space but analogous definitions are also valid in two-dimensional space.

We denote the set of integers by $\mathbb{Z}$, by $\mathbb{N}$ the set of nonnegative integers, by $\mathbb{N}_+$ the set of strictly positive integers and by $\mathbb{R}_+$ the set of strictly positive real values. Moreover, assume that $E = \mathbb{Z}^3$. Let $X \subseteq E$, $\overline{X}$ denote complement of the set X that is $\overline{X} = E \setminus X$. A point $x \in E$ is defined by $(x_1, x_2, x_3)$, with $x_i \in \mathbb{Z}$. For any $x \in E$ we consider the three neighborhoods: $N_6$, $N_{18}$, $N_{26}$ defined by:

$$N_6(x) = \left\{ y \in E : \left|x_1 - y_1\right| + \left|x_2 - y_2\right| + \left|x_3 - y_3\right| \le 1 \right\},$$

$$N_{26}(x) = \left\{ y \in E : \max\left[\left|x_1 - y_1\right|, \left|x_2 - y_2\right|, \left|x_3 - y_3\right|\right] \le 1 \right\}, \quad (14.10)$$

$$N_{18}(x) = \left\{ y \in E : \left|x_1 - y_1\right| + \left|x_2 - y_2\right| + \left|x_3 - y_3\right| \le 2 \right\} \cap N_{26}(x).$$

The set $N_n(x)$ is called the n-neighbourhood of x. We define $N_n^*(x) = N_n(x) \setminus \{x\}$, with n = 6, 18, 26. Any point y of $N_n(x)$ is said to be n-adjacent (n = 6, 18, 26) to x, we also say that y is an n-neighbour of x. We set $N_{18}^+(x) = N_{18}^*(x) \setminus N_6^*(x)$ and $N_{26}^+(x) = N_{26}^*(x) \setminus N_{18}^*(x)$. Two points x and y are said to be strictly n-adjacent (n = 18, 26) if they are $N_n^+$-adjacent.

An object $X \subseteq E$ is said to be n-connected if, for any two points of X, there is an n-path in X between these two points. The equivalence classes related to this relation are the n-connected components of X. Note that, if X is finite, the infinite connected component of $\overline{X}$ (complementary of X) is called the background, the other connected components of $\overline{X}$ are called cavities. The set composed of all the n-connected components of X is denoted by $C_n(X)$. Moreover, if we use an n-connectivity for X we have to use another m-connectivity for $\overline{X}$, i.e., the 6-connectivity for X is associated to the 18- or the 26-connectivity for $\overline{X}$ (and vice versa). This is necessary for avoiding connectivity paradoxes (Kong and Rosenfeld 1989).

We recall now some definitions (Bertrand 1994), which allow one to extract topological characteristics of a point.

*The geodesic n-neighbourhood of x inside $X \subseteq E$ of order k is the set $N_n^k(x, X)$ defined recursively by $N_n^1(x, X) = N_n^*(x) \cap X$ and $N_n^k(x, X) = \cup \{N_n(y) \cap N_{26}^*(x) \cap X, y \in N_n^{k-1}(x, X)\}$.*

*The geodesic neighborhoods $G_n(x, X)$ are defined by $G_6(x, X) = N_6^2(x, X)$, $G_{18}(x, X) = N_{18}^2(x, X)$ and finally by $G_{26}(x, X) = N_{26}^1(x, X)$. We can now give a definition of topological number (Bertrand 1994):*

Let $X \subseteq E$ and $x \in E$, the two *topological numbers* are defined as follows (#X stands for the cardinality of X):

$$T_n(x, X) = \# C_n[G_n(x, X)] \tag{14.11}$$

$$T_m(x, \overline{X}) = \# C_m[G_m(x, \overline{X})] \tag{14.12}$$

The topological characteristic of a point can be established based on the numbers defined above. In particular, these numbers allow one to detect whether a point is simple (Rosenfeld 1970). Informally, a simple point p of a discrete object X is a point, which is "inessential" to the topology of X. In other words, we can remove the point p from X without "changing the topology of X". The notion of simple point is fundamental to the definition of topology-preserving transformations in discrete spaces. Bertrand (Bertrand 1994) proved that $x \in X \subseteq E$ is an n-simple point iff $T_n(x, X) = 1$ and $T_m(x, \overline{X}) = 1$.

The topological numbers allow other types of points to be detected. A point $x \in X$ such that $T_n(x, X) = 0$ is said to be isolated. If $T_m(x, \overline{X}) = 0$, we have an interior point and border points are characterized by $T_m(x, \overline{X}) \neq 0$. Let us consider a point x such that $T_n(x, X) \geq 2$. If we delete x from X, we locally disconnect X. We can say that such a point is a 1D isthmus.

In the same manner, a point such that $T_m(x, \overline{X}) \geq 2$ may be called a 2D isthmus since its deletion locally merges connected components of $\overline{X}$. A point such that $T_n(x, X) = 2$ and $T_m(x, \overline{X}) = 1$ is a simple 1D isthmus. We also have a simple 2D isthmus if $T_m(x, \overline{X}) = 2$ and $T_n(x, X) = 1$.

In (Malandain et al. 1993) one can find interesting characteristic of points in 3D, which can be utilized for the construction of different types of skeletonization algorithms. Visual interpretation of 3D points is presented in Fig. 14.3. The characteristic proposed by Malandain *et al.* is presented in Table 14.2.

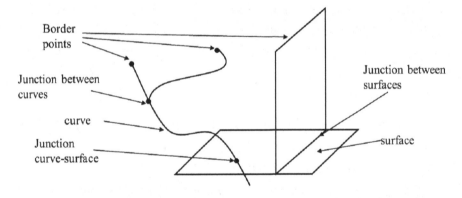

Fig. 14.3 Types of points in 3D space (Malandain et al. 1993)

Let *X, Y* are finite subsets of *E. Y* is a *homotopic thinning of X* if *Y = X* or if *Y* may be obtained from *X* by iterative deletion of simple points. We say that *Y* is an *ultimate homotopic skeleton of X* if *Y* is a homotopic thinning of *X* and if there is no simple point for *Y*. Let *C* be a subset of *X*. We say that *Y* is an *ultimate homotopic skeleton of X constrained by C* if $C \subseteq Y$, *Y* is a homotopic thinning of *X* and if there is no simple point for *Y* in $Y \setminus C$ (see e.g. (Vincent 1991)). The set *C* is called the *constraint set* relative to this skeleton.

Table 14.2 Characteristic of 3D points with the use of topological numbers (Malandain et al. 1993).

| Point type | $T_{26}(x,X)$ | $T_6(x,\bar{X})$ |
|---|---|---|
| Interior point | any | 0 |
| Isolated point | 0 | any |
| Border point | 1 | 1 |
| Curve point (simple 1D isthmus) | 2 | 1 |
| Curve junction | >2 | 1 |
| Surface point (simple 2D isthmus) | 1 | 2 |
| A surface-curve(s) junction | ≥2 | 1 |
| A surface junction | 1 | >2 |

Let $x, y \in E$, we denote by $d^2(x, y)$ the square of Euclidean distance between x and y, that is, $d^2(x, y) = (x_1 - y_1)^2 + (x_2 - y_2)^2 + (x_3 - y_3)^2$. Let $Y \subset E$, we denote by $d^2(x, Y)$ the square of the Euclidean distance between x and the set Y, that is, $d^2(x, Y) = \min\{d^2(x, y) : y \in Y\}$.

Let $X \subset E$ (the "object"), we denote by $D_X^2$ the map from E to $\mathbb{N}$ which associates, to each point x of E, the value $D_X^2 = d^2(x, \bar{X})$. The map $D_X^2$ is called the squared Euclidean distance map of X.

Let $r \in \mathbb{N}_+$, we denote by $B_r(x)$ the ball of (squared) radius r centred on x, defined by $B_r(x) = \{y \in E, d^2(x, y) < r\}$. A ball $B_r(x) \subseteq X \subseteq E$ is maximal for X if it is not strictly included in any other ball included in X. Notice that, for any point $x \in X$, the value $D_X(x)$ is precisely the radius of the maximal ball included in X and centered on x.

The *medial axis of X*, denoted by MA(X), is the set of the centers of all the maximal balls for X. An efficient algorithm has been proposed to extract the exact Euclidean medial axis of a shape, from an exact squared Euclidean distance map and pre-computed look-up tables (Borgefors et al. 1991, Remy and Thiel 2005).

## 14.3.2 *Skeletonization*

One of the most important digital topology applications of object processing concerns skeletonization algorithms. A skeleton is a simple representation of a digital object, which has many important applications e.g.: object matching (Sundar et al. 2003), virtual navigation (Prteux et al. 2004), analysis of human organs (Janaszewski et al. 2009), image registration (Aylward and Bullitt 2002) or morphing (Blanding et al. 2000).

The notion of the skeleton was originally defined by (Blum 1967) through an analogy with a grassfire. If we imagine an object as a field of grass and we set fire to the border of the field, the fire will start propagating inside the field and when fire fronts meet, they vanish. The meeting points of the flame fronts will constitute the skeleton of the object.

Informally, a 3D skeleton can be a set of curves or surfaces or both, which passes through the center of an object. Some simple objects such as a cylinder have only one curve called the centerline. However, there are complicated objects with many branches, tunnels and cavities where a skeleton can be a very complicated structure. Therefore, in the literature different researchers have proposed a set of properties, which are used to define a skeleton. Hilditch gave the first four properties that a skeleton in two-dimensional space should possess (Hilditch 1969), that is: a skeleton should be homotopic to the original object, it should be thin, it should be centered in the original object and making skeletonization of the skeleton should not change anything. Later, such a set of properties was generalized to an $n$-dimensional case and extended by adding more properties, e.g. smoothness, robustness, reliability, cost effectiveness (Cornea et al. 2007, Palagyi 2002). Some of these properties - for example, the property, which states that skeleton should be smooth and reliable - are specially designed and used in particular applications (He et al. 2001, Kang and Ra 2004).

One of the simplest strategy of thinning is to remove simple points sequentially from the input object $X$ until stability, that is, until no more simple points exist. In such a way we obtain the ultimate homotopic skeleton of $X$ (see definition in section 0). Such a strategy results in a skeleton, which has the same topology as an input object $X$, but other important properties are not preserved (for example centricity). Therefore usually more sophisticated strategies are used, which preserve some important points of an input object. In this way one obtains the ultimate homotopic skeleton constrained by a set of points (see definition in section 0). The possibility to decide which points are preserved during thinning gives us a tool to produce different kinds of skeletons. The decision of what kind of skeleton should be used strongly depends on its application. We can distinguish different types of skeletons with respect to the dimension of the result:

- Curvilinear skeleton - is a 1D representation of the object, which preserves both topology and geometry of the object. In order to obtain such kind of skeleton a thinning algorithm has to preserve curve points and curve junction points (see Table 14.2).
- Surface skeleton - in three-dimensional case we can also compute a two-dimensional skeleton. It will provide more geometrical information than

a curvilinear skeleton, which can be useful in some applications. In order to obtain such a kind of skeleton a thinning algorithm has to preserve surface points, surface junction points or 2D isthmuses (see Table 14.2).

Another reason for the generation of constrained skeleton is to preserve centricity property. To do so the skeleton is constrained by the points of the medial axis. The points are by nature centered in the input object, so preservation of them in a skeleton ensures centricity property. Unfortunately, the medial axis is sensitive to small contour perturbations. This is a major difficulty when the medial axis is used in practical applications (e.g. shape recognition). Therefore, it is usually necessary to add a filtering step (or pruning step) to obtain a simplified and more stable representation of an input object.

The simplest strategy to filter the medial axis is to keep only points, which are centers of maximal balls of at least the diameter given. Different criteria can be used to locally threshold and discard spurious medial axis points or branches: see (Attali and Montanvert 1996, Couprie et al. 2007) for methods based on the angle formed by the vectors to the closest points on the shape boundary, or the circumradius of these closest points (Coeurjolly and Montanvert 2007, Hesselink and Roerdink 2008).

Taking into account the above skeletonization algorithm, which preserves important properties described earlier might be presented in the following pseudo-code.

---

**3DSkeleton**(*Input: image X, parameter p; Output S*)

---

01. $Y \leftarrow MA(X)$     //Extraction of medial axis
02. $Y' \leftarrow Filter(Y, p)$ // Medial axis filtering
03. $P \leftarrow ExtractPriority(X)$     //Extraction of priority function for //skeletonization
04. $S \leftarrow UHCSkel(X, Y', P)$     // Ultimate homotopic constrained //skeletonization

---

The first two steps of the algorithm extract and filter the medial axis of an input object, while the third step calculates a priority function for the skeletonization algorithm. One can use the function defined in (Chaussard et al. 2010). During the last step, the algorithm generates an ultimate homotopic skeleton constrained by filtered medial axis $Y'$ and the thinning process is controlled by the priority function $P$. If one would like to generate a surface skeleton the surface points have to be preserved during thinning.

### 14.3.3 *Tunnel closing*

Another example of an algorithm based on digital topology, useful in many applications is tunnel closing presented for the first time for $\mathbb{Z}^3$ in (Aktouf et al. 2002). From the topological point of view the presence of a tunnel in an object is detected whenever there is a closed path, which cannot be transformed into a single point by a sequence of elementary local deformations inside the object (Kong and Rosenfeld 1989, Rosenfeld 1970). Based on this definition a sphere has no tunnel, a torus has one tunnel and a hollow torus has two tunnels. Moreover, it is important to emphasize the difference between tunnels and cavities. The latter are hollows in an object, or more formally, bounded connected components of the background.

The original version of the algorithm is linear in time and space complexity and may be presented as follows: First it computes a bounding box $Y$, which has no cavities and no holes and which contains the input object $X$. Then it iteratively deletes points of $Y \setminus X$ which are border and not 2D isthmuses (see section 14.3.1). If a border point $p$ of $Y \setminus X$ is also a 2D isthmus it can be deleted only if its distance $d(p, X)$ is greater than a predefined parameter *tunnelsize*. The last condition results in an effect that only tunnels of size less or equal to the parameter *tunnelsize* are closed. The deletion process is ordered by a priority function, which is defined as the Euclidean distance from $X$. The algorithm repeats this procedure until stability. The pseudocode of the tunnel closing algorithm (TCA) can be presented as follows:

---

*TCA (**Input** X, tunnelsize, **Output** Y)*

---

*01. Generate a bounding box Y which contains X*
*02. Repeat until no point to delete:*
*03. Select a point p of Y\X such that: $T_m(p,\overline{Y})=1$ or such that: $T_m(p,\overline{Y})>1$*
  *and d(p, X) > tunnelsize*
*04. which is at the greatest distance from X*
*05. Y := Y\p*
*06. Result: Y*

---

An example of the algorithm result when applied to a chain is presented in Fig. 14.4.

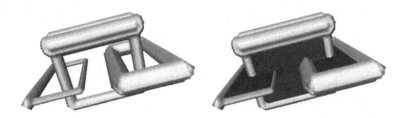

Fig. 14.4 A fragment of a chain in the left hand side. There are three tunnels in the chain. The chain with tunnels filled in the right hand side. The tunnel closing "patches" are presented with dark grey color.

## 14.4 (3D) Image processing dealing with texture

As it is correctly said in the introduction of the book of Petrou and Garcia-Sevilla (Petrou and Garcia-Sevilla 2006), "Texture is all around us, and texture is also on the images we create. Just as variation in what we do allows us to distinguish one day in our life from another, texture allows to identify what we see. And if texture allows us to distinguish the objects around us, it cannot be ignored by any automatic system for vision. Thus, texture becomes a major part of Image Processing, around which we can build the main core of Image Processing research achievements". This comment is particularly true in materials science where microstructure, mainly in metals, presents specific patterns, which can be good candidates for texture analysis. As texture images can be of any kind, such as binary texture, stationary or non-stationary grey texture or even colored-texture images, the author's advice is to refer to the abovementioned book for deeper analysis. Instead, this section will focus on the case or oriented texture images and a set of (potential) methods designed to analyze changes of orientation in patterns and their corresponding boundaries. This will be illustrated by three types of approaches: estimation of local orientation, wavelet transform and its improved extensions, and directional filters.

### 14.4.1 *Estimates of local orientation*

In an image, which contains patterns with marked orientation, a natural approach consists in analyzing the local gradient to estimate the vector field. However, to overcome uncertainties coming from noisy pixels, the analysis of local orientation is usually made over a neighboring domain surrounding each pixel (or voxel in 3D) of the image. Instead of getting the mean value from the neighborhood, a more realistic solution employed in principal component analysis (PCA) consists in calculating the covariance matrix, or more simply stated, the matrix of inertia of the cloud of points and to extract its main axis of inertia by eigenvector decomposition. This method has been used by Jeulin and Moreaud (Jeulin and Moreaud 2008) to segment oriented texture images, such as an image of cellulose cryofracture or the lamellar microstructure of titanium alloy. The latter example has also been presented in (Vanderesse et al. 2008). To do so, they introduced a factor of confidence based on the eigenvalues to detect edges between areas of different main orientations. Indeed, areas with no preferential orientation, such as those close to boundaries, should have very similar eigenvalues (all close to 1/3), while an area with marked orientation should have an eigenvalue much larger than that for the other 2 eigenvectors. The confidence map is further used to estimate zones of well-defined orientation, which are used as markers for a further constrained watershed step to recover boundaries between areas of different orientations. However, their method is very sensitive to noise and it is difficult to predict if boundaries are well recovered. Results are also dependent on the interrogation window used to calculate local orientations.

### 14.4.2 *Discrete wavelet transform (DWT) and its improved extensions*

Since its emergence in the mid 80s by Mallat (Mallat 1988, Mallat 1989), multiresolution analysis for discrete wavelet transform has made a tremendous impact in the signal processing community as the Fast Fourier transform did in earlier times (Polikar 1999). Indeed, it has been exploited with great success in a myriad of signal processing applications and, until now, has been constantly extended. In brief, the DWT replaces the infinitely oscillating sinusoidal basis functions of the Fourier transform with a set of locally oscillating basis functions called wavelets. In the case of 2D images (however what is said hereafter can be easily extended to 3D images), the discrete wavelet transform decomposes the image $f(x,y) \in L^2(R^2)$ in terms of a set of shifted and dilated wavelet functions $\{\Psi^{0o}, \Psi^{90o}, \Psi^{\pm45o}\}$ and scaling function $\Phi(x,y)$

$$f(x,y) = \sum_{k \in Z^2} s_{j_0,k} \Phi_{j_0,k}(x,y) + \sum_{b \in \theta} \sum_{j > j_0} \sum_{k \in Z^2} \omega^b_{j,k} \Psi^b_{j,k}(x,y) \qquad (14.13)$$

with

$$\Phi_{j_0,k}(x,y) = 2^{j_0} \Phi\left(2^{j_0}(x,y) - k\right)$$

$$\Psi^b_{j,k}(x,y) = 2^j \Psi^b\left(2^j(x,y) - k\right)$$

and b $\in \theta = \{0°, 90°, \pm45°\}$. The $0°$, $90°$ and $\pm45°$ denotes the subbands of the wavelet decomposition. A separable 2-D DWT can be computed efficiently in discrete space by convolving the associated 1-D filter bank to each column of the image, then to each row of the resultant coefficients. We can write the 2-D separable transform as follows

$$\Psi^{0°}(x, y) = \Phi(x)\Psi(y)$$

$$\Psi^{90°}(x, y) = \Psi(x)\Phi(y) \qquad (14.14)$$

$$\Psi^{\pm45°}(x, y) = \Psi(x)\Psi(y)$$

where $\Phi$ and $\Psi$ are 1-D scaling and wavelet functions (also known as low-pass and high-pass functions, respectively).

Despite having proven its strength to detect singularities in signals/images and being incorporated, mainly due to the sparsity of the wavelet coefficients, in famous image compression standard JPEG2000, DWT suffers from four fundamental drawbacks: oscillation, shift variance, aliasing and lack of directionality (Selesnick et al. 2005). Indeed, in the latter case, as illustrated in Fig. 14.5a, DWT mainly detect lines along the horizontal and vertical, while for oblique direction, it has a checkerboard appearance since it mixes $+45°$ and $-45°$ orientations. However, there are cases where there are practical applications where ridges, contour or edges in images have to be detected, processed or simulated and more than 3 directions are needed. This is the main reason why in the past 10 years, several new approaches have been developed to solve this drawback such as ridgelet (Donoho 2001), curvelet (Candes et al. 2006), directionlet(Velisavljevic et al. 2006), shearlet (Yi et al. 2009), dual-tree complex wavelet (Selesnick et al. 2005) or the contourlet transform (da Cunha et al. 2006, Do and Vetterli 2005). The frequency domains of the two latter methods are also shown in Fig. 14.5. Note that most of these extensions have mainly been designed for improved image denoising and compression.

Concerning the dual-tree complex wavelet transform (2D-CWT), the mother wavelet is of complex form and composed of two real-valued wavelets, as in the following 1D expression of the high-pass function (similar is the expression for the low-pass function $\Phi$)

$$\Psi(x) = \Psi_h(x) + j\Psi_g(x) \qquad (14.15)$$

where $\Psi_g(x)$ should be approximately the Hilbert transform of $\Psi_h(x)$ (Selesnick et al. 2005). By replacing in the general wavelet expression (14.14) $\Phi$ and $\Psi$ with the expression from (14.15), and taking into account the complex conjugate of $\Phi$ and $\Psi$, one obtains six wavelet subbands, which are strongly oriented in $\theta = \{-75°,-45°,-15°,15°,45°,75°\}$ direction and capture image information in that direction, as illustrated in Fig. 14.5b.

Another example of wavelet extension to compensate the lack of directionality of the DWT is the contourlet method (da Cunha et al. 2006, Do and Vetterli 2005). The authors have proposed another type of strategy to obtain a larger number of subbands than for 2D-CWT, which combines the Laplacian pyramid with the directional filter bank (DFB) proposed by Bamberger and Smith (Bamberger and Smith 1992). The DFB is implemented via an l-level binary tree decomposition that leads to $2^l$ subbands with wedge-shape frequency partitioning. Do and Vetterli (Do and Vetterli 2005) proposed a new construction scheme of the DFB by multiplying a two-channel quincunx filter bank with a fan filter, and the application of a shearing operator to obtain rectangular-like filter with diagonal angles different than $45°$. The modification proposed by Da Cunha et al. (da Cunha et al. 2006) to overcome the problem of shift variance of the multiresolution analysis, despite being more redundant than its counterpart, was based on the nonsubsampled pyramid structure and nonsubsampled directional filter bank (NSDFB), resulting in idealized frequency portioning shown in Fig. 14.5c. Note that this modification was later extended to 3D, known as the surfacelet transform (Lu and Do 2007).

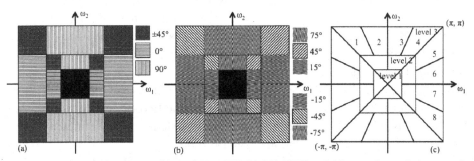

Fig. 14.5 Frequency domain partition of (a) DWT (b) DT-CWT and (c) nonsubsampled contourlet transform (NSCT) showing directional portioning for 3 levels of decomposition.

### 14.4.3 *Directional FIR filter bank*

As mentioned in the previous subsection, the directional filter bank designed in the frequency space has been used in recent wavelet-based methods to mainly denoise, enhance or compress images. A N-D version of the contourlet, called the surfacelet transform (Lu and Do 2007), which uses a multidimensional directional filter bank based on the hourglass shape, is capable of capturing surface-like singularities for further denoising steps. Note that the approach was intended for denoising video sequences rather than segmenting 3D images.

This subsection presents recent approach proposed by the authors to segment patterns in oriented texture images. This has been inspired by the abovementioned method, despite being designed in the discrete space.

In 3D, oriented texture images may look as randomly oriented groups of parallel and equally spaced planes. The detection of these groups of planes requires designing a dedicated directional filter bank (DFB) so as to retrieve the changes of orientation and to delineate boundaries. In that context, the directionality of the features can be retrieved using a 3D filter, which has the following properties:

- cylindrical symmetry
- wedge-shape support

The filter shown in Fig. 14.6a and presented in (Babout et al. 2013) satisfies these conditions. The hourglass filter, which is appropriate for detecting 3D lines, is the complementary part of the presented filter, which is named CHG (Complementary part of Hour Glass filter).

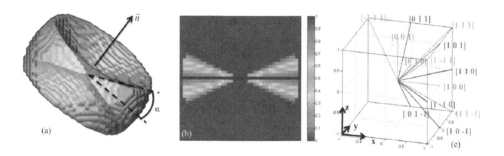

Fig. 14.6 (a) Isosurface of the CHG filter. (b) 2D filter kernel profile. (c) Bank of principal filter normal directions.

In order to have a filter that accepts fluctuations for planes with their normal vector in the vicinity of the vector defining the main direction of the filter, a planar symmetric univariate kernel, which has a similar construction to the Epanechnikov kernel shown in equation (14.3) is used, even though other kernels such as the normal (Gaussian) kernel could be used. This is illustrated in Fig. 14.6b.

The next step concerns the construction of the filter bank. In the approach proposed by (Lu and Do 2007), the checkboard filter bank was used to partition the 3D frequency space into component filters (called surfacelets) with a rectangular-based pyramid shape radiating out from the origin. Back to the spatial domain, the surfacelets have oblate shapes, which indicate a strong filtering response to a specific normal direction of surface-like objects. In the present case, because of the specific shape of the CHG filter, passbands may overlap for directions of neighbor partitioning. This depends on the value of the wedge half-angle $\alpha$. In that context, a slightly different approach is proposed by the authors. A set of normal directions **n** is defined, based on the vertices, edges and faces of the Cartesian cubic grid. More specifically, the family of normal directions <100>, <110> and <111> (following the crystallographic plane and direction notations) regroup 13 independent directions, which are further called the principal directions of the 3D directional filter bank. Their spatial representation is shown in Fig. 14.6c.

The DFB is further used in a texture classification algorithm to locally assign each voxel in the 3D image with the orientation class corresponding to the highest FIR (finite impulse response). We encourage readers to access (Babout et al. 2013) for more details about the proposed algorithm. It is also worth mentioning that this algorithm works for other filter banks, actually in the 2D version, such as the Bamberger DFB (Bamberger and Smith 1992).

## 14.5  Applications

### 14.5.1  *Hole closing: bridge ligament, welding profile*

As mentioned in Section 14.3.3, the tunnel closing algorithm has been thought to repair 3D surfaces by closing tunnels (or holes) lower than a given size. The current section presents two examples that illustrate how this methodology can be very useful to analyse materials science phenomena.

The first example concerns the extraction of bridge ligaments along cracks. In each case of damage accumulation the crack path is interrupted, deflected and bridged. These slow down crack propagation and can increase the materials lifetime. Quantifying the effects of these bridging zones in 3D is therefore becoming increasingly important. Isolating such features from the rest of the microstructure is key to quantifying their morphological properties (e.g. shape, size, orientation) and their microstructural properties (e.g. crystallographic orientation). The reliable extraction of bridging ligaments from tomography data requires an image processing strategy. Standard methods, such as histogram-based segmentation, will not isolate the bridges from the material. Indeed, after segmentation of the crack, bridges correspond to holes (or tunnels) in the crack and are part of the surrounding background. This is in that context that the tunnel closing algorithm has proven to be very useful in extracting these special material features.

An example of such a bridging ligament along a portion of a crack path is shown in Fig. 14.7. This corresponds to the case of an intergranular stress corrosion crack in a sensitized austenitic stainless steel specimen commonly found in the nuclear industry. Results have been published in (Babout et al. 2011). Details of the in-situ X-ray microtomography corrosion experiment can be found in (Babout et al. 2006, King et al. 2008) The reconstructed slice (Fig. 14.7a) shows the crack path (black) is discontinuous in 2D (although continuous in 3D), with unbroken crack bridging ligaments of material. Closed holes in the crack "A" and "B" (i.e. bridging ligaments) are shown in light gray. A clearer visual representation of the bridges is shown in dark gray on the corresponding 3D rendering image of the crack, in light gray (Fig. 14.7b). It can be seen that the bridges tend to have characteristic shapes/locations. For example, bridge C has a trapezoidal-like shape within the grain facet, and bridges A and B appear to be located at the grain facet corners. Therefore, tunnel closing is very useful for further quantitative analysis. For instance, their spatial orientation, their position with respect to the crack front or their crystallographic orientation, obtained by registering the image with data from Diffraction Contrast Tomography images have only been possible using this algorithm, as presented in (Babout et al. 2011, Ludwig et al. 2010) . Note that modifications of tunnel closing algorithm have been proposed to better cope with the complexity of crack morphologies. Janaszewski *et al.* (Janaszewski et al. 2011) have proposed an improved version of the algorithm, which only detects holes from the Euclidean skeleton of the object of interest (i.e. here the crack) and processes the skeleton to remove branches around the hole. This is important since their presence can affect the Euclidean distance map, and consequently, the final extraction of the bridge. Another modification of the algorithm, called hole filling, has also been proposed to propagate the surrounding thickness of the object of interest, defined by its Euclidean skeleton, to the segmented hole (Janaszewski et al. 2010). This had potential interest to estimate the crack opening displacement around the holes that can be indirectly correlated to the bridge failure rate.

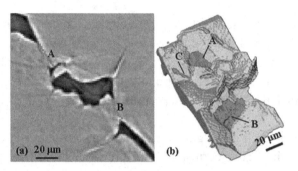

Fig. 14.7 Example of bridge ligament along intergranular stress corrosion crack in stainless steel. (a) 2D view showing crack (black) and position of bridges A and B. (b) 3D rendering of the crack, with bridge A and B filled using tunnel closing algorithm (also presence of another bridge C not present in the 2D view).

Another application of the tunnel closing algorithm is the correction of under-segmented features, due to low contrast/high noise problems in the 3D image. This can occur in X-ray microtomography images, when the density between phases is similar, the phase contrast is weak or one dimension of the feature is below the spatial resolution of the X-ray detector. The latter drawback is illustrated in the following welding example. In the case of friction stir spot welding (FSSW) where a high speed rotating tool is plunged into the upper sheet of a lap joint while an anvil supports the down force, it is very useful to investigate how the material flows during this welding process. To achieve this using X-ray tomography, Bakavos *et al.* (Bakavos et al. 2011) sputtered a thin 50-100nm gold layer on one face of each aluminum lap sheet used for the FSSW. One can see from Fig. 14.8a that the gold layer is visible. This is due to the strong attenuation of gold (~200x larger than that of aluminum at 30 keV). In the present case where a dwell time of 2.5 s is used during the welding process, the gold was not homogeneously distributed, but rather followed a 3D spiral like pattern and a dished shape of the weld interface could be distinguished. However, it is not possible to automatically threshold the layer of gold to visualize the interface position in 3D, as can be seen in Fig. 14.8a. Moreover, manual thresholding can be very burdensome. In the present case, the authors have chosen an intermediate solution, i.e. they have applied a manual segmentation, combined with the tunnel closing algorithm to obtain the 3D profile of the material flow (Fig. 14.8c). The manual segmentation simply uses a "brush tool" to follow the gold layer in few slices of the 3 orthogonal planes, resulting in a 3D grid, as shown in Fig 14.8b, that was finally filled using the tunnel closing algorithm. This strategy, despite not being fully automatic, has proven its usefulness to compare the evolution of the interface displacement as the dwell time (i.e. the time the welding process lasts) increases. More details concerning materials science aspects of the study can be found in (Babout et al. 2010, Bakavos et al. 2011).

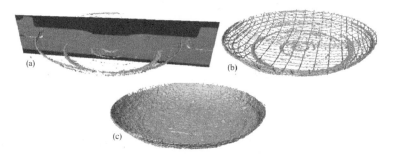

Fig. 14.8 3D visualization of the weld position interface after FSSW of Al alloy sheets using gold layer marker. (a) Sagittal view of the weld with part of the gold marker superimposed after standard histogram thresholding. (b) View of the grid created by manual segmentation (c) 3D rendering of the gold layer after hole closing which clearly shows the dished shape of the material flow.

### 14.5.2 Skeletonization: analysis of open-cell foams

Skeletonization is also another topological approach, which can find many applications in materials science imaging. It can be used for surface objects such as cracks (for instance as a pre-processing step for the hole closing algorithm example presented in the previous subsection) or for structure of curvilinear shape such as open-cell foam. This is illustrated in Fig. 14.9, which shows the isosurface of a polyurethane foam sample with straight ribs and the skeletonization of a similar type of material, but presenting auxetic properties (i.e. bending of the ribs/segments during fabrication process induces auxetic behavior during further loading conditions) (McDonald et al. 2009). The geometrical characterization of the structure is essential for further modeling steps. The skeletonization step helps in that matter, by providing data that can easily be converted to a graph, where vertices (i.e. joints), edges (i.e. ribs), and loops (i.e. open cells) can easily be extracted (see Fig. 14.9b). Moreover, as presented in (Babout et al. 2010, Janaszewski et al. 2011), tunnel closing can be very useful from the visualization point of view. Indeed, it can be used to identify the loops and to represent 3D maps of their deviation from a perfect polygon. This can help to understand how the material has deformed and its potential ability to locally recover its original shape during deformation.

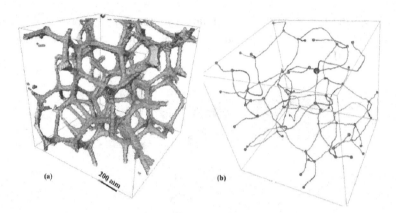

Fig. 14.9 Open cell foam (a) 3D rendering of standard structure (b) Curvilinear skeletonization of an auxetic foam.

### 14.5.3  *Edge preserving smoothing: separation of phases in titanium alloy*

In the process of 3D image segmentation of material a microstructure, it may happen that two or more phases are present and need to be studied separately. In the case of textured images, this might be problematic since one phase can be embedded in a second one. This is the case, for instance, of lamellar microstructure of titanium alloy Ti64, where lamellar colonies of α-phase have grown in β-grains. From the materials science point of view, it is important to extract the β grain boundaries (β-gb) and the lamellar colonies, as presented in (Birosca et al. 2009). In the case of the former feature (the latter will be presented in the next subsection), it is important to "vanish" the presence of the lamellae. Edge preserving smoothing methods presented in Section 14.2 can help in that matter. This is illustrated in Fig. 14.10. One can see that in the present example, the mean shift outperforms smoothing when nonlinear diffusion is used. We can clearly see that α-colonies are better smoothed by the MSS method while preserving the β-gb (Fig. 14.10c). This is also illustrated in the histogram shown in Fig. 14.10d where one can see the apparition of a step in the density profile while the original Gaussian profile of the gray level distribution is narrowed due to smoothing. Unfortunately, not all α-colonies are smoothed and other processing steps are needed towards the accurate segmentation of β-gb.

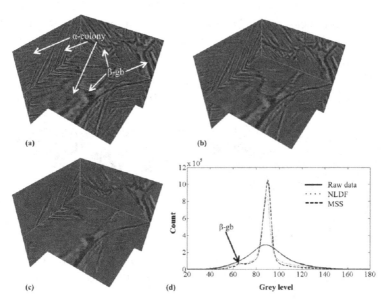

Fig. 14.10 Example of nonlinear filtering preserving edges. (a) Original 3D view of Ti lamellar alloy. (b) Filtering result of nonlinear diffusion filtering (NLDF) with $\sigma$=4.41 (w=15), $\lambda$=3, $\tau$=5 and the number of iteration is 5. (c) Filtering result of mean shift smoothing (MSS) with $h_s$=$h_r$=15. (d) Intensity histogram showing the effect of edge-preserving smoothing of the raw data.

### 14.5.4 *Texture segmentation: case of lamellar titanium alloy*

The previous section has concentrated on the segmentation of $\beta$-grain boundary. This section focuses on the second aspect, i.e. the segmentation of tomographic images of lamellar colonies in Ti alloy. This microstructure can be considered as an exemplar oriented texture image. The result of the comparison between the methods introduced in Section 14.4, i.e. estimation of local orientation, nonsubsampled directional filter bank (NSDFB) and CHG directional filter, is presented in Fig. 14.11 for the 2D case. To facilitate the comparison, the local orientation based on the image gradient (see Section 14.4.1) has been converted to the closest orientation class defined by the NSDFB at level 3. Similarly, 8 principal directions of the 2D version of the CHG filter bank have been chosen to also match the NSDFB, i.e. the <1 0>, <1 1> and <2 1> family of normal directions.

One can see that, of these three methods, segmentation using the estimation of local orientation gives the worst results while the method implementing the CHG filter seems the best, in terms of boundary recovery. The main problem in the latter resides in the presence of regions with low confidence in the local orientation, generating narrow bands at the vicinity of colony boundaries. These are not present in the case of NSDFB and CHG. Still, problems occur for these 2 methods, mainly because of noisy voxels, which interfere with the local detection of orientation, resulting in the segmentation of noisy features within large colonies. However, other steps in the segmentation process can be added to reduce this effect, as presented in (Jopek et al. 2012).

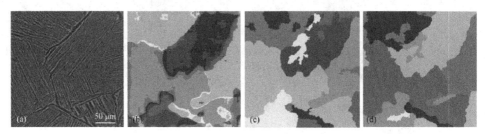

Fig. 14.11 Comparison between methods to segment α-colonies. (a) Method based on estimation of local orientation, w=25. (b) Method based on NSDFB with 3-level decomposition. (c) Method based on CHG filter bank, α=22.5°, w=11.

## 14.6 Conclusion

This chapter has presented the application of advanced image processing methodologies applied to X-ray microtomography imaging of engineering materials. Instead of enumerating a wide list of methods, emphasis has been put on key material problems, which have been solved on the image processing side, but also challenges, which are still under investigation. For instance, it has been shown how the tunnel closing algorithm can be crucial to helping quantify the formation of bridge ligaments along the intergranular stress corrosion crack in stainless steel or reconstructing the interface profile of welded materials. The use of skeletonization to simplify the quantitative analysis of shapes has also been illustrated with its application to open-cell polyurethane foam. Finally, two different groups of approaches, i.e. edge-preserving smoothing as well as directional filter banks used in texture analysis have been applied to separate the segmentation of α and β phases in lamellar titanium alloy.

On the basis of this presentation of key examples, one may apprehend the real potential of materials science imaging to strengthen interdisciplinary collaboration between the materials science and the computer science communities.

# References

Aktouf, Z., Bertrand, G. and Perroton, L. (2002). A three-dimensional holes closing algorithm, Pattern Recog.Lett., 23, pp. 523-531.

Arias-Castro, E. and Donoho, D.L. (2009). Does Median Filtering Truly Preserve Edges Better than Linear Filtering? Annals of Statistics, 37, pp. 1172-1206.

Attali, D. and Montanvert, A. (1996). Modeling noise for a better simplification of skeletons, Proc. International Conference on Image Processing, pp. 13-16.

Aylward, S.R. and Bullitt, E. (2002). Initialization, noise, singularities, and scale in height ridge traversal for tubular object centerline extraction, IEEE Trans.Med.Imaging, 21, pp. 61-75.

Babout, L., Jopek, L. and Janaszewski, M. (2013). A New Directional Filter Bank for 3D Texture Segmentation: Application to Lamellar Microstructure in Titanium Alloy, Proc. 13th IAPR International Conference on Machine Vision Applications.

Babout, L., Janaszewski, M., Bakavos, D., McDonald, S.A., Prangnell, P.B., Marrow, T.J. and Withers, P.J. (2010). 3D inspection of fabrication and degradation processes from X-ray (micro) tomography images using a hole closing algorithm, Proc. IEEE International Conference on Imaging Systems and Techniques (IST), pp. 337-342.

Babout, L., Janaszewski, M., Marrow, T.J. and Withers, P.J. (2011). A method for the 3-D quantification of bridging ligaments during crack propagation, Scr.Mater., 65, pp. 131-134.

Babout, L., Marrow, T.J., Engelberg, D. and Withers, P.J. (2006). X-ray microtomographic observation of intergranular stress corrosion cracking in sensitised austenitic stainless steel, Mater.Sci.Tech.-Lond., 9, pp. 1068-1075.

Bakavos, D., Chen, Y., Babout, L. and Prangnell, P. (2011). Material interactions in a novel pinless tool approach to friction stir spot welding thin aluminum sheet, Metall.Mater.Trans. A, 42, pp. 1266-1282.

Bamberger, R.H. and Smith, M.J.T. (1992). A Filter Bank for the Directional Decomposition of Images - Theory and Design, IEEE Trans.Signal Process., 40, pp. 882-893.

Bertrand, G. (1994). Simple points, topological numbers and geodesic neighborhoods in cubic grids, Pattern Recog.Lett., 15, pp. 1003-1011.

Birosca, S., Buffiere, J.Y., Garcia-Pastor, F., Karadge, M., Babout, L. and Preuss, M. (2009). Three-dimensional characterization of fatigue cracks in Ti-6246 using X-ray tomography and electron backscatter diffraction, Acta Mater., 57, pp. 5834-5847.

Blanding, R., Turkiyyah, G., Storti, D. and Ganter, M. (2000). Skeleton-based three-dimensional geometric morphing, Comput. Geom.-Theory Appl., 15, pp. 129-148.

Blum, H. (1967). Transformation for Extracting New Descriptions of Shape. In: Models for the Perception of Speech and Visual Form, MIT Press, Cambridge, pp. 362-380.

Borgefors, G., Ragnemalm, I. and di Baja, S. (1991). The Euclidean distance transform: finding the local maxima and reconstructing the shape, Proc. 7th Scandinavian Conference on Image Analysis, pp. 974-981.

Candes, E., Demanet, L., Donoho, D. and Ying, L. (2006). Fast discrete curvelet transforms, Multiscale Modeling & Simulation, 5, pp. 861-899.

Charbonnier, P., Blancferaud, L., Aubert, G. and Barlaud, M. (1994). Two Deterministic Half-Quadratic Regularization Algorithms for Computed Imaging, Proc. IEEE International Conference on Image Processing (ICIP-94), pp. 168-172.

Chaussard, J., Couprie, M. and Talbot, H. (2010). Robust skeletonization using the discrete lambda-medial axis, Pattern Recog.Lett., 9, pp. 1-10.

Cheng, Y. (1995). Mean Shift, Mode Seeking, and Clustering, IEEE Trans.Pattern Anal.Mach. Intell., 17, pp. 790-799.

Coeurjolly, D. and Montanvert, A. (2007). Optimal separable algorithms to compute the reverse euclidean distance transformation and discrete medial axis in arbitrary dimension , IEEE Trans.Pattern Anal.Mach.Intell., 29, pp. 437-448.

Comaniciu, D. and Meer, P. (2002). Mean shift: A robust approach toward feature space analysis, IEEE Trans.Pattern Anal.Mach.Intell., 24, pp. 603-619.

Comaniciu, D., Ramesh, V. and Meer, P. (2003). Kernel-based object tracking, IEEE Trans.Pattern Anal.Mach.Intell., 25, pp. 564-577.

Cornea, N., Silver, D. and Min, P. (2007). Curve-skeleton applications, Available: http://www.caip.rutgers.edu/~cornea/CurveSkelApp.

Couprie, M. (2013). Pink, Available: http://www.esiee.fr/~coupriem/Pink/doc/html/.

Couprie, M., Coeurjolly, D. and Zrour, R. (2007). Discrete bisector function and Euclidean skeleton in 2D and 3D, Image Vision Comput., 25, pp. 1543-1556.

da Cunha, A.L., Zhou, J. and Do, M.N. (2006). The nonsubsampled contourlet transform: Theory, design, and applications, IEEE Trans.Image Process., 15, pp. 3089-3101.

Debeir, O., Van Ham, P., Kiss, R. and Decaestecker, C. (2005). Tracking of migrating cells under phase-contrast video microscopy with combined mean-shift processes, IEEE Trans.Med. Imaging, 24, pp. 697-711.

Do, M.N. and Vetterli, M. (2005). The contourlet transform: An efficient directional multiresolution image representation, IEEE Trans.Image Process., 14, pp. 2091-2106.

Donoho, D. (2001). Ridge functions and orthonormal ridgelets, Journal of Approximation Theory, 111, pp. 143-179.

Fukunaga, K. and Hostetler, L. (1975). Estimation of Gradient of a Density-Function, with Applications in Pattern-Recognition, IEEE Trans.Inf.Theory, 21, pp. 32-40.

He, T., Hong, L., Chen, D. and Liang, Z. (2001). Reliable path for virtual endoscopy: Ensuring complete examination of human organs, IEEE Trans.Visual.Comput.Graphics, 7, pp. 333-342.

Hesselink, W. and Roerdink, J. (2008). Euclidean skeletons of digital image and volume data in linear time by the integer medial axis transform, IEEE Trans.on PAMI, 30, pp. 2204-2217.

Hilditch, C.J. (1969). Linear skeletons from square cupboards. In: Machine Intelligence 4, Edinburgh University Press, Edinburgh, pp. 403-420.

Huang, X. and Zhang, L. (2008). An Adaptive Mean-Shift Analysis Approach for Object Extraction and Classification From Urban Hyperspectral Imagery, IEEE Trans.Geosci. Remote Sens., 46, pp. 4173-4185.

Janaszewski, M., Couprie, M. and Babout, L. (2010). Hole filling in 3D volumetric objects, Pattern Recognit, 43, pp. 3548-3559.

Janaszewski, M., Postolski, M., Babout, L. and Kącki, E. (2009). Comparison of several centreline extraction algorithms for virtual colonoscopy, Advances in Soft Computing, 65, pp. 241-254.

Janaszewski, M., Postolski, M. and Babout, L. (2011). Robust algorithm for tunnel closing in 3D volumetric objects based on topological characteristics of points, Pattern Recog.Lett., 32, pp. 2231-2238.

Jarabo-Amores, P., Rosa-Zurera, M., de la Mata-Moya, D., Vicen-Bueno, R. and Maldonado-Bascon, S. (2011). Spatial-Range Mean-Shift Filtering and Segmentation Applied to SAR Images, IEEE Trans.Instrum.Meas., 60, pp. 584-597.

Jeulin, D. and Moreaud, M. (2008). Segmentation of 2d and 3d textures from estimates of the local orientation, Image Analysis and Stereology, 27, pp. 183-192.

Jopek, L., Babout, L. and Janaszewski, M. (2012). A New Method to Segment X-Ray Microtomography Images of Lamellar Titanium Alloy Based on Directional Filter Banks and Gray Level Gradient, Proc. LNCS 7594, International Conference on Computer Vision and Graphics (ICCVG 2012), pp. 105-112.

Kang, D.G. and Ra, J.B. (2004). Automatic path-planning algorithm maximizing observation area for virtual colonoscopy, Proc. SPIE 5367, Medical Imaging 2004: Visualization, Image-Guided Procedures, and Display, pp. 687-696.

Kim, K., Jung, K. and Kim, J. (2003). Texture-based approach for text detection in images using support vector machines and continuously adaptive mean shift algorithm, IEEE Trans.Pattern Anal.Mach.Intell., 25, pp. 1631-1639.

King, A., Johnson, G., Engelberg, D., Ludwig, W. and Marrow, J. (2008). Observations of intergranular stress corrosion cracking in a grain-mapped polycrystal, Science, 321, pp. 382-385.

Kong, T.Y. and Rosenfeld, A. (1989). Digital topology: Introduction and survey, Comput.Vis.,Graph.Im.Proc., 48, pp. 357-393.

Lee, J. and Lee, D. (2005). An improved cluster labeling method for support vector clustering, IEEE Trans.Pattern Anal.Mach.Intell., 27, pp. 461-464.

Liu, T., Zhou, H., Lin, F., Pang, Y. and Wu, J. (2008). Improving image segmentation by gradient vector flow and mean shift, Pattern Recog.Lett., 29, pp. 90-95.

Lu, Y.M. and Do, M.N. (2007). Multidimensional directional filter banks and surfacelets, IEEE Trans.Image Process., 16, .

Ludwig, W., King, A., Herbig, M., Reischig, P., Marrow, J., Babout, L., Lauridsen, E.M., Proudhon, H. and Buffiere, J.Y. (2010). Characterization of polycrystalline materials by combined use of synchrotron X-ray imaging and diffraction techniques, JOM, 62, pp. 22-28.

Malandain, G., Bertrand, G. and Ayache, N. (1993). Topological segmentation of discrete surfaces, Int.J.Comput.Vis., 10, pp. 183-197.

Mallat, S.G. (1989). A theory for multiresolution signal decomposition: the wavelet representation, IEEE Trans.Pattern Anal.Mach.Intell., 11, pp. 674-693.

Mallat, S.G. (1988). Multiresolution representations and wavelet, PhD thesis, University of Pennsylvania.

Mayer, A. and Greenspan, H. (2009). An Adaptive Mean-Shift Framework for MRI Brain Segmentation, IEEE Trans.Med.Imaging, 28, pp. 1238-1250.

McDonald, S.A., Ravirala, N., Withers, P.J. and Alderson, A. (2009). In situ three-dimensional X-ray microtomography of an auxetic foam under tension, Scr.Mater., 60, pp. 232-235.

Palagyi, K. (2002). A 3-subiteration 3D thinning algorithm for extracting medial surfaces, Pattern Recog.Lett., 23, pp. 663-675.

Paris, S. (2008). Edge-Preserving Smoothing and Mean-Shift Segmentation of Video Streams, Proc. 10th European Conference on Computer Vision (ECCV 2008), pp. 460-473.

Perona, P. and Malik, J. (1990). Scale-Space and Edge-Detection using Anisotropic Diffusion, IEEE Trans.Pattern Anal.Mach.Intell., 12, pp. 629-639.

Petrou, M. and Garcia-Sevilla, P. (2006). Image processing dealing with texture, Wiley, 618 p.

Polikar, R. (1999). The story of wavelets, Proc. IMACS/IEEE CSCC'99, pp. 5481-5486.

Prteux, F., Perchet, D. and Fetita, C. (2004). Advanced navigation tools for virtual bronchoscopy, Proc. Conference on Image Processing: Algorithms and Systems III.

Remy, E. and Thiel, E. (2005). Exact medial axis with euclidean distance, Image Vision Comput., 23, pp. 167-175.

Rosenfeld, A. (1970). Connectivity in Digital Pictures, J.ACM, 17, pp. 146-160.

Selesnick, I.W., Baraniuk, R.G. and Kingsbury, N.G. (2005). The dual-tree complex wavelet transform, IEEE Signal Process.Mag., 22, pp. 123-151.

Sundar, H., Silver, D., Gagvani, N. and Dickinson, S. (2003). Skeleton based shape matching and retrieval, Proc. International Conference on Shape Modeling International and Applications, pp. 130-142.

Tao, W., Jin, H. and Zhang, Y. (2007). Image segmentation based on mean shift and normalized cuts, IEEE Trans.Syst.Man Cybern. Part B-Cybern., 37, pp. 1382-1389.

Vanderesse, N., Maire, E., Darrieulat, M., Montheillet, F., Moreaud, M. and Jeulin, D. (2008). Three-dimensional microtomographic study of Widmanstatten microstructures in an alpha/beta titanium alloy, Scr.Mater., 58, pp. 512-515.

Velisavljevic, V., Beferull-Lozano, B., Vetterli, M. and Dragotti, P.L. (2006). Directionlets: Anisotropic multidirectional representation with separable filtering, IEEE Trans.Image Process., 15, pp. 1916-1933.

Vincent, L. (1991). Efficient Computation of Various Types of Skeletons, Proc. SPIE 1445, Medical Imaging V, pp. 297-311.

Weickert, J., Romeny, B. and Viergever, M. (1998). Efficient and reliable schemes for nonlinear diffusion filtering, IEEE Trans.Image Process., 7, pp. 398-410.

Wu, K. and Yang, M. (2007). Mean shift-based clustering, Pattern Recognit, 40, pp. 3035-3052.

Yi, S., Labate, D., Easley, G.R. and Krim, H. (2009). A Shearlet Approach to Edge Analysis and Detection, IEEE Trans.Image Process., 18, pp. 929-941.

Yilmaz, A., Javed, O. and Shah, M. (2006). Object tracking: A survey, ACM Comput. Surv., 38, pp. 13.

Zhou, H., Schaefer, G., Sadka, A.H. and Celebi, M.E. (2009). Anisotropic Mean Shift Based Fuzzy C-Means Segmentation of Dermoscopy Images, IEEE J.Sel.Top.Signal Process., 3, pp. 26-34.

# CHAPTER 15

# SELECTED ALGORITHMS OF QUANTITATIVE IMAGE ANALYSIS FOR MEASUREMENTS OF PROPERTIES CHARACTERIZING INTERFACIAL INTERACTIONS AT HIGH TEMPERATURES

Krzysztof Strzecha, Anna Fabijańska, Tomasz Koszmider, and Dominik Sankowski

*Institute of Applied Computer Science*
*Lodz University of Technology*
*90-924 Łódź, ul. Stefanowskiego 18/22*
*{strzecha, an_fab, t.koszmider, dsan}@kis.p.lodz.pl*

## 15.1 Measurements of the Surface Tension and Contact Angle

In many industrial processes, an important and sometimes a predominant role is played by the phenomena occurring at the interfaces of the liquid, the gas and the solid state. These phenomena occur in welding, the making of composite materials with the participation of the liquid phase, the sintering of powders, the saturation of porous structures, coating, the refining of metals to eliminate non-metallic inclusions, foundry or processes of crystallization from the liquid phase. Thus, the knowledge of physical-chemical processes occurring between the liquid, the gas and the solid is a significant technological problem.

A number of monographs and review articles have been devoted to the significance of phenomena at the interfaces in different fields of technology: (Metcalfe, 1981; Pask and Evans, 1981; Matsunawa and Ohji, 1982; Matsunawa and Ohji, 1983; Matsunawa and Ohji, 1984; Delannay et al., 1987; Baglin, 1988; Nicholas, 1990; Senkara and Windyga, 1990; Mortensen, 1991; Eustathopoulos, 1999; Drzymała, 2001; Butt et al., 2003; de Gennes et al., 2003; Popel, 2003; Rosen, 2004; Deyev and Deyev, 2005; Venables, 2006; Trakhtenberg, 2007). Theoretical foundations of surface phenomena physics and methods of measurement can be found in books: (Woodruff, 1973; Missol, 1974; Murr, 1975; Zangwill, 1988; Adamson and Gast, 1997; Dutkiewicz, 1998; Myers,

1999; Safran, 2003; Hartland, 2004; Bechstedt, 2005; Mittal, 2006; Ibach, 2007), and review papers (Kinloch, 1980; de Gennes, 1985; Kwok and Neumann, 2000; Lam et al., 2001a; Lam et al., 2001b; Xu and Masliyah, 2002).

Among the basic measurable quantities characterizing surface interactions are the surface energy (surface tension) of the liquid phase and the extreme angle of wetting of the base by a liquid. Measurements and calculations performed for the conditions of the thermodynamic equilibrium. For many systems, however, this equilibrium is not reached at all or is reached after too long a time from the technological point of view. This results from processes of diffusion, the dissolution of the base in the liquid, the formation of new chemical compounds or inter-metallic phases. In such cases, crucial information is obtained by measuring dynamic, time-variable quantities of the system, possibly in combination with a structural analysis of the interfacial boundary carried out later.

### 15.1.1 *Review of Selected Solutions*

For a long time, the optical methods, based on the continuous observation of the specimen (or droplet) shape and the manual or photographic recording of changes in its profile as a function of temperature, were the fundamental techniques of determination of the wetting angle and surface tension. The profiles recorded were then analyzed by a qualified specialist who, using graphical methods, measured the basic parameters of the droplet and then calculated the values of the parameters measured. Measuring systems of this type had a number of drawbacks, only to mention an extremely labor-consuming and strenuous measurement calling for the operator's constant undivided attention, as well as an essential effect of the human factor on the measurement results.

In recent years there has been considerable progress in automated techniques for measuring the basic parameters characterizing the interfacial interactions. There are several devices on the market which allow measurement of the extreme wetting angle and the surface tension in systems of liquid metals and solids in protective atmospheres. Primarily, the PR-25/37/45 series of devices produced by the Institute for Tele-and Radio in Warsaw should be mentioned here. These devices are designed for the automatic determination of phase transition temperatures of solids (sintering, softening, melting, melts), as well as for the determination of the wetting angle and surface tension of the melted material as a function of temperature in the range from 700 ° C to 1750 ° C.

Systems enabling fully automated measurement typically only allow the measurement of properties of selected materials in a narrow temperature range.

Additionally, image processing and analysis algorithms implemented in them are not immune to interferences and require the precise positioning of the specimen (Huh and Reed, 1983; Rotenberg et al., 1983; Girault, 1984; Anastasiadis, 1987; Cheng, 1990; Pallas and Harrison, 1990; Hansen and Rodsrud, 1991; Egry et al. 1992; Hansen, 1993; Bachevsky et al., 1994; Song and Springer, 1996; Atae-Allah, 2001; Emelyanenko and Boinovich, 2001; Emelyanenko, 2004; Zuo et al., 2007; Staldera et al., 2010).

The authors of the present chapter were involved in the research projects no. 8 T10C 005 14, within which an automatic device for the measurement of the extreme angle and surface tension in systems of liquid metal and solids was developed. This device, as the first in the world, enabled automated measurement in a wide range of temperatures in a protective atmosphere (Sankowski et al., 2001a; Sankowski et al., 2011). These measurements are based on algorithms for image processing and analysis, and are conducted using a sessile drop technique.

### 15.1.2 *Sessile Drop Method*

The shape of the drops placed on a non-wettable surface results from two types of forces: the surface tension, which tries to give a droplet a spherical shape, and the gravity by which the drop is "flattened". In the case of the spherical drops, it is not possible to determine the surface tension. This becomes possible only when the effects of gravity are comparable with the influence of surface energy, and hence the larger droplets – the equatorial diameter of the metal droplets should be practically at least 0.5 cm. The distortion occurring in a spherical droplet allows one to determine the surface tension on the basis of the droplet dimensions and physical constants. Only symmetric drops could be taken into account (Missol, 1974).

Among the currently used methods for calculating the surface tension three groups can be distinguished:

- the method of the equatorial plane (measured h and d – Fig. 15.1);
- the method of tangents (measured Z and d);
- the method of the total height (measured $z_0$ or z') and the maximum diameter and the diameter of the base (respectively - $x_0$ or x') (Missol, 1974).

The determination of the equatorial plane, according to the literature, is usually performed graphically. After the measurement of h and d, the ratio of

2h/d is calculated. This ratio is essential for further calculations. These can be performed using Porter's equation (15.1) or using Koszewnik's tables (Missol, 1974).

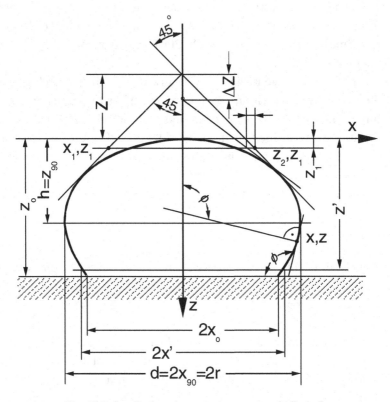

Fig. 15.1 Sessile drop (measured parameters indicated).

$$\frac{\alpha^2}{r^2} = \left(\frac{h}{r}\right)^2 - 0,660\left(\frac{h}{r}\right)^3\left[1 - 4,05\left(\frac{h}{r}\right)^2\right]$$

(15.1)

where:

$\alpha$  –  the capillary constant.

According to the literature, Porter equation is affected by calculation inaccuracy of several percent. For values of $d/2h$ from 2.18 to infinity, it is valid with calculation error no more than 0.2%. Allowing the measurement error 1%, the application of this equation can be extended to the range of $d/2h$ from 1.66 to infinity.

Dorsey, on the basis of Bashforth and Adams tables, empirically developed the following calculation formula:

$$\frac{\alpha^2}{r^2} = \frac{0{,}10400}{\frac{y}{r} - 0{,}41421} - 0{,}24536 + 0{,}0962\left(\frac{y}{r} - 0{,}41421\right) \tag{15.2}$$

which ensures the accuracy of the calculations of about 0.19% in the range of $Z/r$ from 0.515 to 0.66 ($d/2h$ from 1.36 to 2.18) (Missol, 1974).

The determined values of the height and the radius of the droplet allow the determination of a capillary constant. Taking into account the density of the test material, and the gravitational acceleration constant capillary, the surface tension is calculated based on the expression:

$$\sigma = g\Delta\rho\alpha^2 \tag{15.3}$$

where:

$\Delta\rho$ — the density difference, in the case of metal and the gaseous phase: $\Delta\rho = \rho_S$, wherein $\rho_S$ is the density of the metal;

$g$ — the gravitational acceleration.

The determination of the contact angle is a direct measuring method on the basis of the point of contact of the three phases, in accordance with the definitional equation of Young (Fig. 15.2):

$$|\bar{\sigma}_{LV}|\cos\theta + |\bar{\sigma}_{SL}| - |\bar{\sigma}_{SV}| = 0 \tag{15.4}$$

where:

$\theta$ — the wetting angle;

$\bar{\sigma}_{SV}$ — the surface tension on the border of solid-gas;

$\bar{\sigma}_{SL}$ — the surface tension on the border of solid-liquid;

$\bar{\sigma}_{LV}$ — the surface tension on the border of liquid-gas.

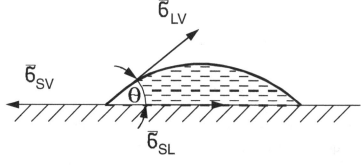

Fig. 15.2 The conditions of the thermodynamic equilibrium.

## 15.2 THERMO-WET Measurement System

Within the framework of the research no 8 T10C 005 14 sponsored by the Polish National Research Committee, carried out from February 1998 until June 2000, The Computer Engineering Department of the Lodz University of Technology in co-operation of the Warsaw University of Technology and the Industrial Electronic Institute (PIE) in Warsaw built a computerized device for the automated measurement of surface phenomena occurring in contact of liquid and solid phases (Sankowski et al., 1999; Sankowski et al., 2000a; Sankowski et al., 2000b; Sankowski et al., 2001a; Sankowski et al., 2001b; Strzecha, 2002; Jeżewski, 2006; Sankowski et al., 2011). The system was named THERMO-WET. The system is capable of measuring:

- the surface tension of a liquid (surface energy);
- the wetting angle of a solid by a liquid (including extreme wetting angle).

These parameters are determined using the sessile drop method in the conditions of thermodynamic equilibrium, and in the case of transient states as a function of time and temperature. These tests may be conducted in a controlled atmosphere, in a temperature range of up to 1800°C, for interfacial solid-liquid systems of two different materials. The measurement process takes place on the basis of specialized, designed specifically for THERMO-WET system algorithms of image processing and analysis. The images processed and analyzed are obtained from a camera observing a sample of the test material placed inside the high-temperature furnace. The measurement results obtained using the presented system are characterized by a much higher accuracy and higher reproducibility compared with those obtained with the previously used, time-consuming methods, depending on the operator's subjective assessments.

### 15.2.1 *Construction of the Measurement System*

The measurement system consists of a two-zone high-temperature electric furnace (1) equipped with a high-precision temperature controller (2), the process gases supply system (3), the system for loading and discharge of the specimen (4), the CCD camera (5) coupled with the computer controlling the measurement process (6), equipped with a specialized programs to image process and analysis, mathematical data processing, editing and archiving of results. A general view of the system THERMO-WET is shown the photograph in Fig. 15.3. A block diagram of the system is presented in Fig. 15.4.

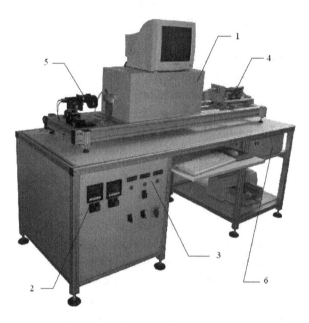

Fig. 15.3 The main view of computerized device for the automated measurement of the surface tension and the wetting angle (1 - the heating chamber of the furnace; 2 – the temperature controllers; 3 - the system of technological gas supply; 4 - the specimen insertion mechanism; 5 – the vision subsystem including CCD camera and infrared filters changer; 6 – the computer controlling measurement process and processing measurement data).

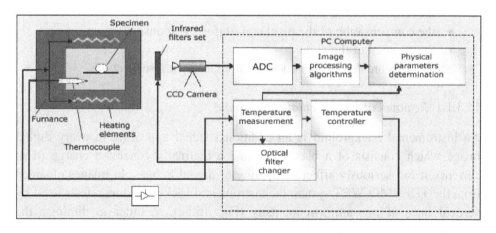

Fig. 15.4 Block diagram of the automated measurement system.

## 15.3 Image Processing and Analysis Algorithms

Image processing and analysis algorithms developed for the THERMO-WET systems can be divided into three groups:

1. image enhancement algorithms;
2. image segmentation algorithms;
3. algorithms for precise measurements of specimen geometrical properties.

These algorithms ensure high quality processing necessary for accurate measurements of surface properties in the measurement system considered.

### 15.3.1 *Image Enhancement*

It should be underlined that there is a strong relationship between the quality of image segmentation and the accuracy of surface property determination. An algorithm computing a wetting angle and surface tension performs its task on the basis of specimen shape analysis, so the stage of image segmentation is to determine the specimen edges and the upper edge of the base plate localization as accurately as possible. However, in the case of the selected segmentation algorithm, accurate image segmentation is only guaranteed by proper image enhancement algorithm selection.

Preprocessing algorithms developed for the THERMO-WET vision system aim at correcting factors arising from the CCD chip imperfections and the specificity of high-temperature measurements. In particular, they include:

- the removal of the instrumental background;
- the compensation of the pixel non-uniform response;
- the removal of the aura.

These steps are described in details in the following subsections.

#### 15.3.1.1 *Removal of Instrumental Background*

An instrumental background is an additional signal appearing in every digital image which consists of a bias signal and a thermally generated charge (dark current). It can seriously affect the quality of digital images. In images obtained from the THERMO-WET system the instrumental background manifests itself by vertical lines of non-uniform intensity (Fig. 15.5a). In order to diminish the influence of the instrumental background, an algorithm constructing its approximate image from a set of dark frames (images obtained with the closed shutter) was developed. A master dark frame – the approximate image of the instrumental background is presented in Fig. 15.5b. The correction is simply the subtraction of the master dark frame from the original image. The result of instrumental background removal is presented in Fig. 15.5c.

Fig. 15.5 Instrumental background elimination; a) original image, copper, $817^0C$; b) master dark - after histogram stretching; c) corrected image.

It should be emphasized that master dark frame construction used in the THERMO-WET system is not simply input dark frames averaging. The developed algorithm operates iteratively. A set of input dark frames is averaged in successive iterations in order to extract dominant components of the instrumental background. The average frame is then subtracted from each of the input frames in order to build an input set for the next iteration. The process is repeated until there are no distortions in the average frame. A block diagram of the algorithm is presented in Fig. 15.6.

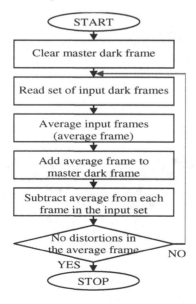

Fig. 15.6 Flow diagram of master dark frame construction algorithm.

### 15.3.1.2 *Compensation of Pixel Non-uniform Response*

For the correction of the non-uniform pixel response, an adaptation of the astronomical method was proposed. The method called flat-fielding is the division of an original image by a flat-field (which is an image of a uniformly illuminated surface) in accordance with the following equation:

$$L' = \bar{L}_{ff} \frac{L - L_b}{L_{ff} - L_b} \tag{15.5}$$

where:

$L'$  -  the corrected image,

$L$   -  the original image,

$L_b$  -  the image of the instrumental background;

$L_{ff}$  -  the flat field (map of pixel sensitivities);

$\bar{L}_{ff}$  -  the average intensity of the flat field.

In order to eliminate random distortions an average of several frames was used as a master flat field. The results of the flat field correction applied to an exemplary image obtained from the THERMO-WET vision system are presented in Fig. 15.7. The comparison is made by means of image segmentation quality. Important details in the specimen shape projection are highlighted by rectangular frames.

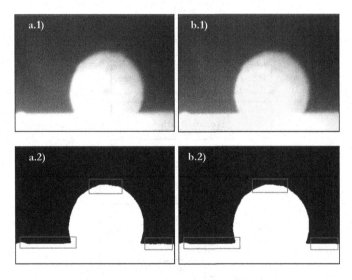

Fig. 15.7 Compensation of pixel non uniform sensitivity; a) original image, copper, $1094^0C$; b) image after correction; 1) image before segmentation; 2) image after segmentation.

The results presented in Fig. 15.7 show that flat-fielding improves the quality of the specimen shape projection due to image segmentation. In the case of original images, the improvement of image quality seems to be rather poor. The actual image quality improvement can be seen in images after segmentation. Preceding the segmentation by flat-fielding significantly increases the accuracy of the specimen shape projection. Objects after segmentation are characterized by smoother and more continuous edges than in the case of no flat-fielded corrected images. It should be stressed that the differences in the accuracy of specimen edges projection are important from the point of view of the specimen shape analysis carried out in the following stage of the measurement process.

### 15.3.1.3 *Automatic Change of Optical Filters*

The specificity of high-temperature measurements requires the use of optical filters. In particular, infrared (IR) and neutral density (ND) filters are applied. They block mid-infrared wavelengths (thermal radiation) and reduce the intensity of light. It should be underlined that only the selection of proper filter allows measurements. The problem is illustrated in Fig. 15.8.

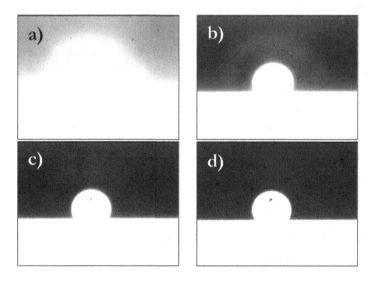

Fig. 15.8   Influence of optical filter selection on image quality; a) original image, palladium, 15640C; b) and c) improper filter selection; d) proper filter selection.

The THERMO-WET vision system enables one to obtain images in the following filter configurations:  FILTER 0 – no filter; FILTER 1 – NG4 3.0 mm (IR filter); FILTER 2 – NG4 4.0 mm+BG38 1.0 mm (IR and ND filter).

During the measurements filters must be changed together with properties of the images obtained. The moment of filter change was determined by the measurement system operator. However, an algorithm of filter automatic change was developed. The algorithm analyzes changes of the image contrast. In Fig. 15.9 a change in the image contrast as a function of time (and in consequence, of temperature) is shown. Images obtained without a filter (FILTER 0) were also considered. Images corresponding to the characteristic points marked on the curve are presented in Table 15.1.

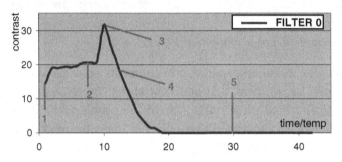

Fig. 15.9  Change of image contrast in function of temperature. Images obtained without filter (FILTER 0) are considered.

It can easily be seen (Fig. 15.8, Tab. 15.1) that together with a temperature growth, the contrast of the images obtained with FILTER 0 increases to reach the maximum and then rapidly falls to a zero value, which corresponds to the saturation of CCD photosensitive elements. The best quality images are obtained for the maximum contrast value. The curves of contrast as a function of time/temperature obtained for FILTER 1 and FILTER 2 have a similar character (see Fig. 15.10).

Moreover, when the contrast of an image obtained with weaker filter decreases after crossing the maximum, the contrast of images acquired with a stronger filter starts to increase. Therefore, a change of the current filter into the stronger one should be made after the maximum contrast is crossed. On this assumption, the proposed algorithm of automatic filter change is based. Experiments proved that in the case of the analyzed images, a 10% decrease in the contrast value means that its maximum was crossed. Block diagrams of the algorithm of automatic filter change and maximum contrast value detection are presented in Fig. 15.11a and 15.11b, respectively.

Table 15.1 Images corresponding with points marked on curve from Fig.8.

| | | | |
|---|---|---|---|
| **1** | | | 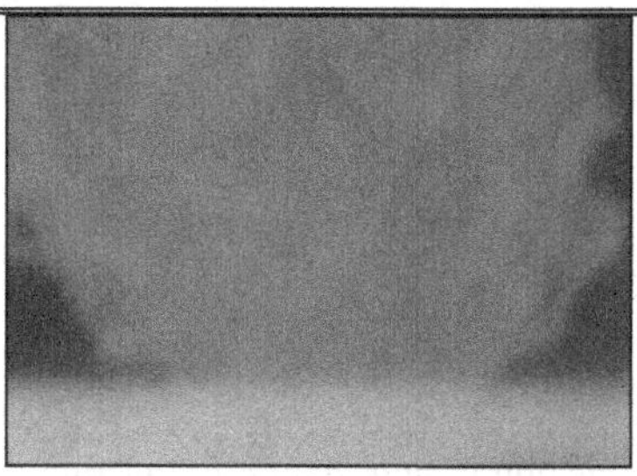 copper, $553^0$C<br>no filter |
| **2** | | | 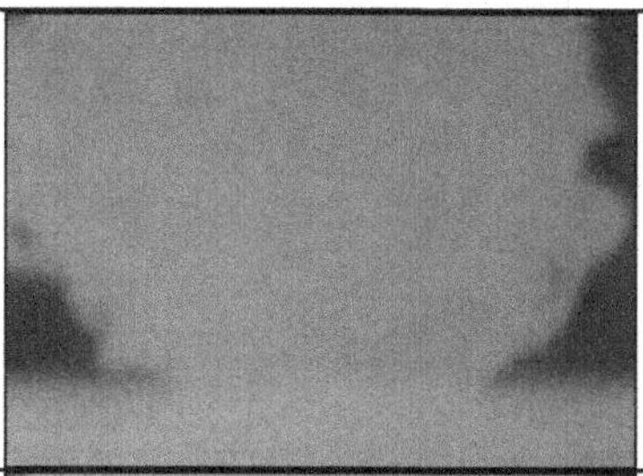 copper, $695^0$C<br>no filter |
| **3** | | | 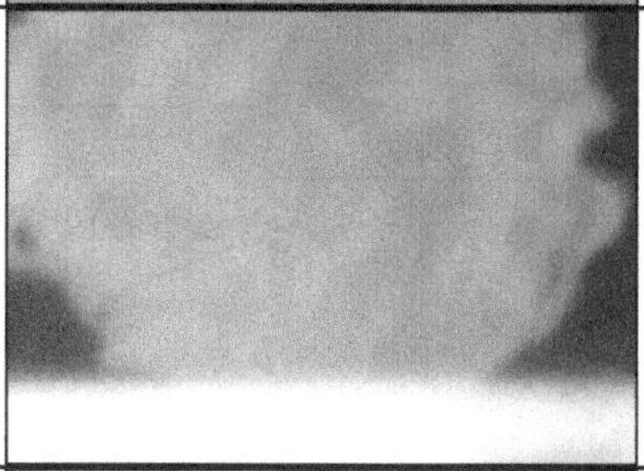 copper, $743^0$C<br>no filter<br>maximum contrast |
| **4** | | | 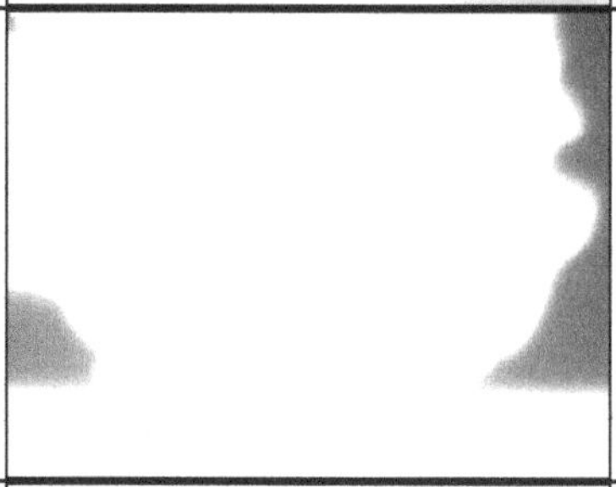 copper, $769^0$C<br>no filter |
| **5** | | | 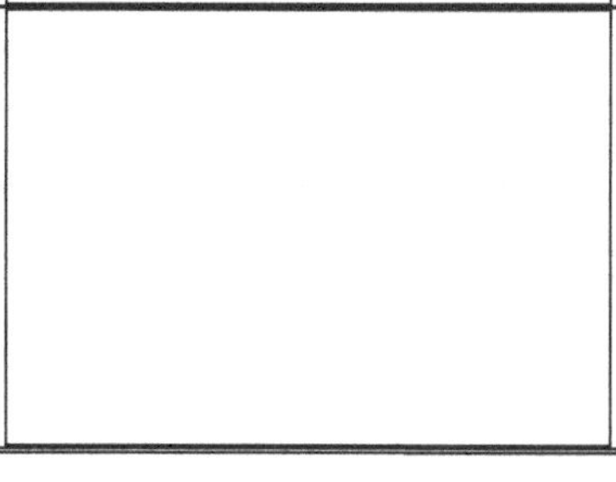 copper, $1300^0$C<br>no filter<br>saturation of CCD<br>photosensitive elements |

352                                   K. Strzecha et al.

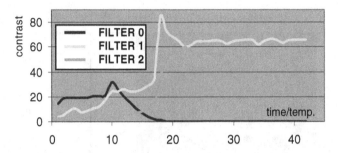

Fig. 15.10. Change of image contrast as a function of temperature for all filters from THERMO-WET vision unit.

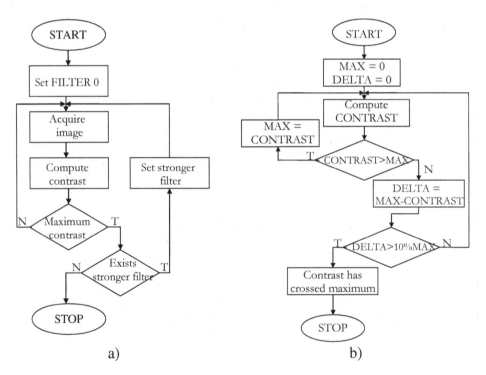

a)                                                b)

Fig. 15.11 Block diagram of algorithm of: a) optical filters automatic change; b) contrast maximum crossing detection.

It should be noted that an automatic change of optical filters eliminates human interference and completely automates measurements made by the THERMO-WET system.

### 15.3.1.4 *Correction of the Distortions Influenced by the Gas Flow*

For most materials, measurements of the surface tension and the wetting angle are carried out in a protective atmosphere. The flow of protective gas introduces significant distortions to the acquired images (Strzecha et al., 2012). Gas, of a temperature of approximately 0°C, is introduced to the furnace chamber from the side of the CCD camera and then, on its way to the specimen it is heated to the current working temperature of, for example, 1500°C.

The absolute index of the light refraction n in gas depends on its density and temperature. For the temperature above 0°C and for the atmospheric pressure, it is given by the equation:

$$\Delta n = \frac{k_1 \gamma}{T}$$

(15.6)

where:

$\gamma$    –    the density of gas;
$T$    –    the absolute temperature;
$k_1$    –    the constant value.

Boyle-Mariotte's law states that for a constant pressure the gas density is inversely proportional to its absolute temperature:

$$\Delta n = \frac{k_2}{T^2}$$

(15.7)

where:

$k_2$    –    the constant value.

Following equation (15.7), in the example above, where the temperature of gas rises from 0°C to 1500°C on its way through the furnace chamber, the value of $\Delta n$ decreases nearly 40 times.

Hence, we can assume that the gas introduced to the furnace should be considered the optical lens, whose index of light refraction smoothly changes between the camera and the specimen.

It should be pointed out that the presented model of "gas lens" is very simplified. In fact, parameters of this lens depend on many different factors, including: parameters of the gas flow, the composition of the gas mixture and the distribution of temperature in the furnace chamber. Additionally, these parameters usually change during measurement experiments. As a result, the lens formed by the protective gas atmosphere becomes an important part of the vision system, introducing significant distortions to the images.

The phenomenon described above is not only related to the flow of protective gas in the heating chamber of the furnace, which has an impact on the quality of the acquired images and, consequently, the accuracy of the measurements.

Fig. 15.12 shows the presumed distribution of isotherms in the gas laminar flowing through the heating chamber of the furnace.

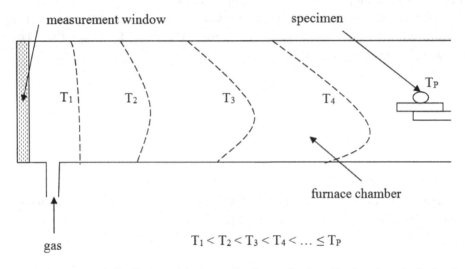

Fig. 15.12   Presumed distribution of isotherms in the gas laminar flowing through the heating chamber of the furnace (Strzecha et al., 2012).

In the area adjacent to the measurement window the gas has a temperature close to room temperature (isotherm T1). Flowing into the furnace chamber, the gas is heated by the furnace walls, but subsequent isotherms are not characterized by axial symmetry because the colder parts of the gas will flow closer to the bottom of the chamber. However, in a horizontal plane such symmetry will occur. Thus, the optical properties of the gas on the way from the measurement window to the specimen will not have axial symmetry. As a result, the vertical and the horizontal transfer of an image differ from each other, which affects the shape of the recorded drop. The higher the temperature at which the measurement is carried out, the greater the deformation of the specimen observed.

Another phenomenon caused by the flow of protective gas, and significantly impedes the measurement of the geometric dimensions of the specimen is the phenomenon of aura shown in Fig. 15.13 (Strzecha et al., 2012).

Fig. 15.13 The phenomenon of aura around a specimen and the surface of the measurement table.

In the image, it is clear that the phenomenon of aura has a significantly lower thickness and clarity on the contour of the drop than the tray table. This observation led to formulate the hypothesis that this phenomenon is the result of a "fata morgana".

On the border between two media with different speed of light, the following relation is fulfilled:

$$\frac{\sin(\alpha_1)}{\sin(\alpha_2)} = \frac{V_1}{V_2} \tag{15.8}$$

where:

$\alpha_1$ — the angle between the direction of incident light and the normal to the boundary surface;

$\alpha_2$ — the similarly, on the other side of boundary surface;

$V_1, V_2$ — the speed of light in the medium 1 and 2, respectively.

Simplifying the problem to the two areas (1 and 2) with the temperatures $T_1$ and $T_2$, where $T_1 > T_2$, in the area 1 the gas density is lower and consequently the speed of light is greater than in the area 2, so the angle $\alpha_1 > \alpha_2$ (Fig. 15.14). In this situation, the rays emitted from the plate areas hidden behind the horizon of the edge falling on the boundary surface of these areas will be broken down and may hit the lens of the camera – hence an aura appears – in fact, a mirage image of the plate surface. In reality, there is a continuous change in the density of the gas, so the rays will run along the convex arc.

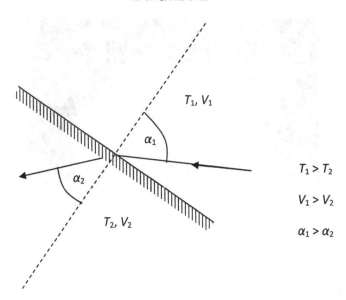

Fig. 15.14 The transition of light rays through the borders of two media with different temperatures (Strzecha et al., 2012).

This hypothesis is further supported by the fact that the aura on the drop contour is very thin. Continuing the argument – this results from the drop curvature – the mechanism described directs the radiation to the camera only from a narrow band adjacent to the contour just beyond the horizon.

### 15.3.1.5 *Correction of the "gas lens" effects*

The full correction of the phenomena described associated with the flow of protective gas in the furnace chamber is not possible. Among them, a particularly strong impact on the quality of the acquired images has the effect of "gas lens". Measurement system developers have concluded that the most effective and the simplest method of correction is to develop and implement an auto-focus algorithm, which will evaluate focus (contrast) of the image during the experiment and, if necessary, carry out complex correction by the appropriate positioning of the camera.

The proposed algorithm is passive, based on the contrast measurement. Its operation can be divided into four basic stages.

### Stage 1: *Definition of the auto-focus point mesh.*

The mesh of auto-focus points defines which image pixels will be considered in determining its focus. The density of the mesh should be chosen to minimize the computational cost, without significantly affecting the quality of processing. The orientation and location of points depend on the edge types found in the image. In the THERMO-WET system two meshes are used:

- with the horizontal auto-focus points, making an assessment of the vertical edges contrast possible (specimen);
- with the vertical auto-focus points, making an assessment of the horizontal edges contrast possible (measurement table).

The proposed mesh of horizontally oriented auto-focus points is presented in Fig. 15.15.

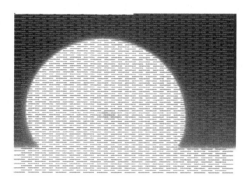

Fig. 15.15 The mesh of horizontal oriented auto-focus points.

*Stage 2: Selecting auto-focus points intersecting edges.*
The second stage of the algorithm is designed to identify the auto-focus points intersected by the edges in the digital image. The following operations are carried out sequentially:

- for all the pixels in the image corresponding to the auto-focus points the gradient is evaluated;
- within each auto-focus point, a pixel with the maximum value of gradient is selected;
- auto-focus points relevant to the assessment of image sharpness are selected; the selection is made on the basis of comparison of the maximum gradient value within an auto-focus point with the experimentally determined threshold (see Fig. 15.16).

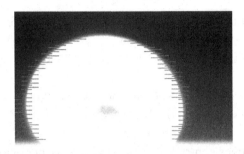

Fig. 15.16 The result of the second stage of auto-focus algorithm for the mesh of horizontal auto-focus points.

*Stage 3: Contrast ratio determination.*

In the third stage the contrast ratio of the analyzed image is determined. It is calculated as the average value of the maximum gradient of auto-focus points selected in the second stage. The designated contrast ratio is a measure defining the edge sharpness in the image.

The results of the algorithm described for a series of images obtained on the THERMO-WET system are shown in Fig. 15.17. Image B is an image with an optimal contrast, image A was acquired for the minimum distance between the specimen and the camera, and image of C for the maximum distance. Images A, B and C are shown in Fig. 15.18.

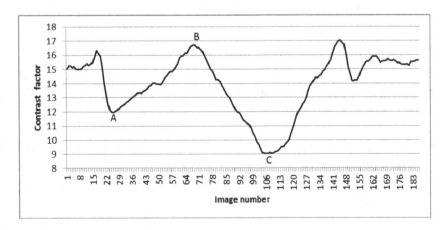

Fig. 15.17 Contrast ratio as a function of distance of the camera from the specimen: A - minimum distance, B - optimal contrast, C - maximum distance.

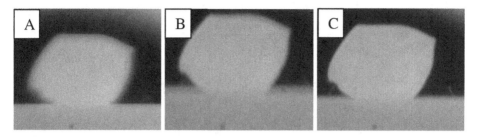

Fig. 15.18 Images obtained during the experimental verification of auto-focus algorithm: A - minimum distance, B - optimal contrast, C - maximum distance.

*Stage 4: Camera positioning.*

The final stage of the algorithm is to correct the position of the camera. It is carried out only when the value of the contrast ratio decreases by more than the assumed threshold. For two images obtained for adjacent camera positions, contrast values are compared and, on this basis, the direction of movement of the camera is determined. The camera is moved until the maximum value of contrast ratio is acquired.

Critical to the quality of the results obtained using the presented algorithm is an adequate selection of a gradient threshold under which the choice of auto-focus points for further analysis is made. Figure 15.19 shows the experimental results for two different values of the threshold. It is clear that if its value is too low, the distinguishing of the contrast in the sequence of images is weaker.

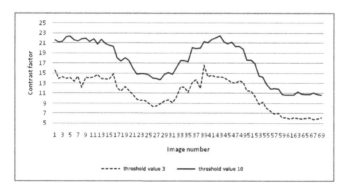

Fig. 15.19 Contrast ratio as a function of distance of the camera from a specimen for two different values of gradient threshold.

The auto-focus algorithm was tested for two different methods of determining the values of gradients, using two and four-point derivative. The graph in Figure 15.20 illustrates the results obtained. Both methods give similar results. Consequently, the final version of the algorithm uses a 2-point derivative as computationally less expensive.

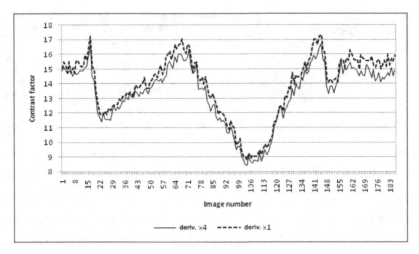

Fig. 15.20 Contrast ratio as a function of distance of the camera from a specimen for two different methods of determining the value of the gradient.

The problem to be solved was the rapid change in the value of the contrast ratio for neighboring images. Simple filtration obtained by averaging the contrast factor values for the four images obtained for neighboring camera positions proved to be an effective solution. The results are shown in Figure 15.21.

The proposed auto-focus algorithm allows such camera positioning during the measurement experiment that images obtained in the entire temperature range are characterized by a similar, high quality, parameters, independent of the actual "gas lens".

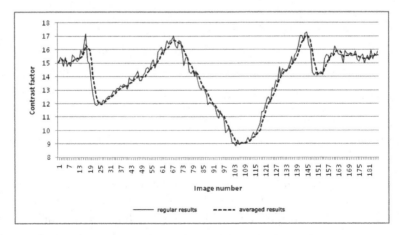

Fig. 15.21 Contrast ratio as a function of distance of the camera from a specimen with and without averaging.

### 15.3.2 *Aura Removal*

Another factor resulting from the specificity of measurements performed by the THERMO-WET system is the aura effect. A specimen heated to a high temperature emits light in the visible part of the spectrum. The illumination causes the saturation of the CCD chip photosensitive elements. In order to diminish mid-infrared (thermal) radiation, infrared filters are used. However, it is impossible to completely eliminate the aura i.e. a glow that forms around the specimen.

Most image segmentation algorithms join the aura with the object, which increases specimen dimensions. Moreover, information about the specimen shape and the upper edge of the base plate localization is lost. Therefore, in order to ensure a high quality of image segmentation and the precision of surface property determination, the aura should be removed from the analyzed scene before the segmentation.

The proposed algorithm of aura removal uses Sobel gradient operators to find areas of an approximately constant intensity value. They are excluded from further analysis. Areas of appropriately high gradient are regarded as an aura and left unchanged. A block diagram of the proposed algorithm of the aura removal algorithm is presented in Figure 15.22. After the aura is extracted, it is simply subtracted from the analyzed image. The image is then subjected to the segmentation process.

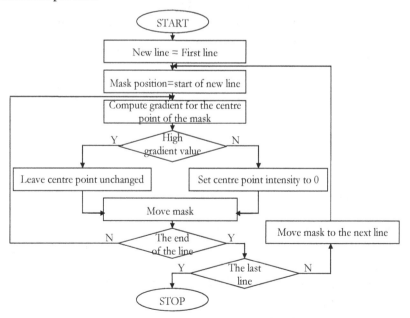

Fig. 15.22 Block diagram of aura extraction algorithm.

The results of applying the aura removal algorithm to an exemplary image are presented in Fig. 15.23. Fig. 15.23 shows the original image (copper at 1239°C). The results of segmentation of the original image are presented in Fig. 15.23b. In Fig. 15.23c the aura-removed image after segmentation can be seen. The horizontal line indicates an approximate localization of the upper edge of the base plate.

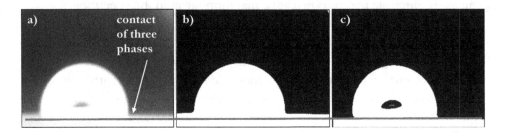

Fig. 15.23 Influence of aura removal process on image segmentation quality; a) original image, copper, 1239°C; b) original image after segmentation; c) aura-removed image after segmentation.

In the case of no aura-removed image after segmentation, the displacement of the base line can easily be seen (Fig. 15.23b). Moreover, important information about the specimen shape and the upper edge of base plate localization (especially in contact of three phases) is lost. It can easily be seen that if image segmentation is preceded by aura removal, the quality of specimen shape determination is significantly improved (Fig. 15.23c). The edges of objects after segmentation are smooth and free from defects. Moreover, significant details of the specimen shape are extracted. The contours obtained provide good estimates of the contours of objects in the original image. The upper edge of the base plate is clear and well defined. Its localization matches the original image. The images obtained can be then successfully used for the further quantitative analysis.

### 15.3.2 Segmentation Algorithms

One of the most difficult and the most important problems of image processing is segmentation. The purpose of segmentation is to divide a set of points on the digital image to disjoint subsets that meet certain criteria for homogeneity (e.g. color, brightness, texture). Each of these subsets has a specific meaning in relation to the characteristics of the observed scene. Although a wide range of different segmentation techniques is known (Fu et al., 1981; Haralick and

Shapiro, 1985; Reed and du Buf, 1993; Li and Gray, 2000; Suri et al., 2002; Yoo, 2004; Nieniewski, 2005; Suri et al., 2005; Sun, 2006; Zhang, 2006), there is no general theory of segmentation.

After a detailed theoretical analysis and experimental verification of various groups of segmentation algorithms, the main assumptions for the segmentation in THERMO-WET system were made:

- developed image segmentation algorithms will be based on edge detection using a gradient filter with dynamically selectable masks;
- algorithms will provide high-quality treatment necessary in measuring systems, understood as giving a precise shape and dimensions of objects in the scene.

### 15.3.2.1 *Algorithm for Fast Specimen Localization*

The task of the algorithm is to provide information on the approximate location of the specimen within the analyzed scene with the minimum possible calculation effort. The resulting location of the sample is described by five location points (Fig. 15.24):

A, E – the upper edge of the tray table;
B, D – the left and the right extreme point of the sample;
C – the top of the drop.

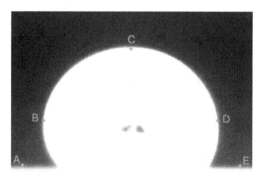

Fig. 15.24 Specimen location points.

The task of the first stage of the fast specimen localization algorithm is to determine the location of the upper edge of the measurement table (A, E). For this purpose, from all points of the columns outermost of 1/20 of the width of the image from its left and right boundary, the pair of pixels distant from each other by the number of pixels representing 2% of the image height, for which the difference of brightness is the greatest, is determined. The greater brightness of the determined pixels will be the brightness value with which the specimen will be identified in the next stage of the algorithm (Fig. 15.25).

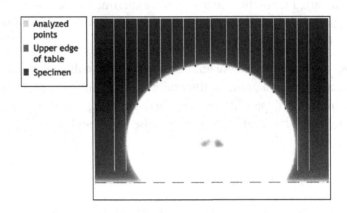

Fig. 15.25  Idea of fast drop localization algorithm.

In the second stage, points D, C and B are searched for in subsequent columns between the first and 15th pixel row above the upper edge of the measurement table. Depending on the point searched for, the algorithm uses one of the following criteria:

- point D: the first point with three following points heaving the brightness higher or equal to the specimen brightness found in the first phase of the algorithm;
- point C: the highest point found with the same criteria as point D, but belonging to a different column;
- point B: the last point found with the same criteria is as point D.

If any of the location points are not found, the algorithm assumes that the image does not represent the specimen or its quality is insufficient for the purposes of further analysis.

The low computational complexity of the discussed algorithm results from the limitation of the number of analyzed points to only 5% of all the pixels. The algorithm examines only the columns separated by a distance equal to 1/20 of the width of the image.

### 15.3.2.2 *Edge detection with dynamic selection of filtration masks*

The algorithm described below is a modified gradient filter which uses directional Sobel masks for 4 base gradient vector directions (Fig. 15.26).

Based on the drop localization points the gradient filtration algorithm specifies areas of drop fragments, for which the gradient vector directions and the axis of ordinates create approximate angles of 0°, 45°, 90° or 135°.

For such defined areas, algorithms select the most appropriate masks (Fig. 15.27):

- the area containing point C: both cross line and horizontal;
- the areas containing points B, D: both cross line and vertical;
- the areas below points A, E: horizontal line mask;
- the remaining areas: both cross line masks.

The edge extraction relies upon the collecting points with the highest value of gradient for the local gradient direction.

| *Cross line masks [3x3]* | | | | | | | |
|---|---|---|---|---|---|---|---|
| 2 | 1 | 0 | | 0 | 1 | 2 | |
| 1 | 0 | -1 | | -1 | 0 | 1 | |
| 0 | -1 | -2 | | -2 | -1 | 0 | |
| *Horizontal and vertical line masks [3x3]* | | | | | | | |
| 1 | 2 | 1 | | -1 | 0 | 1 | |
| 0 | 0 | 0 | | -2 | 0 | 2 | |
| -1 | -2 | -1 | | -1 | 0 | 1 | |

Fig. 15.26 Sobel mask for four basic directions of the vector gradient.

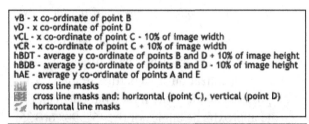

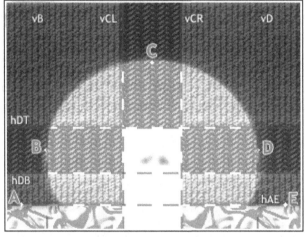

Fig. 15.27 The use of gradient filter masks in different drop areas.

The presented algorithm is characterized by a higher precision of the edge localization by matching the type of masks used for the shape of the object in the analyzed scene. Its computational cost is significantly lower than the classical Sobel method due to the reduction in the number of masks, and the reduction of the number of pixels for which the convolution is performed.

### 15.3.2.3 *Contour tracing*

The last stage of segmentation is responsible for extracting the object from the pre-processed image and presenting it as a list of pixels forming the edge outlining the specimen along with the table. A further analysis of such a list will be possible only if the points it contains will form a continuous edge which:

- is the width of one pixel;
- consists of the points with the maximum of two n-neighbors;
- begins in the first and ends on the last column of the image;

The convolution of an image with a point detection mask gives, as a result, a map of edge points (Fig. 15.28). These points will form a continuous one pixel wide edge if the points in the neighborhood of two consecutive n-neighbors of the concerned edge map element will be deleted. Thus, the obtained list of points represents the binary form of the one-pixel-wide edge.

| *Point detection mask* | | | *N-neighbor detection mask* | | |
|:---:|:---:|:---:|:---:|:---:|:---:|
| -1 | -1 | -1 | 0 | -1 | 0 |
| -1 | 8 | -1 | -1 | 0 | -1 |
| -1 | -1 | -1 | 0 | -1 | 0 |

Fig. 15.28 Point and n-neighbor detection masks.

To verify the continuity of the determined edge simplified contour tracking algorithm based on the analysis of the nearest neighborhood of edge points was used. This algorithm also checks whether the starting point of the edge is in the first, and the final in the last column of the image.

The contour tracking process begins by finding the edge point located in the first column of the image, which initializes the major loop of the algorithm. In this loop n-neighborhood of the current point is tested. Depending on the result of the convolution of the n-neighbor detection mask (Fig. 15.28) with a given point of the image, one of the following actions is taken:

- 1 n-neighbor

deletes all the points added from the last node from the list (more than two n-neighbors) and re-examines its n-neighborhood without the previously considered n-neighbors;

- 2 n-neighbors

considers the n-neighbor, which is not included in the resulting list, as the next point of the edge;

- 3 or 4 n-neighbors

marks the point currently under consideration as a node, and classifies the first encountered n-neighbor, which is not included in the resulting list, as the next edge point, which does not contain a list of the resulting.

The algorithm terminates when the currently considered point belongs to the last column of the image and has only one n-neighbor. However, if the point having one n-neighbor does not belong to the last column of the image, and all n-neighbors of the last node have been processed, the algorithm aborts, indicating that the edge meeting the initial assumption has not been found. Such an image cannot be sent for further analysis.

The final result of the segmentation algorithms developed specifically for the THERMO-WET system is shown in Fig. 15.29.

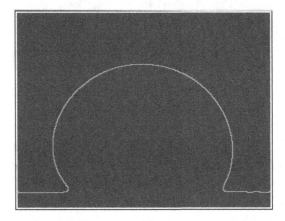

Fig. 15.29 The final result of the segmentation process.

### 15.3.2.4 *Profile correction at the contact point of the three phases*

In the THERMO-WET system the proper determination of contact angles is hindered by a number of phenomena related to the measurements at high temperatures, the most significant of which are:

- an aura;
- a small depth of field of the optical system;
- physicochemical phenomena occurring in the three phases of contact.

The actual shape of the profile of the specimen resulting from the above-mentioned phenomena is shown in Fig. 15.30. The deformation of the specimen profile is clearly visible around the contact point of the three phases.

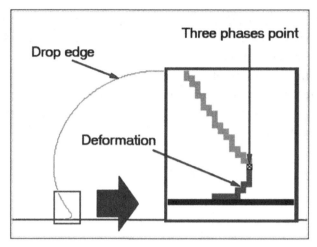

Fig. 15.30 The deformation of the specimen profile at the contact point of the three phases.

In order to avoid the influence of the specimen profile deformation on measurement results there is a need to identify and skip deformation points in further calculations. This operation is performed for both the left and the right profile of the sample. The following operations are carried out sequentially:

- extending the set of the edge points of the specimen in such a way that the contact point of the three phases is not the utmost point;
- approximating the specimen with the fifth or sixth degree polynomial (depending on the shape of the sample) using an extended set of boundary points;
- defining the tangent to the curve approximating the profile of the specimen at the point of three phases contact;

The knowledge of the tangent to the profile of the specimen at the point of three phase contact allows one to determine contact angles on the basis of elementary mathematical relationships.

The algorithm of extending the set of the specimen edge points has a particular significance for the process of determining the contact angles. Its idea is illustrated in Fig. 15.31. For this purpose, first the specimen profile is approximated with a polynomial of the fifth or sixth degree (excluding points in the deformation), next, the tangent to the profile of the specimen at the point of three phase contact is determined. The resulting tangent becomes the axis of ordinates $OY'$ in the local coordinate system with the origin at the three phase contact point.

Extending the set of profile points is simply determining points $A'$ axially symmetric to axis $OX'$ of the local coordinate systems. Points are added to the set of points above the point of three phase contact $A$.

Knowing the equations of lines along the axes $OX'$ and $OY'$ in the local coordinate system and the coordinates of point $A$ in the global coordinate system, it is possible to designate coordinates $x'$ and $y'$ as the distances from the axes $OY'$ and $OX'$, respectively:

$$x' = \frac{|Ax + By + C|}{\sqrt{A^2 + B^2}} \tag{15.9}$$

$$y' = \frac{\left|\dfrac{1}{A}x + By + C\right|}{\sqrt{A^2 + B^2}} \tag{15.10}$$

The determination of extra points is equivalent to the solution of equations (15.9) and (15.10) for the local variables $x'$ and $-y'$.

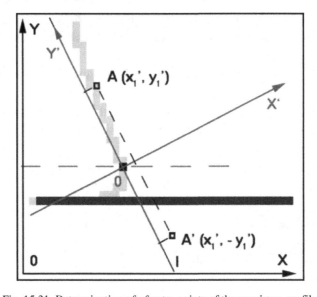

Fig. 15.31 Determination of of extra points of the specimen profile.

The results of contact angle determination in an exemplary image of a low carbon steel specimen at a temperature of 1500°C are shown in Fig. 15.32. Fig. 15.32a shows the original image obtained during the measurement process. In Fig. 15.32b contact angle determination results obtained using the original specimen profile are shown. Finally, Fig. 15.32c presents the results obtained after extending the specimen profile using the proposed method. It is clearly visible that in the case of the considered drop of molten steel, contact angles are below 90°, while ones determined from the original profile are above 90°. Extending the specimen profile using the method described allows one to obtain results with values lower than 90°.

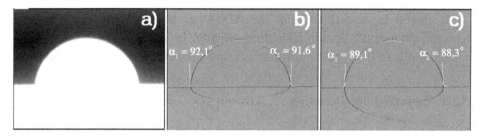

Fig. 15.32 Results of contact angle determination in exemplary image of the low carbon steel specimen at the temperature 1500°C: a) original image; b) results obtained using the original profile; c) results obtained after extending the specimen profile using the presented method.

In Fig. 15.33 the results of contact angle determination in images of copper at temperatures of 1050°C - 1600°C are shown. The results obtained using the old and the new methods are compared with the results obtained using ADSA approach used as a reference.

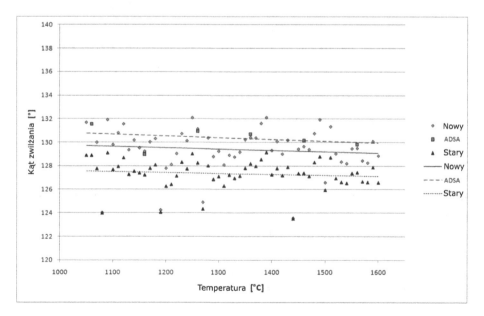

Fig. 15.33 Results of contact angle determination between $Al_2O_3$ and copper at temperature 1050°C-1600°C.

The old method for contact angle determination provides angles characterized by high deviation. Moreover, the determined values of contact angles are significantly smaller than the reference ones provided by ADSA. The introduced method for contact angle determination significantly improves the quality of the results obtained. The determined values are very close to the reference ones and are characterized by smaller deviation.

It should be underlined that the ADSA method used as a reference is characterized by high computational complexity. In addition, it requires a well-defined profile of the specimen, which is usually selected by the system operator.

### 15.3.3 *Subpixel Edge Detection*

In the system considered the precise determination of an object dimension is crucial for the accuracy of measurements. However, due to very intense thermal radiation and aura phenomena, edge detection performed with the traditional accuracy fails to properly define the shape of objects present on the scene. The edge between the object and the background is supposed to be located inside the aura. However, its exact position is not clearly defined. Well-established (derivative-based) methods for edge detection produce different results when applied to the same image. Most commonly, the aura is connected with the object, which increases object dimensions. This, in turn, influences the determined values of surface tension which is calculated based on the characteristic dimensions of the specimen. Therefore, the method for improving the quality of edge detection was introduced. The approach proposed determines edges with sub-pixel accuracy and eliminates weaknesses of the traditional methods for edge detection.

The main concept of the proposed method is to create continuous functions out of the image derivative values provided by the popular approaches to edge detection. Specifically, the Sobel gradient detector is considered. The neighborhood of each coarse edge-pixel provided by the Sobel edge detector is examined in order to build continuous image derivatives in a given direction. The subpixel edge position is indicated by the maximum of the created (i.e. continuous) function.

The introduced approach operates in two main stages, namely:

* the determination of the coarse edge;
* the determination of the subpixel edge;

Sobel masks (the horizontal $h_x$ and the vertical $h_y$) are applied in order to determine the coarse edge. The input image $L$ is convolved with each mask separately in accordance with Equations (15.11) and (15.12). These operations produce images of gradient in horizontal and vertical directions, respectively.

$$|\nabla L_x| \cong |h_x \otimes L| \tag{15.11}$$

$$|\nabla L_y| \cong |h_y \otimes L| \tag{15.12}$$

where: $\otimes$ denotes the convolution.

Next, local intensity maxima are determined along the non-zero pixels of the gradient images $\nabla$Lx and $\nabla$Ly. In the case of image $\nabla$Lx, the search is performed in a horizontal direction and two local maxima (for the left and the right side of the image) are determined in each row. This operation is expressed by Equations (15.13-15.15). Image $\nabla$Ly is searched in a vertical direction and locates one maximum in each column (see Eq. (15.16)).

$$\partial L_x = \partial L_{xl} \cup \partial L_{xr} \tag{15.13}$$

where:

$$\partial L_{xl}(x, y) = \begin{cases} 1 & for\ \nabla L(x, y) = \max_{1 \le x < N/2} (\nabla L(x, y)) \\ 0 & otherwise \end{cases} \tag{15.14}$$

$$\partial L_{xr}(x, y) = \begin{cases} 1 & for\ \nabla L(x, y) = \max_{N/2 \le x \le N} (\nabla L(x, y)) \\ 0 & otherwise \end{cases} \tag{15.15}$$

$$\partial L_y(x, y) = \begin{cases} 1 & for\ \nabla L(x, y) = \max_{1 \le y \le M} (\nabla L(x, y)) \\ 0 & otherwise \end{cases} \tag{15.16}$$

where $M$ and $N$ denote the height and the width of the input image, respectively.

The operations described above produce binary images of one-pixel width coarse edges in horizontal and vertical directions. These binary images are input data for refining the edge location performed in the next step.

Successive steps of the coarse edge detection applied to the exemplary image of heat-emitting steel cylinder at $1020^0$C are presented in Figure 15.34. In particular, Figure 15.34a shows an exemplary image obtained with 8-bit grayscale resolution and spatial resolution $M \times N$ of 240×320 pixels. The edges in the exemplary image are significantly blurred due to intense thermal radiation emitted by the steel cylinder. In Figures 15.34b and 15.34c images of gradient in horizontal and vertical directions are shown. Figures 15.34d and 15.34e present the corresponding edges in vertical and horizontal directions. In Figure 15.34f a complete coarse edge composed from horizontal and vertical components can be seen. Negatives are shown in order to improve the readability of images.

It can easily be seen (Fig. 15.34) that in the case of significantly blurred edges their coarse representation is irregular and discontinuous. There are significant deviations from the main direction of the edge. Therefore, in the next step the coarse location of edges is refined using numerical and mathematical methods.

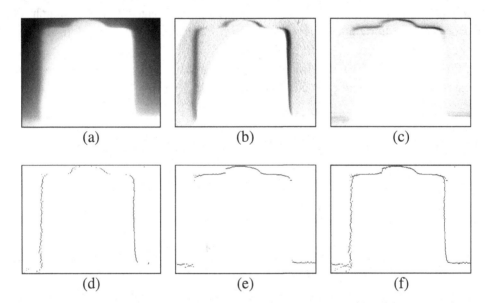

(a)          (b)          (c)

(d)          (e)          (f)

Fig. 15.34 Successive steps of coarse edge determination; (a) exemplary image of steel cylinder at $1020^0$C; (b) horizontal gradient; (c) vertical gradient; (d) coarse vertical edge; (e) coarse horizontal edge; (f) complete coarse edge composed from horizontal and vertical component.

After the coarse edge is found, its location is refined to subpixel position. The accurate location of the edge is approximated using polynomials. Especially, least-square approximation is used. It aims at modeling continuous functions out of the discrete gradient values in the neighborhood of pixels qualified to the coarse edge. The linear neighborhood in the gradient direction is considered; $s$ pixels on each side of the coarse edge pixel are used. The coordinates of the maximum of the approximated gradient function determine the edge location with sub-pixel accuracy. The main idea of refining the edge location to a sub-pixel level is sketched in Figure 15.35.

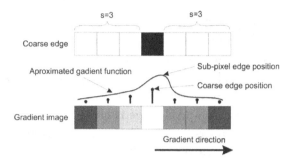

Fig. 15.35 The idea of sub-pixel edge detection using the proposed method.

The experimental results proved that the best (i.e. most regular) edges are obtained when a second-grade polynomial is used. Regardless of the neighborhood size, approximation using Trinomial Square provides the most stable approximation results.

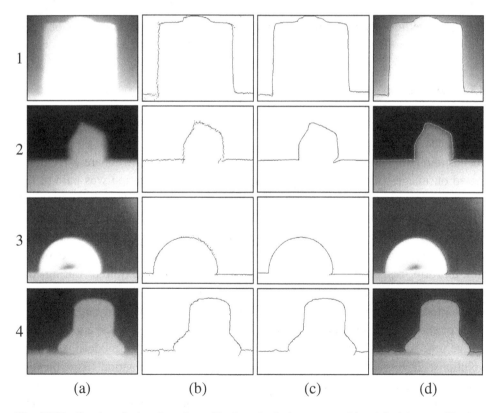

Fig. 15.36 Results of edge detection with the sub-pixel accuracy; (a) original image; (b) edge detected with pixel accuracy; (c) sub-pixel edge; 1- low carbon steel, $1200^0$C; 2- low carbon steel 1, $1000^0$C; 3- silver, $1200^0$C; 4 – low carbon steel, $880^0$C.

The results of edge detection with sub-pixel accuracy in the images of heat-emitting specimens of metals and alloys are presented in Figure 15.36. In particular, Figure 36a presents the input image, Figure 15.36b corresponds with the edge detected with pixel accuracy using Sobel gradient masks and in Figure 15.36c the sub-pixel edge is presented. Finally, the comparison of the input image and the detected (i.e. sub-pixel) edge is given in Figure 15.36d. The input images are denoted by numbers 1 – 4 and present low carbon steel at $1200^0$C, low carbon steel at $1000^0$C, silver at $1200^0$C, and low carbon steel at $880^0$C, respectively. Due to the limitations of the digital raster, the pixel position in the image of the refined edge was rounded to the closest integer value.

The results of applying the method proposed to heat-emitting images prove that sub-pixel processing significantly improves the quality of edge detection. The refined edges are continuous and much more regular than those provided by Sobel gradient masks. The more blurred is the edge, the more significant the difference.

The method proposed was developed for images of heat-emitting objects. However, it can be successfully applied in a wide spectrum of applications.

## 15.4 Conclusions

In this chapter a high temperature image quantitative analysis system was introduced. The system is capable of measuring superficial surface properties of metals and alloys in temperatures of up to $1800^0$C. Measurements are carried out based on images of heat emitting specimens. Particular attention was paid to the selection of digital image processing algorithms. A detailed description of preprocessing methods implemented in the vision unit of the measurement system was given. The algorithms presented consider not only the effect of typical factors arising from the CCD camera electronic components (noise, pixels non-uniform response), but the specificity of images acquired during the measurement process as well. Moreover, precise segmentation algorithms were considered. The application of the algorithm to the images analyzed results in high quality image segmentation. Thanks to the utilization of advanced digital image processing and analysis algorithms, in combination with modern structural solutions, a high precision of surface property determination is achieved. Moreover, the universality of this solution allows the algorithms presented to be successfully used in a wide spectrum of applications in high-temperature image quantitative analysis systems.

# References

Acharya, T., Ajoy, K. (2005). Image Processing: Principles And Application. John Wiley & Sons.

Adamson, A.W., Gast, A.P. (1997). Physical Chemistry Of Surfaces. Wiley-Interscience, New York.

Anastasiadis, S.H., Chen, J.K, Koberstein, J.T, Siegel, A.F., Sohn, J.E., Emerson, J. A. (1987). The Determination Of Interfacial Tension By Video Image Processing Of Pendant Fluid Drops. Journal Of Colloid And Interface Science, Vol. 119, Pp. 55-66.

Atae-Allah, C., Cabrerizo-Vilchez, M., Gomez-Lopera, J.F., Holgado-Terriza, J.A., Roman-Roldan, R., Luque-Escamilla, P.L. (2001). Measurement Of Surface Tension And Contact Angle Using Entropic Edge Detection. Measurement Science And Technology, Vol. 12, No. 3, Pp. 288-298.

Awcock, G.W., Thomas, R. (1995). Applied Image Processing. Mcgraw Hill, New York.

Bachevsky, R.S., Naidich, Y.V., Grygorenko, M.F., Dostojny, V.A. (1994). Evaluation Of Errors In Automatic Image Analysis Determination Of Sessile Drop Shapes. Proc. Int. Conf. High Temperature Capillarity, Smolenice Castle, Pp. 254.258.

Baglin, J.E.E. (1988). Thin Film Adhesion: New Possibilities For Interface Engineering. Material Science And Engineering, Vol. B-1, Pp. 1-7.

Bąkała, M. (2007). Wyznaczanie Wybranych Parametrów Lutowności W Wysokich Temperaturach Z Wykorzystaniem Metody Płytkowej. PhD Thesis. Czestochowa University of Technology, Częstochowa, Poland (in Polish)

Ballard, D.H., Brown, C.M. (1982). Computer Vision. Prentice Hall, New Jersey.

Basu, M. (2002). Gaussian-Based Edge-Detection Methods - A Survey. IEEE Transactions On Systems, Man, And Cybernetics, Part C: Applications And Reviews, Vol. 32, No. 3, Pp. 252-260.

Bauda, I., Kufferb, M., Pfeffer, K., Sliuzasb, R., Karuppannanc, S. (2010). Understanding Heterogeneity In Metropolitan India: The Added Value Of Remote Sensing Data For Analyzing Sub-Standard Residential Areas. International Journal Of Applied Earth Observation And Geoinformation, Vol. 12, Pp. 359–374.

Bauer, N. (2006). Guideline For Industrial Image Processing. Fraunhofer-Gesellschaft.

Baxes, G.A. (1994). Digital Image Processing: Principles And Applications. John Wiley & Sons, Los Altos.

Bechstedt, F. (2005). Principles Of Surface Physics. Springer.

Bergholm, F. (1987). Edge Focusing. IEEE Transactions On Pattern Analysis And Machine Intelligence, Vol. 9, No. 6, Pp. 726-741.

Bovik, A.C. (2000). Handbook Of Image And Video Processing. Academic Press.

Butt, H.J., Graf, K., Kappl, M. (2003). Physics And Chemistry Of Interfaces. John Wiley & Sons, Los Altos.

Canny, J. (1986). A Computational Approach To Edge Detection. IEEE Transactions On Pattern Analysis And Machine Intelligence, Vol. 8, Pp. 679-714.

Castleman, K.R. (1996). Digital Image Processing. Prentice Hall, New Jersey.

Chellappa, R. (1992). Digital Image Processing. IEEE Computer Society Press, New York.

Cheng, P., Li, D., Boruvka, L., Rotenberg, Y., Neumann, A.W. (1990). Automation Of Axisymmetric Drop Shape Analysis For Measurement Of Interfacial Tensions And Contact Angles. Colloid Surfaces, Vol. 43, Pp. 151-167.

Davies, E.R. (2004). Machine Vision: Theory, Algorithms, Practicalities. Morgan Kaufmann.

Davis, L. S. (1975). A Survey Of Edge Detection Techniques. Computer Graphics And Image Processing, Vol. 4, Pp. 248-270.

De Gennes, P., Brochard-Wyart, F., Quere, D. (2003). Capillarity And Wetting Phenomena: Drops, Bubbles, Pearls, Waves. Sprinter-Verlang, Berlin.

De Gennes, P.G. (1985). Wetting: Statics And Dynamics. Reviews Of Modern Physics, Vol. 57, No. 3, Pp. 827-863.

Delannay, F., Froyen, L., Deruyttere A. (1987). The Wetting Of Solids By Molten Metals And Its Relation To The Preparation Of Metal-Matrix Composites. Journal Of Material Science, Vol. 22, No. 1, Pp. 1-16.

Deriche, R. (1987). Using Canny's Criteria To Derive A Recursively Implemented Optimal Edge Detector. International Journal Of Computer Vision, Vol. 1, Pp. 167-187.

Deyev, G.F., Deyev, D.G. (2005). Surface Phenomena In Fusion Welding Processes. CRC Press.

Dougherty, E.R. (Ed.) (1994). Digital Image Processing Methods. CRC Press.

Dougherty, E.R., Astola, J. (1993). Mathematical Non-Linear Image Processing. Kluwer Academic Publishers, London.

Drzymała, J. (2001). Podstawy Mineralurgii. Oficyna Wydawnicza Politechniki Wrocławskiej, Wrocław, Poland (in Polish).

Dutkiewicz, E.T. (1998). Fizykochemia Powierzchni. WNT, Warsaw, Poland (in Polish).

Egry, I., Lohöfer, G., Neuhaus, P., Sauerland, S. (1992). Surface Tension Measurements Of Liquid Metals Using Levitation, Microgravity, And Image Processing. Int. Journal Of Thermophysics, Vol. 13, No 1, Pp. 65-74.

Emelyanenko, A.M. (2004). The Application Of Digital Image Processing To Study Surface Phenomena. Progress In Colloid And Polymer Science, Vol. 128, Pp. 199-201.

Emelyanenko, A.M., Boinovich, L.B. (2001). The Role Of Discretization In Video Image Processing Of Sessile And Pendant Drop Profiles. Colloids And Surfaces A: Physicochemical And Engineering Aspects, Vol. 189, Pp. 197-202.

Erhardt-Ferron, A. (2000). Theory And Applications Of Digital Image Processing. University Of Applied Sciences, Offenburg.

Eustathopoulos, N., Nicholas, M.G., Drevet, B. (Ed.) (1999). Wettability At High Temperatures. Pergamon.

Fabijańska, A. (2007). Image enhancement algorithms for high temperature measurements of surface properties of metals and alloys. PhD Thesis, Lodz University of Technology, Poland.

Forsyth, D.A., Ponce, J. (2003). Computer Vision A Modern Approach. Prentice Hall, New Jersey.

Fu, K.S., Mui, J.K. (1981). A Survey On Image Segmentation, Pattern Recognition Letters, Vol. 13, No. 1, Pp. 3-16.

Girault, H., Schiffrin D,.J., Smith, B.J. (1984). The Measurement Of Interfacial Tension Of Pendant Drops Using A Video Image Profile Digitizer. Journal Of Colloid And Interface Science, Vol. 101, Pp. 257-266.

Gonzalez, R.C., Wintz, P. (1987). Digital Image Processing. Addison-Wesley Publishing, Reading.

Gonzalez, R.C., Woods, R.E. (2007). Digital Image Processing. Prentice Hall, New Jersey.

Hader, D.P. (2000). Image Analysis: Methods And Applications. CRC Press.

Hansen, F.K. (1993). Surface Tension By Image Analysis: Fast And Automatic Measurements Of Pendant And Sessile Drops And Bubbles. Journal Of Colloid And Interface Science, Vol. 160, Pp. 209-217.

Hansen, F.K. Rodsrud, G. (1991). Surface Tension By Pendant Drop. Journal Of Colloid And Interface Science, Vol. 141, Pp. 1-9.

Haralick, R.M. (1982). Zero Crossing Of Second Directional Derivative Edge Operator. Spie Symposium On Robot Vision, 336, Washington, Pp. 91-99.

Haralick, R.M., Shapiro L.G. (1985). Survey: Image Segmentation Techniques. Computer Vision, Graphics, And Image Processing, Vol. 29, Pp. 100-132.

Hartland, S. (Ed.) (2004). Surface And Interfacial Tension: Measurement, Theory, And Applications. CRC Press.

Howell, S.B. (2006). Handbook Of CCD Astronomy, Cambridge University Press, Cambridge.

Huh, C., Reed, R.L. (1983). A Method For Estimating Interfacial Tensions And Contacts Angles From Sessile And Pendant Drop Shapes. Journal Of Colloid And Interface Science, Vol. 9, Pp. 1472-1484.

Hung, M.C., Ridd, M. (2002). A Subpixel Classifier For Urban Land-Cover Mapping Based On A Maximum-Likelihood Approach And Expert System Rules. Photogrammetric, Engineering & Remote Sensing, Vol. 68, Pp. 1173-1180.

Ibach, H. (2007). Physics Of Surfaces And Interfaces. Springer.

Jahne, B. (2002). Digital Image Processing. Springer, Berlin, Heidelberg, New York.

Jahne, B. (2004). Practical Handbook On Image Processing For Scientific And Technical Applications. CRC Press.

Jain, A.K. (1989). Fundamentals Of Digital Image Processing. Prentice Hall, New Jersey.

Jensen, J.R. (2004). Introductory Digital Image Processing. Prentice Hall, New Jersey.

Jeżewski, S., (2006).. High temperature lighting model in issues of image processing of specimens in contact of solid-liquid phases. PhD Thesis. AGH University of Science and Technology, Cracow, Poland.

Jivin, I., Rotman, S.R. (2008). Edge Impact On Subpixel Target Detection In Hyperspectral Imagery. Proceedings IEEE 25th Convention Of Electrical And Electronics Engineers, Pp. 100-104.

Kernco Instruments (1999). Vca 2000 Video Contact Angle Meter. Mat. Kernco Instruments Co. Inc., El Paso.

Kinloch, A.J. (1980). Review: The Science Of Adhesion. Journal On Material Science, Vol. 15, Pp. 2141-2166.

Koszmider, T. (2009). The integrated computer system for determination of geometrical parameters of specimens of metals and alloys in high temperatures. PhD Thesis, Lodz University of Technology, Lodz, Poland.

Koszmider, T., Strzecha, K. (2008). New Segmentation Algorithms Of Metal's Drop Images From Thermo-Wet System. Proc. Iv IEEE Int. Conf. Memstech'2008, Polyana-Lviv, Pp. 81-83.

Kwok, D.Y., Neumann, A.W. (2000). Contact Angle Interpretation In Terms Of Solid Surface Tension. Colloid Surfaces A: Physicochemical And Engineering Aspects, Vol. 161, Pp. 31-48.

Lam, C.N.C, Lu, J.Y., Neumann, A.W. (2001a). Measuring Contact Angle. W: Holmberg, K., Schwuger, M.J., Shsh, D.O. (Ed.) Handbook Of Applied Surface And Colloid Chemistry. Part 5: Analysis And Characterization In Surface Chemistry. Willey Europe.

Lam, C.N.C., Ko, R.H.Y., Li, D., Hair, M.L., Neuman, A.W. (2001b). Dynamic Cycling Contact Angle Measurements: Study Of Advancing And Receding Contact Angles. Journal Of Colloid And Interface Science, Vol. 243, No. 11, Pp. 208-218.

Li, J., Gray, R.M. (2000). Image Segmentation And Compression Using Hidden Markov Models. Springer-Verlag, Berlin.

Lim, J.S. (1990). Two-Dimensional Signal And Image Processing. Prentice Hall, New Jersey.

Lindsay, P.H., Norman, D.A. (1991). Procesy Przetwarzania Informacji U Człowieka. PWN, Warsaw (in Polish)

Marr, D., Hildreth, E. (1980). Theory Of Edge Detection. Proceedings Of Royal Society, B-207, Pp. 187-217.

Materka, A. (1991). Elementy Cyfrowego Przetwarzania And Analizy Obrazów. PWN, Warsaw (in Polish)

Matsunawa, A., Ohji, T. (1982). Role Of Surface Tension In Fusion Welding, Part I. Transactions Of Japan Welding Research Institute, Vol. 11, Pp. 145-154.

Matsunawa, A., Ohji, T. (1983). Role Of Surface Tension In Fusion Welding, Part II. Transactions Of Japan Welding Research Institute, Vol. 12, Pp. 123-130.

Matsunawa, A., Ohji, T. (1984). Role Of Surface Tension In Fusion Welding, Part III. Transactions Of Japan Welding Research Institute, Vol. 13, Pp. 147-156.

Metcalfe, A.G. (1981). Interfaces In Metal-Matrix Composites. Academic Press, New York.

Missol, W. (1974). Energia Powierzchni Rozdziału Faz W Metalach. Wydawnictwo Śląsk, Katowice (in Polish).

Mittal, K.L. (2006). Contact Angle, Wettability And Adhesion. Vsp.

Mortensen, A. (1991). Interfacial Phenomena In The Solidification Processing Of Metal-Matrix Composites. Material Science And Engineering Vol. A-135, Pp. 1-11.

Murr, L.E. (1975). Interfacial Phenomena In Metals And Alloys. Addison-Wesley Publishing, Reading.

Myers, D. (1999). Surfaces, Interfaces, And Colloids: Principles And Applications. Wiley-Vch.

Nicholas, M.G. (Ed.) (1990). Joining Of Ceramics. Chapman & Hall, London.

Nieniewski, M. (2005). Segmentacja Obrazów Cyfrowych. Metody Segmentacji Wododziałowej, Akademicka Oficyna Wydawnicza Exit, Warsaw.

Nikolaidis, N., Pitas, I. (2001). 3d Image Processing Algorithms. Wiley-Interscience.

Pallas, N.R., Harrison, Y.R. (1990). An Automated Drop Shape Apparatus And The Surface Tension Of Pure Water. Colloids Surfaces, Vol. 43, Pp. 169-194.

Pask, J., Evans, A. (Ed.) (1981). Surfaces And Interfaces In Ceramic And Ceramic-Metal Systems. Plenum Publishing Co., New York.

Pavlidis, T. (1987). Grafika And Przetwarzanie Obrazów. WNT, Warsaw.

Pitas, I. (2001). Digital Image Processing Algorithms And Applications. Wiley-Interscience.

Popel, S.I. (2003). Surface Phenomena In Melts. Cambridge International Science Publishing.

Pratt, W.K. (2001). Digital Image Processing. John Wiley & Sons.

Putiatin, E., Awierun, S. (1990). Obrabotka Izobrażjeni W Robotjechnikje. „Maszinostrojenie", Moskwa.

Reed, T.R., Du Buf, J.M.H. (1993). A Review Of Recent Texture Segmentation And Feature Extraction Techniques. Computer Vision, Graphics And Image Processing, Vol. 57, Pp. 359-372.

Rosen, M.J. (2004). Surfactants And Interfacial Phenomena. Wiley-Interscience, Los Altos.

Rotenberg, Y., Boruvka, L., Neumann, A.W. (1983). Determination Of Surface Tension And Contact Angle From The Shapes Of Axisymmetric Fluid Interfaces. Journal Of Colloid And Interface Science, Vol. 93, Pp. 169-183.

Russ, J.C. (2002). The Image Processing Handbook. CRC Press.

Safran, S.A. (2003). Statistical Thermodynamics Of Surfaces, Interfaces, And Membranes. Westview Press.

Sankowski, D., Mosorow, W., Strzecha, K. (2011). Przetwarzanie And Analiza Obrazów W Systemach Przemysłowych. Wybrane Zastosowania. PWN, Warsaw.

Sankowski, D., Senkara, J., Strzecha, K., Jeżewski, S. (2001a). Automatic Investigation Of Surface Phenomena In High Temperature Solid And Liquid Contacts. 18th IEEE Instrumentation And Measurement Technology Conference, Budapest, Pp. 346.249.

Sankowski, D., Senkara, J., Strzecha, K., Jeżewski, S. (2001b). Image Segmentation Algorithms In High Temperature Measurements Of Physical Properties Using Ccd Camera. Proc. 18th IEEE Instrumentation And Measurement Technology Conf., Budapest, Pp. 346-249.

Sankowski, D., Strzecha, K., Janicki, M., Koszmider, T. (2006). Thermowet: Case Study Of Control Application Design. Selected Problems Of Computer Science, Warsaw, Pp. 392-401.

Sankowski, D., Strzecha, K., Jeżewski, S. (2000a). Image Processing In Physical Parameters Measurement. Proc. 16th Imeko World Congress, Vienna, Pp. 277-283.

Sankowski, D., Strzecha, K., Jeżewski, S. (2000b). Digital Image Analysis In Measurement Of Surface Tension And Wettability Angle. Proc. IEEE Int. Conf. On Modern Problems In Telecommunication, Computer Science And Engineers Training, Lviv, Pp. 129-130.

Sankowski, D., Strzecha, K., Jeżewski, S., Senkara, J., Łobodziński, W. (1999). Computerised Device With Ccd Camera For Measurement Of Surface Tension And Wetting Angle In Solid-Liquid Systems. Proc. IEEE Instrumentation And Measurement Technology Conf., Venice, Pp. 164-168.

Senkara, J., Windyga, A. (1990). Podstawy Teorii Procesów Spajania. Wydawnictwo Politechniki Warszawskiej, Warsaw.

Senthilkumaran, N., Rajesh, R. (2009). Edge Detection Techniques For Image Segmentation – A Survey Of Soft Computing Approaches. International Journal Of Recent Trends In Engineering, Vol. 1, No. 2, Pp. 250-254.

Shapiro, L., Rosenfeld, A. (1992). Computer Vision And Image Processing. Academic Press, Boston.

Shapiro, L., Stockman, G. (2000). Computer Vision. Prentice Hall, New Jersey.

Sharma, G. (2002). Digital Color Imaging Handbook. CRC Press.

Song, B., Springer, J. (1996). Determination Of Interfacial Tension From The Profile Of A Pendant Drop Using Computer-Aided Image Processing: 2. Experimental. Journal Of Colloid And Interface Science, Vol. 184, No 1, Pp. 77-91.

Sonka, M., Hlavac, V., Boyle, R. (2007). Image Processing, Analysis, And Machine Vision. Thomson-Engineering.

Staldera, A.F., Melchiorb, T., Müllerb, M., Saged, D., Bluc, T., Unserd, M. (2010). Low-Bond Axisymmetric Drop Shape Analysis For Surface Tension And Contact Angle Measurements Of Sessile Drops. Colloids And Surfaces A: Physicochemical And Engineering Aspects, Vol. 364, Pp. 72-81.

Strzecha, K. (2002). Application of image processing and analysis in high temperature measurements of surface properties of selected materials. PhD Thesis, Lodz University of Technology, Lodz, Poland.

Strzecha, K., Bąkała, M., Fabijańska, A., Koszmider, T. (2010). New Ideas In High Temperature Computerized Measurements Of Surface Properties. Proc. 6th Int. Conf. Perspective Technologies And Methods In Mems Design. Lviv-Polyana, Pp. 81-84.

Strzecha, K., Koszmider, T. (2008). Drop Shape Analysis For Measurements Of Surface Tension And Wetting Angle Of Metals At High Temperatures. Proc. Iv IEEE Int. Conf. Memstech'2008, Polyana-Lviv, Pp. 57-59.

Strzecha, K., Koszmider, T., Zarębski D., Łobodziński W. (2012). Passive auto-focus algorithm for correcting image distorsions caused by gas flow in high-temperature measurements of surface phenomena. Image Processing & Communications, Vol. 17, No. 4, Pp. 379-384.

Strzecha, K., Sankowski, D., Janicki, M., Koszmider, T. (2006). Control Application Design Of Thermo-Wet System. Proc. Viii Imeko World Congress, Rio De Janeiro.

Sun, J. (2006) Edge Detection, Image Segmentation And Their Applications In Microarray Image Analysis. Proquest/Umi.

Suri, J.S., Setarehdan, S.K, Singh, S. (Ed.) (2002). Advanced Algorithmic Approaches To Medical Image Segmentation: State Of The Art Applications In Cardiology, Neurology, Mammography And Pathology. Springer-Verlag, Berlin.

Suri, J.S., Wilson, D., Laxminaryan, S. (2005). Handbook Of Biomedical Image Analysis: Volume 1: Segmentation Models, Springer.

Tadeusiewicz, R. (1992) Systemy Wizyjne Robotów Przemysłowych. WNT, Warsaw, Poland (in Polish)

Teuber, J. (1993) Digital Image Processing. Prentice Hall, New Jersey.

Trakhtenberg, L.I., Lin, S.H., Ilegbusi, O.J. (Ed.) (2007). Physico-Chemical Phenomena In Thin Films And At Solid Surfaces. Academic Press.

Umbaugh, S.E. (2005) Computer Imaging: Digital Image Analysis And Processing. CRC Press.

Venables, J.A. (2006) Introduction To Surface And Thin Film Processes. Cambridge University Press.

Watkins, C.D., Sadun, A., Marenka, S. (1995). Nowoczesne Metody Przetwarzania Obrazu. WNT, Warsaw.

Weng, Q.H., Lu, D.S. (2009). Landscape As A Continuum: An Examination Of The Urban Landscape Structures And Dynamics Of Indianapolis City, 1991–2000, By Using Satellite Images. International Journal Of Remote Sensing, Vol. 30, Pp. 2547-2577.

Woodruff, D.P. (1973). The Solid-Liquid Interface. Cambridge University Press.

Woods, J.W. (2006). Multidimensional Signal, Image, And Video Processing And Coding. Academic Press Inc. Orlando.

Woodward, R.P. (Ed.) (1996). Two Dimensional Contact Angle And Surface Tension Mapping. First Ten Angstroms, Portsmouth.

Woźnicki, J. (1996) Podstawowe Techniki Przetwarzania Obrazu. WKŁ, Warsaw, Poland (in Polish)

Xu, Z., Masliyah, J.H. (2002). Contact Angle Measurement On Oxide And Related Surfaces. W: Hubbar, D.A., (Ed.) Encyclopaedia Of Surface And Colloid Science. Marcel Dekker, New York, Pp.1228-1241.

Yoo, T.S. (Ed.) (2004). Insight Into Images: Principles And Practice For Segmentation, Registration, And Image Analysis, Ak Peters.

Young, T., Gerbrands, J.J., Van Vliet, L.J. (1998). Fundamentals Of Image Processing. The Netherlands At The Delft University Of Technology, Delft.

Zangwill, A. (1988). Physics At Surfaces. Cambridge University Press.

Zhang Y.J. (Ed.) (2006). Advances In Image And Video Segmentation. IRM Press.

Ziou, D., Tabbone, S. (1998). Edge Detection Techniques - An Overview. International Journal Of Pattern Recognition And Image Analysis, Vol. 8, Pp. 537-559.

Zuech, N. (2000) Understanding And Applying Computer Vision. Marcel Dekker Inc., New York.

Zuo, Y.Y., Do, C., Neumann, A.W. (2007). Automatic Measurement Of Surface Tension From Noisy Images Using A Component Labeling Method. Colloids And Surfaces A: Physicochemical And Engineering Aspects, Vol. 299, Pp. 109-116.

# CHAPTER 16

## THEORETICAL INTRODUCTION TO IMAGE RECONSTRUCTION FOR CAPACITANCE PROCESS TOMOGRAPHY

Radosław Wajman, Krzysztof Grudzień, Robert Banasiak, Andrzej Romanowski, Zbigniew Chaniecki, and Dominik Sankowski

*Institute of Applied Computer Science*
*Lodz University of Technology*
*90-924 Łódź, ul. Stefanowskiego 18/22*
*{rwajman, k.grudzien, rbanasi, a.romanowski z.chaniecki, dsan}@kis.p.lodz.pl*

### 16.1 Introduction

Industry processes, which are characterized by high dynamics or permit only limited access to their installation systems, require effective and reliable measurement techniques. An example of such robust measurement methods is process tomography. Tomography is considered a measurement technique which allows one to visualize the state of dynamic processes – taking place in reactors, pipelines or boilers during extraction or production – as 2D/3D images. Process tomography originates from medical tomography, however, in the medical field there are usually living organisms to be examined, where dynamic changes, as those in industrial processes, are not observed. Industry processes, on the other hand, are of a rapidly changing dynamic character, in such cases as liquid, solid or multiphase flows. An additional advantage of process tomography, as a modern computerized technology, is the noninvasive visualization of industrial processes ongoing in non-transparent industrial installations, in contrast to classical solutions which are usually based on invasive methods disturbing the process monitored in various ways, with the necessity of use of expensive apparatus. Process tomography offers, to scientists and engineers, unique opportunities to measure the structure of materials, liquids, and the speed of their movement in real time. The information about the process, obtained with the use of process tomography, can be useful for developing new hardware systems, or the proper monitoring of production or extraction in an industry environment.

The description of a theoretical background to image reconstruction methods presented in the chapter focuses on Electrical Capacitance Tomography (ECT); however, the issues discussed are shown in a broad aspect of the inverse problem in electrical process tomography. Before the description of the main part of the chapter, information is given about the main components of the measuring system and the methodology followed for data gathering. This knowledge allows a better understanding of the problem of image reconstruction algorithms.

## 16.2 Principles of tomography measurement

Tomography (from Greek: *tomos* – to slice and *graph* – to write) means producing an image of the cross-section of an object of interest, based on the measurements acquired from its boundaries (Williams and Beck, 1995; Scott and McCann, 2005). The common approach to the tomographic measurement data analysis is their processing in order to obtain a reconstructed image of the distribution of a required physical property in the examined cross section area. In the case of ECT, this property is the electric permittivity ε. Therefore, ECT tomography constitutes a unique, non-intrusive measurement tool for the imaging of non-conducting material distribution in the form of a 2D cross-sectional plane or 3D volume during ongoing industrial processes under investigation. The visualization of the interior of the investigated body is obtained on the basis of the acquired capacitance measurements. Electric permittivity distribution $\varepsilon$ is revealed from the measured capacitance data $C$. A schematic diagram of a typical, single plane, electrode layout arrangement, which is attached around the pneumatic conveyor pipe, is shown in Fig. 16.1. In contrast to other modalities, for example, X-ray imaging (which uses the principle of radiation intensity attenuation), ECT allows visualization of two-phase flows using the property of different characteristics of electrical permittivity ($\varepsilon_1$, $\varepsilon_2$) of flow phases (Dyakowski, 1996; Huang et al., 1988; Plaskowski et al., 1995; Yang and Liu, 2000). The permittivity distribution $\varepsilon$ could then be related directly to the material concentration present in the sensor space.

The number of independent measurement data $m$ is defined as:

$$m = N(N-1)/2 , \tag{16.1}$$

for the $N$ electrode system. Both for the simplification of the theoretical discussion and increasing the data acquisition speed, it is assumed that the capacitance, which is measured between two electrodes $l_p$ and $l_q$ (where p represents an active electrode and q represents a grounded electrode), is the same as for the opposite direction.

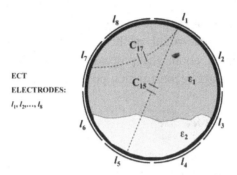

Fig. 16.1 ECT electrodes sensor cross-section diagram, $\varepsilon_1, \varepsilon_2$ indicate different electric permittivity distributions of distinct flow phases.

ECT imaging is characterized by a limited spatial resolution (compared to other tomography modalities) but its acquisition rate is relatively high. A typical measurement data acquisition speed is within a range of 50-200 frames per second (the number of sets of measurement data per second)and 1000+frames per second as reported by(PTL; York et al., 2006). Other important features of capacitance tomography are the simplicity and robustness of the tomography system. These advantages constitute a very important reason to choose ECT tomography as a mature measurement tool for dynamic flow monitoring systems.

The electrical capacitance tomography system can measure a capacitance change between two electrodes using a charge-discharge capacitance measuring circuit or an AC-based principle measurement technique (Yang, 1999; Yang et al., 2001; Brzeski et al., 2003). In both types of ECT systems the signal is taken from a set of electrodes installed around the pipe circumference. Another common feature is that both systems can measure an electrical signal which is proportional to capacitance values. Therefore, it is possible to calculate the capacitance, using circuit analysis. A typical measurement sequence for the charge-discharge system, starts from voltage application to the first (active) electrode $l_1$. The remaining electrodes ($l_2...l_n$) are all grounded (typically named detector electrodes). The capacitance between the active and the detector electrodes is measured. This process is repeated, while the next electrode is active, and afterwards the procedure is repeated for all the other electrodes in the cycle. There are a number of variations of the measurement protocol that allow grouping a number of electrodes into subsets, and performing the voltage excitation process in pairs or, for example, of four electrodes at the same time.

The main advantage of such a solution is an increase in the number of independent measurements, which can improve image reconstruction quality and enhance capacitance measurement sensitivity. However, this implies the limitation of a measurement acquisition rate.

In order to better understand methods of the inverse problem solution, information is presented about the processing of measurements data usually conducted before the image reconstruction procedure.

### 16.2.1 *The calibration procedure*

As in other imaging techniques, an electrical process tomography system needs to be calibrated. The calibration process usually consists of two steps. In the first step the pipe with the sensor volume is filled with a low electrical permittivity material. For most cases the air with the permittivity value equal to one is used. The same procedure is carried out for a high permittivity value material to provide a contrast. It is worth emphasizing that the key element for the correct calibration procedure is the proper selection of lower and higher permittivity values of materials. Materials prepared for sensor fill should be selected just below and just above the level of the lowest and the highest expected permittivity values to be measured. This should be done to leave a small margin for situations when a small porosity change, or a packing density rise may result in extending the measurement range (within [0, 1] or [1, 2] relative ranges). For an "empty pipe" step, the pipe should be carefully emptied, which in industrial circumstances is difficult to achieve. As a result of the measurement, the vector of the measured capacitances $C_{min}$ is given.

In the next step, the sensor volume must be filled with a material with a possibly higher permittivity. In the case of solid material in a granular form, it is difficult to fill the sensor only with a solid material. Materials in the form of porous grains are usually mixed with air or water. Therefore, the real value of permittivity depends on the concentration and the humidity of the material. After the pipe is filled, the vector of capacitances for the maximal permittivity values is measured and saved as $C_{max}$.

### 16.2.2 *The normalization of data vector*

The elements $c_i$ of inter-electrode capacitance values are normalized in an online mode, using related $c_{i\,min}$ values (elements of $C_{min}$) and $c_{i\,max}$ values (elements of $C_{max}$), which are both taken during the calibration procedure. In the literature a different approach to data normalization can be found (Yang and Byars, 1999; Xie and Huang, 1992; Zhiheng, 2011). Typically, the $c_{i\,norm}$ value is obtained based on the parallel model:

$$c_{i\,norm} = \frac{c_i - c_{i\,min}}{c_{i\,max} - c_{i\,min}} \tag{16.2}$$

The calibration process is performed to correctly prepare the ECT equipment to measure and visualize material distribution inside the sensor. This is a fundamental assumption of electrical capacitance tomography (ECT) technique measurements. This calibration should be done both for the sensor and the measurement unit and should precede the measuring process proper for a given object.

### 16.2.3 *ECT measurements sensor*

The proper design of the mechanical and electrical structures of capacitance 2D/3D sensor is a complex task. It includes determining the proper surface area of the electrodes and their shape, the inter-electrode horizontal and vertical spacing, the width of the inter-electrode and the boundary screen and also the thickness of protective insulation. The most commonly used layout of electrodes, their shapes and geometric dimensions are determined according to the useful range of capacitance tomography measurement system sensitivity and according to the expected range of dielectric constants of the test objects. The sensitivity of the ECT system requires both the measured values and the values changes to have a certain minimum. This enforces the use of electrodes with a sufficiently large surface area. In the case of the ECT imaging technique, the measured capacitance range can vary from tens to hundreds picofarads and a change in capacitance caused by a change in the dielectric permittivity distribution can be in a range of femtofarads. In addition, the development of an analogous two-sensor concept for pipes of different diameters (but with the same kind of processes – the same range of the dielectric permittivity values) is not a matter of simple geometry "rescaling". The electric field is known to be highly non-linear, and the choice of ECT sensor geometry parameters must be optimized through the simulation or experimentally by using an ultra-precise LCR meter. During this experimental work the inner space of the test sensor should be filled with a medium; first, of a high dielectric constant and then, of a low dielectric constant. Static capacitance measurements should be performed. The analysis of experimentally collected capacitance data allows adjusting optimal key parameters of the sensor design for an ECT tomography sensor measuring range.

A typical two- or three-dimensional capacitance sensor comprises an array of conducting plate electrodes, which are mounted on the outside of a non-conducting pipe (e.g. PVC polyvinyl chloride), and surrounded by an electrical shield. For a metal wall pipe or vessel, the sensing electrodes must be mounted internally, with an insulation layer between the electrodes and the metal wall and using the metal wall as an electrical shield. A measurement sensor (see Fig. 16.2.) is usually made of copper foil, in the case of an outer sensor, folded out on the pipeline. Additionally, in the case of fluid flow, an inner PVC dielectric shield (3) should be installed to protect it from an unfavorable effect of this fluid. The outer shield is also made of copper and acts as radial screening (1) for the electrodes set being connected to the ground. This screen is extremely important in proper securing the sensor against the exterior environment changes. It is possible to install an additional set of screens (4) which separate the electrodes and are known in the literature as driven guards (Yan et al., 1999). The driven guards and main screening security make it possible to make measurements in very similar electrical conditions. The PVC dielectric shield protects the electrodes against the corrosion effect of fluids and ensures constant conditions for the tomography system.

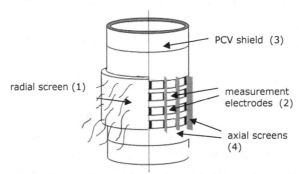

Fig. 16.2 The scheme of the 2D/3D ECT sensor.

## 16.3 2D or 3D – proper choice of visualization

The description explained above concerns the 2D visualization of a process in the form of the 2D cross-section of the object investigated, e.g. a pipeline. Such an approach is very useful when the dynamics of a process is definitely high and the time analysis of measurement data or images is crucial for the proper control of the process. However, the process has a strongly three-dimensional nature and it should not be reduced to a two-dimensional space. If time-consuming data acquisition and their processing are not crucial, a better solution is to apply a 3D sensor for the visualization of the process in the entire measurement space. 3D visualization gives full information about the material distribution inside the sensor.

Three-dimensional visualization can be achieved by interpolating a series of 2D slices from different planes into a 3D image, as in many medical applications. This approach used to be known as 2.5D tomography. In fact, it is not a real 3D reconstruction but only a rough approximation. When evaluating the concept of capacitance tomography, we can see that it is based on obtaining a series of measurements from sensors which are placed on the selected cross-section planes. Various image reconstruction techniques can then be applied to obtain the distribution of dielectric permittivity on these cross-section planes. The same idea can be used for directly obtaining the distribution of a given material parameter in 3D space.

In 2D tomography the natural three-dimensional properties of soft-field modalities interfere with obtaining good quality images. An example of this is an electric field in capacitance tomography which spreads into 3D space following Maxwell's equations. This means that some serious assumptions have to be made about the electric field distribution. As a result, the sensor has to incorporate additional electrodes which reduce the effect of three-dimensional character of the electrical field. It seems to be obvious to transform inconveniences of three-dimensional character of the electric field phenomenon into the benefit of 3D tomography.

In the past few years there has been a great deal of interest in using volumetric ECT both in 3D (Soleimani et al., 2009; Solemani, 2006; Wajman et al., 2006; Warsito and Fan, 2003; Warsito et al., 2007) and in a dynamic imaging mode of 4D ECT. The electrodes at the surface of the volume apply an electrical potential into the surrounding area and the emergent charge is measured at the boundary. The measured data is then used in an electrostatic model to determine the dielectric permittivity properties of the object. Three-dimensional ECT imaging is likely to become an important tool in industrial imaging for process monitoring.

In the case of 3D capacitance tomography, the basic structure of the sensors and the measurement concept are the same as in 2D tomography (see Fig. 16.3). It is a set of electrodes, each with a relatively high active surface area. The difference lies in their layout. In 2D capacitance tomography, with its planar layout of electrodes, some inhomogeneities (i.e. objects) cannot be distinguished and properly located in 3D space as presented in Fig. 16.3b. The inhomogeneity of the examined process should be understood as an area with a different value of the electrical permittivity than that of its neighborhood. In the case of 2D or 2.5D capacitance tomography, the measurements are obtained only for one or some cross-sections, therefore any inhomogeneity (i.e. important for given process object) can exist outside the investigation zone and then cannot affect the measurement values. Thus, they will not be reconstructed. In a 3D capacitance tomography approach, the measurements are also made between the electrodes from different layers and therefore, any inhomogeneity will certainly affect the measurement values and will be distinguished in the final image.

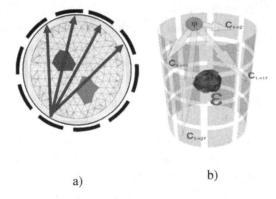

a)                                              b)

Fig. 16.3  2D (a) and 3D (b) tomography sensor.

## 16.4 Inverse problem solution in ECT systems

The obtained vector of measurements is the first step in the image reconstruction procedure. The next step is the retrieval of information on the internal state of the process under examination. Solving an inverse problem is nothing else but looking for a relationship of the measurement (result) with the source data (cause), that is to say, searching for the cause, which gives a measurement.

Reconstruction methods have to be based on recovering information about several hundred/thousand pixels from just tens of measurements. The second issue, which makes the reconstruction in the process tomography system more difficult, is the soft-field effect. This means that the electrical field inside the sensor is distorted by the material with the inhomogeneous distribution of any electrical properties (i.e. permittivity). Due to this fact, the image reconstruction process for capacitance tomography needs to be considered as a non-linear issue (Warsito et al. 2003; Wajman et al. 2006; Soleimani et al. 2007). In order to better understand methods of the inverse problem solution is to analyze the electrical field distribution inside the ECT sensor. The electrical field inside the ECT sensor can be calculated using Poisson's equation which is given by:

$$\nabla^2 \varphi = \nabla \cdot \left( \varepsilon(\vec{x}) \nabla \varphi(\vec{x}, t) \right) = -\rho(\vec{x}) \tag{16.3}$$

where $\nabla^2$ is the Laplace's operator, $\varphi(\vec{x}, t)$ is the electrical potential and $\varepsilon(\vec{x})$ is the electrical permittivity. The vectors $\vec{x}$ are the coordinates in the volume of the sensor. Typically, alternating current of a fixed frequency is employed in ECT systems. Because the frequency is sufficiently small, the problem can be treated as quasi-static and the magnetic field can be neglected, and then the right side of equation (16.3) becomes zero. Using equation (16.3) from a given permittivity distribution $\varepsilon(\vec{x})$ and boundary conditions for the potential $\varphi$ at the electrodes and the shield, the potential field in the sensor can be calculated numerically. If the potential field $\varphi(\vec{x})$ is known, the induced charge $Q_R$ at the grounded electrode $g$ and the capacitance $C_g$ can be calculated using Gauss's law:

$$C_g = \frac{Q_g}{V} = -\frac{1}{V} \oiint_{S_g} \varepsilon(\vec{x}) \nabla \varphi(\vec{x}, t) \, d\vec{S} \tag{16.4}$$

where $S_g$ is the area of the electrode $g$ and $V$ is the potential value set on the excited electrode. In equation (16.4) another relationship can be found (Yang et al. 2003). The permittivity distribution $\varepsilon(\vec{x})$ affects the potential field $\varphi(\vec{x})$ and the capacitance can be considered as a functional of permittivity distribution.

$$C = \xi(\varepsilon) \tag{16.5}$$

The change of capacitance in response to a perturbation of the permittivity distribution is given by

$$\Delta C = \frac{d\xi}{d\varepsilon}(\Delta\varepsilon) + O((\Delta\varepsilon)^2) \tag{16.6}$$

where the term $d\xi/d\varepsilon = J$ is the sensitivity of the capacitance to changes in the permittivity distribution and $O((\Delta\varepsilon)^2)$ represents the $(\Delta\varepsilon)^2$ and higher order terms. In ECT applications the values of the terms of the second order and higher are usually small and by neglecting the term $O((\Delta\varepsilon)^2)$ equation (16.6) can be simplified to its linear form:

$$\Delta C = J\Delta\varepsilon \tag{16.7}$$

## 16.5 Mathematical aspects of image reconstruction

In order to find the image of the electrical permittivity distribution inside the ECT sensor, equation (7) has to be discretized by dividing the sensor area into $n$

elements (image elements). However, for an ECT sensor of N electrodes, the $m = N(N-1)/2$ independent measurements can be obtained. Then, equation (16.7) can be written as:

$$\underset{m\times 1}{\Delta\mathbf{Cm}} = \underset{m\times n}{\mathbf{J}} \ \underset{n\times 1}{\Delta\varepsilon} \tag{16.8}$$

where $\mathbf{J}$ is the Jacobian matrix, i.e. sensitivity distribution matrix given as a set of sensitivity maps for every pair of electrodes.

The aim of the image reconstruction process for ECT is to find the electrical permittivity distribution of the material inside the sensor. This task in an iterative concept is complicated and consists in the measurement of a forward and an inverse problem as presented in the following figure:

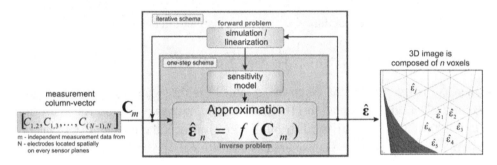

Fig. 16.4 The flow diagram of the iterative image reconstruction schema.

The step called an Inverse Problem is responsible for the direct calculation of the permittivity distribution $\Delta\varepsilon$ from a set of the measured capacitances $\Delta\mathbf{Cm}$ (usually derived from equation 16.8) and from the numerical point of view is under-determined. During this step the inverse of $\mathbf{J}$ has to be found. This task is very difficult because $\mathbf{J}$ is not a square matrix, and instead of having its inverse $\mathbf{J}^{-1}$, which does not exist, the pseudo-inverse $\mathbf{J}^{+}$ must be found as, for instance, an approximated solution. This is due to the fact that the number of variables $n$ (i.e. image elements) is significantly greater than the number of equations $m$ (i.e. measurements). This causes more than one solution to exist. In addition, the solution is ill-conditioned and is very susceptible to measurement noise. The relationship between the electrical permittivity and the measured capacitance is strongly non-linear. This is mostly called a "soft-field" effect and makes the solution very sensitive to small changes in the measured capacitances.

The forward problem, in turn, is of great importance in an iterative reconstruction process. In every step for the current reconstructed image a new calculated capacitance vector should be determined and compared with the measured values to obtain an error, which is minimized during the next iterative step. Therefore, the forward problem is defined as obtaining the measurement vector for the assumed material parameter distribution. From the mathematical point of view solving the forward problem in a 3D or 2D space is almost the same – all equations are based on Gauss equations (16.4). In most cases, a forward problem is solved by numerical methods of which the most popular is the finite elements method (FEM). The linear forward projection (LBP) is also used but – due to its low accuracy – rather as a part of iterative reconstruction algorithms. The FEM method provides both the measurement vector and the electric field distribution from which a Jacobian matrix is usually calculated.

### 16.5.1 *Sensitivity model*

The aim of an image reconstruction process consists in possibly the best approximation of the permittivity distribution of the material on the basis of a set of the measured capacitances from the sensor electrodes. The capacitance sensors (both 2D and 3D) belong to the type of sensors characterized by the "soft-field" effect. The electrical field distribution as well as the set of the measured capacitance changes $C$ strongly depends on the changes of the electrical permittivity distribution $\varepsilon$ of the material inside the sensor and this dependence is non-linear. In order to be able to directly describe this dependence, it is necessary to carry out the linearization process. The linearization in all electrical tomography systems can be performed by applying the Jacobian matrix (also called a sensitivities matrix).

Before the physical meaning of the Jacobian matrix is explained, some information about the types of ECT visualization (2D and 3D) needs to be given. Basically, the structure of the sensors and the measurement concept are the same in both types of ECT. This is a set of electrodes, each with a relatively high area of the active surface. The difference lies only in their layout and in the case of 3D, the methods for sensor modeling have to take the third dimension into consideration. Due to these similarities, 3D capacitance tomography can employ the same image reconstruction and Jacobian calculation methods as the classical 2D tomography. Only the additional third dimension in visualization requires the Jacobian matrix to have the spatial information about the sensitivity changes in the whole sensor volume.

The Jacobian matrix **J** consists of $m$ sensitivity distributions (known as sensitivity maps) for each possible measurement electrode pair (without repetitions). The sensitivity of a specific location in the discretized sensor volume is defined as a change in capacitance as a result of the change in permittivity in that particular location. In Fig. 16.5 the projection areas between the opposite electrode pairs are drawn. The width of each projection is much smaller than the total volume of the sensor. These projections can be of different sensitivity values and, in addition, are superimposed on each other. At the same time, there can be areas with a negative sensitivity value. This is because in those areas the increasing permittivity values affect the reduction of the measured capacitances.

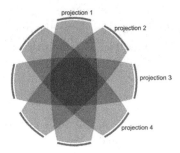

Fig. 16.5 The manner of creating and superimposition sensitivities areas.

As mentioned before, the electric field distribution as well as the vector of changes of measurement values depends non-linearly on the changes in the permittivity distribution. If a function $\xi$ represents this dependence, the change in the permittivity $\delta\varepsilon$ for the $n^{th}$ image element and the change in the capacitance $\delta C$ are related to each other according to the equation:

$$dC_i = \xi(d\varepsilon_n) \qquad (16.9)$$

where $C_i$ is the measured capacitance between the $i^{th}$ measured electrode pair. This relation can be determined as a derivative described using the equation below:

$$\mathbf{J} = [J_{i,n}] = \left[ \frac{dC_i}{d\varepsilon_n} \Big|_{\varepsilon_n = 0} \right] \qquad (16.10)$$

Calculating all of the values of $J_{i,n}$ determined for each image element $n$ and for each measurement electrode pair $i$ gives a matrix which approximates the sensitivity of capacitance changes to the changes in the permittivity distribution.

The sensitivity matrix is a very important part in the image reconstruction process. Its main task is to approximate the electric field inside the ECT sensor. Usually, it is calculated only ones, right before the complete image reconstruction process. In this case, such a matrix does not take into consideration the current permittivity distribution being inspected by the ECT measurement. Neglecting this fact in solving an inverse problem significantly deteriorates the quality of the final image. However, non-linear iterative image reconstruction methods use numerical analysis to solve a forward problem within the iteration. The basic parameters of the electric field (such as a potential distribution inside the sensor ECT) are determined according to the current image of permittivity distribution. This in turn can be used to update the Jacobian matrix. The updated matrix better reflects the nature of the electric field for a given measurement vector and causes a smaller error of approximation in the image reconstruction process.

The Jacobian matrix can be determined using a method based on the measurements or can be calculated as well. One of the most commonly applied methods in applications of ECT imaging is an algorithm based on the electrical field energy distribution analysis. The sensitivity map for the ECT sensor, in this case, can be calculated according to the known electrical field distribution obtained by any numerical methods (Lionheart 2001). Irrespective of the type of ECT measurements (2D or 3D), the Jacobian matrix depends on the electrical field distribution $\vec{E}$ inside the sensor:

$$J = \frac{1}{U^2} \int_{\Omega} \vec{E}^2 d\Omega \qquad (16.11)$$

According to the above, the sensitivity of the $n^{\text{th}}$ element of the discretized measuring volume for the $i^{\text{th}}$ measured electrode pair $(e1:e2)$ can be calculated using the equation:

$$J_{i(e1,e2),n} = -\int_{\Omega_n} \frac{\vec{E}_{e1}}{V_{e1}} \frac{\vec{E}_{e2}}{V_{e2}} d\Omega, \qquad (16.12)$$

where $\vec{E}_{e1}$ and $\vec{E}_{e2}$ are the vectors of the electrical field in the $n^{\text{th}}$ element when the potentials $V_{e1}$ on the electrode $e1$ and $V_{e2}$ on the electrode $e2$ are set. The $\Omega_n$ is the volume of the $n^{\text{th}}$ element.

Before using the Jacobian matrix throughout the image reconstruction process, its values need to be normalized. As the result of the normalization process, the sum of all of the elements of the sensitivity map is equal to 1.

$$\sum_{n=1}^{N} J_{i,n} = 1 \qquad (16.13)$$

The symbol $i$ is the index of the considered sensitivity map for the $i^{\text{th}}$ measuring electrode pair while the $n$ is the image element.

## 16.6 Reconstruction methods

In the literature, one can come across many methods for the solution of image reconstruction which allow one to show the material distribution inside the sensor space. Some of them are adopted directly from the hard field tomography system, used on a large scale in medicine. Such an approach, using the methods borrowed, yields images of low quality; however, the time of reconstruction in this case is very short. On the other hand, is it possible to apply iterative, non-linear, methods which solve an inverse problem with images of a much higher quality. The fact that they are time consuming does not allow such images to be reconstructed in real time if the implementation of the algorithms is conducted on the CPU. Such a situation is significant for 3D reconstruction methods when the number of voxels is, very often, of the order of tens of thousands. The chapter contains an overviewof methods for tomographic image reconstruction in ECT systems.

### 16.6.1 *Linear back-projection*

The most common direct image reconstruction method dedicated for ECT is the one-step linear back-projection method (LBP). It finds the relationship between the measured capacitances and the permittivity distribution through the linear approximation. The LBP method provides ECT systems with a fast manner of image reconstruction from both the measured and simulated data. The algorithm considered is mostly applied in prototype diagnostic systems with real-time imaging.

The main goal of the method is the linearization of an inverse problem. The linearization assumes that the total change in the measured capacitance for any electrode pair is equal to the sum of the particle capacitance changes. The particle change is directly proportional to the change in the electrical field energy in the image element caused by the change in the material permittivity value in this element. So the three-dimensional forward problem defined can be expressed by the following matrix equation:

$$\mathbf{Cm} = \mathbf{J}_{m \times n} \cdot \hat{\varepsilon}, \tag{16.14}$$

where $\mathbf{Cm}$ is the column-vector with normalized measurement data for $m$ measurement pairs of $N$ electrodes, considering the volumetric inter-plane measurements, i.e. measurements between the electrodes from different layers (Banasiak et al. 2004). $\mathbf{J}_{m \times n}$ determines the normalized sensitivity matrix (Jacobian) calculated for $n$ image elements and m measurement data. The column-vector $\hat{\varepsilon}$ corresponds to the three-dimensional image, which approximates the permittivity distribution into the grid voxels using any color scale. Therefore, when using the LBP algorithm, the image is created by multiplying the normalized measurement column-vector **Cm** and normalized pseudo-inverse sensitivity matrix $\mathbf{J}^{+}_{m \times n}$. Practically, the matrix $\mathbf{J}^{+}_{m \times n}$ is approximated by the transpose matrix (Yang et al., 2002).

$$\mathbf{J}^{-1}_{m \times n} \approx \mathbf{J}^{T}_{m \times n} \tag{16.15}$$

Converting equation (16.8), the linear back-projection formula is as follows:

$$\hat{\varepsilon} = \mathbf{J}^{T} \cdot \mathbf{Cm} \tag{16.16}$$

### 16.6.2. *Tikhonov regularization*

The idea behind of the linear back projection can be improved by incorporating a better pseudo inverse sensitivity matrix. One of the examples of such an approach is Tikhonov regularization (Tikhonov, 1977; Penget al., 2000) which incorporates a pseudo inverse matrix with a regularization parameter as presented below:

$$\hat{\varepsilon}_{Tik} = \left( \mathbf{J}^{T} \cdot \mathbf{J} + \alpha^{2} \mathbf{I} \right)^{-1} \cdot \mathbf{J}^{T} \cdot \mathbf{Cm} \tag{16.17}$$

In equation (17) $I \in \Re^{N \times N}$ determines a singular matrix. The inverted matrix is now well conditioned, symmetric and always exists. The reconstruction result is much better at the expense of the more complex calculation of the inverse matrix, which, fortunately, can be done once before any reconstruction occurs.

### 16.6.3 *(Non-) linear iterative reconstruction methods*

This method is known as the Landweber iterations technique (Li and Yang, 2008). This algorithm is still fast but, having some internal optimization features, is typically more accurate which can be sufficient if the high image resolution and small objects visibility in a gas-liquid mixture are not important. This method is defined as follows:

$$\hat{\varepsilon}(k+1) = \hat{\varepsilon}(k) - \alpha \mathbf{J}^{T} \left( \mathbf{J}\hat{\varepsilon}(k) - \mathbf{Cm} \right), \tag{16.18}$$

where $k$ is the number of the current iteration, $\hat{\varepsilon}(k+1)$ is the newly reconstructed image, $\hat{\varepsilon}(k)$ is the previously reconstructed image, **Cm** is the one-column vector with the measurement data, $\alpha$ is the relaxation parameter for convergence rate control that can be selected empirically or by calculation.

The sensitivity matrix that is determined by using a forward solution can be used in the next stage of the image reconstruction process – an inverse problem. The most complicated form of the inverse solution takes into consideration the non-linearity of the electrical field and uses the computer simulation of the complete model of the ECT sensor. It can be formulated using matrix equation (16.18) where the image $\hat{\varepsilon}$ is iteratively reconstructed from the measurement data **Cm**:

$$\hat{\varepsilon}(k+1) = \hat{\varepsilon}(k) + \alpha \mathbf{J}^T (\mathbf{V}(k))(\mathbf{Cm} - FEM(\hat{\varepsilon}(k))) \qquad (16.19)$$

where $\mathbf{J}^T(\mathbf{V}(k))$ is the updated transposed sensitivity matrix for $k-th$ iteration (it addresses the non-linearity nature of the electrical field), $FEM(\hat{\varepsilon}(k))$ states for the computer simulation of the ECT system based on the previous image $\hat{\varepsilon}(k)$.

The initial step of the whole non-linear reconstruction process is to find a starting approximation of the flow volumetric interior $\hat{\varepsilon}(0)$ by using the experimental data **Cm** and the threshold-type linear back-projection method with an initial Jacobian for $k = 0$. For the initial calculation of starting Jacobian, the simulated homogeneous permittivity distribution was applied. It should be selected for the simulation of homogeneous fill according to the known maximum and minimum values of permittivity. The next step is to find a nonlinear forward solution $\mathbf{Cc}(0)$ and $\varphi_0$ using 3D ECT forward modeling *FEM* for the initial flow image $\hat{\varepsilon}(0)$, where $\mathbf{Cc}(0)$ is the calculated measurement data and $\varphi(0)$ is the electrical field distribution. At this stage a new updated Jacobian can be calculated. The $k^{th}$ image $\hat{\varepsilon}(k)$ can be found using conventional iterative back-projection optimization. Every $k-th$ step of the whole 3D image reconstruction process includes a new nonlinear forward problem solution for $\varphi(k)$ and both $\mathbf{Cc}(k)$ and the Jacobian $\mathbf{J}(\varphi(k))$ are updated based on the current $k-th$ reconstructed image. The convergence of image reconstruction can be adjusted by the relaxation factor $\alpha$. The stopping criterion is aided by a dynamic change in the relaxation parameter and performed by tracing the progress in capacitance residual decrementing. The process is stopped if the progress is not sufficient and is controlled by a fixed value.

### 16.6.4 *Area- and point-based reconstruction (voxels vs. nodes)*

One of the most significant tasks related to the deterministic algorithms of image reconstruction is the necessity of dividing the sensor volume into a finite number of elements. For the cross-sectional imaging as the image elements, triangles or rectangles were chosen. In the case of 3D imaging, mostly tetrahedrons (so-called voxels) as image elements are considered. Then, in the process of forward problem calculation in each image voxel the permittivity value is set. However, in the inverse problem each element obtains an appropriate color value. This value corresponds to the permittivity value of the monitored process.

The process of voxel-based image reconstruction in the case of 3D ECT is very time consuming. This is due to a large number of image elements which are repeatedly analyzed during the algorithm lifetime. The mentioned algorithms involve the matrix computation. A commonly known "voxel" approach uses matrices which significantly absorb the computer memory. The size of the sensitivity matrix $\mathbf{S}_{m \times n}$ for the 3D ECT sensor may use even hundreds of megabytes of the operating memory. The problem is made worse in the case of iterative algorithms where there is a need of storing two similar matrices (the $\breve{\mathbf{S}}_{m \times n}$ and his pseudo-inverse $\breve{\mathbf{S}}_{m \times n}^{+}$). The size of the sensitivity matrix determined for a 32-electrode ECT sensor, for example, in a function of the complexity of the image is shown Fig. 16.5a. Additionally, in Fig. 16.5b the time necessary for solving a forward problem is shown as a function of the image elements count.

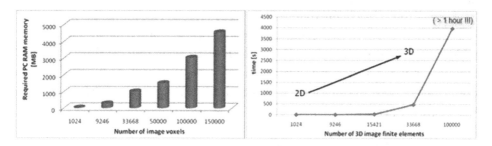

Fig. 16.5   Time and memory complexity in terms of the number of voxels in 3D reconstruction methods.

The sensor mesh with a reduced number of elements lowers the efficiency of object shape detection and additionally increases the numerical error of the forward problem solution. Moreover, the voxel-based reconstructed image has to be approximated to the node-based representation to show it on a computer screen. This is due to the 3D image analysis and visualization algorithms, and the 3D graphics acceleration hardware used, which normally handles the images in non-volume pixels (i.e. nodes). During the approximation process some of the important data are irretrievably lost.

The solution for this inadvertence could be a node-based sensitivity matrix. Moreover, the node-based sensitivity matrix is able to significantly reduce the computation time necessary for 3D image reconstruction processing. This reduction is directly proportional to the ratio the number of mesh voxels to the number of nodes. The final 3D image is determined directly in the nodes of the mesh and now does not need to be approximated.

### 16.6.4 *Stochastic techniques based on Bayesian theory*

If the measurement system provides sufficiently many measurements, then high confidence in the solution of an inverse problem can be achieved. The reality of image reconstruction in the process tomography system is different. There usually are tens of measurements. On this basis, the analysis aims to reach an acceptable image resolution of about 1000 pixels for 2D and much higher for 3D visualization. Since this is an ill-posed inverse problem, in order to achieve this goal and ensure the solution stability, a range of different regularization schemes is used. The regularization techniques presented above allow one to "improve" the ill-posed problem. However, the boundaries of the different permittivity (ECT), resistivity or conductivity (ERT, EIT) regions could then be masked by over-blurring. Such effects could irreversibly destroy the valuable information hidden in the measurement data. Hence the need to incorporate expert knowledge into regularization methods is then obvious, especially that such knowledge is now available. Let us consider a physical model of the tomography system describing the relation between the electrical parameters of the process being investigated and the measurement data from a data acquisition device (see Fig. 16.6). The set of source parameters is denoted as an event A and the resulting observations by event B.

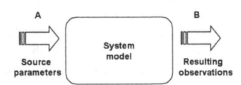

Fig. 16.6 Process tomography theoretical model.

The forward probability density in the case above is the conditional probability of the result occurrence given known cause: $p(B|A)$. The unknown is the inverse probability density, that is to say, determining the conditional probability of a potential cause given the result: $p(A|B)$. Since the forward relationship is usually derived from the theory, the inverse one is missing. Bayes' theorem helps one to relate these two quantities. It combines joint and conditional probabilities in the following form:

$$p(A,B) = p(A \mid B)p(B) = p(B \mid A)p(A) \tag{16.20}$$

and,

$$p(A \mid B) = \frac{p(B \mid A)p(A)}{p(B)}. \tag{16.21}$$

One can see that the formula above describes the unknown – the inverse probability, e.g. the state of the investigated object on the basis of the conducted measurement result. The term p(A) is called the prior probability, since it is associated with the information about possible causes without (or before) considering measurements, whereas the posterior probability term corresponds to the inverse probability. The remaining factors are required for transition from a prior to a posterior probability. As the p(B|A) term was described above to be associated with theoretical knowledge (forward probability), it can be shown as a function of cause set A for a fixed result B. In this case, we call it a likelihood function. The last term in the formula is derived from the fact that the sum of all possible events must be equal to one, e.g. the sum of the posterior probability function over all potential causes equals one.

The exploration of the Bayesian approach, in particular with MCMC sampling, is more flexible and convenient in terms of controlling regularization (Kaipio et al. 2000) and incorporating prior knowledge in image reconstruction (West et al. 2003, 2004). As stated earlier, the Bayesian approach is about producing the posterior density of the required parameter distribution on the basis of the measurements. It could be permittivity, resistivity, conductivity or whatever is the quantity of interest. In order to compose an appropriate relationship, the prior probability density and likelihood is necessary. The posterior can be obtained by combining both of them.

The prior information can involve temporal or spatial dependencies. There is a distinction between low and high level modeling of image data that can be applied to tomography. Low levels refer to basic operations such as noise or blur removal. It is usually associated with the application of models for images that operate at pixel level (Winkler 2003). The high level usually refers to more complex operations, for example, object recognition.

In most applications the spatial permittivity distribution is generally relatively smooth and can be modeled by homogenous Markov random field (Winkler

2003). The adjacent pixels (permittivity values) have a similar value. To promote this, locally linear smoothing prior of the following form can be applied:

$$\pi(\mathbf{x}) \propto \exp\{-\beta \left\| \mathbf{x} - \overline{\mathbf{x}} \right\|_p^p\}; \beta > 0, 0 \le p \le 2. \tag{16.22}$$

The $\beta$ is the smoothing parameter controlling the degree of correlation between the neighboring pixels, hence the smoothness of the image. The above prior with the 1-norm (p=1) leads to a total variation prior and with 2-norm (p=2) expresses the Gaussian prior. The total variation prior leads to sensible smoothing for plain regions and will not oversmooth boundaries. The choice of a particular prior is dependent on the phenomenon or process investigated. The Gaussian distribution is a common choice (West *et al.* 2003, 2004).

The modeling of the likelihood for electrical tomography is associated with the physical properties of the electromagnetic field responsible for phenomena present in the measurement space. If the permittivity distribution $\varepsilon$ (for ECT) is known, the boundary capacitance can be calculated. This forward problem is non-linear, yet well-posed, and allows calculation of capacitances by the finite element method (FEM) with a high accuracy. The measurements (inter-electrodes capacitance $C$) are subject to errors:

$$C = C(\varepsilon) + \delta \tag{16.23}$$

where $\delta$ is the vector of identically distributed, independent Gaussian errors with a variance $\tau^2$. Therefore, the likelihood density function (the conditional distribution of the capacitance for a given permittivity distribution) has:

$$p(\mathbf{C} | \varepsilon) = \frac{1}{(2\pi\tau^2)^{n/2}} \exp\{-\frac{1}{2\tau^2} \left\| \mathbf{C} - \mathbf{C}(\varepsilon)^* \right\|^2\} \tag{16.24}$$

In the case of considering spatial modeling, there is a single permittivity taken into account for the calculation of capacitance (forward problem solution) at each time point.

The final formula for the reconstruction is given below on the basis of the prior (4) and the likelihood (6):

$$p(\varepsilon | C) \propto p(\mathbf{C} | \varepsilon) * \pi(\varepsilon). \tag{16.25}$$

The posterior probability density function represents the complete state of knowledge of process parameters in terms of a Bayesian perspective given the measurements. An estimate of the parameters is usually sufficient, instead of providing the entire state of knowledge. There is a whole range of various estimates such as MAP (maximum a posteriori), a mean of a posterior probability. The choice of particular one is dependent on the situation. In Fig. 16.7 an example of a reconstructed image based on iterative algorithms and the posterior mean is presented.

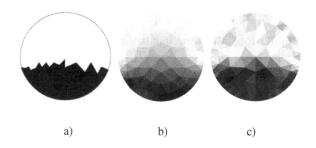

a)        b)        c)

Fig. 16.7 Images of a) simulated truth, b) SIRT reconstruction c) posterior mean.

## 16.7 Parallel computing

The reconstruction methods presented in the previous subsection require a lot of computation. If the process tomography diagnosis is used in the control unit and the analysis of images is conducted to estimate the process control parameters, the reconstructed images must be obtained in real time to have a possibility to react to the process state. These reconstruction methods take a much longer time in the case of 3D visualization and the success of the application of 3D ECT to control the process depends on the real time reconstructed procedure.

Three-dimensional ECT imaging is likely to become an important tool in industrial process imaging where the cross-sectional 2D ECT systems provide inaccurate information about the process (Yang and Peng, 2003; Romanowski et al., 2006). The problem of ECT image reconstruction is therefore two-fold; the model to describe the electrical field distribution within the area must be accurate and the inverse problem must be reliable and computationally efficient at estimating the electrical properties within the imaging area (Banasiak et al., 2009). The computational efficiency of the inverse problem is the one of the major issues in 3D capacitance tomography. There are typically hundreds of thousands of numbers to be processed while a 3D image is being built and optimized and the time of reconstruction may reach several minutes or even hours (Banasiak et al., 2009). This is a one of the important drawbacks of 3D

ECT image computation and it can seriously limit potential industrial applications. A new approach has been developed to achieve a significant acceleration of the image reconstruction process. The proposed idea assumes that all tomography inverse problem computations can be performed in parallel using modern fast graphics processors.

GPGPU (General-Purpose computing on Graphics Processing Units) is a technique of using graphic cards (GPUs), which normally handles graphics rendering for computations that are usually handled by a processor (CPU – Central Processing Unit). A growing interest in GPU computations started with the difficulties to clock CPUs above a certain level, because of the limitations of silicon based transistor technology and a constant demand for improvements. Due to this interest in multi-core technology and parallel computing started to emerge, even though sequential programs are easier to understand and more natural. There is a limit though of how fast such a program can run. Sequential speed up is now based on architecture improvements in the CPU rather than higher clocks, but even this approach has limitations. Parallel programming is not a new idea though until only recently it was reserved for high performance clusters with many processors and the cost of such a solution was extremely high. This changed with the introduction of many-core processors to the mainstream market. Graphics processors not only fit well in that trend but even take it to a higher level as well. Compared to the CPUs which today have a maximum of 2 to 12 cores, GPUs consist of dozens, and even hundreds of smaller, simpler cores designed for high-performance massive computing. CPUs are built and designed to execute single thread, no matter how unpredictable, diverse or complicated it may be, as fast as possible. On the other hand, GPUs mostly take care of data computations that are much simpler in their nature and for that reason their execution units, or cores, can be much simpler and, as a result, smaller. Thanks to this there can be a great number of them on a single chip with numbers reaching dozens or even hundreds. Additionally, because graphics computations can be easily parallelizable, cores in the GPU work on the basis of the SIMD (Single Instruction, Multiple Data), or more precisely the SIMT (Single Instruction, Multiple Thread) architecture, where one instruction is applied to a big portion of independent data. This translates into a much greater number of operations per second than can be achieved on the CPU. GPU based systems can run hundreds or even thousands of threads at once, compared to only a few on the CPU. Many computations that were dramatically time consuming can now be made in a near real-time scheme with a small investment in hardware compared to the cost of achieving the same results using established methods. Yet, a real breakthrough in ease-of-use of the GPGPU came with the introduction of the Nvidia CUDA (Compute Unified Device Architecture) in 2007. Programmers could then use C language, with some specific extensions and easily create programs and tools that could utilize the potential of the massively parallel Graphics Processing Units and increase the speed of many algorithms. However, GPU computations are not the golden means for everything. They have many limitations. One of the biggest is that they were designed for parallel performance and high computing power and as such will not do well on sequential tasks, or ones that involve a lot of branching. This is still the domain of CPUs as they were designed to cope well with such processing operations.

The 3D nonlinear inversion GPU acceleration development was performed using the AccelerEyes Jacket Matlab toolbox. Nowadays, a Jacket GPU library is a commercial and mature solution that enables one to write and run a code on the GPU in the native M-Language used in MATLAB. Jacket accomplishes this by automatically wrapping the M-Language into a GPU CUDA (CUBLAS and CUSPARSE) compatible form. By simply casting input data to Jacket's GPU data structure, MATLAB functions can easily be transformed into GPU functions. Jacket also preserves the interpretive nature of the M-Language by providing real-time, transparent access to the GPU compiler which increases the efficiency of a parallel computing code. There is also a new GPU computing library available, embedded in the standard Matlab engine but currently in its early stage of development and relatively problematic. There are two equivalent nonlinear IBP based algorithm examples, presented in Fig. 16.8, using the Matlab sequential CPU M-code and the GPU Jacket parallel M-code.

The implemented 3D ECT inverse problem can be considered as a mixed parallel data-oriented and parallel task-oriented image reconstruction algorithm. Due to single GPU computations the parallel data-oriented scheme was applied in the Matlab code. There are three important steps (1), (2) and (3) where CUDA computations may dramatically help to reduce computational time. Step (1) is the FEM forward modeling that requires solving a large-scale set of linear equations. As a result of step (1), a spatial electric potential distribution and capacitance vector is computed. For the purpose of the paper, Jacket implementation of LU factorization with partial pivoting was used to solve the 3D ECT FEM phase in both, the GPU parallel and CPU sequential scheme. Step (1) is considered to be the most time-consuming part of the 3D ECT forward problem solution. Step (2), that is sensitivity matrix computation, is based on the step (1) solution. Step (2) can be time consuming when it is performed using the sequential CPU scheme (even optimized by Matlab vectorization routines). Due to the fact that sensitivity matrix computation for 3D ECT is mainly the data parallel-oriented process, it can easily be parallelized using CUDA. Therefore, step (2b) is much faster when CUDA computing is used, because all the sensitivity matrix values can be computed in parallel by separate CUDA threads. step (3) requires the step (1) and step (2) solution. Step (3) is an image optimization function which aims an iterative update of the image using the previous image, FEM modeling and sensitivity analysis from the previous iteration. This step is also well-parallelized and the usage of CUDA is efficient by using intensive linear algebra computations.

```
mask = find(tril(ones(size(elecgnd,1),
                 size(elecgnd,1)),-1));

pic = SInverse * Cm;
NdEltrd = NdEltrd * voltage;

for i=1:nb_iters
[Y] = ECT3D_FEMStiffnessMatrix(
                 vtx,simp,picFEM.*80);

RhsKnown = -Y(NdIndUnknown,NdIndKnown)
            * NdEltrd(NdIndKnown,:));

Vfr = full(NdEltrd);

Vfr(NdIndUnknown,:) =
3DECT forward on CPU(Y,RhsKnown);     (1)

EltrdCur =   (Y*Vfr)' * NdEltrd;
CcF = (-1)*EltrdCur(mask);
Cc = (CcF - Cl)./(Ch - Cl);

list = find(Cm <0);
Cc (list)=0;
list = find(Cm >1);
Cc (list)=1;

[SInverse]=ECT3D_FEMSMatrixOnCPU(Vfr);   (2)

pic = pic + RF * SInverse *(Cm - Cc);    (3)

list = find(pic <0);
pic (list)=0;
list = find(pic >1);
pic (list)=1;

end
```
a)

```
clear gpu_hook

C_GPU = gsingle(Cm);
SInverseGPU = gsingle(SInverse);
vtxGPU = gsingle(vtx);
simpGPU = gsingle(simp);
pic_srGPU=gzeros(size(simpGPU,1),1);
RF_GPU = gsingle(RF);

Cl_GPU = gsingle(Cl);
Ch_GPU = gsingle(Ch);

mask = find(tril( ones(size(elecgnd,1),
                 size(elecgnd,1)),-1 ));

pic_GPU = SInverseGPU * C_GPU;
NdEltrd = NdEltrd * voltage;

for i=1:nb_iters
picFEM = double(pic_GPU .* 80);
[Y] = ECT3D_FEMStiffnessMatrix(vtx,simp,picFEM);
YGPU = gsingle(full(Y(NdIndUnknown,NdIndUnknown)));
RhsKnownGPU = gsingle(-Y(NdIndUnknown,NdIndKnown)
                 *NdEltrd( NdIndKnown, :));
VfrGPU = gsingle(full(NdEltrd));
VfrGPU(gsingle(NdIndUnknown),:) =
3DECT forward on GPU(YGPU,RhsKnownGPU);   (1)
Vfr=double(VfrGPU);
EltrdCur =   (Y*Vfr)' * NdEltrd;
Cc = (-1)*EltrdCur(mask);
Cc_GPU = gsingle(Cc);
Cm_GPU = (Cc_GPU-Cl_GPU)./(Ch_GPU-Cl_GPU);
list = find(Cm_GPU<0);
Cm_GPU(list)=0;
list = find(Cm_GPU>1);
Cm_GPU(list)=1;
[SInverseGPU]=ECT3D_FEMSMatrixOnGPU(VfrGPU);   (2)
pic_GPU = pic_GPU + gsingle(RF) * SInverseGPU
                 *(C_GPU - Cm_GPU);  (3)
list = find(pic_GPU<0);
pic_GPU(list)=0;
list = find(pic_GPU>1);
pic_GPU(list)=1;
end

pic = single(pic_GPU);
```
b)

Fig. 16.8 Example M-code listing for a) CPU and b) GPU 3D nonlinear inverse solution.

The solution proposed was verified using a high performance workstation running under Windows 7 and Matlab 2010b with the AccelerEyes Jacket GPU Toolbox 1.7.1. The system was equipped with the Intel Core i7 CPU at 2.80 GHz and 12 GB of RAM and the Tesla C2070 CUDA computing system with 6GB of RAM. The experimental verification was done using single precision for Matlab sequential and Jacket parallel computations. Table 16.1 contains average times of nonlinear problem computations using various densities of the 3D ECT mesh. There were three representatives of small, middle and large scale three-dimensional meshes with 20499, 60896, 125232 image voxels, respectively. The nonlinear reconstruction algorithm was run with 10 iterations. The FEM solution and sensitivity matrix update were computed for all 10 iterations. The results achieved show 9÷11-time speedup for the 3D ECT nonlinear inverse solution depending on the mesh tested. A small improvement in performance is also visible when the scale of the mesh complexity is increasing.

Table 16.1 Average times of nonlinear problem computations using various densities of 3D ECT mesh.

| 3D nonlinear inversion | Intel Core i7 CPU [s] | Tesla C2070 GPU [s] | Speedup [x] |
|---|---|---|---|
| 20499 image voxels | 49.75 | 4.99 | 9.96 |
| 60896 image voxels | 179.25 | 17.07 | 10.50 |
| 125232 image voxels | 637.29 | 55.12 | 11.56 |

Table 16.2 compares three different reconstruction technique results according to their imaging performance, using CPUs and CUDA GPU based systems. The middle scale mesh (60896) was used for this test.

Table 16.2 The imaging performance of three image reconstruction algorithm representatives in CPU and GPU computing scheme.

| 3D nonlinear inversion | Intel Core i7 CPU | Tesla C2070 GPU |
|---|---|---|
| 60896 image voxels - LBP Alg. | 91.74 image/s | 1666,66 image/s |
| 60896 image voxels - Linear Landweber Alg. | 0.55 images/s | 4 images/s |
| 60896 image voxels – nonlinear IBP Alg. | 0.001 images/s | 0.02 images/s |

The results obtained demonstrate two important facts. Nowadays, three-dimensional capacitance tomography GPU CUDA computations are sufficiently fast to visualize industrial processes near real time, using linear iterative techniques (Landweber's algorithm) – 4 images/s were obtained using the Tesla C2070 hardware. For the nonlinear inversion, CUDA computations still require relatively small-scale 3D ECT meshes to meet industrial real-time imaging requirements. However, the performance of graphics processor based computations is incomparably better than the commonly used CPU sequential image reconstruction code. Recent advances came with novel, dedicated algorithms developed for a forward problem using GPU, based on Cholesky-Banachiewicz matrix decomposition (Matusiak et al, 2013). First tests proved 6-7 times speed-up in case of 2D image reconstruction, hence it is a very promising direction for further research in terms of 3D reconstruction as well as advanced data processing algorithms (Romanowski et al, 2006, Grudzien et al, 2006).

## 16.8 Conclusions

The chapter covers a field of process tomography, in particular measurement methodology and visualization techniques of material distribution in sensor space. Theoretical considerations are presented on the basis of Electrical Capacitance Tomography; however, as already mentioned, they can be applied for other electrical tomography techniques (Electrical Impedance or Electrical Resistance Tomography). The review of image reconstruction algorithms, from Linear Back Projection to statistical methods based on the Bayesian approach, shows a range of options for choosing the proper algorithm. The main criterion is the time of reconstruction. If the process is characterized by slow changes of the state, the application of 3D visualization seems to be a better option than the 2D visualization used. In the case of some part of the process characterizing high dynamic changes the simple LBP algorithms can be only one alternative.

In order to increase the information quality visible in a reconstructed image new algorithms are developed, which implement parallel computing. The modern methods of applying GPU CUDA computations presented herein allow one to significantly increase the speed of image reconstruction; however, non-linear methods still require improvements, both in the elaboration of parallel algorithms and their implementation in the GPU.

In addition to the time of reconstruction, the aspect of speed of the acquisition system has to be considered. Sometimes, the use of real-time algorithms does not bring success in process monitoring. Such a situation can take place when, during the gathering of measurement vectors, the state of the process is changed. It is necessary to select the proper speed of the acquisition unit in order to minimize the spread of the measurements around the time point.

# References

Banasiak, R., Wajman R., Sankowski D., Soleimani, M. (2010). Three-dimensional nonlinear inversion of electrical capacitance tomography data using a complete sensor model, *Progress in Electromagnetics*, Research PIER, 100, pp. 219-234.

Banasiak, R., Wajman, R., Betiuk, J., Soleimani, M. (2008). An efficient Nodal Jacobian method for 3D electrical capacitance image reconstruction, INSIGHT, pp. 1354-2575

Brzeski, P., Mirkowski J., Olszewski, T., Plaskowski, A., Smolik, W., Szabatin, R. (2003). Multichannel capacitance tomograph for dynamic process imaging, *Opto-Electronics Review*, 11, pp. 175-180.

Gilks, WR., Richardson, S., Spiegelhalter, D.J. (1995). Markov Chain Monte Carlom Practice, *London: Chapman & Hall*, pp. 75-88.

Grudzien, K., Romanowski, A., Williams, R.A. (2006). Application of a Bayesian approach to the tomographic analysis of hopper flow, *Part. Syst. Charact*, 22, 4, pp. 246-253

Huang, X.M., Dyakowski, T., Plaskowski, A.B., Beck, M.S. (1988). A tomographic flow imaging system based on capacitance measuring techniques, *In Proc: IX International Conference on Pattern Recognition*, Rome, Italy

Li, Y, Yang, W.Q. (2008). Image reconstruction by nonlinear Landweber iteration for complicated distributions, Meas. Sci. Technol., 19 094014

Lionheart, W.R.B. (2004). Review: Developments in EIT reconstruction algorithms: pitfalls, challenges and recent developments, *Physiol. Meas.*, 25, pp. 125-142.

Loser, T., Wajman, R., Mewes D. (2001). Electrical capacitance tomography: image reconstruction along electrical field lines, Meas. Sci. Technol., 12, pp. 1083-1091.

Matusiak, B., Romanowski, A., Sankowski, D., Grudzień, K. (2013). Forward problem solution acceleration with graphical processing units (GPU) for electrical tomography imaging in proceedings of 7th World Congress on Industrial Process Tomography, *WCIPT7*, 2-5 September 2013, Krakow, Poland

Plaskowski, A.B., Beck, M.S., Thorn, R., Dyakowski, T. (1995). Imaging Industrial Flows, Institute of Physics Publishing, Bristol

PTL, http://www.tomography.com/pdf/TFL%20R5000%20Flow%20Measu-rement.pdf

Romanowski, A., Grudzien, K., Williams R.A. (2006). Analysis and Interpretation of Hopper Flow Behaviour Using Electrical Capacitance Tomography Part. Part. Syst. Charact., 22, 3-4 pp. 297-305

Romanowski, A., Grudzien, K., Aykroyd, R.G., Williams, R.A., (2006), Advanced Statistical Analysis as a Novel Tool to Pneumatic Conveying Monitoring and Control Strategy Development, *Part. Part. Syst. Charact.*, 23, 3-4, ISSN: 0934-0866, pp. 289-296.

Scott, D.M., McCann H. (2005). Process Imaging for automatic control, Taylor and Francis Group, pp. 439

Soleimani, M. (2006). Three-dimensional electrical capacitance tomography imaging, Insight, Non-Destructive Testing and Condition Monitoring, 48, 10, pp. 613-617.

Soleimani, M., Mitchell, C. N., Banasiak, R., Wajman, R., Adler, A., (2009) Four-dimensional electrical capacitance tomography imaging using experimental data, *Progress In Electromagnetics Research-PIER*, 90, pp. 171-186.

Wajman, R., Banasiak, R., Mazurkiewicz, Ł., Dyakowski, T., Sankowski, D., (2006). Spatial imaging with 3D capacitance measurements, *Measurement Science and Technology*, 17, 8, pp. 2113-2118;

Warsito, W., Fan, L-S (2003). Development of 3-Dimensional Electrical Capacitance Tomography Based on Neural Network Multi-criterion Optimization Image Reconstruction, *Proc. of 3rd World Congress on Industrial Process Tomography (Banff)*, pp. 942-947;

Warsito, W., Marashdeh, Q., Fan, L-S., (2007). Electrical capacitance volume tomography, IEEE Sensors Journal, 7, 3-4, pp: 525-535;

West, R.M., Aykroyd, R.G., Meng, S., Williams, R.A.. (2004). Markov Chain Monte Carlo Techniques and Spatial-Temporal Modelling for Medical EIT, *Physiological Measurements*, 25, pp. 181-194.

Williams, R.A., Beck M.S. (1995). Process Tomography – Principles, Techniques and Applications, Butterworth-Heinemann, Oxford

Winkler, G. (2003). Image Analysis, Random Fields and Markov Chain Monte Carlo: A Mathematical Introduction Berlin, Heidelberg, Springer-Verlag, pp. 404.

Xie, C. G., Huang S.M. et al., (1992). Electrical capacitance tomography for flow imaging-system model for development of image reconstruction algorithms and design of primary sensors," IEE Proc. G, 139, 1, pp.89-98.

Yang, W.Q., Peng L. (2003). Image reconstruction algorithms for electrical capacitance tomography, *Measurement Science and Technology* , 14,pp. R1-R13

Yang, W.Q. (2010). Design of electrical capacitance tomography sensors, Meas. Sci. Technol. 21

Yang, W.Q., Byars M. (1999). An Improved Normalisation Approach for Electrical Capacitance Tomography, *1st World Congress on Industrial Process Tomography*, Buxton UK, pp. 215-218.

York, TA., Phua, T.N, Reichelt, L., Pawlowski, A., Kneer, R. (2006). A miniature electrical capacitance tomograph, Meas. Sci. Technol. 17, pp. 2119-2129

Zhiheng, G. (2011). New Normalization method Of Imaging Data for Electrical Capacitance Tomography, *2011 International Conference on Mechatronic Science, Electric Engineering and Computer*, August 19-22, 2011, Jilin, China

# CHAPTER 17

# INFRA-RED THERMOVISION IN SURFACE TEMPERATURE
# CONTROL SYSTEM

Jacek Kucharski, Tomasz Jaworski, Andrzej Frączyk, and Piotr Urbanek

*Institute of Applied Computer Science*
*Lodz University of Technology*
*90-924 Łódź, ul. Stefanowskiego 18/22*
*{jkuchars, tjaworski, a.fraczyk, p.urbanek}@kis.p.lodz.pl*

## 17.1 Measurement and control system for the induction heating of a rotating cylinder

Thermal processes play a pivotal role in many industrial technologies of ever growing requirements, concerning the accuracy of the temperature value and its distribution. Among many areas in which temperature is an essential quantity, papermaking and textile industries are those where specific requirements concerning temperature distribution on the surface of moving bodies are formulated. Generally, rotating steel cylinders are used in such cases as the heated objects, thus they can be efficiently heated under operation with the use of electrical induction. As the information on the temperature of the cylinder surface must be captured during the process, non-contact infrared thermovision is one of the best methods of data collection. In order to fulfill technological requirements, there should exist a feedback from temperature to the heating power signal, which constitutes the closed loop system, in which the temperature signal has the form of 2D image. Thus, some image processing and analysis methods must be included in measurement and control algorithms.

At the Institute of Applied Computer Science at Lodz University of Technology, the semi-industrial set up of induction heated rotating steel cylinder, equipped with a thermovision camera and a computer-based control system has been designed and developed [Fraczyk *et al.*, 2008]. A general block diagram of the system is given in Fig. 17.1.

The rotating steel cylinder is the central part of the experimental set up. It is heated by means of six mobile inductors supplied by dedicated high-frequency generators. In the multi-loop computer-based measurement and control system, three main functional channels can be distinguished:

1.  the cylinder surface temperature measurement and control channel, which is the main loop of the entire system, governing other subsystems to some extent,
2.  the stabilized power generation channel,
3.  the cylinder rotation speed measurement and control channel.

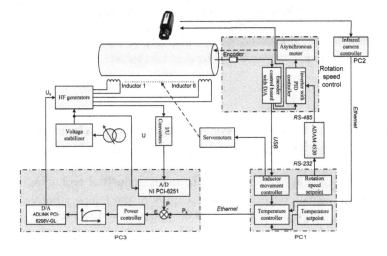

Fig. 17.1. General block diagram of the measurement and control system of induction heated rotating steel cylinder

Three computer units (PC1, PC2, PC3), communicating via Ethernet and other digital buses, are the main hardware nodes of the applied distributed data acquisition and processing system. PC1 unit governs the main temperature control loop together with PC2, which forms the input signal based on the thermal image acquired from an FLIR A615 infrared (IR) camera and with PC3 being in charge of giving the proper, stabilized value of the heating power to the inductor. Moreover, PC1 controls the location of inductors and their velocity under the cylinder surface as well as its angular velocity. Within the temperature measurement channel, the information on the current angular position of the cylinder, provided by the encoder, is fed to the IR camera through the frame synchronization module. It enables one to measure the temperature of the selected areas of the rotating object, which is essential for the temperature map reconstruction algorithm.

## 17.2 Thermal image processing algorithms

In this subsection, the set of algorithms for reconstructing a full map of the cylinder's surface temperature is presented. In further sections, the map obtained is used as a source signal for temperature control. This algorithm can be considered as a set of steps that are executed one after another, in order to achieve the final outcome (the map). Each of them can also be considered as an independent image processing step (fig. 17.2).

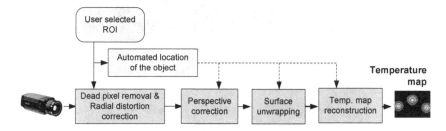

Fig. 17.2. An overview of the algorithm presented in the paper

The image processing steps, connected in series, have their own inputs and outputs, hence the image resulting from e.g. the perspective correction block is the source for a surface unwrapper. Finally, when the surface unwrapping and reconstruction block [Jaworski and Kucharski, 2013a] collects a sufficient amount of data, it provides an output temperature map of the cylinder's surface to the temperature control algorithms described in subsection 17.4. ROI selection and automated object location are not involved in the main pipeline. They use IR images, when experiment begins; however in a normal process they only provide information to the main processing blocks. Moreover, the images received from the IR camera must be corrected due to the existence of so called *dead pixels*. A dead pixel in a sensor is a measurement point which yields values that are constant or vary within a small range, with no or little regard to the observed value (i.e. the observed temperature) [Cho *et al.*, 2011]. The removal is acquired by applying a median filter on the selected dead pixels with a mask with 3x3.

All the operations are performed on a selected region of interest (ROI) of the acquired thermal image. Since every image, regardless of its content, should be processed by the radial correction block, the ROI selection is performed after the first processing step. The coordinates of ROI can be user-selected, and additionally, a special localization algorithm [Jaworski and Kucharski, 2013b] was developed to ensure algorithm's efficiency in various operating conditions, especially when the cylinder is at ambient temperature.

### 17.2.1 *Automated location of the object in ambient temperature*

To obtain the correct surface temperature map, the reconstruction algorithm should take into account the spatial location and orientation of the cylinder. This poses a problem with the proper calibration and placement of the camera. The most important is to correctly set the yaw/pitch/roll angles of the camera on a stand in order to obtain a video stream that would be easy to process.

Fig. 17.3. Input image with calibration markers and unwanted objects visible as hotter areas

Such a task is easy when the object is hot and clearly visible in the infrared. However, when the whole heating process is to be registered, starting from the cold object, it is mostly indistinguishable from its background due to the low gradient value on the edges and thermal noise. To enable proper registration, a set of boundary markers was placed on the object monitored (Fig. 17.3) and an algorithm to indentify them was automatically developed. The purpose of the above-mentioned markers is to calibrate the infrared camera position properly in order to register the whole heating process and to recalibrate the camera in case it is moved or displaced during the experiment.

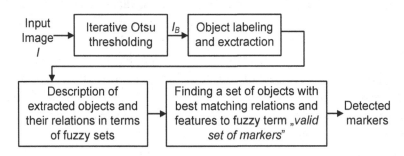

Fig. 17.4. Marker identification algorithm

A sample input image with a cylinder and a set of hot markers is shown in Figs 17.3 and 17.6. There are three markers visible: two vertical along the diameter and one horizontal, under the object. All of them are made of a resistive wire and are of approximately the same length. When a need for calibration arises, they are powered up by a power supply of 24V and 1A DC current.

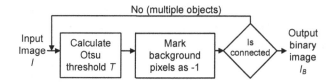

Fig. 17.5. Iterative version of the Otsu algorithm, used in Fig. 17.4

The total amount of power dispatched by the markers (~75W) affects the total time of calibration, which is crucial to avoid any temperature rise on the object's surface. The IR camera applied is able to measure up to 50FPS, hence one second of the calibration process is sufficient. A set of 50 frames, obtained from a steady object, is then averaged to attenuate the thermal noise and arrive at the final input image used in the calibration algorithm described, presented in Fig. 17.4.

The input image obtained, shown in Fig. 17.3 is initially processed by the Otsu algorithm [Otsu, 1979] in order to extract hot areas, clearly visible over the background. It is executed iteratively (as shown in Fig. 17.5) in order to obtain the connected background area.

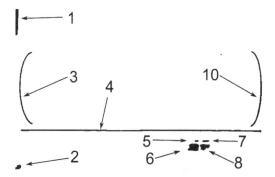

Fig. 17.6. A 8-connected input image with calibration markers and unwanted objects visible as resulting objects for further processing

After the algorithm from Fig. 17.5 is executed, the background pixels (with the temperature lower than threshold value T) are removed from further iterations so that the Otsu histogram calculation routine will ignore them. In the final decision block, the set of removed background pixels is checked for connectedness [Klette and Rosenfeld, 2004]. If all the pixels belong to one 8-connected object, the algorithm terminates. An example of a final image $I_B$ is presented in Fig. 17.6. Afterwards, a set of 8-connected objects is extracted and labeled. It can be noticed that three of them (labeled 3, 4, 10) are desired markers and others objects are unwanted hot areas from e.g. central heating pipes or human heat reflected by glossy components in the room.

Generally, as shapes, distances, orientations and mutual relations between all the extracted objects cannot be expressed and described precisely, a set of features and relations between each of them is calculated on the basis of the fuzzy logic theory [Zadeh, 1965]. The proposed fuzzy assessment of features and relations in the image, leading to the proper detection of the markers, is given below.

The **distance** between two objects [Rosenfeld, 1985] is calculated using Eq. (17.1):

$$d(R,A) = \inf_{\substack{p \in R \\ q \in A}} d(p,q) \qquad (17.1)$$

where: $R$, $A$ – the objects (set of points) extracted from $I_B$, (objects R and A, when in a relation, can be read as **R**eference and **A**rgument), $d(p,q)$ – the Euclidean distance between two points $p$ and $q$.

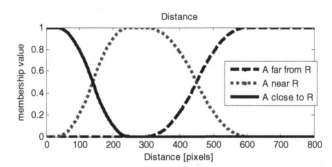

Fig. 17.7. Fuzzy labels for assessing distance between two objects

Eq. (17.1) can be interpreted as the shortest distance between any of two points belonging to objects $R$ and $A$. Such a value of the distance is then fuzzified using fuzzy variable *Distance* as shown in Fig. 17.7, in order to obtain degrees of truth of expressions "$A$ is far from $R$" and "$A$ is close to $R$".

Potentially, a histogram of distances [Hudelot *et al.*, 2008] could be used in this case, however there was a problem encountered with the distance relation between object 3/10 and 4. From their point of view the object 4 (horizontal line) is simultaneously close, near and far. This issue justifies the use of (17.1) and the fuzzy description for dealing with the *near/far* relation in situations that can be considered both ambiguous and imprecise.

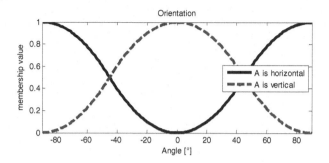

Fig. 17.8. Fuzzy labels for assessing orientation of an object

The **orientation** of the object *R* is recognized by discovering its major axis with the help of the Principal Component Analysis (PCA) method, [Jackson, 1991] which in fact rotates a given set of points to be as horizontal as possible (first coordinate variance maximization). The implementation produces an eigenvectors matrix $V \in \mathfrak{R}^{2 \times 2}$ in which is the scaling part of 2D rotation transform matrix [Jahne, 2002] and as such, allows the rotation angle $\alpha$ to be calculated. Finally, knowing that the object *R* has to be rotated by angle $-\alpha$ in order to become vertically oriented, one can fuzzify the angle $-\alpha$ to obtain the fuzzy variable *Orientation*. This can be done by a set of fuzzy labels depicted in Fig. 17.8.

The **shape** of the object *R* can be calculated, since one knows the object is oriented vertically. In this case, eq. (17.3) can be considered as an object's *squareness* measures.

$$S(R) = \frac{\min\limits_{p \in R}\left( \text{stddev } p_x, \text{stddev } p_y \right)}{\max\limits_{p \in R}\left( \text{stddev } p_x, \text{stddev } p_y \right)} \qquad (17.3)$$

where: $p_x$, $p_y$ – X and Y coordinate of a point *p*. The S(R) values for objects in Fig 17.6 were calculated and shown in Table 17.1.

Table 17.1. Results for squareness measure

| Object | 1 | 2 | 3 | 4 | 5 | 6 | 7 | 8 | 9 |
|---|---|---|---|---|---|---|---|---|---|
| S(R) | 0.085 | 0.629 | 0.163 | 0.006 | 0.677 | 0.364 | 0.549 | 0.250 | 0.126 |

Eq. (17.3) is based on the standard deviation in $\mathfrak{R}^2$ with two coordinates (X,Y). The value yielded by (17.3) can be used to distinguish between three shapes:

- R is a square – S(R) is close to 1 (or equal)
- R is a line – S(R) is close to 0 (but never equal)
- R is a rectangle – S(R) $\in$ (0,1)

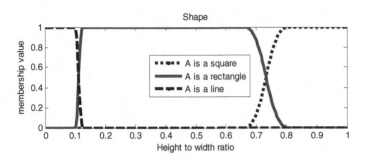

Fig. 17.9. Fuzzy variable for shape description

Due to the inherently imprecise boundary between the square, the rectangle and the line (especially in the case of noisy image), the shape feature can be considered as the most appropriate to be fuzzified and used as such. For this purpose, three fuzzy labels were proposed, appropriate for shape assessment in order to arrive at fuzzy variable *Shape*, as depicted in Fig. 17.9. They were selected as the answer to the question "*When does a square become a rectangle and when does a rectangle become a line?*"

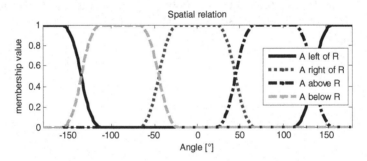

Fig. 17.10. Fuzzy variable for describing directional spatial relation

A ***direction*** relation of two objects, the reference object R and the argument object A, was obtained by the Histogram of Angles method, also called the Compatibility Method, which was introduced by Miyajima and Ralescu [Miyajima and Ralescu 2004a, Miyajima and Ralescu, 2004b] and analyzed further in [Bloch and Ralescu, 2003]. Its main idea is based on the similarity comparison between fuzzy labels describing the relation, the membership function of which is given by one of the plots in Fig. 17.10 and the histogram interpreted as an unlabeled fuzzy set. More detailed information about directional spatial relation is available in [Jaworski and Kucharski, 2010].

Having a set of features describing every object and relations between every pair of objects, a fuzzy inference about the content of the input image can be carried out. In order to complete this task, it is necessary to describe desired relations in the scene by a set of fuzzy rules. As previously mentioned, the goal is to detect and distinguish a set of markers on the thermal image. For this reason, the following fuzzy description was used for the term *the valid set of markers*:

A **set of markers**, composed of three objects: A1, A2 and B is **valid** if and only if

| |
|---|
| **A1** is a *rectangle* and **A2** is a *rectangle* and **B** is a *line* and |
| **A1** is oriented vertically and **A2** is oriented vertically and **B** is oriented horizontally and |
| **A1** is *on the left* of **A2** and **A2** is *on the right* of **A1** or **A1** is *on the right* of **A2** and **A2** is *on the left* of **A1** and |
| **A1** is *close to* **B** and **A1** is *not near* **B** and **A2** is *close to* **B** and **A2** is *not near* **B** and **A1** is *far* from **A2** |

then the selected object triple (X,Y,Z): X as A1, Y as A2 and Z as B is *the valid set of markers*

The above set of rules was used to obtain the membership of truth for the *"valid set of markers"* term and was applied to every triple of objects, assuming that X, Y and Z are unique in the selected triple. Fuzzy operators: product (*and*) and sum (*or*) were defined by *min* and *max*, respectively.

The algorithm presented was applied to the input image, depicted in Fig. 17.4. There were 10 objects found in the image, yielding 720 triples of unique objects. The following Table 17.2 presents the obtained results of the algorithm. Only results with $\mu_{valid}$ greater than zero are shown.

Table 17.2. Obtained results for a given input image and 10 objects

| B | 4 | 4 | 4 | 4 | 4 | 4 | 4 | 4 | 4 | 4 |
|---|---|---|---|---|---|---|---|---|---|---|
| A1 | 2 | 2 | 2 | 2 | 2 | 3 | 3 | 3 | 3 | 3 |
| A2 | 5 | 6 | 7 | 8 | 10 | 5 | 6 | 7 | 8 | 10 |
| $\mu_{valid}$ | 0.0092 | 0.002 | 0.0191 | 0.0016 | 0.2008 | 0.0092 | 0.002 | 0.0191 | 0.0016 | 0.9626 |

| B | 4 | 4 | 4 | 4 | 4 | 4 | 4 | 4 | 4 | 4 |
|---|---|---|---|---|---|---|---|---|---|---|
| A1 | 5 | 5 | 6 | 6 | 7 | 7 | 8 | 8 | 10 | 10 |
| A2 | 2 | 3 | 2 | 3 | 2 | 3 | 2 | 3 | 2 | 3 |
| $\mu_{valid}$ | 0.0092 | 0.009 | 0.002 | 0.002 | 0.0191 | 0.0191 | 0.002 | 0.0016 | 0.2008 | 0.9626 |

Two columns were marked gray: objects 4,3,10 and objects 4,10,3 based on the highest level of membership value $\mu_{valid}$. When comparing them with objects shown in Fig. 17.5, one can easily see that only these two columns contain valid identifiers of objects, which can be considered as valid markers. The selection of object triple (4,3,10) vs (4,10,3) depends on whether object A1 will be called left- or right-most marker. Similar positive results were obtained for various positions of the IR camera.

### 17.2.2 *Temperature map reconstruction algorithm for cylindrical surface*

After the spatial orientation of the cylinder and IR camera are determined, the system must provide useful information of the surface temperature to the control algorithms. In order to feed these algorithms with temperature information about the entire surface of the cylinder rather than a cut currently observed by the camera, as shown in Fig. 17.11, an algorithm for reconstructing the full temperature map of the rotating steel cylinder was developed.

Fig. 17.11. A cut of cylinder surface in the camera viewport (height 120px)

To arrive at the temperature map of a half-infinite cylindrical surface, the proposed algorithm utilizes two edge-triggered synchronization signals, generated by a quadrature encoder (mounted to the cylinder's shaft): the first is triggered once per cylinder's revolution and the second is triggered N-times per revolution. This allows the IR camera to sign individual frames during the acquisition process by two bit flags each time the camera's input stage detects a rising edge in one of the signals. Fig. 17.12 shows the distribution of angular synchronization points: **RS** synchronization is issued once every revolution, while **IS** is the intermediate synchronization issued N times per revolution.

In the design and implementation of the algorithm, the cylinder surface was divided into 8 images with one indirect synchronization pulse **IS** per image. Each image is shifted against the previous one by 45°. Such a number of areas, along with the camera viewport height, provide a sequence of images high enough to overlap with neighbors and to avoid using parts of input images affected by the Lambert's cosine law [Litwa, 2010]. The overlays are shown in Fig. 17.12 as dotted, rectangular areas. Finally, sequences consisting of eight consecutive frames are obtained. A sample sequence is shown in Fig. 17.14.

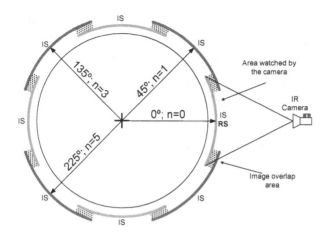

Fig. 17.12. Image synchronization points on the circumference

The reconstruction algorithm, using the above-described sequence of images, consists of two steps. The first one is to eliminate the influence of surface natural cylindrical curvature on the object observed. This step is carried out by the so-called *texture unwrapping* from the input image. This process can be interpreted as opposed to texturing, known from all popular 3D modeling packages. The algorithm employs a three-dimensional model of a cylinder, given by expression (17.4). Its goal is to calculate a point in the three-dimensional model space coordinate system for each point of unwrapped texture, located at coordinates $u$ and $v$, in texture space (2D). The model is located at the origin of the coordinates, denoted as $Q(u, v) = |0\ 0\ 0|^T$.

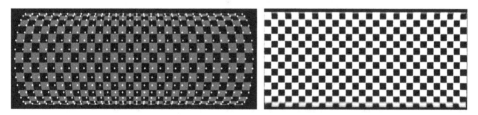

Fig. 17.13 a) A 3D model of a cylinder with a superimposed map of points; b). Texture unwrapped by applying the 3D model

$$Q(u,v) = \left| \frac{uC_L}{M_W} - \frac{1}{2}C_L; C_R \sin\left(2\pi\frac{v}{M_H}\right); C_R \cos\left(2\pi\frac{v}{M_H}\right) \right|^T \tag{17.4}$$

where $u$, $v$ – the coordinates of a point in the final unwrapped texture of width $M_W$ and height $M_H$, $C_L$ – the length of the cylinder, $C_R$ – the radius of the cylinder.

From all the points calculated by (17.4) only the points visible to the camera were selected, as an angle between the vector normal to the cylinder surface at the point $Q(u, v)$, and the camera direction vector, coinciding with the optical axis, oriented on the point $Q(u, v)$. Each visible point $Q(u, v)$ was then mapped to the input image (from an IR camera) by perspective projection (17.5) [Cyganek, 2002]:

$$P^{input}(u,v) = \left| \frac{fQ_X(u,v)}{f+Q_Z(u,v)} \quad \frac{fQ_Y(u,v)}{f+Q_Z(u,v)} \right| \qquad (17.5)$$

where: $P^{input}$ – the coordinates in the input image, $f$ – the camera focal length, subscripts $X$, $Y$ and $Z$ are components of the vector $Q(u, v)$ given by (17.4).
Fig. 17.13a shows a sample visualization of the model given by (17.4) with only visible points $Q(u, v)$ marked. Each of them has a corresponding $P^{input}$ point. The result of the unwrapping algorithm, performed for a synthetic input image is depicted in Fig. 17.13b. To determine the final value of each point in unwrapped texture, a bilinear interpolation was employed.

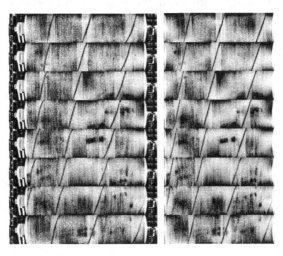

Fig. 17.14 a). A sequence of 8 images input sequence arranged from the oldest to the youngest (left); b) Unwrapped and corrected textures (right)

Fig. 17.14b shows the result of the unwrapping algorithm performed for an arbitrarily selected sequence of eight input images (Fig. 17.14a). To better illustrate the result of the developed algorithm, the presented images show an additional marker wrapped around the cylinder (the black line).

After unwrapping the texture from each image of the sequence, the last step –
the reconstruction algorithm, is carried out. Its task is to combine a set of textures
(Fig. 17.14b) into a complete cylinder's surface temperature distribution map
(Fig. 17.16). The algorithm assumes that the surface area seen by the camera is
high enough for successive textures in the sequence to overlap each other at their
upper and lower parts (as illustrated in Fig. 17.12). These folds can be used in
two ways, as shown in Fig. 17.17.

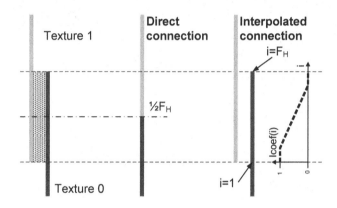

Fig. 17.15. Two methods for consecutive textures connecting

**Direct connection**: for a series of textures of the same height, the heights of
folds are equal. This allows the folds to be spilt in half their height and then the
two neighboring textures to be connected.
**Interpolated connection**: two textures can be connected by a fold which allows
one texture to pass into the other. For this case, the algorithm uses linear
interpolation, guaranteeing a smooth transition between two consecutive textures.
Interpolation is performed using a transfer function $I_{coef}(i)$, shown in Fig. 17.14
and used in expression (17.6).

$$M(i) = T_0(T_H - F_H + i)I_{coef}(i) + T_1(T_H - i)\left(1 - I_{coef}(i)\right) \tag{17.6}$$

where: $T_x(\bullet)$ – the row of texture $x$, $T_H$ – the height of the texture, $F_H$ – the height
of the fold.

The result of the reconstruction algorithm executed for sample data from Fig.
17.14b is shown in Fig. 17.16 as a *complete temperature distribution*. Such a
map can be used in the temperature control to ensure a desired profile of
temperature on the surface of the cylinder.

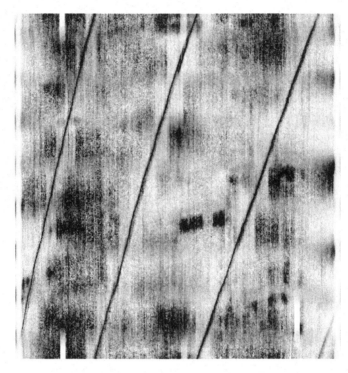

Fig. 17.16. The final temperature distribution map of the rotating steel cylinder after surface reconstruction

## 17.3 Identification of dynamic properties of heated rotating cylinder

### 17.3.1 *Transfer function model*

The designing of a temperature control algorithm requires some knowledge of dynamic properties of the plant under control. In the case of a heated rotating cylinder the input signal of such a plant results from the heating power generated in the cylinder mantle by moving inductors. On the other hand, the processed image acquired by an IR camera plays the role of the output signal. Typically, the dynamic properties of the plant can be expressed in the form of a transfer function, illustrating the relation between input and output. One of the possible structures of a transfer function model of the heating system analyzed is presented in Fig. 17.17.

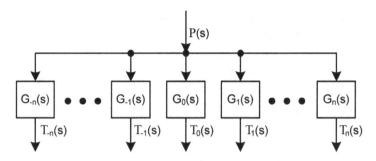

Fig. 17.17. The structure of the transfer function model.

In this model, the cylinder was divided into 2n+1 zones along x-axis and direct thermal couplings between the heating power and the temperature of each section was taken into account. For the purpose of such modeling the average temperature along x-axis T(x) was calculated basing on the thermal image of the cylinder surface, given by (17.7):

$$T(x) = \frac{\sum_{y=1}^{N} I(x, y)}{N} \tag{17.7}$$

where: N – the number of pixels along y-axis

and the temperature of zone i ($T_i$) was calculated as an average temperature of pixels in zone i.

The first order inertia transfer function with time delay (17.8), typically used for the modeling of various thermal systems [Sankowski *et al.*, 1993, Michalski *et al.*, 2001], can be applied as every single component of the model in Fig. 17.17:

$$G(s) = \frac{K \cdot e^{-s \cdot L}}{1 + s \cdot N} \tag{17.8}$$

where:

K – the model gain,
N – the time constant,
L – the delay time.

### 17.3.2 Step input response identification

The dynamic parameters of the transfer function model (Fig.17.17 and Eq. (17.8)) can be determined experimentally by applying one of typical excitation

signals [Bohlin, 2006] to the cylinder surface and analyzing its output –
temperature images. The step input response method can easily be realized by
heating up the rotating cylinder using a single static inductor and acquiring a
sequence of the resulting thermal images of its surface. An example of such an
image during the experiment with a step change of heating power from 0 to
700W supplied to the rotating cylinder is shown in Fig. 17.18.

Fig. 17.18. Sample image of a cylinder surface obtained from an IR camera during the step-input
identification process

Basing on the recorded sequence of thermal images the time dependent step
input responses of the cylinder sections can by determined using (17.7).
Examples of such signals are presented in Fig. 17.19. Due to the symmetry of the
model in Fig. 17.17 only half of the sections can be taken into account. The
results of the least square estimation of transfer functions parameters, using the
criterion defined by (17.9) are collected in Table 17.3 and illustrated in Fig.
17.19.

$$E = \sum_{i=0}^{n} \left( \frac{T_i - T_{mi}}{T_i} \right)^2 \cdot 100\%$$

(17.9)

where:
$T_i$  – the response of the object in measured time step i,
$T_{mi}$ – the response of the transmittance model in measured time step i,
n  – the number of time steps.

Some intuitive relations between the parameters of particular blocks can be
seen in Table 17.3: the more distant the heated region, the lower the model gain
and the higher time constant and time delay. These results also show that the L/N

quotient is very low for the most important sections of the model so that the on-off control algorithm can be used for temperature control purposes [Michalski *et al.*, 1981].

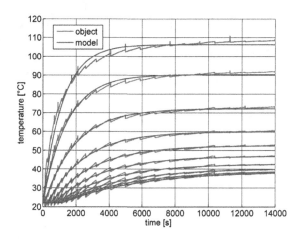

Fig. 17.19. Comparison of the measured data obtained and the cylinder transfer function model

Table 17.3 Parameters of the transfer function model determined using a step-input response

|  | $G_0$ | $G_1$ | $G_2$ | $G_3$ | $G_4$ | $G_5$ | $G_6$ | $G_7$ | $G_8$ | $G_9$ |
|---|---|---|---|---|---|---|---|---|---|---|
| K [°C/W] | 0.121 | 0.098 | 0.072 | 0.054 | 0.044 | 0.036 | 0.030 | 0.026 | 0.025 | 0.024 |
| N [s] | 1176 | 1473 | 1916 | 2230 | 2557 | 2917 | 3300 | 3733 | 4194 | 4466 |
| L [s] | 0.0 | 0.0 | 81.9 | 291.6 | 494.3 | 675.7 | 856.7 | 979.3 | 1080.8 | 1172.8 |

## 17.4  Control algorithms for cylinder surface temperature control and inductors movement

The temperature distribution at the cylinder surface is shaped in the system analyzed by heating power generated by moving inductors. Thus, the general idea of a temperature control system embraces both the manipulation of the heating power of every single inductor and the inductors movement along the cylinder axis. This leads to a system in which heating power and the movement of the inductor are controlled separately in two feedback loops, on the basis of the instant temperature distribution yielded by image processing algorithms, as shown in Fig. 17.20.

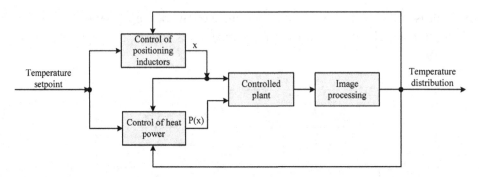

Fig. 17.20. Structure of closed loop temperature control system

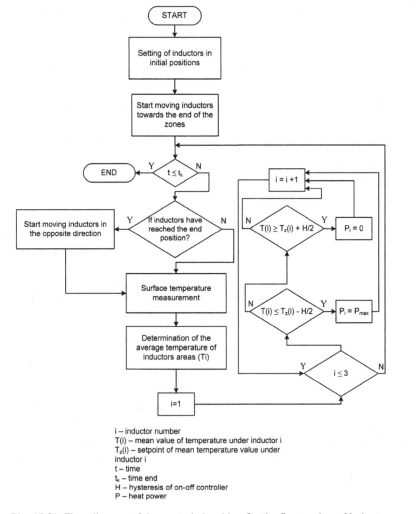

i – inductor number
T(i) – mean value of temperature under inductor i
$T_z(i)$ – setpoint of mean temperature value under inductor i
t – time
$t_k$ – time end
H – hysteresis of on-off controller
P – heat power

Fig. 17.21. Flow diagram of the control algorithm for the first section of inductors

The control accuracy in such a system depends inherently on the algorithms applied in both control loops. Since the most general aim of the control system considered is to reach as close as possible to the required temperature profile along the cylinder axis, the control strategy lies in feeding the heating power to these parts of the cylinder where the difference between the required and actual temperature value proves the most significant. This means that during each control cycle the inductor should be moved to the place where the heating power is the most needed while the value of the heating power is calculated by the on-off algorithm.

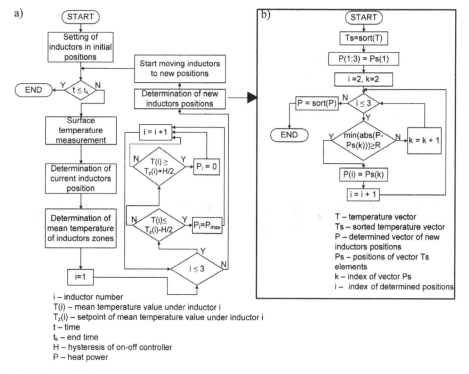

Fig. 17.22. Flow diagram of the control algorithm for the first section of inductors (a) and the detailed algorithm of "Determination of new positions of the inductors" (b).

Considering that a semi-industrial laboratory stand is equipped with two independent sections of 3 mobile inductors each, realizing two different tasks.

The first section is responsible for the uniform heating of the cylinder surface according to so called "lift algorithm" shown in Fig. 17.21. The inductors of the second section can move freely along the cylinder axis, assisting those of the first section so that they are capable of eliminating the local non-uniformity of temperature distribution, which can occur during the system operation. The flow chart of this algorithm is shown in Fig. 17.22. The detailed algorithm of the block named "Determination of new positions of the inductors" in Fig. 17.22 (a) is shown in Fig. 17.22 (b).

## 17.5 Experimental verification of the temperature control system

The effectiveness of algorithms presented was confirmed experimentally. Both the heating-up process and disturbance attenuation were analyzed. For comparison purposes, the quality index was defined as follows:

$$J = \sum_{k=1}^{m} \sum_{x=1}^{n} (T_{x,k} - T_z)^2 / 10^6 \tag{17.10}$$

where: m – the number of time steps, n – the number of pixels along y-axis, $T_z$ – the setpoint of the temperature value.

The heating-up of the cylinder surface from ambient temperature (20°C) to the required temperature $T_z$ (60°C) is shown in Fig. 17.23. The results obtained for two control algorithms applied for the inductor movement along with the on-off algorithm governing the heating power are presented in the following figure.

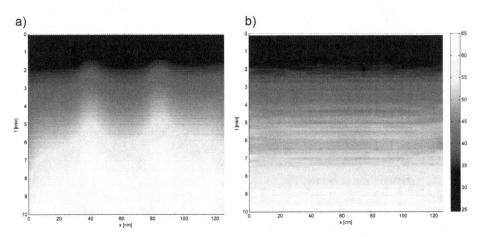

Fig. 17.23. Temperature distribution at the cylinder surface recorded during the heating-up process using the algorithm presented in Figs 17.21 (a) and 17.22 (b). The quality index J is equal to 121 and 118, respectively.

It can be noticed that six inductors, moving according to the "lift" algorithm (given in Fig. 17.21), do not ensure the acceptable uniformity of temperature at the cylinder surface (Fig. 17.23a). Particularly, the edges of the cylinder proved slightly colder while its other regions are overheated. The temperature accuracy improves significantly when the inductors are divided into two sections and three of them move according to the algorithm presented in Fig. 17.22. The inductors of this section, dynamically located in places of the lowest temperature (Fig. 17.22b), compensate, for example, for different heat exchange conditions. It can be noticed, when analyzing the histogram of their locations along the cylinder axis given in Fig. 17.26a, that the cylinder edges are the most frequently occupied positions since this is the region of the highest temperature losses. Fig. 17.23b demonstrates the high uniformity of temperature distribution at the cylinder surface during the entire heating-up process when the two-section control algorithm is used.

Another important issue in the temperature control of the cylinder surface is the ability of the control system to attenuate thermal disturbances that can occur during a technological process. In the developed system various types of such disturbances were generated and both control algorithms discussed above were examined. The chosen examples of the results are shown in Fig. 17.24.

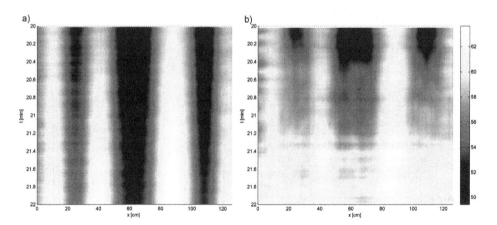

Fig. 17.24. Temperature distribution at the cylinder surface during the local thermal disturbance when the algorithm presented in Figs 17.21 (a) and 17.22 (b) is used. The quality index J is equal to 3.24 and 0.82, respectively.

The application of two sections of inductors moving according to the algorithms given in Figs 17.21 and 17.22 proved to be a very efficient way of disturbance attenuation. The analysis of the histogram given in Fig. 17.26b shows that freely moving inductors, besides the heating-up of cylinder edges, are mainly located in the cylinder regions of the deepest temperature drop. In addition, the trajectories of the inductors, shown Fig. 17.25, recorded during the disturbance attenuation confirm the appropriate reaction of the inductors after the disturbance has occurred. They immediately follow the coldest cylinder regions heating them up and, after reducing the temperature non-uniformity, they return to their regular action. The advantage of the solution proposed (seen in Fig 17.24b) can quantitatively be expressed using the quality index (17.10). The value of this index, in the case of the two-section control algorithm (Fig. 17.24a), outnumbers its value for six inductors moving according to the "lift" algorithm (Fig. 17.24b) by 4 to 1.

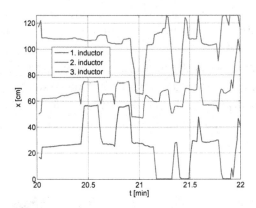

Fig. 17.25. Trajectories of the movement of inductors during the disturbance shown in Fig.17.24

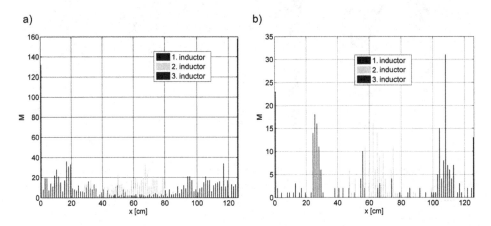

Fig. 17.26. Histogram of the location of freely moving inductors during the cylinder heating-up process (a) and thermal disturbance attenuation (b)

Besides keeping the required temperature value at the cylinder surface highly uniform, the control system should also be able to set and change dynamically any temperature profile at the surface. As an example, the results of such an experiment are shown in Fig. 17.27. After the heating-up process and steady state temperature control at a level of 40°C (lasting for approximately 40 min) the required temperature profile along the cylinder axis was changed to the linear one from 35 to 45°C. Fig. 17.27 proves that the control algorithms fulfilled this task successfully and ensured the high accuracy of temperature control in various operating conditions.

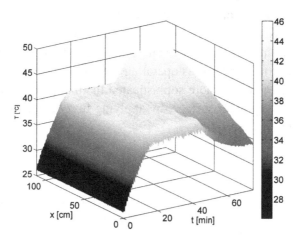

Fig. 17.27. System operation during the change of the required temperature profile along the cylinder axis.

## 17.6 Summary

In the present chapter the computerized system for the surface temperature control of the rotating cylinder has been presented. Such a device is used in many industrial technological processes, largely in the papermaking and textile industries. The induction heating of the cylinder with a set of moving inductors is recognized as an efficient and ecological solution, but the temperature control of the cylinder surface requires some efforts. In view of high accuracy requirements as well as the expected flexibility of the system in terms of various technology-resulted required temperature profiles fulfilled at the cylinder surface, the use of an infra-red camera in a temperature control system is necessary. In the solution presented, an IR camera works as a source of information about the thermal state of cylinder surface for the rest of the system. The necessary image processing algorithm as a basis for a control algorithm has been developed. First, an algorithm for the automatic localization of the cold cylinder by an IR camera using a set of thermal markers and a fuzzy recognition algorithm was developed. It enables one to calibrate the system properly in various operating conditions. As the cylinder rotates permanently, different parts of its surface are exposed and can be captured by an IR camera so that a special reconstruction algorithm was proposed. The algorithm does not only reconstruct the temperature map for the entire cylinder surface from the sequence of acquired partial thermal images, but also eliminates the influence of the natural cylindrical curvature of the object observed. So, the obtained image of the cylinder surface was used for the identification of dynamic properties of the plant and a control algorithm was designed on this basis. The image processing and developed control algorithms were verified experimentally using a semi industrial laboratory set-up and their efficiency was confirmed in various operating conditions. The current research is devoted to the recognition of more sophisticated thermal patterns in the acquired thermal image of the cylinder surface, by applying fuzzy clustering methods and a fuzzy spatial relation description.

## References

Bloch, I. and Ralescu, A. (2003). Directional relative position between objects in image processing: A comparison between fuzzy approaches, Pattern Recognition, 36, nr 7, pp. 1563–1582

Bohlin, T. (2006). Practical Grey-box Process Identification. Theory and Applications. Springer. Germany.

Cho, C.-Y., Chen, T.-M., Wang, W.-S. and Liu C.-N. (2011). Real-Time Photo Sensor Dead Pixel Detection for Embedded Devices Digital Image Computing, International Conference on Techniques and Applications (DICTA), pp. 164 -169

Cyganek B. (2002). Komputerowe przetwarzanie obrazów trójwymiarowych (in Polish), Exit.

Fraczyk, A., Urbanek, P. and Kucharski J. (2008). Computer-based system for non-contact temperature measurement of high-glittering induction-heated rotating steel cylinder. IV International Conference of Young Scientists "Perspective Technologies and Methods in MEMS Design". Polyana, Ukraine.

Hudelot, C., Atif, J. and Bloch, I. (2008). Fuzzy spatial relation ontology for image interpretation, Fuzzy Sets and Systems, 159, nr 15, pp. 1929–1951

Jackson, J. E. (1991). A User's Guide to Principal Components, John Wiley and Sons.

Jahne, B. (2002). Digital Image Processing, Springer Verlag Berlin Heidelberg New York, ISBN 3-540-67754-2

Jaworski, T. and Kucharski, J. (2010). The use of fuzzy logic for description of spatial relations between objects, Automatyka - Zeszyty Naukowe AGH, 14, nr 3/1, pp. 563–580

Jaworski, T. and Kucharski, J. (2013a). An algorithm for reconstruction of temperature distribution on rotating cylinder surface from a thermal camera video stream, Przegląd Elektrotechniczny, ISSN 0033-2097, R. 89 NR 2a/2013, pp. 91-94

Jaworski, T. and Kucharski, J. (2013b). An algorithm for finding visual markers in an infrared camera images based on Fuzzy Spatial Relations, Przegląd Elektrotechniczny, ISSN 0033-2097, R. 89 NR 2b/2013, pp. 276-279

Klette, R. and Rosenfeld, A. (2004). Digital Geometry: Geometric Methods for Digital Picture Analysis, Morgan Kaufmann.

Litwa M. (2010). Influence of Angle of View on Temperature Measurements Using Thermovision Camera, IEEE SENSORS JOURNAL, VOL. 10, NO. 10, pp. 1552-1554

Michalski, L., Kuźmiński, K. and Sadowski, J. (1981). Regulacja temperatury Urządzeń elektrotermicznych (in Polish). WNT, Warszawa.

Michalski, L., Eckersdorf, K., Kucharski J., McGhee J. (2001). Temperature Measurement. Second Edition. John Wiley & Sons, LTD.

Miyajima, K. and Ralescu, A. (2004a). Spatial organization in 2D segmented images: Representation and recognition of primitive spatial relations, Fuzzy Sets and Systems, 65, nr. 2-3, pp. 225–236

Miyajima, K. and Ralescu, A. (2004b). Spatial organization in 2d images, in of Proceedings of the 3rd IEEE Conference on Fuzzy Systems 1, pp. 100–105

Otsu, N. (1979). Threshold selection method from gray-level histograms, IEEE Trans Syst Man Cybern, SMC-9, nr 1, pp. 62–66

Rosenfeld, A. (1985). Distances between fuzzy sets, Pattern Recognition Letters, 3, nr 4, pp. 229–233

Sankowski, D., Mc Ghee, J., Henderson, I., Kucharski, J. and Urbanek, P. (1993). Application of Multifrequency Binary Signals for Identification of Electric Resistance Furnaces. Chapter 8 of book "Perturbation Signals for System Identification. Editor Keith Godfrey. Prentice Hall.

Zadeh, L. (1965). Fuzzy sets, Information and Control, 8, pp. 338-353

## Part 4

# Medical and Other Applications of Computer Vision

# CHAPTER 18

# THE COMPUTER EVALUATION OF SURFACE COLOUR CHANGES IN CULTIVATED PLANTS INFLUENCED BY DIFFERENT ENVIRONMENTAL FACTORS

Joanna Sekulska-Nalewajko and Jarosław Gocławski

*Institute of Applied Computer Science*
*Lodz University of Technology*
*90-924 Lodz, ul. Stefanowskiego 18/22*
*{jsekulska, jgoclaw}@kis.p.lodz.pl*

Nowadays, biologists are especially interested in founding non-destructive, fast and accurate methods of observation and quantification of abnormal plant development. Therefore, imaging techniques and image processing methods are well established as tools for visualization and measurement of changes in plant physiology and morphology at the level of single plant parts. There are many visual symptoms of plant growth distortions caused by different factors operating in the environment. Some of them concern the plant shape and development. For example tobacco or tomato leaves can be wrinkled, smaller or completely stunted after the tobacco mosaic virus infection (*Nicotiana virus*) as well as roots of other plants (e.g. wheat) could have similar symptoms because of the higher concentration of heavy metals in soil. The second group of symptoms is associated with colour changes enhancing the surface or subsurface part of leaves, roots, steams, fruits or even seeds. The origin of these changes also can be very different. For example pathogen infections, toxic compounds and adverse growth conditions can induce:

- damage in surface and internal structures of leaf subsequently leading to necrosis (death of cells in a large area),
- production and accumulation of secondary metabolites, which are mainly reactive oxygen species (ROS) and salicylic acid (SA), as a consequence of plant defence mechanism,
- breakdown of photosynthetic pigments (when it becomes apparent it is called chlorosis).

Generally each of the phenomenon has its specific colour reaction and in the case of well recognized and clear visible with a naked eye symptoms, such as spots of necrosis or chlorosis, the image processing algorithms can be applied directly to the traditional image. In the case of ROS detection the histochemical dying must be used before the computer analysis. But we are still aware of the importance of founding the methods of fast diagnosis of plant condition before the clear visible symptoms appear. There are non-destructive techniques such as thermal [Chaerle *et al.* (1999)] and fluorescence imaging [Chaerle and Van Der Straeten (2000); Chaerle *et al.* (2004)] or optical coherence tomography [Changho *et al.* (2012)] able to visualize pathogen infections at the early stage (before damage appeared), but they are not the subject of this chapter.

Here are presented selected image processing methods aimed at the identification and measurement of changed plant regions of a few cultivated plant species: wheat (*Triticum aestivum*), cucumber (*Cucumis sativus*) and pumpkin (*Cucurbita maxima*). Particular issues represent different approach in terms of plant group, plant body, factor inducting the color change and the proposed measurement algorithm. Considered plants were cultivated in experimentally modified and controlled conditions, such as the presence of pathogens or higher concentration of heavy metals, draught and salinity. Laboratory culture conditions were provided in special growth chambers at the Department of Plant Physiology and Biochemistry of University of Lodz. Plants in the changed conditions were subjected to experiments, designed to determine the impact of certain factors on the growth and development of plants and the discovery of the relationship between the simultaneous influence of several factors.

The selected algorithms discussed in this chapter deal with three main computational problems. The first concerns wheat root discolouration analysis and it is solved based on fuzzy segmentation and transformations colour space. The second method based on the application of hybrid neural network enables identification of ROS reaction regions of the stained leaves of cucumber and pumpkin. The last method compares green colour in wheat leaves from control group against leaves of other groups treated with nickel (heavy metal) and selenium.

## 18.1 Software solutions for automatic detection and classification of plant morphological abnormalities

Right from the beginning the scientists were highly motivated by potential applications of pattern recognition in natural sciences. But as opposed to medical science, where computer image analysis plays an important role, the life sciences does not have many solutions based on computer applications that are considered common and standard. Since 1990's have been developed only few commercial software applications that are dedicated directly to botany and agriculture. But the potential of these applications indicates that semi-automatic and automatic image processing methods over the next few years may replace coarse and expensive visual inspection of plant surfaces or biochemical treatment, that do not allow obtain certain spatial data of studying phenomena. There is a set of specialized applications developed by

Regent Instruments Inc. demonstrating the variety and complexity of the problems, who need this type of solution. The specific purposes of these application are: morphology analysis of leaves and disease/damage quantification (*WinFOLIA*), root measurement (*WinRHIZO*), seed morphology and disease analysis (*WinSEEDLE*), tree-ring, stem, wood density analysis and measurement (*WinDENDRO*), canopy and solar radiations analysis (*WinCANOPY*) and wood cells analysis and quantify the changes in wood structure over annual rings (*WinCELL*). In the most cases these products comprises a computer program and hardware for image acquisition (scanner or digital camera and accessories). Another software product of Regent Instruments Inc. is *WinCAM NDVI*, that uses image analysis to quantify areas of different colour that falls into specified ranges. It is not as specialized for a group of objects as the previous, but it has a wide variety of applications which may overlap with those. For example, it can measure the diseased area on fruits, leaves, flowers or any other material that shows a minimum of colour contrasts. *WinCAM NDVI* is very sensitive and can detect very small color differences. Among the software mentioned above the disease, necrosis and insect damage quantification can be made with the *WinFOLIA Pro* version, which has colour analysis. It allows to choose the region of the image to analyse as well. The *Basic*, and *Regular WinFOLIA* versions measure only the leaf morphological features: area, length and width in a simple manner as done with scanner and camera devices (in the horizontal and vertical directions), perimeter, holes area and some other measurements. Regular version can also differentiate the blade from the petiole.

Directly to plant disease quantification the software *Assess 2.0* is dedicated (e.g. [Bade and Carmona (2011)]). There can be analysed single leaves with different symptoms of disease exposed on the artificial background as well as aerial images containing natural vegetation. For the first category of images the user can operate on the three threshold plates: automatic, manual or classic, choosing colour spaces as well as appropriate methods of diseased regions thresholding. The second category of images can be analysed using the agronomist mode, where for example the ground cover of the vegetation can be calculated.

A lot of automated methods of leaf analysis are based on common software packages. For example quantifying the symptoms of fungus infection (*Colletotrichum destructivum*) in *Nicotiana benthamiana* and other plant species was done by the freely available *Scion Image* application [Goodwin and Hsiang (2010); Wijekoon *et al.* (2008)]. But his popular, general purpose image analyser requires time-consuming operations or substantial knowledge for appropriate macros building to extract the diseased regions. In the case of cited studies only grey-level images were considered to identify bright leaf blade profiles and dark diseased areas inside of them by simple, global thresholdings.

There are many highly advanced imaging systems for scientific and industrial applications, that are useful for biology. One of them is *Matrox Inspector*, which has many tools for image enhancement and processing (thresholding, edge extraction, model finder or pattern-matching tools), that can be successfully used for measure lesion areas in leaves e.g. cucumber leaves attacked by dangerous fungus *Colletotrichum orbiculare* [Kwack *et al.* (2005)].

In many cases dealt with issues related to the development of plants custom approach to extracting relevant information from the image is required. In therm of multiple classes' classification that involves more than two classess, the researchers should use a number of methods to deal with this common scenario asuch as one-against-one method. Specially for this purpose are designed systems which are often characterized by a hybrid structure. And so, for example intelligent system for automated pomegranate disease detection and the stage of infection recognition uses fuzzy logic, back propagation neural network and support vector machines [Sannakki *et al.* (2011)].

## 18.2   Image aquisition and measurement system

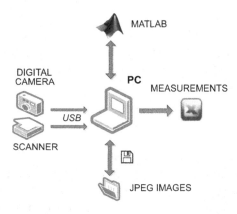

Fig. 18.1   The block diagram of measurement system for plant tissue colour changes; PC: personal computer with the *Windows* operating system, MEASUREMENTS: evaluation results saved in an *Excel* worksheet, JPEG IMAGES: the folder of plant root or leaf images acquired from an input scanner or digital camera using firmware.

The proposed measurement system, shown in Fig. 18.1, consists of the following parts:

- a standard flat-bed scanner with 300/600 *dpi* resolution and 24 *bit/pixel* colour depth, equipped with appropriate software package controlling the scanning process,
- a personal computer with dual core processor 2 *GHz*, 4 *GB* RAM and operating system *Windows 7 32/64 bit*,
- the *MATLAB* environment and *Visual Studio Express 2008 C++* compiler to build hybrid applications from *M* language scripts and *DLL* libraries included

in *MEX* files, which provide the evaluation of required changes in plant tissues using image analysis methods,

* *Excel* application to preserve a series of measurement results.

The tested plant leaves or roots are typically scanned in *A*4 format at $300/600\,dpi$ and $24bit/pixel$ resolutions. To facilitate image analysis the objects are exposed on a highly saturated and uniformly coloured background. The background colour should be possibly different from each colour appearing inside of the tested objects. After scanning, the images are saved as disk files in *JPEG* format before reading them into *MATLAB* environment. The developed *MATLAB* programs provide further processing and analysis of plant images with the help of appropriate *M* scripts and *C++* functions stored in *MEX* files. The programs use several library functions from *Image Processing Toolbox* [The Mathworks Inc. (2012b)]. For computers without *MATLAB* environment they can be compiled as standalone applications including built in *MATLAB* run-time libraries.

## 18.3    Plant tissue material and tested stress symptoms

The segmentation methods presented in this chapter are related to the roots or leaves, wich can be newly detached from wheat seedlings or specifically stained to expose plant stress reaction. They include three following cases:

(A) image segmentation of wheat root systems with dark discolourations influenced by heavy metals (HM),
(B) image segmentation of cucurbits leaves with colour symptoms of plant stress,
(C) the comparative assessment of green pigment content in wheat leaves under HM stress conditions.

### 18.3.1    *Wheat roots material for discolouration assessment*

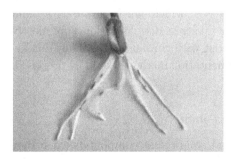

Fig. 18.2 Exemplary photo of Ni-treated wheat root system with coloured areas (dark places, originally red) after staining for lignin detection.

Fig. 18.3 Exemplary photo of Ni-treated wheat root system with coloured areas (dark places, originally blue) after staining with Evans blue for dead cells detection.

The analysis of plant root systems (method (A)) provides quick, coarse information about plant's health, the presence of possible diseases or growth distortions. It also helps to estimate plant tolerance to the influence of chemical agents like HM [Terry *et al.* (2000); Jung *et al.* (2004); Seregin and Kozhevnikova (2009)]. Heavy metals are absorbed by plants mainly via the root systems. Therefore, their toxic effects are clearly manifested in root growth inhibition and morphological disturbances. Some inhibitory effects of HM can be suppressed by the presence of some metal and non-metal ions e.g. calcium [Seregin and Kozhevnikova (2009)] or selenium respectively [Cartes *et al.* (2010)].

The authors have considered roots of young wheat cultivated in the presence of higher concentration of nickel. The plant material have been examined after seven days culturing in water and subjecting abiotic stress. Healthy wheat roots should be properly developed and be uniformly white after washing. Roots reaction to HM is usually expressed by their limiting growth and the appearance of toxicity symptoms like greyish or brownish discolourations. After histochemical or vital staining also blue, yellow or red discolouration can be observed (Fig. 18.2 and Fig. 18.3).

Wheat roots always form fibrous root systems [Weaver (1926)], with a bundle of first order roots growing out of the seed. To minimise overlapping in the zone around origin, all first order roots should be separated from each other and laid out vertically side by side in front of the scanner glass (Fig. 18.5). The image background made of plastic sheet with colour approximating pure red or blue *RGB* components has been applied. The selected background colour is intended to provide maximal hue distance from expected chromatic discolourations of roots.

Despite the existence of many basic segmentation methods using edge detection, thresholding, watersheds, textures or image graphs [Gonzalez and Eddins (2004); Gonzalez (2008); Vincent and Soille (1991)] the considered root image class needs individual approach. The previous known methods of root image analysis are mostly based on global thresholding of grey level images with roots initially stained to improve their contrast over background [Smit *et al.* (2000); Glasbey and Horgan (1995)]. Commercial programs like *WinRHIZO* [Regent Instruments Inc. (2008)] include basic methods determining colour classes, which can be grouped around selected colour hues. The considered heavy metal stress discolourations, are mostly achromatic, have variable brightness and coexist with chromatic discolouration areas of vital staining. In this case a special segmentation method should be proposed.

### 18.3.2 *Plant defence mechanisms assessment*

Plants produce a wide spectrum of organic and non organic compounds when they are attacked by herbivores, pathogens or in response to many other biotic and abiotic stresses, e.g. drought or salt. In most cases it was recognized as a part of a plant defence mechanism, altering the photochemical cycles, preventing oxidation process [Loreto and J. (2010)] or attracting natural enemies of the attacking herbivores [Huang *et al.* (2010b)]. Some synthesized compounds are emitted in the form of volatile blends thus there are not easy to observe [Loreto and J. (2010)], but many others substances are cumulated in a plant tissues where play a relevant role e.g. in antioxidant action. However, the cumulated substances are hardly detected using traditional techniques of visualization. Among these compounds is salicylic acid (SA), produced as a signal in defence against pathogens. SA induces metabolic heating mediated by alternative respiration in flowers of thermogenic plants, and, when exogenously applied, increases leaf temperature in non-thermogenic plants (due to stomatal closure). The latter phenomenon would be detectable in the early stage of infection (before any disease symptoms became visible) when SA is synthesized locally in plant leaves. For these purposes the thermographical imaging has been adapted [Chaerle *et al.* (1999)]. In turn of reactive oxygen species (ROS) which are generated in the presence of a variety of biotic and abiotic stressors - pathogens, drought or salinity [Mahajan and Tuteja (2005); Pitzschke *et al.* (2005); Bolwell and Daudi (2009)], the histochemical dying in order to enable ROS quantification is used. The ROS species and free radical, including superoxide ($O_2^-$), hydrogen peroxide ($H_2O_2$) and hydroxyl radicals ($\cdot HO$), are generated in photosynthesis during over-reduction of components within the electron transport chain and transferring the electrons to oxygen. These ROS need to be scavenged by the plant as they may lead to photo-oxidation and damage of cell and organellas membranes (cause extensive peroxidation and de-esterification of membrane lipids), as well as lead to protein denaturation and mutation of nucleic acids. For those reasons for a long time, ROS have been considered only as dangerous molecules, whose levels need to be kept as low as possible. Now it has been realized that they play important roles in the defence against pathogens, in plant development and in regulation of gene expression [Pitzschke *et al.* (2005)]. Redox signals are like a forewarning for the plant, controlling the energy balance of the leaves. For example $H_2O_2$ acts as a signal for the closure of leaf stomata, acclimation of leaf to high irradiation and the induction of heat shock proteins [Mahajan and Tuteja (2005)]. Therefore, it is necessary for cells to control the level of ROS tightly, but not to eliminate them completely.

Traditional methods of detection and determination of the cumulative amount of oxygen compounds have great limitations. The quantitative biochemical and chemiluminescence methods developed for ROS determination do not provide information on their localization in plant tissues and the assessment by microscopy is limited to small tissue sections. Increased production of ROS is visible in leaf tissues as specific colour regions only after histochemical detection. These histochemical methods allow obtaining detailed data on in situ ROS distribution and accumulation in different leaf parts, thus enabling better comparison of various treatments. However, the histochemical detection of ROS imposes limitations on their quantitative analysis. The methods based on dye extraction from the stained tissue are time- and labour-consuming. The interpretation of results often relies on subjective visual estimation of stained areas yielding only qualitative or semi-quantitative data [Huang et al. (2010a)]. Recently, image analysis methods have been developed and applied to the quantification of the products of ROS-mediated histochemical reactions in plant tissues [Soukupova and Albrechtova (2003)]. The authors have been faced with the identification of the sites of ROS generation in pumpkin and cucumber leaves subjected to abiotic stresses (drought and salinity) and infected with a pathogen. The subjects of quantification were regions of accumulation of two ROS species: superoxide anion radical and hydrogen peroxide, visible after leaf staining as blue or red-brown spots, respectively. In histochemically stained and then cleared (chlorophyll free) leaves, these regions differ from the intact leaf tissues by colour hue and saturation values. The colour features and multidimensionality of the feature space suggest using the colour space instead of grey-levels and a formal classifier, e.g., with an ANN (Artificial Neural Network) instead of thresholding. So far, LVQ (Linear Vector Quantization) type neural networks have been successfully applied to many classification problems like blood cell recognition [Tabrizi et al. (2010)] or sea floor acoustic images segmentation [Tang et al. (2007)]. The authors propose the use of a slightly modified self-clustering WTA (Winner Takes All) network [Kohonen (1997, 2001)] concatenated with a linear perceptron layer type [Widrow et al. (1988); Hagan et al. (2009)]. Using a sufficient number of clusters, the network can recognise all visible leaf staining colours and then combine them in two groups: intact blade areas and the concentration regions of the stress reaction. For a network of such a type, these groups do not need to be assumed as linearly separated, which is not guaranteed in the examined populations. The capabilities of using Kohonen networks in image segmentation in the $L^*u^*v^*$ colour space have already been studied [Ong et al. (2002)].

The material for image analysis consisted of a set of five-week-old plants subjected to abiotic stresses. Plants were cultivated in water deficit (drought stress) or irrigated with $50\,mM$ NaCl (salt stress) for seven days. The second group of plants were not treated with abiotic factors. Then each group was divided into two subsets: control and inoculated with the pathogenic fungus Erysiphe cichoracearum. The plants were analysed five days after inoculation. Detached leaves were examined for superoxide anion radical ($O_2^-$) visualisation according to [Unger et al. (2005)] and for hydrogen peroxide ($H_2O_2$) detection according to [Thordal-Christensen et al. (1997)]. After staining and clearing in ethanol, leaves became almost white, as a result of chlorophyll removal, and colour products of histochemical reactions of $O_2^-$ and $H_2O_2$ were visible as blue and red-brown spots, respectively.

### 18.3.3  Wheat leaves green pigment evaluation

A significant content of chlorophylls in photosynthetic organisms is responsible for their green colour. In higher plants mainly chlorophylls a and b occur, which deal 85% of total amount of photosynthetic pigments. Chlorophyll content in leaves provides valuable information about plant condition, because it can change with plant physiological status depending on growth stages and environmental conditions. There are several physical and chemical methods for chlorophyll content determination. The most of them are destructive for blades and time-consuming, because need a procedure of pigment extraction. One of non-destructive method in this group is leaf reflectance and transmittance measurement in $700 - 1000\,nm$ wavelength [Daughtry *et al.* (2000)]. But in that case we lose also useful information about chlorophyll distribution in leaves, which is sometimes as important as quantitative amount. Because of the drawbacks of existing chlorophyll measurement processes mentioned above there is a need to develop a fast and approximate method for the comparative assessment of green pigment changes in leaves.

The authors proposed a new solution to the problem of pigments assessment in young wheat leaves cultivated with the presence of nickel (HM) and selenium. In many plant species photosynthetic pigment decreases with increase nickel concentration due to oxidative stress [Dubey and Pandey (2011)]. Toxic effects of nickel may be ameliorated by nutrition of plants with selenium [Yao *et al.* (2009)]. One group of seven days old wheat seedlings was subjected to different concentrations of nickel sulphate ($NiSO_4$) every alternate day. In the same time other two groups were treated with different concentrations of sodium selenite pentahydrate ($Na_2O_3Se \bullet 5H_2O$) and with both compounds. All groups were then compared with control group. We expect a relationship between chlorophyll concentration and the degree of leaf greenness, but in the case of this particular study and results of biological experiment only visual inspection of leaves and chlorophyll distribution was taken into account.

## 18.4 Separating of plant tissue samples from the image background

### 18.4.1 *Plant root extraction*

All segmentation methods discussed in the chapter apply images acquired from scanner device as described above in Sec. 18.2. In the case of the segmentation (A) white roots with grey discolouration areas are achromatic in the contrary to the applied background of uniform and highly saturated red or blue colour. Therefore to obtain the binary mask of visible root objects image chroma component is to find. One possible solution is the transformation of $RGB$ into $HCI$ (Hue-Chroma-Intensity) colour space [Lambert and Carron (1999)]). This selection ensures that equal distances in the tristimulus space correspond to equal perceptual differences between colours and enables minimization of non-uniform noise sensitivity accompanying non-linear transforms. Besides, more than one from output image components can be useful in the segmentation process. The $HCI$ space is defined as the combination of:

- a linear transform of $RGB$ to $YC_1C_2$ components:

$$\begin{bmatrix} Y \\ C_1 \\ C_2 \end{bmatrix} = \begin{bmatrix} 1/3 & 1/3 & 1/3 \\ 1 & -1/2 & -1/2 \\ 0 & -\sqrt{3}/2 & \sqrt{3}/2 \end{bmatrix} \times \begin{bmatrix} R \\ G \\ B \end{bmatrix}, \tag{18.1}$$

- a nonlinear transform providing $HCI$ components:

$$I = Y, \tag{18.2}$$

$$C = \sqrt{C_1^2 + C_2^2}, \tag{18.3}$$

$$H = \begin{cases} \arccos{(C_1/C)} & \text{if } C_1 > 0 \\ 2\pi - \arccos{(C_1/C)} & \text{otherwise} \end{cases}. \tag{18.4}$$

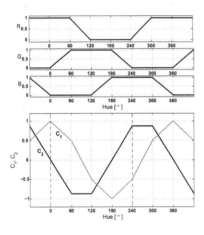

Fig. 18.4 a) The relationship between $RGB$ coordinates and hue values, b) The plot of C1, C2 chroma components along hue axis.

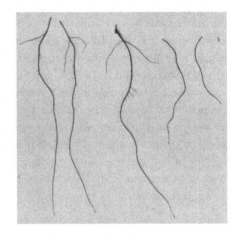

Fig. 18.5 A scanned chroma image $\mathcal{I}_C$ with the set of healthy roots split from one seedling.

The arrays of $C(x,y)$ and $I(x,y)$ for each pixel $(x,y)$ represent grey level chroma and intensity images denoted as $\mathcal{I}_C$ and $\mathcal{I}_I$ respectively. Hue component formally describing the colour itself as an angle $[0, 360°]$ is useless for its high sensitivity to noise at low saturations (chroma). High saturation root colourations observed in our experiments were "red-like" type, which includes tones of red, brown and yellow. Therefore background colour was selected as blue. The colour space transform presented $RGB \rightarrow HCI$ outputs two chroma components $C_1$ and $C_2$ of red and blue primaries, respectively (Fig. 18.4). The complex chroma image $\mathcal{I}_C$ provides good visual separation of achromatic root parts from their chromatic background and can be thresholded to identify the achromatic elements (Fig. 18.5). Applying global thresholding level is undesirable because the intensity histogram of $\mathcal{I}_C$, with roots covering few percent in the whole image area, includes single, narrow peak at the brightness of background. The searching for a common intensity threshold in the histogram valley is ill conditioned. Additionally achromatic root pixels can change their brightness, because the root tissue filled with water can be locally semi-transparent. The authors applied the relatively fast and efficient Bernsen

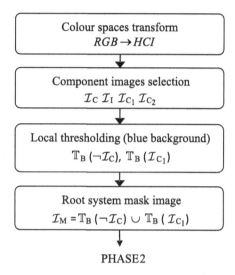

Fig. 18.6    Flow diagram of the extraction of a root system binary image. $\mathbb{T}_B$ -Bernsen thresholding with bilinear interpolation.

method [Bernsen (1986)] of local thresholding with properly adjusted square window and local contrast limitation. Additionally the image of local threshold values was smoothed with bilinear interpolation. In the case of coexistence of achromatic and chromatic root discolourations (e.g. staining for lignin detection), chromatic regions differing in colour from the background can be also detected in a proper image chroma component. Then root system mask binary image $\mathcal{I}_M$ can be evaluated as a logical sum of thresholded inversion of chroma image $\neg\mathcal{I}_C$ and either red $\mathcal{I}_{C_1}$ or blue $\mathcal{I}_{C_2}$ chroma component.

$$\mathcal{I}_M = \mathbb{T}_B(\neg\mathcal{I}_C) \ \cup \ \mathbb{T}_B(\mathcal{I}_{C_1}) \quad \text{or} \quad (18.5)$$
$$\mathcal{I}_M = \mathbb{T}_B(\neg\mathcal{I}_C) \ \cup \ \mathbb{T}_B(\mathcal{I}_{C_2}),$$

where $\mathbb{T}_B(\cdot)$ means Bernsen thresholding operator. Root image processing steps described above are depicted in Fig. 18.6 as a flow diagram. The steps represent the first phase of the whole algorithm and provide only the extraction of root system mask.

### 18.4.2   *Leaf blade extraction*

In the case of method (B) (Sec. 18.3) input images of stained cucurbits leaves used for segmentation include single leaf blades occupying substantial part of the view field. Therefore the global manipulations of $RGB$ colour components are appropriate to detect leaf objects. The selection of background colour contrasting with in-blade colours is necessary. The main image processing steps of leaf blade

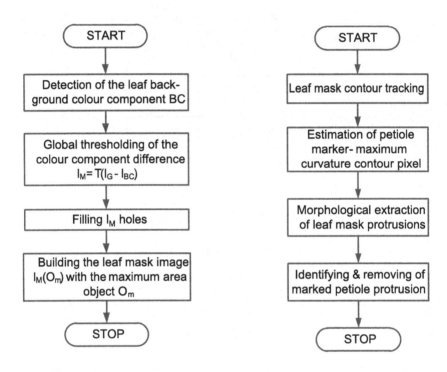

Fig. 18.7  Flow diagram of the algorithm providing the binary mask of leaf image.

Fig. 18.8  Flow diagram of the algorithm eliminating a leaf petiole.

extraction are depicted in Fig. 18.7 as a flow diagram. One of the $R, B$ colour components dominating in the image upper left corner $W[50 \times 50]\ px$ is assumed as the image background colour $BC$. This testing corner must be always free from any foreground object. Therefore the difference of image colour components $I_{\mathrm{BC}}$ and $I_G$ (green) exposes highly saturated background colour $BC$ and makes the resulting grey-level image independent from potential background intensity variations. The global thresholding of $I_{\mathrm{BC}} - I_G$ provides the inversion of a binary leaf blade mask image $I_M$:

$$I_M = \neg\mathbb{T}(I_{\mathrm{BC}} - I_G), \qquad (18.6)$$

where $\mathbb{T}$ denotes Otsu thresholding operator [Otsu (1979)]. After the thresholding given in Eq. (18.6)the binary image $I_M$ can be considered as the set of white objects $\{O_i\}$ of 8-adjacent pixels on a black background. All potential "holes" in the leaf mask object should be removed by a flood-fill operation on 4-adjacent background pixels. In the $MATLAB$ environment this process is represented by the function *imfill* from [The Mathworks Inc. (2012a)]. Only one object $O_m$ from the set $\{O_i\}$ with the greatest area $A_m$ is preserved as a leaf blade mask. Objects with smaller areas are regarded as spurious data and must be eliminated as shown in Eq. (18.7).

$$(x, y) \in O \wedge A_O < A_m \Rightarrow I_M(x, y) = 0 \qquad (18.7)$$

In the tested population of stained leaves a petiole is usually the place of high concentration of dye, but biologists ignore this part of leaf during the visual estimation of plant's stress. Therefore in the presented algorithm the petiole is removed from the leaf mask image $I_M$ by the method shown in Fig. 18.8. Based on the observation of leaf mask contours the authors noticed that the petiole is always the most protruding part of every leaf located at the leaf base. They formulated the hypothesis that petiole tip can be distinguished as a leaf contour point with the highest curvature value. The hypothesis was successfully verified in the tested population of leaves. Fig. 18.9 shows the example image with marked white contour

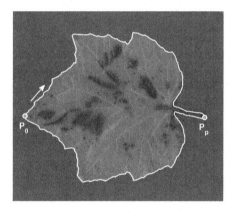

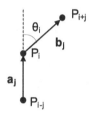

Fig. 18.9 Example leaf image with the white contour overlapped on the edge of a leaf mask, $P_0$ - the contour starting pixel, $P_p$ - the petiole tip pixel.

Fig. 18.10 The illustration of idea for the curvature computation of a leaf blade contour. Letter symbols explained in the text.

found around the leaf mask. For each contour point $P_i$, $i \in [0, \ldots, N-1]$ the local curvature is represented as the bending angle $\theta_i$ computed by [Du Buf and Bayer (2002)]:

$$\theta_i = \frac{1}{M} \sum_{j=1}^{M} \arccos \frac{\mathbf{a}_j \cdot \mathbf{b}_j}{|\mathbf{a_j}| \cdot |\mathbf{b_j}|}, \qquad (18.8)$$

where $\mathbf{a}_j = P_{i-j}P_i$ and $\mathbf{b}_j = P_iP_{i+j}$ are vectors as shown in Fig. 18.10, $\mathbf{a}_j \cdot \mathbf{b}_j$ is the vector inner product and $M$ - the half size of an averaging mask. This mask represents built in low-pass filter smoothing the curvature values $\theta_i$ evaluated along a leaf mask contour. This filtering is necessary due to the high sensitivity of $\theta_i$ to any contour ripples. The value of $M$ has been chosen experimentally in relation to the leaf contour length $N$ as:

$$M = 2.5/100 \times N. \qquad (18.9)$$

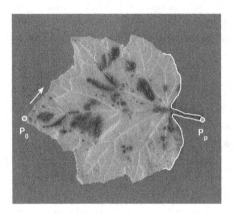

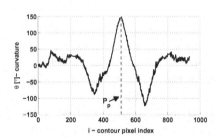

Fig. 18.11  Example leaf image with the contour part used to calculate the curvature $\theta_i$, $P_0$ - the contour starting pixel, $P_p$ - the petiole tip pixel.

Fig. 18.12  The example curvature plot $\theta(i)$ along the section of the leaf contour shown in Fig. 18.11.

Proper selection of the $M$ value plays a key role in the detection of petiole protrusion. Each of the scanned leaves is assumed to be placed horizontally in the field of view with its tip on the left side (Fig. 18.11). Then leaf mask contour tracing begins at the leaf's tip $P_0$ and runs through the petiole tip $P_p$ located close to half the length of the contour $N/2$. Therefore only the curvature of 30% of the contour pixels around $P_{N/2} \in [P_{i_1}, P_{i_2}]$ is considered. The petiole tip pixel $P_p$ is determined as in Eq. (18.10).

$$P_p = arg\ max\ \theta_i, \quad i \in [i_1, i_2]. \tag{18.10}$$

An example curvature plot with the global maximum corresponding to the contour part in Fig. 18.11 is depicted in Fig. 18.12. The petiole tip marks the petiole region belonging to the set of protrusion objects. These objects are extracted by the subtraction of the leaf blade mask $I_M$ from its morphological opening and by additional separative dilation with appropriate structuring circles. Finally the petiole region is xor'ed from the mask image $I_M$.

In the problem (C) listed in Sec. 18.3 yellow-green or pure green wheat leaves are put on the blue background. The difference of selected $RGB$ colour components easily extracts leaf blade profile. The images are initially divided into the parts including single leaf objects Fig. 18.13a. The idea of leaf blade extraction consists in thresholding the difference of colour components $I_R$ and $I_B$ of the image $I_{\text{RGB}}$ as given below:

$$I_M = \neg\ T(I_B - I_G), \tag{18.11}$$

where $I_B, I_G$: blue and red leaf image components, $I_M$: binary image of leaf blade mask, $\neg$: binary image negation symbol, $T(\cdot)$: Otsu thresholding operation [Otsu (1979)]. Because of using difference operator global image thresholding is fully acceptable even in the presence of foreground object shadows which are in fact minimal for naturally flat leaves.

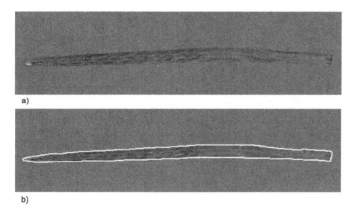

Fig. 18.13 The images of an example wheat leaf a) before segmentation, b) after segmentation with visible boundaries of detected leaf mask contained in the image $I_M$. Original image background is blue; leaf blade is green with yellow-green regions (brighter places).

## 18.5 Identification of requested tissue regions in the image colour space

### 18.5.1 *Wheat roots discolourations — the method of identification*

Dark type discolourations in root images are detected as local intensity suppressions in relation to their neighbourhoods due to the fluctuations of chroma (translucency) and brightness along the roots. A sequence of square windows around selected sample points has been proposed as the local neighbourhoods in segmentation process. The sample points for local neighbourhoods are located in regular distances $\Delta l$ measured along roots medial axes (skeleton lines) in the in binary image $\mathcal{I}_M$ from Eq. (18.5) as shown in Fig. 18.14. Satisfactory skeletoning result has been obtained for a series of these images using an algorithm of morphological thinning proposed by Lam [Lam *et al.* (1992)], which is also included in *MATLAB Image Processing Toolbox* [The Mathworks Inc. (2012b)].

Problems in the extraction of dark discolouration regions in the image $\mathcal{I}_I$ include ambiguity of their exact locations due to variable brightness of neighbouring pixels in healthy tissue, relatively small size, unsharp borders and the presence of noise. For image segmentation uncertainty problems a suitable tool can be the Fuzzy C-Means method (*FCM*) applying fuzzy set theory[Bezdek (1981); Horvath (2006); Mohamed *et al.* (1999); The Mathworks Inc. (2012a)]. The method is an iterative procedure which minimizes the membership uncertainty of feature space data to a predefined number of clusters. It is realized by minimizing the objective function

$$J_m = \sum_{i=1}^{N} \sum_{j=1}^{M} u_{ij}^{m} \|d_i - v_j\|^2, \qquad (18.12)$$

where $m \geqslant 1$ is "fuzzification" parameter, $u_{ij}$ is the degree of membership of the $d_i$ data in the cluster j, $v_j$ is the cluster centre and $\| \cdot \|$ is any norm expressing the

PHASE 1

Root mask $\mathcal{I}_M$ thinning and skeleton
sampling at pixels $S_i$

$U_i$ = **fcm** ( $D_i$ )
Local FCM segmentation of dataset $D_i$
inside windows $W_i$ ( $S_i$ )
$D_i$ = { $\mathcal{I}_I$ ( $W_i$ ) : $\mathcal{I}_M$ ( $W_i$ ) > 0 }

End of $W_i$ ( $S_i$ ) ?     NO

YES

Root discolourations image mask
$\mathcal{I}_{C0}$ = **defuzz** { $U_i$ }
with maximum membership

Elimination of semi-transparent root
pixels under $\mathcal{I}_{C0}$ mask

STOP

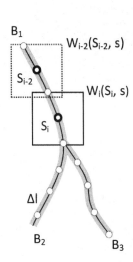

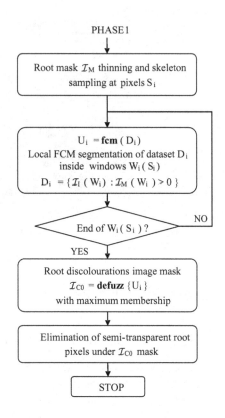

Fig. 18.14 Illustrative fragment of a root system with skeleton branches $B_i$ (black medial axes), sample points $S_i$ and segmentation square windows $W_i$ with side $s$.

Fig. 18.15 Flow diagram with final steps of the algorithm accomplishing the detection of dark discolourations.

distance between measured data and the cluster centre. The case of the Euclidean distance and two cluster centres of dark and bright pixels ($M = 2$) with intensity values $D_i = [d_i]$, $i \in \{1, 2, ...N\}$ as input data is taken into account. At the $k$-th iteration step, the membership values $u_{ij}$ and the cluster centres $v_j$ are updated as follows:

– calculate the new centres vectors $V^{(k)} = [v_j]$ with $U^{(k-1)} = [u_{ij}]$

$$v_j = \frac{\sum_{i=1}^{N} u_{ij}^m \cdot d_i}{\sum_{i=1}^{N} u_{ij}^m}, \tag{18.13}$$

– update $U^{(k)}$ with the new $V^{(k)}$

$$u_{ij} = \frac{1}{\sum_{k=1}^{M} \left( \frac{\|d_i - v_j\|}{\|d_i - v_k\|} \right)^{2/(m-1)}}, \tag{18.14}$$

This iteration will stop at the $k$-th step when

$$\|U^{(k)} - U^{(k-1)}\| < \varepsilon \ \vee \ k > k_{\max}, \tag{18.15}$$

where $\varepsilon$, $k_{\max}$ mean termination criteria. The *FCM* method converges only to a local minimum or a saddle point of $J_m$.

To cluster image $\mathcal{I}_I$ data of the roots ($\mathcal{I}_M > 0$) into groups of bright or dark pixels the authors applied fuzzy c-means clustering described by Eqs. (18.13)–(18.15) to each window $W_i$ centred at the skeleton sample point $S_i$ (Fig. 18.14):

$$V_i, U_i, J_{mi} \leftarrow FCM(D_i, M, m, \varepsilon, k_{\max}), \tag{18.16}$$

where

$$D_i = \{\mathcal{I}_I(W_i) : \mathcal{I}_M(W_i) > 0\}, \tag{18.17}$$

$D_i$ is the dataset of image intensity values in selected region, $M = 2$ is the number of clusters, $m = 2$ is the "fuzzification" parameter, $\varepsilon$ and $k_{\max}$ stand for iterations stopping criteria, $V_i$ is the array of final cluster centres, $U_i$ is the final membership array, $J_{mi}$ denotes values of the objective function during iterations.

For every window $W_i$, the membership dataset $U_i$ obtained in Eq. (18.16) was mapped into image raster domain $\mathcal{I}_M^{-1} = X \times Y$ as the array $P_i$ (Eq. (18.18)).

$$\underset{i}{\forall} \ \{U_i, x_i, y_i\} \to (P_{yx})_i. \tag{18.18}$$

The local membership values $(p_{yx})_i$, for each pixel $(x, y)$, descended from overlapping windows $W_i$ should be averaged :

$$\underset{n_{yx}>0}{\forall} \quad P_{yx} = \frac{1}{n_{yx}} \sum_{i=1}^{n_{yx}} (p_{yx})_i, \tag{18.19}$$

where $n_{yx}$ is an entry of two-dimensional $N_{yx}$ array with $\mathcal{I}_M^{-1}$ domain and represents the count of windows $W_i$ overlapping in a pixel $(x, y)$. The entries $p_{yx}$ of the array $P_{yx}$ are pairs of calculated clusters membership values completing to one. During programming in *MATLAB* environment partial cluster memberships and their counts are cumulated only in two arrays: $P_{yx}(Y \times X \times 2)$ and $N_{yx}(Y \times X)$, stored as sparse types. Calculations of local membership sets $U_i$ in windows $W_i$ (Eq. (18.16)) can give inappropriate segmentation results when cluster centres are too close or the number of pixels in the dataset $D_i$ (Eq. (18.17)) is too small to get reliable membership values $U_j$ (Eq. (18.16)). In these situations local arrays $U_i$ for the sets of points $\{x_i, y_i\}$ must be substituted with equivalent values of global membership array $U$ created through global *FCM* clustering. The employed defuzzification procedure of fuzzy partition array $P_{yx} = [P_{yx}(0), P_{yx}(1)]$ assigns every image pixel $(x, y)$ inside the root mask ($\mathcal{I}_M > 0$) to class $\mathcal{C}_{k\max}$ with the highest membership. The binary result of the proposed *FCM* segmentation is the image:

$$\mathcal{I}_{\mathcal{C}_0} = \begin{cases} 1 & \text{if } \mathcal{C}_{k\max} = \mathcal{C}_0 \\ 0 & \text{otherwise} \end{cases}, \tag{18.20}$$

where $\mathcal{C}_0$ is the class of dark pixels in the segmentation region.

Some pixels of root objects can be misclassified in Eq. (18.20) as discolourations relaying only on the values of the intensity image $\mathcal{I}_I$. Lower intensities of these pixels come from the background colour (e.g. blue) passing through thin roots, which are slightly coloured from background when filled with water. This effect is eliminated by removing from discolourations the root pixels with background hue. Binary discolouration regions are finally processed with morphological functions to eliminate spurious objects with areas below a certain limit value $A_{\min}$. Remaining objects are smoothed by morphological opening with disk structuring element of experimentally adjusted radius.

### 18.5.2 *Wheat roots discolourations — experimental results*

The proposed segmentation method was applied to the population of several images of wheat root systems prepared as described in Sec. 18.4.1. The acquired images included normally developed roots without discolourations (Fig. 18.5) as well as those heavy metal-treated, topologically reduced root systems with dark type discolourations (Fig. 18.16(a)). In the first case, the segmentation process only extracts roots from the background, in the second – it should also separate dark regions of roots. Figure 18.16(a) shows in greys an original colour image $\mathcal{I}_{\mathrm{RGB}}$ of the root system, where the background is blue and the roots are white with dark spots (grey-brownish). Colour space transformation (Eqs. (18.1)-(18.4)) produces the chroma image $\mathcal{I}_C$ visible in Fig. 18.16(b). From Figs. 18.16(c) and (d) we can see that global Otsu thresholding of $\mathcal{I}_C$ can "loose" some parts of root masks compared to the local thresholding using the Bernsen method. Standard, hand-labelled segmentations were carried out for the same series of images as in the automatic method. Then the quality of segmentation with the proposed image processing method can be expressed by the relative error:

$$\delta_{D_I} = \frac{D_I^I - D_I^M}{D_I^M}. \qquad (18.21)$$

The symbols $D_I^I = A_D^I/A_R^I$ and $D_I^M = A_D^M/A_R^M$ mean discolouration indexes found by the proposed image processing and visual inspection methods respectively. Similarly $A_D^M$ ($A_R^M$) and $A_D^I$ ($A_R^I$) represent areas of discolouration (root mask) estimated manually and by image segmentation. All of the areas are calculated as the numbers of image pixels. The data evaluated manually are regarded as true, reference values. Figure 18.17 shows discolouration index errors for a series of $N = 10$ images. An average error in this set marked as $\bar{\delta}_{D_I}$ is equal $-2.15\%$ and its standard deviation is $7.4\%$.

The most important parameters of the algorithm are: the size of Bernsen local window $W_B$, the size of the local *FCM* window $W_i$, local *FCM* algorithm stopping criteria, the minimum allowable distance between two cluster centres, the discolouration areas lower limit $A_{\min}$. The sensitivity of the algorithm to the pa-

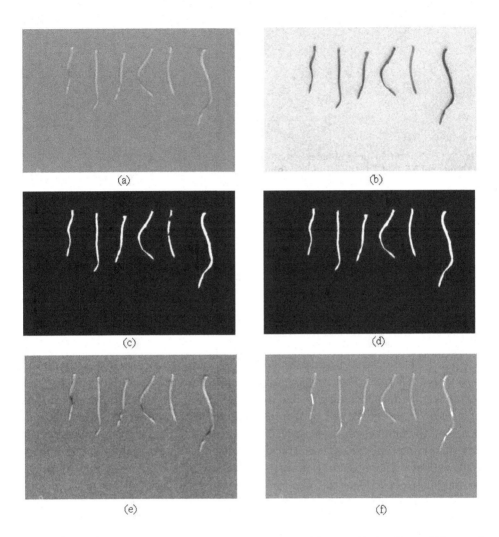

Fig. 18.16  Example stages of root system image processing: (a) original image $\mathcal{I}_{\mathrm{RGB}}$ of Ni-treated roots on blue background, (b) chroma image $\mathcal{I}_C$, (c) binary root mask image after global Otsu thresholding, (d) binary root mask image $\mathcal{I}_M$ after local thresholding, (e) intensity image $\mathcal{I}_I$ from $HCI$ space corresponding with (a), (f) results of discolourations detection with our method – marked white regions overlapped on $\mathcal{I}_{\mathrm{RGB}}$.

rameters is discussed in [Gocławski *et al.* (2009)]. The method was executed in the measurement system presented in Sec.18.2. Example execution times are presented in Table 18.1. Data in the first row are related to the image of healthy roots in Fig. 18.5. The second and third rows include processing parameters of the same roots from Fig. 18.16(a), but for different image margins. In the proposed method, execution time of the first phase depends on the area of processed image $\mathcal{I}_C$. Colour space transformations, filtering and local thresholding refer to each im-

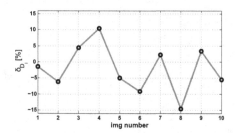

Fig. 18.17   The discolouration index errors $\delta_{D_I}$ of automatic segmentation method in comparison with manual method, for the set of tested images.

Table 18.1   Comparison of image sizes, roots features and execution times for the proposed method.

| image size [px×px] | roots length [px] | area [px²] | execution time phase 1 [s] | phase 2 [s] |
|---|---|---|---|---|
| 1470×1440 | 6682 | 38004 | 2.92 | 2.46 |
| 690×460 | 1065 | 7159 | 0.48 | 0.34 |
| 1380×920 | 1065 | 7159 | 1.66 | 0.75 |

age pixel. The same is with the second phase regarding morphological and filtering operations. Therefore, before processing, scanned images should be limited to their rectangular regions including visible roots plus reasonable margins. Execution time of local *FCM* procedure and semi-transparency elimination during second phase depends on the number of pixels inside the roots mask $\mathcal{I}_M$. The total length of roots determines the times of skeleton tracking and sampling. Fortunately, contaminated root systems with discolourations are significantly reduced in area and length. The second phase of the algorithm is normally unused for images with healthy roots like this one shown in Fig. 18.5 and described in the first row of Table 18.1. In the future it is planned to study biochemical reactions of wheat seedlings to heavy metal stress. The statistical dependency will be searched between the data obtained by image analysis and those acquired by biochemical analysis.

### 18.5.3   *Symptoms of ROS reaction in cucurbits leaves — the region identification with neural network*

The observation of leaf blade images stained for ROS detection (Sec. 18.3.2) leads to the conclusion that ROS accumulation areas can be distinguish from other leaf regions by their colour features. Depending on ROS type ($H_2O_2$ or $O_2^-$) and chemical agents used at staining these areas appear as red-brown or blue. The rest of the leaf blade has a background colour mixed with white while passing through chlorophyll free semi-translucent leaf tissues except for opaque leaf veins. Red-brown stained locations in a tested leaf population have usually high saturation (Fig. 18.23) while blue areas can be medium or even low saturated (Fig. 18.21). Most of low saturated areas cannot be properly distinguished in grey levels. Therefore the image feature space for colour regions segmentation (classification) has been built from two image colour components: hue $H$ and saturation $S$ obtained by colour space transform from $RGB$ colour space [Smith (1978)] as

$$\{I_H, I_S, I_V\} = HSV(I_{\mathrm{RGB}}).  \tag{18.22}$$

The proposed classification of leaf colour regions in $(H, S)$ space is based on a hybrid artificial neural network described below. The traditional hex-cone HSV colour model was applied to follow the *HSB (HSV)* transformation of Corel Photo Paint 12, which helps specialists in the manual classification procedure using color masks. This procedure provides pattern results for training and error estimation of the proposed automatic classification.

The image for network training can be selected intuitively, by visual assessment of images or using heuristic formula of training ability factor, which applies the weighted variances of a leaf blade histogram in the $(H, S)$ space. The formula has been proposed proposed in the paper [Gocławski *et al.* (2012)]. In the visual method only single leaf images with the greatest variability and colour changes inside of leaf blade are selected as the representatives of their classes. Despite of selecting single images there seems to be no danger of network's overfitting in the learning process because of the large number of training feature vectors derived from pixel colours inside of a leaf blade mask. The scanned leaf images typically have a size about $1000 \times 1000$ pixels and a leaf blade occupies about half of the image area. The first network layer proposed in the next section with two inputs and 8-12 neurons has only 16-24 vector weights. The second layer with one neuron has the same amount of weights as the number of first layer outputs, what gives a total of maximum 36 weights to learn.

The proposed model of the classifier applies a two-layer neural network of the counter propagation type [Hecht-Nielsen (1987)] shown in Fig. 18.18. The first layer is a self-organising Kohonen network with MWTA (*Modified Winner Takes All*) learning accepting two-dimensional input vector $\mathbf{x} = [x_1, x_2]^T$ of $I_H$ and $I_S$ pixel values masked by $I_M$. It has been developed, because there is no guarantee that the regions of ROS generation and other leaf parts are linearly separable in the $(H, S)$ space and always can be extracted by a single layer network or other linear classifier. Using the modified WTA learning method aims to achieve more reliable classification result and will be discussed below. Input image data of hue and saturation physically restricted to the domain $[0, 1]$ are mapped to the range $[-1, 1]$ optimal for Kohonen network [Kohonen (1997); Masters (1993)].

All components of the input vector are connected to each of the $M$ output neurons $\mathbf{y} = [y_1, \ldots, y_M]^T \in [0, 1]^M$ representing the centres of clusters (Fig. 18.18). The output neuron $y_i$ of the first layer is defined by the weight vector $\mathbf{w}_i^{(1)} = [w_{i1}^{(1)}, w_{i2}^{(1)}]^T$. The neuron $y_c(\mathbf{w}_c^{(1)})$ which weights $\mathbf{w}_c^{(1)}$ are closest to the data vector $\mathbf{x}$ becomes the 'winner' (Eqs. (18.23), (18.24)) and its weights are slightly updated

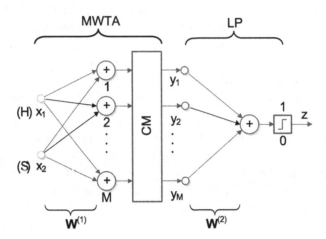

Fig. 18.18   The structure of 2-layer neural network applied for the segmentation. $M$ - the number of clusters, $\mathbf{W}^{(1)}, \mathbf{W}^{(2)}$ - the weight arrays of the first and second network layer respectively, $CM$ - data competition module.

towards the input vector by the WTA learning rule.

$$c = arg\ \min_i \rho(\mathbf{x}, \mathbf{w_i}^{(1)}), \tag{18.23}$$

$$y_j = \begin{cases} 1 & \text{if } j = c, \\ 0 & \text{otherwise}, \end{cases} \tag{18.24}$$

where $\rho$ - distance between the input $\mathbf{x}$ and i-th weight vector $\mathbf{w}_i^{(1)}$, $c$ - index of the 'winner' weight vector closest to the input vector. The second, linear perceptron type layer contains 1 neuron with the activation step function (Eq. (18.26)) not used in learning rule. This layer merges the initially clustered pixel data at $M$-dimensional input in two final classes: with and without ROS colour features, which are expressed respectively as zeros and ones at single output. The weight vectors of the output neuron $\mathbf{w}^{(2)} = [w_1^{(2)}, \dots, w_M^{(2)}]^T$ (Eq. (18.25)) must be calculated by matching output pattern (supervised learning) prepared by specialists.

$$u = \sum_{i=0}^{M} w_i^{(2)} y_i, \tag{18.25}$$

$$z = \begin{cases} 1 & \text{if } u > 0.5, \\ 0 & \text{otherwise}. \end{cases} \tag{18.26}$$

where $y_0$ is the offset adjusted in the output layer.

The goal of the MWTA training modification is to minimise some drawbacks accompanying standard WTA algorithm [Kohonen (2001)]. All changes have been made between the consecutive epochs of standard WTA learning as shown in Fig. 18.19. The WTA training procedure begins with the random initialisation of the weight vectors $\mathbf{w}_i^{(1)}$ representing initial class centres, collected into the weight array $\mathbf{W}^{(1)}$. They are selected by uniform, no-replacement, random sampling from indexes of the $Q$ data samples entered in one epoch (Eq. (18.27)).

$$\mathbf{w}_i^{(1)}(0) = rand(\{1, \ldots, Q\}), \ i \in \{1, \ldots, M\}. \tag{18.27}$$

For the proposed MWTA method its start is extended to the set $\{\mathbf{W}^{(1)}\}$ of several weight vector arrays handled independently according to the WTA rule during one epoch. WTA learning cycles minimise the quantisation error $E(\mathbf{W}^{(1)})$ given in Eq. (18.28), which occurs at the approximation of all input data vectors $\mathbf{x}$ by M weight vectors (or neurons) shown in Fig. 18.18.

$$E\left(\mathbf{W}^{(1)}\right) = \max_c \frac{1}{Q} \sum_{q=1}^{Q} \left\|\mathbf{x}(q) - \mathbf{w}_c^{(1)}(q)\right\|^2, \tag{18.28}$$

where Q is the number of data vectors in one epoch, $\mathbf{w}_c^{(1)}(q)$, $c \in \{1, \ldots, M\}$ is the weight of neuron winning at the presentation of the $\mathbf{x}(q)$ vector.

WTA learning process ensures only the convergence to a local minimum, when the learning rate $\alpha^{(1)}$ experimentally adjusted is small enough ($\alpha^{(1)} = 0.002$). During each epoch of WTA training all valid sets (arrays) of weight vectors

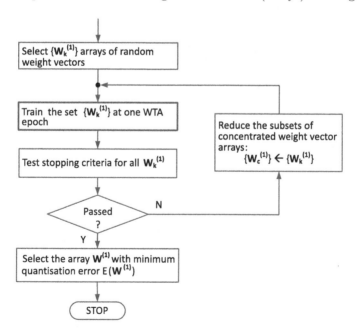

Fig. 18.19   Flow diagram of the proposed MWTA algorithm for training the first layer of the neural network. $\mathbf{W}_k^{(1)}$, $k = 1, \ldots, N_k$ - the $k$-th set of neuron weights, $\mathbf{W}_c^{(1)}$ - the centroids array of any $\mathbf{W}_k^{(1)}$ set.

$\mathbf{W}_{\mathbf{k}}^{(1)} = [\mathbf{w}_{\mathbf{i}}^{(1)}]_k$, $k = 1 \dots N_k$, $i = 1 \dots M$ are independently moved towards the nearest class centres minimizing the error function in Eq. (18.28). Each of them also comes under the stopping criterion of minimum change of weights norm or the iterations limit [Masters (1993)]. After stopping, currently fixed weight sets are stored for final solution analysis (Fig. 18.19). All of still shifted sets are tested for their possible concentrations in the input data space. Sufficiently concentrated groups of weight arrays are replaced by the concentration centres before a next training epoch. It is expected that with the successive training epochs the weight arrays $\mathbf{W}_{\mathbf{k}}^{(1)}$ will concentrate around the different local minima of $E(\mathbf{W}_{\mathbf{k}}^{(1)})$ existing in the considered optimisation task.

The network's linear layer is learned by Widrow-Hoff method to minimise the error [Kohonen (2001); Osowski (2006)]:

$$E(\mathbf{W}^{(2)}) = \frac{1}{Q} \sum_{q=1}^{Q} \| \boldsymbol{\delta}(q) \|^2, \qquad (18.29)$$

$$\boldsymbol{\delta}(q) = \mathbf{t}(q) - \mathbf{W}^{(2)}\mathbf{y}(q),$$

as a quadratic function of weight matrix $\mathbf{W}^{(2)}$, where $\mathbf{t}(q)$ symbolises the $q$-th vector of image pattern, $\mathbf{y}(q)$ is the $q$-th output vector of the self-organised layer, $\boldsymbol{\delta}(q)$ is the error between $q$-th pattern and output vectors, $q \in \{1, \dots, Q\}, Q$ is the number of data vectors per one epoch. To verify network generalisation ability 5-fold cross-validation has been carried out on the vectorised hue, saturation of leaf blade data $\mathbf{x} = [x_1, x_2]^T$ in several training images with manually labelled patterns. The folding routine implements the *MATLAB* function *crossvalind*, which returns randomly generated indices for a K-fold cross-validation of $Q$ data items. As a result of the cross-validation confusion matrices are computed for the classified data from trained images. The matrices include separate counts of all possible classification results. The confusion matrix based factors selected to estimate classification method quality

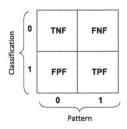

Fig. 18.20  The structure of confusion matrix for classification into two classes, $TNF$ - true negative fraction, $FNF$ - false negative fraction, $FPF$ - false positive fraction, $TPF$ - true positive fraction.

are:

$$PE = \frac{FPF}{TNF + FPF}, \qquad (18.30)$$

$$NE = \frac{FNF}{FNF + TPF}, \qquad (18.31)$$

$$ER = \frac{FNF + FPF}{\sum F}, \qquad (18.32)$$

where $\sum F$ denotes the sum of all fraction counts, $PE$ is the false positive error rate, $NE$ is the false negative error rate, $ER$ is the classifier error rate. The last factor has been used for the classifier validity.

### 18.5.4 *Symptoms of ROS reaction in cucurbits leaves - experimental results of identification*

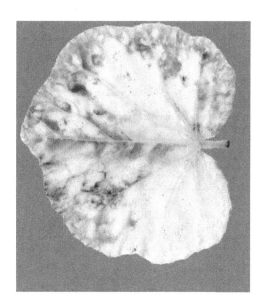 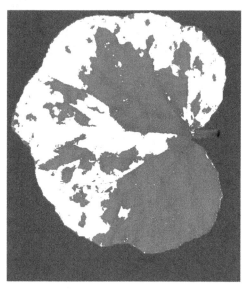

Fig. 18.21 Example of a pumpkin leaf image undergoing segmentation, with blue stained regions of ROS reaction products visible as darker pixels. Image size: 844 × 952 $px$, depth: 24 $bit/px$, leaf blade area: 498473 $px$.

Fig. 18.22 Segmentation result for the image from Fig. 18.21 with the class of ROS stained pixels shown as white regions.

The classifier error rates $ER$ at 5-fold cross-validation for the training images from Fig. 18.21 (Fig. 18.22) and Fig. 18.23 (Fig. 18.24) are visualised in Fig. 18.27. Small values of the errors in the range [0.9, 1.6]% confirm the proper training of the classifier, free of overfitting phenomenon.

The classification errors at the segmentation of 12 example images have been listed

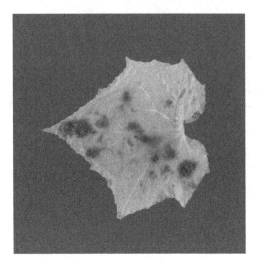 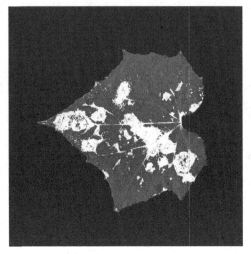

Fig. 18.23   Example of a cucumber leaf image before segmentation with red-brown stained regions of ROS reaction products visible as darker pixels. Image size: 1180 × 1182 *px*, depth: 24 *bit/px*, leaf blade area: 387693 *px*.

Fig. 18.24   Segmentation result for the image from Fig. 18.23 with the separated class of ROS stained pixels shown as white regions.

Table 18.2   The list of classification errors derived from confusion matrix.

| Image number | False Positive Error Rate $PE$ [%] | False Negative Error Rate $NE$ [%] | Classifier Error Rate $ER$ [%] |
|---|---|---|---|
| 1 | 2.39 | 0.01 | 1.36 |
| 2 | 2.25 | 0.01 | 1.77 |
| 3 | 1.79 | 0.18 | 1.59 |
| 4 | 3.04 | 0.19 | 2.77 |
| 5 | 0.08 | 8.64 | 0.24 |
| 6 | 0.75 | 0.03 | 0.56 |
| 7 | 1.22 | 0.00 | 0.95 |
| 8 | 1.45 | 1.83 | 1.52 |
| 9 | 0.10 | 3.47 | 0.97 |
| 10 | 1.60 | 6.22 | 1.94 |
| 11 | 1.20 | 5.30 | 1.39 |
| 12 | 1.87 | 3.78 | 1.92 |

in Table 18.2. The classification was preceded by three network trainings with images representing differently stained leaf groups. Each of the trainings involved the $H, S$ feature data of all leaf blade pixels. Classification errors were evaluated for all these images, whose binary patterns were manually labelled for this purpose (Fig. 18.25 and Fig. 18.26). The computed error rates $ER$ vary from 0.24% to

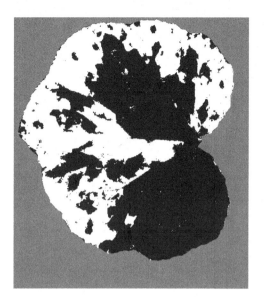

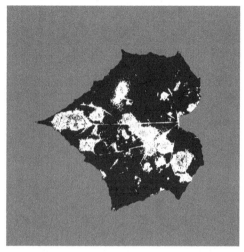

Fig. 18.25 Manually labelled pattern for the image from Fig. 18.21 with the class of ROS stained pixels shown as white regions and a grey image background.

Fig. 18.26 Manually labelled pattern for the image from Fig. 18.23 with the class of ROS stained pixels shown as white regions and a grey image background.

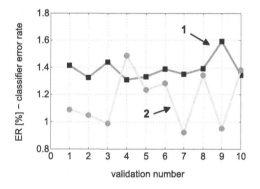

Fig. 18.27 The example plots of the error rates $ER$ computed according to Eq. 18.32 at the 5-fold cross-validation for images in Fig. 18.22 (plot 1) and Fig. 18.24 (plot 2).

2.77% (mean 1.42%), what means good compliance with the patterns. False positive errors (mean 1.49%) are regarded as less important than false negative errors (mean 2.30%), because further studies of dye concentration in the stained areas enable detection of this type errors. It should be emphasised that sometimes the manual identification of stained foliar areas can be ambiguous, what gives different possible patterns for one image. In such cases the pattern variant closest to the automatic classification results was taken into account. The accuracy of final clas-

*J. Sekulska-Nalewajko & J. Gocławski*

Table 18.3 Example learning times for the images from Fig. 18.22 and Fig. 18.24 respectively. The learning times of layer 1 are calculated for $N_k = 10$ initial neuron weight sets.

| Image number (validation) | NN learning time layer 1 [s] | layer 2 [s] |
|---|---|---|
| 1(1) | 29.95 | 0.97 |
| 1(2) | 43.18 | 0.76 |
| 1(3) | 27.63 | 0.61 |
| 1(4) | 33.61 | 0.55 |
| 1(5) | 30.08 | 0.63 |
| 2(1) | 18.90 | 3.94 |
| 2(2) | 17.69 | 2.70 |
| 2(3) | 17.58 | 3.44 |
| 2(4) | 19.84 | 4.23 |
| 2(5) | 15.86 | 4.46 |

sification results has been accepted by the specialists identifying the sites of ROS accumulation in stained leaves.

The execution times of MWTA self-clustering and linear perceptron supervised learning registered at the example cross-validation are listed in Table 18.3 for $M = 8$ MWTA layer neurons and $N_k = 10$ weight vector sets. The average clustering times computed from time data shown in Table 18.3 are about $33s$ and $18s$ respectively for the training images from Fig. 18.21 and Fig. 18.23. The times can be various because of different leaf areas and $(H, S)$ distributions as well as the random starting values of MWTA initial weight vectors. The MWTA training with 10 initial weight sets lasts an average of $4 \div 5$ times longer than the classic WTA learning of the same tested image population. This is the cost of increasing the chance to achieve global minimum of clustering error at the first layer. The MWTA ($N_k = 10$) classification errors shown in Table 18.2 are of the same order as in the case of applying classic WTA ($N_k = 1$) when each of these methods stops at the same weight array $\mathbf{W}^{(1)}$ indicating the global minimum of $E(\mathbf{W}^{(1)})$ given in Eq. (18.28). If the WTA method reaches only a local, but not the global, minimum of $E(\mathbf{W}^{(1)})$ the error rates of classification can be relatively high. It can happen for some types of error functions with different local and global minima, which shapes are explicitly unknown and a single initial weight array in the WTA algorithm is randomly selected. The training of linear perceptron layer with the maximum allowable learning rate is much faster ($0.7s$ and $3.8s$ on average). It should be remembered that the learning process refers only to the small number of training images and the rest of each image population is intended for the classification, typically $10 - 20$ times faster than the learning. The algorithm parameters are discussed in the paper [Gocławski *et al.* (2012)].

### 18.5.5    *Wheat leaf greenness comparative assessment*

To assess greenness of the tested leaves the original image $I_{\mathrm{RGB}}$ with tested leaf object is converted into $HSV$ colour space [Gonzalez and Eddins (2004); The Mathworks Inc. (2012b)] using the function $rgb2hsv(\cdot)$ available in *MATLAB Image Processing Toolbox*. In the successive steps of the proposed algorithm presented in Fig. 18.28 only the hue image component $I_H$ is considered. The authors pro-

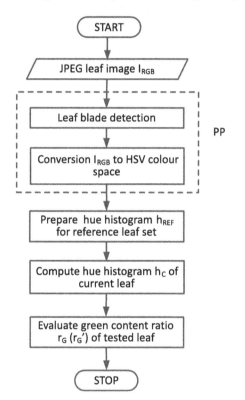

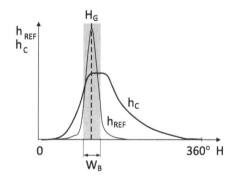

Fig. 18.28   Flow diagram of the algorithm for the proposed leaf greenness assessment method, PP - image preprocessing module.

Fig. 18.29   The hue histograms: $h_C$ - of a tested leaf, $h_{\mathrm{REF}}$ - of a referenced group of leaves. $H_G$ - the hue of green.

pose two approaches (methods) to the greenness evaluation of wheat leaves. Both of them apply a control group of leaves, which carries information of a reference green colour. For this group the normalised colour hue histogram $h_{\mathrm{REF}}$ is computed according to Eq. (18.33), which serves as a reference of green pigment distribution.

$$h_{\mathrm{REF}}(H) = \frac{1}{N_R} \sum_{i=1}^{N_R} \frac{1}{|M_i|} \sum_{p \in M_i} \left( I_H^{(i)}(p) = H \right), \qquad (18.33)$$

where: $p$: the symbol of an image pixel, $M_i$: the set of leaf blade mask pixels in the leaf mask image $I_M^{(i)}$ with saturation values greater than an assumed minimum,

$I_H^{(i)}(p)$: the value of hue at a pixel $p$ in the image $I_H^{(i)}$ with number $i$, $N_R$: the cardinality of reference image set. It is well known from the definition of *HSV* colour components model that for very low saturation values the colour hue is highly uncertain. Therefore all hue histograms are calculated for image pixels p with the saturation value $I_S(p)$ above some minimum value $S_{MIN}$.

Method 1 applies the hue mean value based on the histogram $h_{REF}$, to determine the reference hue $H_G$ of the green colour computed according to Eq. (18.34).

$$H_G = \frac{1}{S_h} \sum_i i \times h_{REF}(i), \tag{18.34}$$

where $S_h$: the sum of $h_{REF}$ histogram bins. Symmetric hue band $B_G = [H_G - W_B/2, H_G + W_B/2]$ of width $W_B$ is then introduced around $H_G$ as a representation of green hues. Therefore greenness ratio can be defined as in Eq. (18.35):

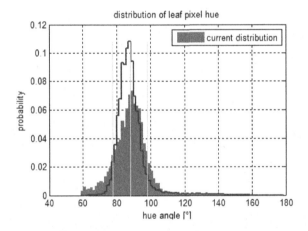

Fig. 18.30  Example histogram of leaf pixel hue distribution; stairstep plot– reference set distribution, bar plot - current leaf; plot for nickel treated (70 *mg/l*) leaf and narrow hue band used; outer vertical lines - predefined narrow band $B_G$ for the purpose of $r_G$ calculation.

$$r_G = \frac{\sum\limits_{H=H_G-W_B/2}^{H_G+W_B/2} h_C(H)}{\sum\limits_{H=H_G-W_B/2}^{H_G+W_B/2} h_{REF}(H)}, \tag{18.35}$$

where $h_C(H)$, $h_{REF}(H)$ hue histograms of current leaf and reference group of leaves respectively. The bandwidth $W_B$ used for computation of $r_G$ should be relatively narrow, e.g. about $20°$ in the full hue range $[0, 360°]$ because the appropriate histogram counts are averaged inside of $W_B$ as stated in Eq. (18.35). This idea of $r_G$ ratio calculation has been depicted in Fig. 18.29. It is assumed that the average of $h_{REF}$ sum located in the denominator of Eq. (18.35) is always greater than a given minimal value $h_{MIN}$.

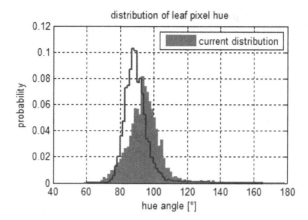

Fig. 18.31  Example histogram of leaf pixel hue distribution; stairstep plot - reference set distribution, bar plot - current leaf; plot for selenium treated (89 $mg/l$) leaf and wide hue band used.

In method 2 the greenness ratio $r'_G$ defines mean conformity between the hue histogram of a reference leaves group and the hue histogram of a currently tested leaf. The hue range taken into account is typically full ($W_B = 360°$) or at least widely spread around $H_G$ in comparison with the method 1. The proposed hue similarity measure between a current leaf and the reference leaf group is given in Eq. (18.36). It includes the proportions $f_H$ of minimum to maximum counts for every bin $H$ of the normalised hue histograms $h_{REF}$ and $h_C$. Additionally the ratios $f_H$ can be provided with different weights $w_H$ incorporating different importance of bin count dissimilarities closer to the reference hue $H_G$ (Eq. (18.37)). Currently all $w_H$ are assumed equal to $1/N_B$, where $N_B$ means the number of bins pairs with maximum count in each pair greater than a certain value $h_{MIN}$.

$$r'_G = \sum_{H \in B_G} w_H \, f_H, \tag{18.36}$$

$$\sum_{H \in B_G} w_H = 1, \quad f_H = \frac{\min\left(h_C(H), h_{REF}(H)\right)}{\max\left(h_C(H), h_{REF}(H)\right)} \tag{18.37}$$

Figure 18.30 and Fig. 18.31 show example histograms of leaf hue distribution for method 1(narrow band) and method 2(wide band) respectively. They can be visualised in *MATLAB* application implementing the algorithm.

The algorithm for greenness assessment has been applied to the four groups of leaves including one reference group and three other groups treated with different combination of heavy metals. The numerical results are included in [Sekulska-Nalewajko and Gocławski (2012)]. Both the reference group of leaves and other groups have been represented by a set of 5 images each containing single leaf object. Due to the activity of nickel or selenium within the period of wheat cultivation chlorosis (spots or scratches with yellow-green or washy green colour) appeared in the leaf blades. The algorithm confirms observed colour changes by the reduction averaged green assessment ratios in plants treated with nickel down to the value of $r_G \approx 0.87$ (ver. 1) and $r'_G \approx 0.51$ (ver. 2). The treatment of wheat seedlings with the combination of both nickel and selenium at the same time changes the $r_G$ ratio to above 0.89 and 0.61 for each method respectively and therefore seems to have less impact on the loss of chlorophyll. The expected correlation between computed $r_G$ ratios and associated heavy metal concentrations is unfortunately unnoticed. The reason for this may be to small range of these concentrations. The results obtained by the proposed method should be compared with the measurements of chlorophyll content provided by biochemical methods applied to the same plant groups.

# References

Bade, C. I. A. and Carmona, M. A. (2011). Comparison of methods to assess severity of common rust caused by _Puccinia sorghi_ in maize, _Tropical Plant Pathology_ **36**, 42, pp. 264–266.

Bernsen, J. (1986). Dynamic thresholding of gray-level images, in _International Conference on Pattern Recognition_ (Paris), pp. 1251–1255.

Bezdek, J. C. (1981). _Pattern Recognition with Fuzzy Objective Function Algorithms_ (Plenum Press, New York, NY).

Bolwell, G. and Daudi, A. (2009). Reactive oxygen species in plantpathogen interactions, in _Reactive Oxygen Species in Plant Signaling_ (Springer-Verlag, Berlin, Heidelberg), pp. 113–133.

Cartes, P., Jara, A. A., Pinilla, L., Rosas, A. and Mora, M. L. (2010). Selenium improves the antioxidant ability against aluminium-induced oxidative stress in ryegrass roots, _Annals of Applied Biology_ **156**, pp. 297–307.

Chaerle, L., Hagenbeek, D., De Bruyne, E., R., V. and Van Der Straeten, D. (2004). Thermal and chlorophyll-fluorescence imaging distinguish plant-pathogen interactions at an early stage, _Plant and Cell Physiology_ **45**, 7, pp. 887–896.

Chaerle, L., Van Caeneghem, W., Messens, E., Lambers, H., Van Montagu, M. and Van Der Straeten, D. (1999). Presymptomatic visualization of plant-virus interactions by thermography, _Nature Biotechnology_ **17**, 8, pp. 813–816.

Chaerle, L. and Van Der Straeten, D. (2000). Imaging techniques and the early detection of plant stress, _Trends in Plant Science_ **5**, 11, pp. 495–501.

Changho, L., Seung-Yeol, L., Hee-Young, J. and Jeehyun, K. (2012). The application of optical coherence tomography in the diagnosis of marssonina blotch in apple leaves, _Journal of the Optical Society of Korea_ **16**, 2, pp. 133–140.

Daughtry, C. S., Walthal, l. C. L., Kim, E., M. S.and Brown de Colstoun and McMurtrey, J. E. (2000). Estimating corn leaf chlorophyll concentration from leaf and canopy reflectance, _Remote Sensing of Environment_ **74**, pp. 229–239, http://naldc.nal.usda.gov/download/26536/PDF.

Du Buf, H. and Bayer, M. (eds.) (2002). _Automatic Diatom Identification_ (World Scientific Publishing, New York/London/Singapore/Hong Kong).

Dubey, D. and Pandey, A. (2011). Effect of nickel (Ni) on chlorophyll, lipid peroxidation and antioxidation enzymes activities in black gram (_Vigna mungo_) leaves. _International Journal of Science and Nature_ **2**, 2, pp. 395–401, http://scienceandnature.org/IJSN_Vol2(2)J2011/IJSN-VOL2(2)-45.pdf.

Glasbey, C. A. and Horgan, G. W. (1995). _Image Analysis for the Biological Sciences_ (John Wiley & Sons, New York, NY).

Gocławski, J., Sekulska-Nalewajko, J., Gajewska, E. and Wielanek, M. (2009). An automatic segmentation method for scanned images of wheat root systems with dark discolourations, *International Journal on Applied Mathematics and Computer Science* **19**, 4, pp. 679–689.

Gocławski, J., Sekulska-Nalewajko, J. and Kuźniak, E. (2012). Neural network segmentation of images from stained cucurbits leaves with colour symptoms of biotic and abiotic stresses, *International Journal on Applied Mathematics and Computer Science* **22**, 3, pp. 669–684.

Gonzalez, R. E., E. R.and Woods (2008). *Digital Image Processing* (Prentice Hall, Upper Saddle River, NJ).

Gonzalez, R. E., E. R.and Woods and Eddins, S. L. (2004). *Digital Image Processing Using MATLAB* (Prentice Hall, Upper Saddle River, NJ).

Goodwin, P. H. and Hsiang, T. (2010). Quantification of fungal infection of leaves with digital images and Scion Image software, *Methods in Molecular Biology* **638**, pp. 125–135.

Hagan, M. T., Demuth, H. B. and Beale, M. H. (2009). *Neural Network Design* (University of Colorado, Denver, CO), http://www.personeel.unimaas.nl/westra/ Education/ANO/10widrow_hoff.pdf.

Hecht-Nielsen, R. (1987). Counterpropagation networks, *Applied Optics* **26**, 23, pp. 4979–4984.

Horvath, J. (2006). Image segmentation using fuzzy c-means, http://bmf.hu/ conferences/sami2006/JurajHorvath.pdf.

Huang, C. S., Liu, J. H. and Chen, X. J. (2010a). Overexpression of PtrABF gene, a bZIP transcription factor isolated from *Poncirus trifoliata*, enhances dehydration and drought tolerance in tobacco via scavenging ROS and modulating expression of stress-responsive genes, *BMC Plant Biology* **10**, 230, http://www.biomedcentral. com/1471-2229/10/230.

Huang, M., Abel, C., Sohrabi, R., Petri, J., Haupt, I., Cosimano, J., Gershenzon, J. and Tholl, D. (2010b). Variation of herbivore-induced volatile terpenes among *Arabidopsis* ecotypes depends on allelic differences and subcellular targeting of two terpene synthases, TPS02 and TPS03, *Plant Physiology* **153**, pp. 1293–1310.

Jung, V., Olsson, E., Caspersen, S., Asp, H., Jensen, P. and Alsanius, B. W. (2004). Response of young hydroponically grown tomato plants to phenolic acids, *Scientia Horticulturae* **100**, (1-4), pp. 23–37.

Kohonen, T. (1997). The self-organising map, *Proceedings of IEEE* **78**, 9, pp. 1464–1479.

Kohonen, T. (2001). *Self-organizing Maps*, 3rd edn. (Springer-Verlag, Berlin/ Heidelberg/New York).

Kwack, M. S., Kim, E. N., Lee, H., Kim, J.-W., Chun, S.-C. and D., K. K. (2005). Digital image analysis to measure lesion area of cucumber anthracnose by *Colletotrichum orbiculare*, *Journal of General Plant Pathology* **71**, pp. 418–421.

Lam, L., Lee, S. and Suen, C. Y. (1992). Thinning methodologies –a comprehensive survey, *IEEE Transactions on Pattern Analysis and Machine Intelligence* **14**, 9, pp. 869–885.

Lambert, P. and Carron, T. (1999). Symbolic fusion of luminance-hue-chroma features for region segmentation, *Pattern Recognition* **32**, 11, pp. 1857–1872.

Loreto, F. and J., S. (2010). Abiotic stresses and induced BVOCs, *Trends in Plant Sciences* **15**, 3, pp. 154–166.

Mahajan, S. and Tuteja, N. (2005). Cold, salinity and drought stresses: An overview, *Archives of Biochemistry and Biophysics* **144**, pp. 139–158.

Masters, T. (1993). *Practical Neural Network Recipes in C++* (Academic Press Inc., San Diego, CA).

Mohamed, A. N., Ahmed, M. N. and Farag, A. (1999). Modified fuzzy c-means in medical image segmentation, in *Proceeding of IEEE International Conference on Acoustics, Speech, and Signal Processing*, Vol. 6 (Phoenix, AZ, USA), pp. 3429–3432.

Ong, S., Yeo, N., Lee, K., Venkatesh, Y. and Cao, D. (2002). Segmentation of color images using a two-stage selforganizing network, *Image and Vision Computing* **20**, 4, pp. 279–289.

Osowski, S. (2006). *Neural Networks for Information Processing* (Warsaw University of Technology Press, Warsaw).

Otsu, N. (1979). A threshold selection method from grey-level histograms, *IEEE Transactions on Systems, Man, and Cybernetics* **9**, 1, pp. 62–66.

Pitzschke, A., Fornazi, C. and Hirt, H. (2005). Reactive oxygen species signalling in plants, *Antioxidants and Redox Signaling* **8**, pp. 1757–1764.

Regent Instruments Inc. (2008). *WinRHIZO For Root Morphology And Architecture Measurement*, http://www.regent.qc.ca/products/rhizo/Rhizo.html.

Sannakki, S. S., Rajpurohit, V. S., Nargund, V. B., Arun Kumar, R. and P.S., Y. (2011). A hybrid intelligent system for automated pomegranate disease detection and grading, *Interantional Journal of machine Intelligence* **3**, 2, pp. 36–44.

Sekulska-Nalewajko, J. and Gocławski, J. (2012). An image analysis method for the evaluation of green pigment reduction in wheat leaves under the influence of heavy metals, in *IEEE Proceedings of the VIII-rd International Conference MEMSTECH* (Lviv-Polyana, Ukraine), pp. 000–000.

Seregin, I. V. and Kozhevnikova, A. D. (2009). Enhancement of nickel and lead accumulation and their toxic growth inhibitory effects amaranth seedlings in the presence of calcium, *Russian Journal of Plant Physiology* **56**, 1, pp. 80–84.

Smit, L. A., Bengough, A. G., Engels, C., Van Noordwijk, M., Pellerin, S. and Van de Geijn, S. C. (2000). *Root Methods: A Handbook* (Springer Verlag, Heidelberg).

Smith, A. (1978). Color gamut transform pairs, *Computer Vision and Image Understanding* **12**, 3, pp. 12–19.

Soukupova, J. and Albrechtova, J. (2003). Image analysis – Tool for quantification of histochemical detection of phenolic compounds, lignin and peroxidases in needles of Norway spruce, *Biologia Plantarum* **46**, 4, pp. 595–601.

Tabrizi, P. R., Rezatofighi, H., S. and Yazdanpanah, M. J. (2010). Using PCA and LVQ neural network for automatic recognition of five types of white blood cells, in *Engineering in Medicine and Biology Society* (Annual International Conference of the IEEE, Teheran, Iran,), pp. 5593–5596.

Tang, Q. H., Liu, B. H., Chen, Y. Q., Zhou, X. H. and Ding, J. S. (2007). Application of LVQ neural network combined with the genetic algorithm in acoustic seafloor classification, *Chinese Journal Geophysics* **50**, 1, pp. 291–298.

Terry, N., Zayed, A. M., de Souza, M. and Tarun, A. (2000). Selenium in higher plants, *Annual Review of Plant Physiology and Plant Molecular Biology* **51**, pp. 401–432, http://www.plantstress.com/articles/toxicity_i/selenium.pdf.

The Mathworks Inc. (2012a). *Fuzzy logic toolbox user's guide*, http://www.mathworks.com/help/fuzzy/.

The Mathworks Inc. (2012b). *Image processing toolbox user's guide*, http://www.mathworks.com//help/toolbox/images/.

Thordal-Christensen, H., Zhang, Z., Wei, Y. and Collinge, D. (1997). Subcellular localization of $H_2O_2$ in plants. $H_2O_2$ accumulation in papillae and hypersensitive response during the barley-powdery mildew interaction, *Plant Journal* **11**, 6, pp. 1187–1194.

Unger, C., Kleta, S., Jandl, G. and Tiedemann, A. (2005). Suppression of the defence-related oxidative burst in bean leaf tissue and bean suspension cells by the necrotrophic pathogen *Botrytus cinerea*, *Journal of Phytopathology* **153**, 1, pp. 15–26.

Vincent, L. and Soille, P. (1991). Watersheds in digital spaces: An efficient algorithm based on immersion simulations, *IEEE Transactions on Systems, Man, and Cybernetics* **13**, 6, pp. 583–598.

Weaver, J. E. (1926). *Root Development of Field Crops* (McGraw-Hill Book Company, New York, NY/London).

Widrow, B., Winter, R. G. and Baxter, R. A. (1988). Layered neural nets for pattern recognition, *IEEE Transactions on Acoustics, Speech and Signal Processing* **36**, 7, pp. 1109–1118.

Wijekoon, C. P., Goodwin, P. H. and Hsiang, T. (2008). Quantifying fungal infection of plant leaves by digital image analysis using Scion Image software, *Journal of Microbiological Methods* **74**, (2-3), pp. 94–101.

Yao, X., Chu, J. and Wang, G. (2009). Effects of selenium on wheat seedlings under drought stress, *Biological Trace Element Research* **130**, 3, pp. 283–290, `http://naldc.nal.usda.gov/download/26536/PDF`.

# CHAPTER 19

# VARIOUS APPROACHES TO PROCESSING AND ANALYSIS OF IMAGES OBTAINED FROM IMMUNOENZYMATIC VISUALIZATION OF SECRETORY ACTIVITY WITH ELISPOT METHOD

Wojciech Bieniecki and Szymon Grabowski

*Institute of Applied Computer Science*
*Lodz University of Technology*
*90-924 Łódź, ul. Stefanowskiego 18/22*
*{wbieniec, sgrabow}@kis.p.lodz.pl*

## 19.1 Introduction

Enzyme linked immunospot assay (ELISPOT) (Gebauer et al., 2002; Hricik et al., 2003) is a powerful technique used as a non-invasive diagnostic tool for the prediction of long term renal allograft function and the early detection of chronic graft rejection process markers (Koscielska-Kasprzak et al., 2009). It focuses on the detection and quantification of antigen specific immunological responses at a single cell level. The ELISPOT experiment (Fig. 19.1) is performed in a membrane bottomed 96-well plates with each well bottom (6 mm diameter) coated with antibodies specific against each of the cytokines analyzed.

During the experiment the suspension with a known number of recipient lymphocytes is placed in the wells. To selected wells inactivated donor splenocytes are added. Cytokines secreted by activated recipient lymphocytes are captured in the immediate vicinity of the cells by the specific antibody bound to the membrane. After the removal of the cells, the presence of the exuded cytokine is detected with the use of the enzyme immunoassay reaction. As a result of the enzyme-catalyzed conversion of chromogen to an insoluble stain, spots, seen in the image, are formed in places prior to test the presence of cytokine-secreting cells.

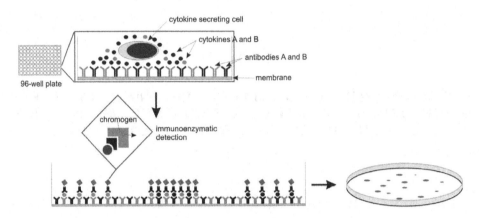

Fig. 19.1  ELISPOT procedure.

ELISPOT assay can be used for the simultaneous analysis of two different cytokine secretions, resulting in the two-color images consisting of spots of the following colors:

- purple-blue, as a result of the presence of the reaction of the substrate BCIP/NBT catalyzed by alkaline phosphatase;
- bricky-red derived from the oxidation of the AEC substrate by peroxidase in the presence of $H_2O_2$;
- transient color.

The procedures for preparing cells for two-color ELISPOT test did not differ from those for a standard single color tests. Special optimization requires only an examination, which is based largely on commercial sets of antibodies and reagents.

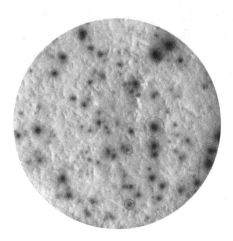

Fig. 19.2  ELISPOT image.

## 19.2 Characteristics of the images

Viewed under a microscope, an image was photographed with one of the two available acquisition devices:

- a digital camera Nikon E4500 giving a resolution of 2272×1704 pixels and JPEG image format;
- a motion digital camera Nikon DS5M-U1 giving a resolution of 2560×1920 pixels for BMP and JPEG formats.

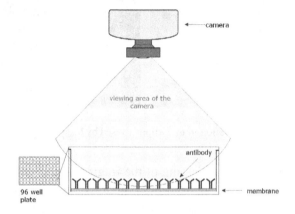

Fig. 19.3 ELISPOT image acquisition.

For both devices, acquiring the images entails some inconvenience that may have a significant effect on the correct operation of the image processing algorithm.

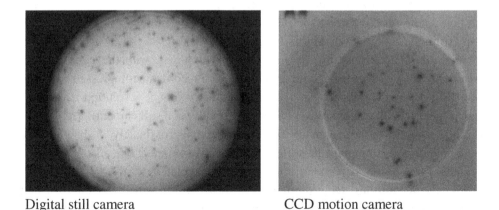

Digital still camera          CCD motion camera

Fig. 19.4 The ELISPOT acquired with two different devices.

A still camera allows recording only in JPG format, which, despite a fairly high-resolution, introduces errors due to lossy compression.

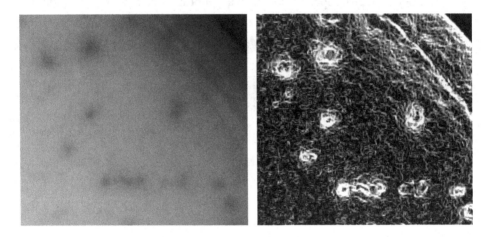

Fig. 19.5 Visualization of distortions caused by lossy compression.

In Fig. 19.5 the result of processing of the fragment of an image with an egde filter is depicted. Consequences of JPEG compression are visible as horizontal and vertical brighter lines. This distorts the process of detecting spots' shapes and boundaries, which affects the accuracy of measurement of their size.

Another disadvantage is the unequal exposure of the image because the device was not designed for this type of lighting. This defect may have an adverse effect on the proper identification of dichroic colors in images.

The image has a relatively good resolution thanks to the effective focal length of the lens control. However, the depth of field is very poor, which means that it is not possible to set the correct focus around the area of interest. Fragments further away from the center are darker and slightly blurred.

A motion camera captures an image evenly exposed, with the correct white balance, uniform sharpness and allows it to record uncompressed images. However, due to the optical system used, an effective image resolution is twice as low as that of a still camera (3600 DPI vs 7200 DPI for a still camera). This means that the spot of a diameter of 0.03 mm has approximately 5 pixels. The segmentation algorithm must be sufficiently sensitive to detect objects of this size.

Other observable defects of the images are related to the production of ELISPOT preparations. Blue pigments can produce a much better contrast and color saturation than the brown pigment which hinders the simultaneous detection of both stainings.

## 19.3 Challenges of the image processing algorithm

The image processing algorithm should:
-   detect the bounds of the bottom of the well as a circle which will be the region of interest in the image. Knowing that its diameter is 6 mm it is possible to obtain the scale

If we define a color image I as a weighted graph:

$$I = (D, E, f) \text{ where } D : [1,...,M] \times [1,...,N] \subset N \times N \tag{19.1}$$

where E defines the neighbourhood of its points and $f : D \rightarrow N \times N \times N$, the region of interest is defined as:

$$ROI = \left\{ (x,y) : (x,y) \in I \wedge (x_0 - x)^2 + (y_0 - y)^2 \leq R^2 \right\} \tag{19.2}$$

-   distinguish from the image the spots of types A, B and C (three colors) as connected components, and the background

$$Bg, I_{A1}, I_{A2}...., I_{sA}, I_{B1}, I_{B2}...., I_{sB}, I_{C1}, I_{C2}...., I_{sC}$$

$$I = \bigcup_{k=1}^{s} I_k \cup Bg \tag{19.3}$$

$$\forall i, j \in \{1,...,s\}, i \neq j \Rightarrow I_i \cap I_j = \varnothing$$

-   the spots are approximately round (AR>0.4) where AR is defined as follows:

$$AR_I = \frac{\min\limits_{p \in I \setminus \text{int}(I)} \left\{ \rho(C_I, p) \right\}}{\max\limits_{p \in I \setminus \text{int}(I)} \left\{ \rho(C_I, p) \right\}} ; \ 0 < AR_I \leq 1 ; \tag{19.4}$$

C – "gravity center" of the component

-   the area of the individual spot is in the range [0.03 mm² – 0.1 mm²].

The segmentation may be hampered by the fact that the spots have fuzzy edges. Below (Fig. 19.6) are graphs of intensity distribution of the dye in the cross-section of the spot.

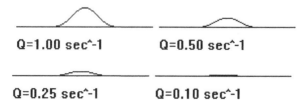

Q=1.00 sec^-1          Q=0.50 sec^-1

Q=0.25 sec^-1          Q=0.10 sec^-1

Fig. 19.6 Exemplary sizes of the spots and distributions of dye intensity.

The subject of image analysis, and more specifically – quantitative analysis – is to obtain the area distribution of individual spots in the visual field and present the result in the form of histograms, separately for spots A, B and C.

The area of the spot is an approximation of quantity of the secreted dye, but not the only one. To improve the accuracy, an additional measure, called the spot weight was introduced. Below is the definition for type A.

Let $\{I_{A1}, I_{A2},..., I_{Ak}\}$ be the set of all segmented A-type spots. Denote by $l$ the dye intensity of the image pixel. For monochrome images, it will be an inverse of the luminance, and for color images e.g. the color saturation. Let us denote $L_{\min} = \min_{(x,y)\in \cup I_i} \{l(x,y)\}$ and $L_{\max} = \max_{(x,y)\in \cup I_i} \{l(x,y)\}$ as extreme values of the intensity. Then the spot weight is defined by:

$$W_k := \sum_{(x,y)\in I_k} \frac{L_{\max} - l(x,y)}{L_{\max} - L_{\min}} = \frac{n_k \cdot L_{\max} - \sum l(x,y)}{L_{\max} - L_{\min}} \tag{19.5}$$

The equation allows one to scale the value in this way, which would be comparable to the area of the spot (W=S in the case of even intensity and perfectly sharp edges) (Fig. 19.7).

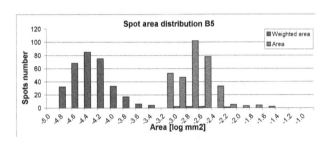

Fig. 19.7 Spot area vs. weighted area distributions for a sample image.

## 19.4 Preprocessing algorithms

### 19.4.1. *White balance correction*

This operation is very important in the case of digital images. Color space transformation is linear, so there is no loss of image information. Let $I = (I_R, I_G, I_B)$ be an input color image. Denote:

$$\bar{R} = \{\bar{p} : p \in I_R\}, \quad \bar{G} = \{\bar{p} : p \in I_G\}, \quad \bar{B} = \{\bar{p} : p \in I_B\} \tag{19.6}$$

$$MaxR = \max\{p \in I_R\}, MaxG = \max\{p \in I_G\}, MaxB = \max\{p \in I_B\}$$

$$MinR = \min\{p \in I_R\}, MinG = \min\{p \in I_G\}, MinB = \min\{p \in I_B\}$$

$$MaxI = \max\{MaxR, MaxG, MaxB\}, MinI = \min\{MinR, MinG, MinB\},$$

We have the following cases. If
$\bar{R} = \min\{\bar{R}, \bar{G}, \bar{B}\}$ then

$$WR = \frac{256}{1 + MaxR - MinI}; \quad WG = \frac{256 \cdot \bar{R}}{(1 + MaxR - MinI) \cdot \bar{G}};$$
$$WB = \frac{256 \cdot \bar{R}}{(1 + MaxR - MinI) \cdot \bar{B}}. \tag{19.7}$$

If $\bar{G} = \min\{\bar{R}, \bar{G}, \bar{B}\}$ then

$$WR = \frac{256 \cdot \bar{G}}{(1 + MaxG - MinI) \cdot \bar{R}}; \quad WG = \frac{256}{1 + MaxG - MinI};$$
$$WB = \frac{256 \cdot \bar{G}}{(1 + MaxG - MinI) \cdot \bar{B}}. \tag{19.8}$$

If $\bar{B} = \min\{\bar{R}, \bar{G}, \bar{B}\}$ then

$$WR = \frac{256 \cdot \bar{B}}{(1 + MaxB - MinI) \cdot \bar{R}}; \quad WG = \frac{256 \cdot \bar{B}}{(1 + MaxB - MinI) \cdot \bar{G}};$$
$$WB = \frac{256}{1 + MaxB - MinI}. \tag{19.9}$$

Otherwise (if none of the listed cases applies), all the mean values are equal:

$$WR = WG = WB \frac{256}{(1 + MaxI - MinI)} \tag{19.10}$$

The output image $I' = (I'_R, I'_G, I'_B)$ is generated as follows:

$$I'_R = \{p' = WR \cdot p - MinI; p \in I_R\}; \quad I'_G = \{p' = WG \cdot p - MinI; p \in I_G\};$$
$$I'_B = \{p' = WB \cdot p - MinI; p \in I_B\} \tag{19.11}$$

### 19.4.2 *Red color correction*

The aim of the operation is the correction of the intensity of red spots, because its intensity is worse than in the case of blue spots. The input parameters will be

*W. Bieniecki & S. Grabowski*

reference hue of red $H_0$, the tolereance of the hue $T_H$ and the gain amplitude $A$. The steps are as follows:

Convert the image to HSV color space:

$$U:(I_R,I_G,I_B)\rightarrow(I_H,I_S,I_V)$$
(19.11)

Select pixels which are not background pixels, which means:

$$I_T=\{p\in(I_H,I_S,I_V):S>0.28\wedge V<0.8\}$$
(19.12)

For each pixel $p$ denote:

$$dH=\begin{cases}||H-H_0|| & if\quad |H-H_0|<180\\360-|H-H_0| & otherwise\end{cases}$$
(19.13)

if $dH<T_H$ then

$$V'=V\cdot\left(1-\exp\left(-\frac{dH^2}{10\cdot T_H}\right)\cdot\frac{A}{10}\right)$$
(19.14)

Convert the image back to RGB. For the parameters set experimentally $H_0 = 5$, $T_H = 30$ and $A = 3$ the algorithm corrects the red spots. It will not affect the hue of any pixel of the image.

### 19.4.3. *Uneven exposure correction*

The algorithm requires that the round region of interest (2) is yet specified. The steps of the algorithm:

Paint the rest of the image black

$$I_1=\{p:p\in ROI\cup(0,0,0):p\in I\setminus ROI\}$$
(19.15)

For the obtained image do the averaging (a convolution filter using square mask $M$ with all 1s):

$$I_2=I_1*M\ .$$
(19.16)

Make the subtraction of the images:

$$I_3=I_2-I_1.$$
(19.17)

Paint the region outside the ROI white:

$$I_4=\{p:p\in ROI\cup(255,255,255):p\in I_3\setminus ROI\}.$$
(19.18)

The size of the square mask *M* is set experimentally to $0.25 \cdot R$ (one fourth of the ROI radius).

## 19.5 Segmentation of monochrome and color images using color thresholding

The simplest method of segmentation of ELISPOT images is the selection of image pixels that belong to spots from all the image pixels. This type of segmentation is called pixel classification. The classification may be carried out by comparison of its color, intensity, brightness or any other property to a threshold value τ. (Fig. 19.8). For instance, after thresholding the gray level image is converted to a binary one. There exist algorithms that use more than one threshold values (multithresholding), which enables one to assign pixels to one of a few classes instead of only two. Threshold value(s) may be entered manually or automatically. A thorough survey of automated image thresholding selection can be found in (Mehmet and Bulent, 2004). For the purpose of our research we selected three methods for the automatic selection of τ value for ELISPOT images. Originally intended for grayscale images, the method was adapted for color images.

The first of them is Bernsen's method (Bersen, 1986), while two others are based on the analysis of a pixel color histogram. Bernsen's method was applied with only slight modifications, while histogram-based methods had to be reimplemented.

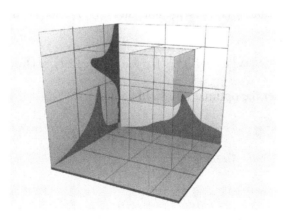

Fig. 19.8 Thresholding in RGB color space.

### 19.5.1 *Bernsen algorithm*

In the Bernsen algorithm each pixel $(i, j)$ is considered with its surrounding (usually square) window W

$$W_{ij} = \{(x, y) \in I : |y - i| < b \wedge |x - j| < b \qquad (19.19)$$

As a local threshold the mean value of the maximum and the minimum pixel intensity within W is taken:

$$T(i, j) = 0.5[\max{}_w (I(i + m, j + n)) + \min{}_w (I(i + m, j + n))]$$
$$= 0.5[I_{high}(i, j) + I_{low}(i, j)] \qquad (19.20)$$

Additionally, the contrast for the search window for pixel $(i, j)$ is defined:

$$C(i, j) = I_{high}(i, j) - I_{low}(i, j) \qquad (19.21)$$

If $C(i, j)$ is not sufficient (the experimentally set threshold value – usually for images of global contrast 255, it is set as 15) the threshold value from (20) is replaced by a global value $\tau_0$. This value must be set a priori. Although it may be set manually, we decided to obtain it by simple image statistics (SIS) (Kittler et al., 1985), the algorithm for adaptive threshold selection.

Let us assume that the perfect image presents objects with intensity a over the background with intensity b, and because of some light and material non-uniformity, the intensity values are to some extent distorted – the noise is introduced to image pixels. Despite this, the best threshold that discriminates the objects from the background is $\tau = (a + b)/2$.

For each image point $p(i, j)$ with intensity $l(i, j)$, let us define its gradient module:

$$e(i, j) = |\nabla p| = \max\{|l(i - 1, j) - l(i + 1, j)|, |l(i, j - 1) - l(i, j + 1)|\} \qquad (19.22)$$

Then we may set the optimal threshold value as

$$\tau_0 = \frac{\displaystyle\sum_{i=1}^{m}\sum_{i=1}^{n} |l(i, j) \cdot e(i, j)|}{\displaystyle\sum_{i=1}^{m}\sum_{i=1}^{n} |e(i, j)|} \qquad (19.23)$$

The Bernsen algortithm has some drawbacks. If the image contains some impulse noise (Fig. 19.9) (isolated pixels of a different color) applying formula (19.20) can give unexpected results.

Fig. 19.9 "Hot" pixel.

In such a case the estimated contrast may be rather high and the region uder the moving window is supposed to be nonhomogeneous. In fact, all the pixels of this fragment of the image belong to the background. But the Bernsen algorithm considers this bright pixel as a background, and its neighborhood as an object. To avoid this phenomenon, the standard deviation was used as a measure of color homogeneity.

$$\sigma = \sqrt{\frac{\sum_{i=1}^{N}(l(i,j)-\mu)^2}{N}} \quad (19.24)$$

Regarding the manner of local threshold selection, Bernsen's method has another significant drawback. The window $W$ size should fit the size of expected objects (shapes of fairly uniform intensity) in order to perform the correct segmentation; thus it must be set experimentally before running the actual Bernsen routine. If the image contains objects of a wide size range or the object scale is unknown, the segmentation fails.

With too large a window the method works slowly and tends to misclassify small objects with low contrast to the background, especially when they are located near bigger objects with higher contrast. The small window range can cause incorrect border detection of large objects and incorrect identification of large objects.

To eliminate this phenomenon in our research, we introduced a multipass version. In each iteration, the size of W is increased so that both small and larger spots are detected.

### 19.5.2. *Histogram analysis*

Both examined histogram-based methods that have been tested against ELISPOT images analyze the extreme values of a brightness distribution function, which are maxima (peaks) and minima (valleys).

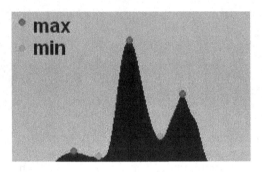

Fig. 19.10  Maxima and minima in the brightness distribution function.

The algorithm searches for a sequence of three points: maximum - minimum - maximum. The location of such groups suggests that there are two classes of objects with different brightness and gives a hint how to set a threshold value, for instance:

$$\tau = \frac{\max_1 + \max_2}{2} \tag{19.25}$$

The first of these methods is called "Peaks & Valleys (valley)" because the value of the threshold is set as a point where a local minimum is found. In the second method "Peaks & Valleys (half)", the threshold is set according to formula (25)

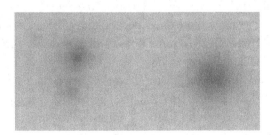

Fig. 19.11  An examined fragment of the image.

In Fig. 19.11 a fragment of an exemplary ELISPOT image is presented and in Fig. 19.12 the histograms of its brightness.

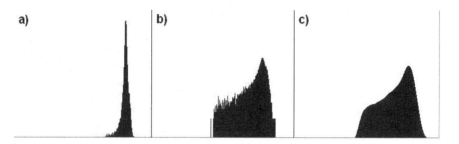

Fig. 19.12 Histograms of the image a) normal b) logarithmic c) smoothed logarithmic.

For this fragment of the image none of peaks&vallys variant is effective.

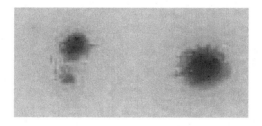

Fig. 19.13 The image after initial filtering (uneven exposure correction algorithm).

A solution may be filtering the image using algorithms from section 4.

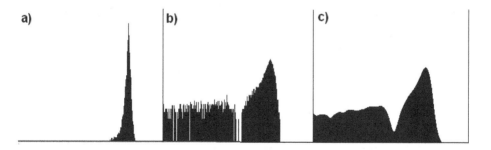

Fig. 19.14 Histograms of the filtered image a) normal b) logarithmic c) smoothed logarithmic.

### 19.5.3. *Tests of the algorithms*

The results were developed for all three methods and for three images. The images test1, test2 and test3 are filtered with an uneven exposure correction filter.

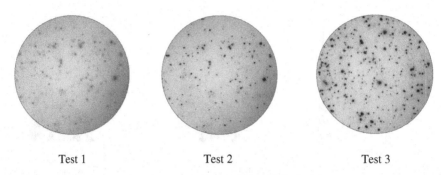

Test 1                        Test 2                        Test 3

Fig. 19.15  Test images (2272 x 1704 pixels).

For all three algorithms the window size was set to 15 pixels, 75 pixels and a two-pass version was run.

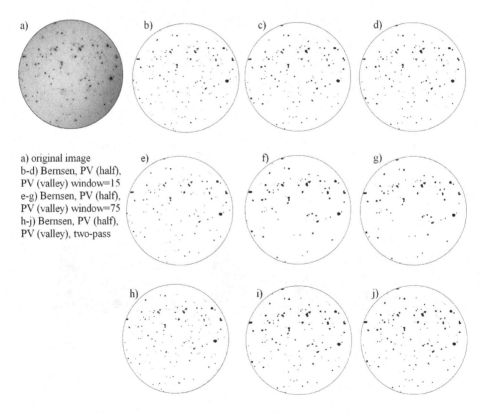

a) original image
b-d) Bernsen, PV (half),
PV (valley) window=15
e-g) Bernsen, PV (half),
PV (valley) window=75
h-j) Bernsen, PV (half),
PV (valley), two-pass

Fig. 19.16  test2 image segmented with different algorithms.

All three methods (if appropriate parameters have been set) give very similar and satisfactory results. The resulting images helped to see that the use of a small window (15x15) is suitable for small spots, while the bigger ones are of poor quality.The use of a larger window (75x75) resulted in a significant improvement of this deficiency; however, this window coped worse with the detection of very small objects. A trade-off is to carry out two runs with two windows. Unfortunately, this operation is time-consuming.

## 19.6 Segmentation of color images using color clustering

Clustering (Hanson and Riseman, 1978) is a technique which requires defining a feature space and mapping the image pixels to vectors in this space. With the use of statistical methods, the feature space is split into connected regions (clusters), which implies grouping the pixels of similar color to a number of classes.

One of the most commonly used clustering algorithms is $k$-means (MacQueen, 1967). The most important parameter for this algorithm is $k$, the number of clusters in a feature space (here: color space) to be found. A fixed value of $k$ may be a drawback of the algorithm if the predicted number of classes is unknown. The algorithm must start from initializing $k$ centers of the clusters – the initial values may be set manually or chosen randomly. In each iteration, each point is attracted to the nearest centroid and the centroids are updated (as a mean value of all points belonging to one class). The algorithm presented below uses clustering and thresholding techniques.

The algorithm operates in four phases: initialization (Fig. 19.17), teaching, grouping and finalization.

---

1. Get segmentation params and the coordinates of ROI

2. Initialize the class matrix for image pixels

3. Load the image

4. Set the ROI

---

Fig. 19.17 The initialization phase.

An image to be processed is divided into square regions using specified masks (Fig. 19.18).

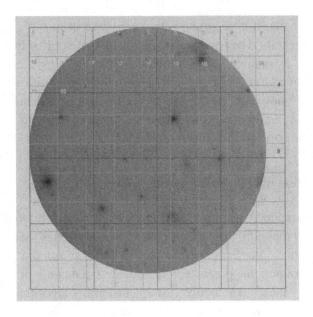

Fig. 19.18 The masks for segmentation algorithm.

This helps to cope with uneven lighting and varying focal conditions in the entire image. The mask for teaching phase is bigger. Teaching and grouping phases work separately on the square regions.

The following parameters are set by the user at the start:

- the number of classes, diameters of scanning masks (separately for teaching and grouping phases);
- the gray-level thresholds (the level of homogeneity which controls the sensitivity of the algorithm and the level of spot which affects the spot sharpness);
- the method of initialization of centroids;
- optionally – post segmentation median filtering.

In this stage, a class matrix for the image pixels is allocated.

In the next – teaching phase – the gravity centers for each class are initialized and pixels are initially classified. Those centers are generated on the basis of one of three methods: global gray-level, global feature level, local feature level.

The global gray level is a method where the centers are initialized based on the maximum and minimum levels found in the whole region of interest (Table 19.1).

Table 19.1  Centroid initialization – global gray level.

| centroid | R | G | B |
|---|---|---|---|
| 0 | maxgray | maxgray | maxgray |
| 1 | maxgray | mingray | mingray |
| 2 | mingray | mingray | maxgray |
| 3 | mingray | maxgray | mingray |
| 4 | maxgray | maxgray | mingray |
| 5 | mingray | maxgray | maxgray |
| 6 | maxgray | mingray | maxgray |

The global feature level method (Table 19.2) assumes that the initial positions of centroids are constructed of feature values (R, G, B) of a pixel with the darkest (min) and the brightest (max) within the region of interest.

The last variant, the local feature level (Table 19.3), is analogous in operation to the previous one, but the values are evaluated individually for each square region.

Table 19.2. Centroid initialization – global feature level.

| centroid | R | G | B |
|---|---|---|---|
| 0 | max | max | max |
| 1 | max | min | min |
| 2 | min | min | max |
| 3 | min | max | min |
| 4 | max | max | min |
| 5 | min | max | max |
| 6 | max | min | max |

Table 19.3. Centroid initialization – local feature level.

| centroid | R | G | B |
|---|---|---|---|
| 0 | maxR | maxG | maxB |
| 1 | maxR | minG | minB |
| 2 | minR | minG | maxB |
| 3 | minR | maxG | minB |
| 4 | maxR | maxG | minB |
| 5 | minR | maxG | maxB |
| 6 | maxR | minG | maxB |

The thresholds: homogeneity level and spot level are used as follows.

The homogeneity level (*hl*) is a factor used in the expression of homogeneity of the analyzed square:

$$\left(localMaxGray - localMinGray\right) > \left(Maxgray - Mingray\right) \cdot hl \cdot 0.01 \qquad (19.26)$$

where Maxgray and Mingray are the levels defined in Table 19.1, while localMaxgray and localMinGray are the minimum and maximum pixel levels within an individual quare.

The spotlevel (*sl*) is a factor in the decision rule if the pixel is assigned as a spot pixel. The rule is:

$$pixelGray > MinGray + |MaxGray - MinGray| \cdot sl \cdot 0.01 \qquad (19.27)$$

As a result of the teaching phase, the matrix of pixel classes is filled. The algorithm is presented in Fig. 19.19.

---

1. Initialize the matrix of gravity centers (for each mask region)

2. If the region is homogenous (homogeneity level is not achieved) set all region pixel class as background and proceed to the next region. Otherwise, go to 3.

3. Get the next pixel in the region.

4. If the gray-level is less than the spot level, set the pixel class as background, otherwise assign the pixel class to the nearest class and go to 5.

5. If all region pixels are set, go to 6 else go to 3.

6. Update the matrix of gravity centers.

7. if the gravity center matrix has been changed go to step 3, else proceed to the next region.

---

Fig. 19.19  The teaching phase.

In the grouping phase the region of interest is split using a mask of a smaller diameter. This is the final step of pixel classification; all the pixels are assigned to some class. For each mask region:

---

1. Initialize the gravity center matrix based on the class matrix.

2. Evaluate the mean gray-level in the region.

3. Get the next pixel.

4. If the gray level of the pixel is less than the mean value, assign the pixel as a background and go to 6 else assign it to the nearest class and go to 5.

5. If all pixels are classified go to 6 else go to 3.

6. Update the gravity center matrix.

7. Once again assign all pixels to nearest centers.

---

Fig. 19.20  The grouping phase.

In the last – finalization – phase, the image of classes is generated. The part which is not in the Region of Interest is colored gray, the background pixels are white, and the other pixels have the color corresponding to the adequate gravity center value. If the median filter option is on, a filtering procedure to remove the image artifacts and to smooth the ragged object edges is carried out.

The algorithm was tested to prove the reliability and high performance of the segmentation. The followimg aspects were tested:

- the impact of the number of gravity centers upon the convergence of the algorithm (Tables 19.4–19.5),
- the impact of mask sizes upon the speed and quality of the segmentation (Table 19.6),
- the impact of the homogeneity and sensitivity thresholds upon the segmentation accuracy (Table 19.7).

Table 19.4 Convergency tests for various centroid initialization methods (part 1).

| Centroid initialization method | Iteration 0 Centroids coords. | | | D | Iteration 1 Centroids coords. | | | D |
|---|---|---|---|---|---|---|---|---|
| | R | G | B | | R | G | B | |
| Global gray level | 169 | 169 | 169 | 24 | 160 | 152 | 143 | 8 |
| | 52 | 52 | 169 | 70 | 72 | 87 | 112 | 3 |
| | 169 | 52 | 52 | 73 | 135 | 94 | 86 | 14 |
| | 52 | 169 | 52 | 184 | 0 | 0 | 0 | 0 |
| | 169 | 169 | 52 | 83 | 147 | 116 | 104 | 37 |
| | 52 | 169 | 169 | 87 | 104 | 116 | 131 | 25 |
| | 169 | 52 | 169 | 245 | 0 | 0 | 0 | 0 |
| Global feature level | 172 | 172 | 172 | 29 | 160 | 152 | 143 | 8 |
| | 35 | 35 | 172 | 87 | 66 | 81 | 108 | 6 |
| | 172 | 35 | 35 | 96 | 131 | 89 | 81 | 22 |
| | 35 | 172 | 35 | 179 | 0 | 0 | 0 | 0 |
| | 172 | 172 | 35 | 100 | 144 | 110 | 99 | 45 |
| | 35 | 172 | 172 | 103 | 97 | 110 | 128 | 34 |
| | 172 | 35 | 172 | 246 | 0 | 0 | 0 | 0 |
| Local feature level | 180 | 177 | 174 | 36 | 160 | 153 | 144 | 7 |
| | 35 | 54 | 174 | 79 | 76 | 91 | 114 | 14 |
| | 180 | 54 | 55 | 76 | 138 | 99 | 90 | 8 |
| | 35 | 177 | 55 | 189 | 0 | 0 | 0 | 0 |
| | 180 | 177 | 55 | 86 | 150 | 121 | 109 | 31 |
| | 35 | 177 | 174 | 105 | 104 | 118 | 134 | 20 |
| | 180 | 54 | 174 | 256 | 0 | 0 | 0 | 0 |

The results shown in Tables 19.4–19.5 indicate that the global gray level approach is the most convergent and the initial positions of centroids correspond to the colors: red, green, blue, cyan, magenta, yellow and white. The other methods require more iterations to achieve the final positions but the initial position is set better, which may be an advantage when the maximum number of iterations is limited. All three methods are convergent to the same centroid positions.

Table 19.5 Convergency tests for various centroid initialization methods (part 2).

| Centroid initialization method | Iteration 2 Centroid coords. | | | D | Last iteration | Last iteration Centroid coords. | | |
|---|---|---|---|---|---|---|---|---|
| | R | G | B | | | R | G | B |
| Global gray level | 162 | 155 | 146 | 4 | | 163 | 157 | 149 |
| | 65 | 81 | 107 | 7 | | 70 | 85 | 110 |
| | 133 | 91 | 83 | 18 | | 140 | 104 | 94 |
| | 0 | 0 | 0 | 0 | 14 | 0 | 0 | 0 |
| | 152 | 127 | 115 | 20 | | 157 | 141 | 129 |
| | 118 | 127 | 136 | 6 | | 123 | 130 | 138 |
| | 0 | 0 | 0 | 0 | | 0 | 0 | 0 |
| Global feature level | 161 | 154 | 145 | 5 | | 163 | 157 | 149 |
| | 62 | 78 | 105 | 12 | | 70 | 85 | 110 |
| | 130 | 87 | 80 | 24 | | 140 | 104 | 94 |
| | 0 | 0 | 0 | 0 | 15 | 0 | 0 | 0 |
| | 150 | 123 | 111 | 26 | | 157 | 141 | 129 |
| | 115 | 124 | 135 | 10 | | 123 | 130 | 138 |
| | 0 | 0 | 0 | 0 | | 0 | 0 | 0 |
| Local feature level | 162 | 155 | 146 | 4 | | 163 | 157 | 149 |
| | 66 | 81 | 108 | 1 | | 67 | 82 | 108 |
| | 135 | 96 | 87 | 13 | | 141 | 105 | 95 |
| | 0 | 0 | 0 | 0 | 17 | 0 | 0 | 0 |
| | 154 | 130 | 118 | 17 | | 157 | 142 | 130 |
| | 119 | 127 | 137 | 3 | | 121 | 129 | 137 |
| | 0 | 0 | 0 | 0 | | 0 | 0 | 0 |

Mask size adjustments affect the algorithm speed and the segmentation quality. The experiments carried out on the images showed that the mask in the teaching phase should be smaller than in the grouping with the ratio about 1:3. The suggested mask sizes for the processed resolution are 20 for teaching and 60 – 10 for grouping. Too small masks defect the weak fuzzy spot detection and too large masks dramatically increase the computation time.

Table 19.6  Segmentation quality for various segmentation mask sizes.

| Centroid initialization method | Mask size | | Time [ms] | Fract. of properly detected spots [%] |
| | Teaching phase | Grouping phase | | |
| --- | --- | --- | --- | --- |
| Global gray level | 5 | 5 | 1210 | 65 |
| | | 10 | 1122 | 68 |
| | | 20 | 1109 | 80 |
| | 20 | 10 | 1192 | 78 |
| | | 60 | 1333 | 81 |
| | | 100 | 1241 | 92 |
| | 200 | 100 | 1728 | 43 |
| | | 200 | 1699 | 72 |
| | | 400 | 1953 | 62 |
| | | 800 | 2346 | 66 |
| Global feature level | 5 | 5 | 1217 | 35 |
| | | 10 | 1240 | 38 |
| | | 20 | 1225 | 53 |
| | 20 | 10 | 1364 | 37 |
| | | 60 | 1402 | 56 |
| | | 100 | 1204 | 66 |
| | 200 | 100 | 1675 | 43 |
| | | 200 | 1807 | 64 |
| | | 400 | 1808 | 67 |
| | | 800 | 2246 | 64 |
| Local feature level | 5 | 5 | 1240 | 67 |
| | | 10 | 1107 | 68 |
| | | 20 | 1273 | 80 |
| | 20 | 10 | 1258 | 84 |
| | | 60 | 1214 | 96 |
| | | 100 | 1339 | 88 |
| | 200 | 100 | 1653 | 43 |
| | | 200 | 1670 | 68 |
| | | 400 | 1778 | 64 |
| | | 800 | 2274 | 65 |

The homogeneity and spot levels have a very high impact on the quality of segmentation. The values must be adjusted experimentally and may differ when the camera is replaced or the reagents are changed.

The experiments showed that the proper adjustment of the segmentation parameters enables segmentation of all the tested ELISPOT images.

The method with local feature levels offers the highest segmentation quality. Despite its worst convergence, it is also the fastest because a segmentation mask may be relatively small. Its drawback is high sensitivity to threshold factor adjustment, which is a serious weakness for unattended processing.

Table 19.7 Segmentation quality as a function od thresholding factors.

| Centroid initialization method | hl [%] | sl [%] | Time [ms] | Spot count | False spot count |
|---|---|---|---|---|---|
| Global gray level | | 1 | 1302 | 91 | * |
| | 1 | 10 | 1331 | 86 | 15 |
| | | 75 | 1578 | 92 | 10 |
| | 10 | | 1453 | 96 | 6 |
| | 25 | 100 | 1175 | 93 | 10 |
| | 75 | | 971 | 22 | 78 |
| Global feature level | | 1 | 1640 | 95 | 12 |
| | 1 | 10 | 1706 | 81 | 19 |
| | | 75 | 1763 | 83 | 17 |
| | 10 | | 1689 | 81 | 19 |
| | 25 | 100 | 1545 | 83 | 17 |
| | 75 | | 1235 | 23 | 77 |
| Local feature level | | 1 | 1806 | * | * |
| | 1 | 10 | 1807 | * | * |
| | | 75 | 2719 | * | * |
| | 10 | | 2759 | * | * |
| | 25 | 100 | 1663 | 96 | 4 |
| | 75 | | 1220 | 24 | 76 |

\* means very poor segmentation

The global feature level method gives high-quality segmentation without the user having to take excessive care of the parameters, but it requires large segmentation masks, which increases the computation cost.

The method of global gray level gives stable results without modifying the clustering parameters, but during the segmentation the weakest and smallest spots are often omitted.

## 19.7 Monochrome image segmentation using Hit-Miss transform

The hit-miss transform is a basic binary morphological operation. It is most frequently used to search for specific structures that form the object and the background pixels. Its idea was thoroughly described in (Fisher et al., 2004). In practice, with this transformation, it is possible to search for objects of various shapes, or of the same shape, but changing orientation. In this chapter, HMT has been adapted for greyscale and color images. Since the objects to be found are circular spots of certain colors and size ranges, it is possible to design suitable hit-and-miss conditions.

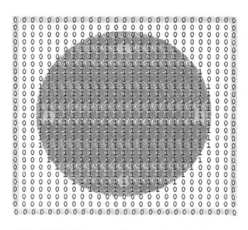

Fig. 19.21 The mask for scanning the image.

To detect spots of different diameters, a multi-pass algorithm is used. In each cycle, the image is scanned with a decreasing size of the mask.
After the setup phase of the algorithm, the mask array is filled as shown in Fig. 19.21.

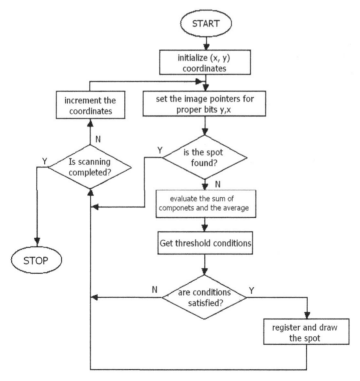

Fig. 19.22 The flowchart of the main function of the algorithm.

The array consits of the region representing the object ('ones') and the background ('zeros'). After filling the mask, the main loop of the algorithm is run (Fig. 19.22)

In each pass of the function, a new size of the mask is set and the mask is filled. Then, scannig for each pixel of the image starts. The implementation uses a pointer to optimize the algorithm performance. Initially, for each inspected pixel it is checked whether there is a spot at these coordinates (possibly found earlier). If so, the point is omitted. If no, the negihborhood of the pixel determined by the mask is examined. Pixels which belong to the pink region (see Fig. 19.21) and markcd "1" are supposed to belong to the object, while those which belong to the cyan region are supposed to belong to the background.

It was experimentally determined that the color of the background pixels differs significantly from the object pixels in each color component. In Table 19.8 there are model values of colors for spots and the background presented.

Table 19.8 Model colors for background and spot pixel.

| Color component | Background color | Spot color |
|---|---|---|
| Red | 249 | 167 |
| Green | 204 | 105 |
| Blue | 113 | 84 |

For each position of the mask we calculate:

- an average value of R, G, B for pixels marked pink;
- an average value of R, G, B for pixels marked cyan;

These to values are compared with threshold values to classify the examined point. The threshold values are calculated locally for each position of the mask using the calculated averages of R, G, and B. This mitigates the influence od inequal exposure of the image. In Table 19.9 threshold values are presented

Table 19.9 Threshold values for the spot and the background.

| Value of S=avg(R+G+B) | | Threshold value | |
|---|---|---|---|
| from | to | Object | Background |
| 0 | 180 | 0 | 0 |
| 180 | 250 | S - 12 | S + 8 |
| 250 | 300 | S - 15 | S + 10 |
| 300 | 350 | S - 17 | S + 10 |
| 350 | 400 | S - 18 | S + 12 |
| 400 | 460 | S - 20 | S + 12 |
| 460 | 510 | S - 24 | S + 12 |
| 510 | 570 | S - 22 | S + 10 |
| 570 | 630 | S - 22 | S + 10 |
| 630 | | S - 24 | S + 10 |

Apart from comparing the sums of colors with the thresholds, we use specific control points. In Fig. 19.21 they are "1" marked cyan. The necessary condition of the possitive detection of a spot is that the values of the pixels must be less than the threshold value incresed by 6% (the margin for the control points).

If the detection is positive, a spot is registered, otherwise we move to the next pixel. The algorithm terminates when all the pixels have been examined.

### *Algortithm benchmarks*

The algorithm was tested to evaluate its efficiency. Figures 19.23 and 19.24 present the fragments of segmented images from two different cameras and exposure conditions. An uneven exposure is, to a certain extent, not an obstacle in the operation of the algorithm.

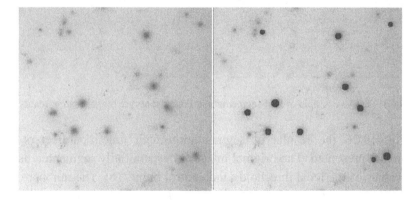

Fig. 19.23 Exemplary image and found spots (even exposure).

The results from these images were obtained using low thresholds for the colors of the objects and the background and a rather high margin of the control points.

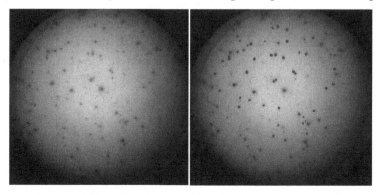

Fig. 19.24 Exemplary image and found spots (uneven exposure).

Although the sensivity of the algorithm is sufficiently good to find all the actual spots it has a little drawback. The radii of the recognized spots are larger than in reality. (Fig. 19.25)

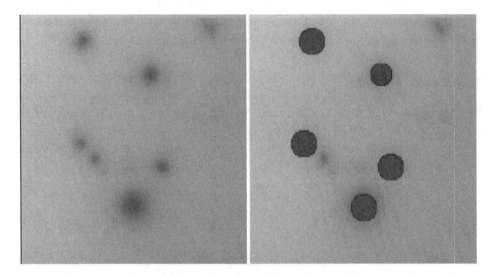

Fig. 19.25 An example of HM segmentation. Found spots are bigger than in reality.

In Fig. 19.26 the results of segmentation with tuned parameters of the algorithm are presented. The original image (a) was initially segmented using the default values of the local thresholds and control points (b). The number of spots found is 85. In (c) the local thresholds were decreased and the control margin was decreased to 104%. The number of spots found is 94.

In (d) the result of segmentation with default thresholds and the margin of control points increased to 108% is shown. The number of spots found is 93 and the effect of too large a size of spots found is minimal.

The experiments show that decreasing the thresholds and increasing the margins enables catching spots of low contrast, but then the "strong" objects have a false size. Increasing thresholds but decreasing the control margin gives the result of losing small objects. The best results were obtained for the thresholds given in Table 19.9 and the control point margin 108% of the thresholds.

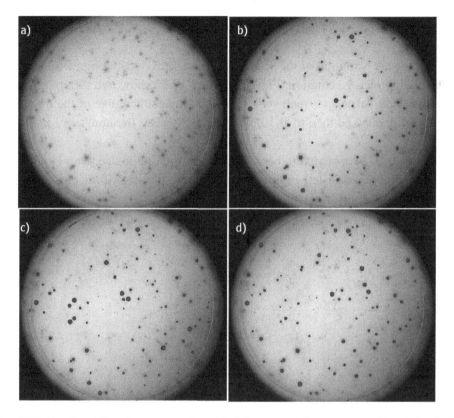

Fig. 19.26 Results of Hit-miss segmentation with different sensivity parameters. Description in the text.

Simple image preprocessing (noise reduction and contrast enhancement) may significantly increase the efficiency of the method (Fig. 19.27).

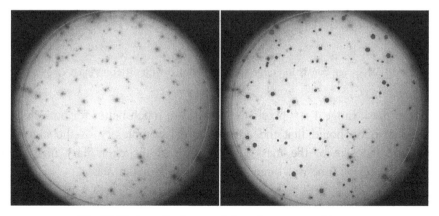

Fig. 19.27 A result of segmentation of preprocessed image.

The number of detected spots increased by 10% compared to the original image.

The range of variation of the scanning mask size has also a significant impact on the quality of segmentation. Figs. 19.28 and 19.29 show two different images processed: with large and small spots. For each image the range of the mask was set to 6–24 pixels and then to 1–10 pixels. In both cases the number of iterations was 10.

The range 6–24 appeared to be good only for the first image, with large spots. When the image consisted of many small spots, the algorithm misinterpreted clusters of small spots as larger spots.

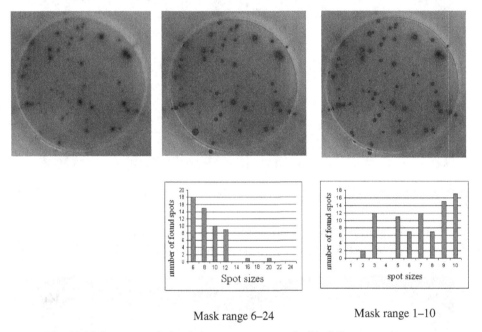

Mask range 6–24                    Mask range 1–10

Fig. 19.28 Image consisting large spots segmented with different mask ranges.

For the mask range 1–10 pixels the results are much better for both images. It should also be noted that the computational complexity of the algorithm depends on the size of the mask. In the case of the exemplary images the algorithm execution time was three times faster for smaller masks.

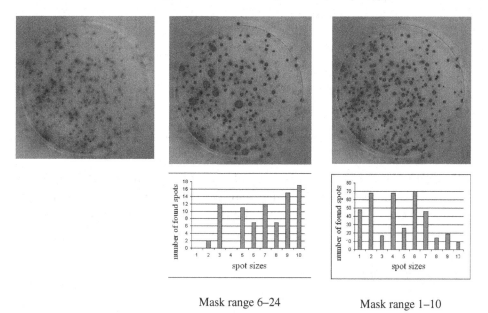

Mask range 6–24          Mask range 1–10

Fig. 19.29  Image consisting small spots segmented with different mask ranges.

## 19.8  Monochrome image segmentation using edge detection and geometrical fitting of the shape

The routines examined in the previous chapters allow successful detection of rough location and size of the spots as well as recognition of their color.

But the main problem was how to precisely measure the amount of the staining, which translates to the area of individual spots. The reason why it is difficult to determine is blurry edges of the spots. All the approaches listed above use thresholding at least in their initial stage and thus are very sensitive to the selection of the threshold; in other words, changing the threshold slightly may severely change the measured area of a given object. In the current chapter we present an algorithm based on edge detection which appears to yield more stable results and the returned spot contours quite closely correspond to real ones.

Tracing edges on an image has a potential of being more accurate in returning local object boundaries, but there are two problems concerning it: an easy one is to fill the interior of the contour of the spot, and a harder one is to obtain a closed contour instead of several non-connected arcs. We follow this approach in the current description.

In general, the algorithm performs the operations outlined in Figure 19.30.

1. resolution normalization;

2. noise reduction using low-pass filtering;

3. boundary emphasize using sharpening filtering;

4. Canny curve detection;

5. circle-fitting procedure applied to arcs;

6. reconstruction of circular spot boundaries.

Fig. 19.30  The general outline of the algorithm.

We claim that the algorithm is less vulnerable to variable image properties: image resolution, out-of-focus zones and improper exposure (non-uniform lighting). Even if the detected spot borders are severely fragmented and incomplete, the fitting procedure can evaluate the spot circles.

### *Image preprocessing*

In this phase we attempt to prepare the image for the Canny procedure, which performs best if the border to be traced is approximately 3–5 pixels wide. In the examined images obtained using a Nikon Camera (1182 x 1280) the ROI has approximately 750 pixels of diameter, and the diameter of the spots varies from 10 to 50 pixels. To obtain the best performance, the dimensions of the picture must be doubled (using bilinear interpolation).

On the resized (resampled) image two procedures are imposed:
- the Gaussian smoothing filter with a mask diameter of 5 pixels.
- the Laplacian as a sharpening filter with a mask diameter of 3 pixels.
- the linear adjustment of the image intensity histogram (for contrast equalization).

Fig. 19.31  Gaussian smoothing and Laplace sharpening masks.

### *Boundary detection*

There are many approaches to edge and boundary detection. For this purpose the Canny procedure (Canny, 1986; Grigorescu, 2004) was chosen because of its benefits listed below:

- after the proper preprocessing it generates stable results,
- it is invulnerable to image rotation,
- thanks to the "threshold hysteresis", it can detect and trace weak or fading lines.

The algorithm is carried out in three phases:

1.  The image is filtered with a vertical and horizontal Sobel mask in order to estimate the gradient magnitude and angle.

| 1 | 2 | 1 |
|---|---|---|
| 0 | 0 | 0 |
| -1 | -2 | -1 |

| -1 | 0 | 1 |
|---|---|---|
| -2 | 0 | 2 |
| -1 | 0 | 1 |

Fig. 19.32 Sobel gradient masks.

2.  If the intensity gradient magnitude is greater than the threshold value $t_1$, the procedure of line tracing is started, and continued until the line direction changes or its strength falls below the $t_2$ threshold ($t_2 < t_1$),
3.  The algorithm usually finds line zones (several perpendicular neighboring lines). The proper lines are extracted with the aid of the non-maximum suppression procedure.

The Canny procedure generates a binary image, which is subsequently segmented into connected components (individual lines), which are then regarded as candidates for arcs of the reconstructed circles.

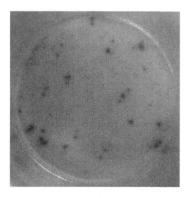 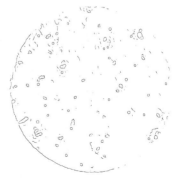

Fig. 19.33 The input image and after Canny operator.

## Arc approximation

In this stage each connected component is regarded as a tabulated function and its analytical representation has to be approximated. The circle-finding procedure introduced in (Gander, 1994) runs in two steps:

1. the fitting procedure using the linear least squares method (the implementation from (Press et al., 2002));
2. the iterative fine-fitting using geometric distance (the implementation from (Press et al., 2002));

Consider $x^{\mathrm{T}}$ and $y^{\mathrm{T}}$ as $m$-length vectors of coordinates of the component points. Let us denote:

$$[A] = \begin{bmatrix} x_1^2 + y_1^2 & x_1 & y_1 & 1 \\ x_2^2 + y_2^2 & x_2 & y_2 & 1 \\ \vdots & \vdots & \vdots & \vdots \\ x_m^2 + y_m^2 & x_m & y_m & 1 \end{bmatrix} \tag{19.28}$$

Assuming that $m$, the number of points in each connected component, is more than 4, the matrix $[A]$ is rectangular and singular. To deal with $[A] \cdot x = b$ linear equation in this case, the singular value decomposition method must be applied. The idea of decomposition of the matrix $[A]$ (MxN) is the creation of a column orthogonal matrix $[U]$ (MxN), a diagonal matrix $[W]$ (NxN) with zero or non-zero elements and square, and an orthogonal matrix $[V]$ (NxN).

$$[A] = [U] \cdot \begin{bmatrix} w_{11} & & 0 \\ & \ddots & \\ 0 & & w_{nn} \end{bmatrix} \cdot [V]^T \quad \text{where} \quad [U]^T \cdot [U] = [V]^T \cdot [V] = [1] \tag{19.29}$$

In the further processing of the fitting algorithm, the square matrix $[V]$ (4x4) is used and the circle parameters are calculated as follows:

$$x_0 = \frac{-v_{24}}{2 \cdot v_{14}}; \quad y_0 = \frac{-v_{34}}{2 \cdot v_{14}}; \quad r_0 = \sqrt{\frac{v_{24}^2 + v_{34}^2}{4 \cdot v_{14}^2} - \frac{v_{44}}{v_{14}}} \tag{19.30}$$

When initial $(x_0, y_0, r_0)$ values are calculated, the iterative procedure is launched.

The aim of the procedure is to minimize the square error

$$\varepsilon = \sum \left( \left\| (x, y) - (x_0, y_0) \right\| - r \right)^2 \tag{19.31}$$

Let us denote:

$$f = \begin{bmatrix} \sqrt{(x_1 - x_0)^2 + (y_1 - y_0)^2} - r_0 \\ \sqrt{(x_2 - x_0)^2 + (y_2 - y_0)^2} - r_0 \\ \vdots \\ \sqrt{(x_m - x_0)^2 + (y_m - y_0)^2} - r_0 \end{bmatrix} \qquad (19.32)$$

and

$$[J] = \begin{bmatrix} \dfrac{x_1 - x_0}{\sqrt{(x_1 - x_0)^2 + (y_1 - y_0)^2}} & \dfrac{y_1 - y_0}{\sqrt{(x_1 - x_0)^2 + (y_1 - y_0)^2}} & 1 \\ \dfrac{x_2 - x_0}{\sqrt{(x_2 - x_0)^2 + (y_2 - y_0)^2}} & \dfrac{y_2 - y_0}{\sqrt{(x_2 - x_0)^2 + (y_2 - y_0)^2}} & 1 \\ \vdots & \vdots & \vdots \\ \dfrac{x_m - x_0}{\sqrt{(x_m - x_0)^2 + (y_m - y_0)^2}} & \dfrac{y_m - y_0}{\sqrt{(x_m - x_0)^2 + (y_m - y_0)^2}} & 1 \end{bmatrix} \qquad (19.33)$$

The algorithm solves the equation:

$$[J] \cdot h = f \qquad (19.34)$$

with the help of the QR decomposition method.

$$[J] = [Q] \cdot [R] \quad \Rightarrow \quad [R] \cdot h = [Q]^T \cdot f \qquad (19.35)$$

where $[Q]$ is orthogonal ($[Q]^T \cdot [Q] = [1]$) and $[R]$ is triangular (upper). In each iteration, the circle is updated

$$(x_0, y_0, r_0) = (x_0, y_0, r_0) + h \qquad (19.36)$$

and the error

$$\delta = \frac{\|h\|_\infty}{\|(x_0, y_0, r_0)\|_\infty} \qquad (19.37)$$

is estimated. The iteration repeats until $\delta < \delta_0 = 10^{-5}$ or the number of iterations exceeds 100 (then, the algorithm is not convergent – the line is not an arc).

### Spot border reconstruction

The Canny algorithm finds the following lines:

- fragments of the spot borders (sometimes more than 1 for each spot);
- fragments of the spot interiors (caused by non homogenous spot staining);
- the well boundaries;

Initially, we leave only the circles for which $r \subset (0.05 \cdot D; \quad 0.1 \cdot D)$ where $D$ is the diameter of ROI. Then, we delete the circles that cross the ROI circle.

It is very likely that the remaining connected components are parts of the spot contours.

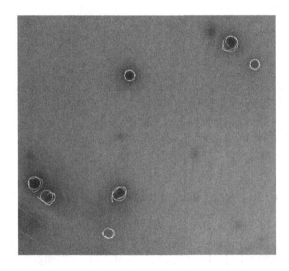

Fig. 19.34 The circles approximated for found spot contours.

It sometimes happens that the spot contour is fragmented. In this case, the same number of circles is produced for each spot. The circles which overlap, or in other words, for which the following condition is satisfied:

$$\rho\big((x_1, y_1), (x_2, y_2)\big) \leq \min(r_1, r_2), \tag{19.38}$$

represent the same contour and the respective arcs should be merged.

The merging algorithm works in three phases.

Phase 1: For each arc, find the neighboring arcs based on (38). Let us look at Fig. 19.35 (the artificial spots).

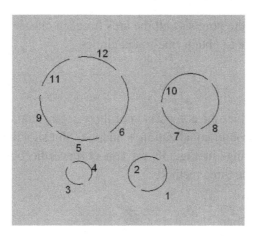

Fig. 19.35 The exemplary image: enumeration of found arcs.

For this example, the following neighbors are found (Table 19.10).

Table 19.10 Object parameters for the exemplary image.

| Object number | Circle $(x, y, r)$ | Found neighbors |
|---|---|---|
| 1 | (360.1, 256.1, 30.9) | 2 |
| 2 | (367.2, 254.2, 43.0) | 1 |
| 3 | (224.8, 260.7, 22.7) | 4 |
| 4 | (228.3, 259.8, 21.2) | 3 |
| 5 | (235.2, 406.5, 85.6) | 6; 9; 11; 12 |
| 6 | (248.5, 397.8, 71.1) | 5; 9; 11; 12 |
| 7 | (435.9, 391.3, 52.3) | 8; 10 |
| 8 | (444.0, 389.1, 47.1) | 7; 10 |
| 9 | (224.1, 398.4, 71.9) | 5; 6; 11; 12 |
| 10 | (438.8, 390.3, 55.8) | 7; 8 |
| 11 | (238.1, 390.4, 87.8) | 5; 6; 9; 12 |
| 12 | (236.4, 389.1, 87.7) | 5; 6; 9; 11 |

Phase 2: After the process of traversing the adjacency graph, we build a list of clusters (Table 19.11).

Table 19.11 Clusters for the exemplary image.

| Cluster number | Arcs numbers |
|---|---|
| 1 | 1; 2 |
| 2 | 3; 4 |
| 3 | 5; 12; 11; 9; 6 |
| 4 | 7; 10; 8 |

In the last phase, the points of all the arcs for each cluster are submitted to the Find-Circle procedure to obtain fine-tuned circles.

### *Results*

The experiment involves the comparison of three segmentation methods: Bernsen thresholding, Background equalization + adaptive thresholding, and Canny edge detector + arc merging. In Fig. 36 we can observe the borders of segmented objects with use of the three methods.

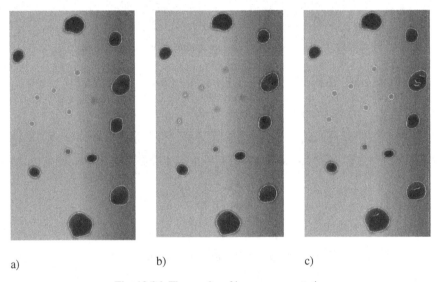

a)  b)  c)

Fig. 19.36 The results of image segmentation.

For this image the basic geometry properties: area, perimeter, Aspect Ratio and Gamma coefficient were calculated. The results for individual spots are shown in Table 19.12. From Fig. 19.36 and Table 19.12 we can see that the adaptive thresholding missed some small spots. This happens because small spots have weak intensity. Both Bernsen's algorithm and Canny's operator were able to detect the spots. Comparing the area value obtained for the spots for different methods, we easily notice that edge detecting gives larger spots than thresholding (in this case). Using another thresholding algorithm or changing its parameters will yield give different areas. Changing the thresholds in Canny's method may introduce some irrelevant lines (edges) or suppress some important fragments of the contours, but the contours will always appear in the same place. This guarantees that the sizes of the objects are more stable. Proper contour detection enables accurate measurement of the object shape descriptors, which is also visible in Table 19.12.

Table 19.12 Object parameters for different segmentation types: A – adaptive, B – Bernsen, C – Canny.

| Obj. | Area | | | Perimeter | | | AR | | | Gamma | | |
| meth | A | B | C | 117 | B | C | A | B | C | A | B | C |
|---|---|---|---|---|---|---|---|---|---|---|---|---|
| 1 | 1290 | 1127 | 1283 | 62 | 123 | 132 | 0.78 | 0.77 | 0.78 | 0.93 | 0.93 | 0.92 |
| 2 | 1085 | 874 | 1021 | 61 | 108 | 117 | 0.84 | 0.81 | 0.79 | 0.91 | 0.93 | 0.93 |
| 3 | 2684 | 222 | 303 | 60 | 53 | 62 | 0.68 | 0.86 | 0.86 | 0.88 | 0.97 | 0.98 |
| 4 | - | 226 | 300 | 205 | 54 | 61 | - | 0.88 | 0.92 | - | 0.96 | 0.98 |
| 5 | - | 214 | 292 | 64 | 50 | 60 | - | 0.90 | 0.88 | - | 1.00 | 1.00 |
| 6 | 2684 | 2503 | 3000 | 62 | 189 | 205 | 0.68 | 0.67 | 0.66 | 0.88 | 0.88 | 0.89 |
| 7 | 230 | 257 | 323 | 254 | 56 | 64 | 0.87 | 0.88 | 0.88 | 1.00 | 1.00 | 0.98 |
| 8 | - | 196 | 295 | 61 | 49 | 62 | - | 0.83 | 0.85 | - | 0.99 | 0.96 |
| 9 | 3892 | 3862 | 4572 | 99 | 237 | 254 | 0.82 | 0.82 | 0.84 | 0.88 | 0.86 | 0.89 |
| 10 | - | 194 | 297 | 62 | 49 | 61 | - | 0.84 | 0.88 | - | 1.00 | 0.98 |
| 11 | 733 | 627 | 720 | 147 | 93 | 99 | 0.69 | 0.65 | 0.68 | 0.91 | 0.91 | 0.91 |
| 12 | - | 216 | 294 | 203 | 52 | 62 | - | 0.84 | 0.85 | - | 0.98 | 0.95 |
| 13 | 1122 | 1313 | 1593 | 152 | 135 | 147 | 0.85 | 0.78 | 0.80 | 0.91 | 0.90 | 0.92 |
| 14 | 1211 | 1373 | 2969 | 209 | 139 | 203 | 0.79 | 0.73 | 0.76 | 0.91 | 0.88 | 0.90 |
| 15 | 1926 | 2486 | 1642 | 172 | 188 | 152 | 0.74 | 0.78 | 0.68 | 0.81 | 0.88 | 0.89 |
| 16 | 2027 | 2655 | 3101 | 173 | 196 | 209 | 0.69 | 0.67 | 0.70 | 0.84 | 0.86 | 0.89 |

## 19.9 Conclusions

In this chapter we have shown that segmentation of ELISPOT images is not a trivial task and requires combination of standard and specially prepared image processing algorithms. A very important phase in the process of segmentation is proper image preprocessing, which allows the main algorithm to work smoothly.

Despite this, it is possible to use at least two main approaches for the segmentation process:

- pixel classification: the thresholding of the grey level or color, the clustering of the color space
- shape analysis: matching of the shapes and sizes of the objects or finding their contours.

All the described algorithms yielded accurate results on condition that their parameters were tuned. It is unlikely that the design of an algorithm running for each image completely without user supervision is possible.

# References

Gebauer, B. S. et al. (2002). Evolution of the enzyme-linked immunosorbent spot assay for post-transplant alloreactivity as a potentially useful immune monitoring tool, *Am. J. Transplant*, 2, pp. 857–866.

Hricik, D. E. et al. (2003). Enzyme linked immunosorbent spot (ELISPOT) assay for interferon-gamma independently predicts renal function in kidney transplant recipients, *Am. J. Transplant*, 3, pp. 878–884.

Koscielska-Kasprzak, K. et. al. (2009). Pretransplantation Cellular Alloreactivity Is Predictive of Acute Graft Rejection and 1-Year Graft Function in Kidney Transplant Recipients, *Transplantation Proceedings*, 41, 8, pp. 3006–3008.

Mehmet, S., Bülent, S. (2004). Survey over image thresholding techniques and quantitative performance evaluation, *Journal of Electronic Imaging*, 13, 1, pp. 146–168.

Bernsen, J. (1986). Dynamic thresholding of grey-level images, *Proceedings 8th International Conference on Pattern Recognition*, pp. 1251–1255.

Kittler, J., Illingworth, J., Foglein, J. (1985). Threshold selection based in a simple image statistic, *Computer Vision, Graphics and Image Processing*, 30, pp. 125–147.

Hanson, A. R., Riseman, E. R. (1978). Segmentation of natural scenes. In Hanson and Riseman, editors, *Computer Vision Systems*, pp. 129–163, Academic Press, NJ.

MacQueen, J. (1967). Some methods for classification and analysis of multivariate observations. 5th Berkeley Symposium on Mathematics, Statistics and Probability, University of California Press, Berkeley, CA, USA, 1, pp. 281–296.

Fisher, R. B., Perkins, S., Walker, A., Wolfart, E. (2004). Hypermedia Image Processing Reference, John Wiley & Sons LTD.

Canny, J. F. (1986). A computational approach to edge detection, *IEEE Trans. Pattern Analysis and Machine Intelligence*, 8, 6, pp. 679–698.

Grigorescu, C., Petkov, N., Westenberg, M. A. (2004). Contour and boundary detection improved by surround suppression of texture edges, *Image and Vision Computing*, 22, 8, pp. 609–622.

Gander, W., Golub, G. H, Strebel, R. (1994). Least-Squares Fitting of Circles and Ellipses, *BIT Numerical Mathematics*, Springer.

Press, W. H., Teukolsky, S. A., Vetterling, W. T., Flannery, B. P. (2002). Numerical recipes in C. *Cambridge University Press*, Cambridge.

# CHAPTER 20

# IMAGE PROCESSING AND ANALYSIS ALGORITHMS FOR ASSESSMENT AND DIAGNOSIS OF BRAIN DISEASES

Anna Fabijańska and Tomasz Węgliński

*Institute of Applied Computer Science*
*Lodz University of Technology*
*90-924 Łódź, ul. Stefanowskiego 18/22*
*{an_fab, tweglinski}@kis.p.lodz.pl*

Medical imaging is one of the fastest growing fields in medicine. The influence and impact of digital images on the way of diagnosis and treatment of various diseases is tremendous. The rapid development of imaging devices caused the significant progress in computerized image reconstruction and computer-aided diagnosis. These in turn led the medical image processing to one of the most important scientific fields in today's computer science.

## 20.1 Introduction

Radiology is a branch of medicine which uses imaging technology to diagnose lesions and pathological conditions that affect the human body. Recently, medical imaging is widely applied for the monitoring and assessment of various diseases. As a result, many of them are detected at their early stages. Additionally, radiological examinations allow accurate detection and location of lesions. This in turn is often proves crucial in making a decision to open surgery and provides useful information about the effectiveness of treatment in the postoperative control of the patient.

There are several radiological techniques used for medical imaging. Most popular ones include computed tomography (CT), magnetic resonance imaging (MRI), positron emission tomography (PET) and ultrasonography (USG). Recently, these methods are widely applied, inter alia, for the diagnosis and assessment of various pathological conditions affecting the brain, such as tumors, cerebrospinal fluid disorders, craniosynostosis, cerebral atrophy or strokes.

Radiological techniques, including computed tomography and magnetic resonance imaging, allow one to acquire three dimensional (3D), volumetric images of the brain. With the use of device-integrated software, the diseased regions can be visualized and analyzed from different axes and angles. Yet, despite the capabilities of imaging devices, the assessment of lesions and diseased regions is often mainly visual. This, however, does not allow for a precise, reliable and objective analysis. Manual measurements of the key characteristics used to assess the disease progress are often too cumbersome and too time-consuming for everyday clinical routine. Therefore, many recent studies in medical image processing are focused on the development of computer-aided tools and software dedicated to automatic detection and analysis of various pathological changes within the brain.

Recent studies in brain image processing follow in many directions. In particular, there are researches concerned with the enhancement and pre-processing of input CT or MRI data (Xue et al., 2003; Nappi et al., 2008; Boussion et al., 2008), segmentation of brain and lesion (Zhang et al., 2001; Ahmed et al., 2002; Prastawa et al., 2005; Artaechevarria et al., 2009) brain modeling (Maniadakis and Trahanias 2009, Maniadakis and Trahanias 2005; Chen and Zhong, 2008), linear, planar and volumetric analysis of the diseased regions (Sun et al., 2010; Pustkova et al., 2010; Butman and Linguraru, 2008) or blood and liquid flow analysis (Zhang et al., 2007; Lo et al., 2008; Noda et al., 2008).

Due to the aim and scope of the ongoing research performed by the authors, this chapter focuses on the description of the main steps of the brain image processing and analysis. The methods and results presented are mainly examples from the authors' research aimed at the detection and quantitative analysis of hydrocephalus in children. However, the use of the presented methods may be extended to other problems and applications in the field of medical image processing.

## 20.2 The main steps of image processing and analysis

Image processing in medicine basically aims at the extraction of all relevant information from the entire information contained within images. Then, a quantitative description of the key characteristics related to this information is performed. Operations applied to certain medical images usually depend on a particular application and may vary according to specific needs. In the case of brain images, two basic processing steps can be distinguished, namely:

(1) Image Processing including pre-processing, segmentation and post-processing;

(2) Image Analysis — including linear, planar or volumetric measurements.

Input data for image processing is usually a volumetric dataset provided by a medical imaging device. The result of image processing in most cases is a binary or labeled image which represents the lesion (or lesions) being investigated. These serve as input data for image analysis, which produces numerical values describing important properties of the diseased regions. Figure 20.1 presents a diagram of the main steps related to brain image processing and analysis. The brief description of each step is given below.

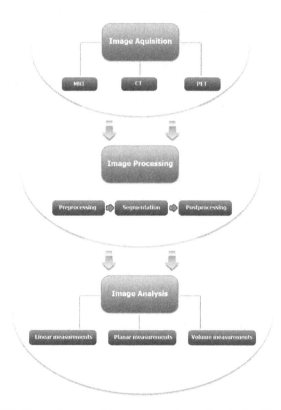

Fig. 20.1 The main steps of image processing and analysis of the brain.

Usually, after acquisition, the image is subjected to pre-processing. The goal of this step is improvement of the image quality by enhancing image information connected with the diseased region, which is relevant to further analysis and interpretation of image content. Simultaneously, redundant information is attenuated. Particular operations applied in this step strongly depend on the application and cover a wide range of problems. In the case of brain images, they commonly include: the removal of noise and artifacts caused by imperfections of imaging devices (del Fresno et al., 2009; Mohamed et al., 1999), the correction of the partial volume effect (Boussion et al., 2005) or histogram processing and intensity adjustment (Zang et al., 2010). It should be remembered that in the case of medical images the result of image enhancement is very subjective. There are also no common methods that allow one to qualify the resulting image as good or bad. However, the reliable and efficient image enhancement algorithm is crucial for further processing, as it significantly influences the precision of image segmentation, which is usually the next step of medical image processing.

Image segmentation is the most important step in many medical applications. It aims at the extraction of specific information from the image, important from the point of view of the application. Generally, image segmentation divides it into the object of interest (most commonly, the lesion or the diseased regions) and the background (i.e. remaining non-significant information). However, in some applications image segmentation can be seen as the process of partitioning an image into multiple segments, called ROI - regions of interest. These usually correspond with the lesion and other important regions used during the assessment of the disease (e.g. the whole area of the diseased organ). The precise segmentation of the lesion is crucial to further reliable analysis, measurements or diagnosis of the disease. This step is usually accompanied by the significant reduction of image information, because it produces a binary (or labeled) image, representing the particular object (or objects) extracted from the image.

The last step of image processing is called post-processing. The objective of this step is to improve the binary image after segmentation. Post-processing is often based on a-priori knowledge of the required shape and smoothness of the segmented ROI. In this step, the binary image obtained from the segmentation is processed with basic morphological operations (such as opening or closing), median filtering or smoothing algorithms. However, this step is optional and should be performed carefully as ineffectual application of morphological processing may result in the loss of data. In many cases (such as 3D shape modeling), post-processing plays an indisputable role and significantly influences the accuracy of the results.

Image processing is followed by image analysis which aims at the numerical description of the lesion under investigation. The input data for analysis algorithms is a binary (or labeled) image provided by the segmentation process. Depending on a particular application, the aim of the image analysis step is different. Commonly, having the binary data and a-priori knowledge about the image resolution and scale, the 2D linear and planar measurements can be performed. On the other hand, three-dimensional data is usually accompanied by information about the size of the gap between the consecutive slices (slice thickness), so that the volume of segmented ROI can be calculated.

The following chapters briefly describe the methods and results of image processing and analysis performed on a large set of DICOM images from both MRI and CT scanners. These are the results of the authors' research on the assessment of hydrocephalus and brain tumors.

## 20.3 Image processing

This section contains the description of image processing methods and algorithms commonly used to process volumetric image data of the brain. It was divided into three subsections, corresponding with the main processing steps, namely: image enhancement, image segmentation and post-processing. The methods and results presented were collected from the previous and ongoing researches conducted by the authors. The solutions presented were tested and verified on a large set of image data from MRI and CT scanners.

### 20.3.1 *Image enhancement*

Image enhancement is the process of highlighting significant image information. It also aims at the removal of redundant information from the image. As far as brain image processing is concerned, it aims at improving the quality and contrast of the image and removing the unnecessary information related to the skin, eyes or nose.

Algorithms applied during preprocessing may vary depending on the considered image data and application purposes. In the case of brain images, they strongly depend on imaging devices used for acquisition.

As can be seen from Fig. 20.2, brain images produced by the MRI scanners have better input contrast than the CT scans. Therefore, in most cases, the MRI images require much less preprocessing than CT datasets. The low input contrast of CT images requires histogram manipulations to demonstrate or enhance the various structures within the brain, important from the viewpoint of a given application. In the simplest case, windowing is performed to adjust the displayed range of intensities to Hounsfield units connected with the objects of interest. According to the DICOM specification, every pixel in the CT image is scaled in accordance with the following formula:

**MRI**

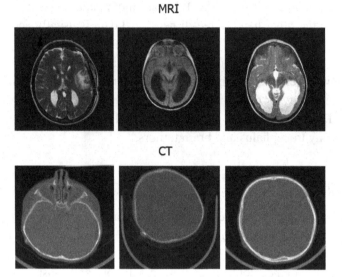

**CT**

Fig. 20.2 Exemplary brain images; top panel – MRI examinations; bottom panel – CT examinations.

$$g(x, y) = f(x, y) \cdot R_s + R_i \qquad (20.1)$$

where:

$g(\cdot)$ – the output intensity;
$f(\cdot)$ – the input intensity;
$R_s$ – the rescale slope;
$R_i$ – the rescale intercept;
$x, y$ – the pixel coordinates;

Another popular transformation is the normalization of intensities to a 12 bit range (see Eq. 20.2). The range of intensity values obtained (0 - 4095) fully covers the Hounsfield scale, so (contrary to windowing) there is no data lost.

$$g'(x, y) = \begin{cases} g_{min} & for\ g(x, y) \leq a \\ g_{max} & for\ g(x, y) > b \\ h(x, y) & for\ a < g(x, y) \leq b \end{cases} \qquad (20.2)$$

where:

$$a = c - 0.5 \cdot (2 - w) \qquad (20.3)$$

$$b = c - 0.5 \cdot w \qquad (20.4)$$

$$h(x, y) = \frac{g(x, y) - (c - 0.5)}{(w - 1) + 0.5} \cdot (g_{max} - g_{min}) + g_{min} \qquad (20.5)$$

and:

$g'(\cdot)$ – the input pixel intensity;

$g(\cdot)$ – the output pixel intensity;

$g_{min}$ – the requested minimum output intensity;

$g_{max}$ – the requested maximum output intensity;

$w$ – the window width;

$c$ – the window center;

The values of $g_{min}$ and $g_{max}$ are set to 0 and 4095, respectively. The parameters $R_s$, $R_i$ (in Eq. 20.1), $w$ and $c$ (in Eq. 20.2) are stored in the header of each DICOM file.

The normalization of the image intensities to 12 bits with respect to DICOM specification emphasizes the information connected with the lesion in the central part of the brain and the information related to the skull. After preprocessing, the maximum intensity value (i.e. 4095) represents the tissues with the highest calcium content, which, in the case of the brain images, belongs to the skull. Some elements of the CT tube were emphasized too (see Fig. 20.3b).

Since pixels belonging to the skull and the CT tube are irrelevant from the point of view of brain image segmentation and the assessment of the lesion, they can also be removed during preprocessing. As these pixels have the highest intensity value after normalization they can be removed by simple thresholding, in particular – by setting all the pixels of the maximum intensity to zero. In this way the redundant information connected with the skull and the CT tube is removed. The results of thresholding applied to an exemplary image from Figure 20.3b are presented in Figure 20.3c.

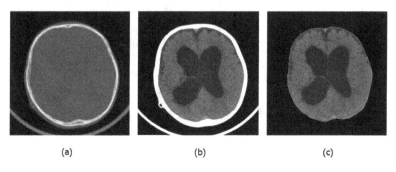

(a)             (b)             (c)

Fig. 20.3 Results of image enhancement (a) original image; (b) image after brightness and contrast adjusting; (c) results of skull removal.

A special group of image enhancement algorithms dedicated to CT brain scans is connected with recent requirements and restrictions related to the manner of CT imaging. In order to prevent the harmful effects of radiation to a human body and long-term consequences caused by ionizing radiation, sometimes the dose of x-rays used during image acquisition is reduced to the minimum. This is especially important in the case of children, which are more sensitive to harmful radiation. Reduced doses of radiation significantly lower the quality and contrast of the produced CT images and introduce so called film grain artifacts. These cause limitations in the accuracy of visual assessment, but also influence the precision and accuracy of image processing algorithms.

One of possible solutions to improve the quality of low-dose CT images is to apply the traditional low-pass image enhancement filters such as: the median filter, the Gaussian filter or the averaging filter. However, having in mind the side effects of these methods (e.g. additional blurring), in most cases the results obtained are not satisfactory.

On the other hand, the lost pixel values can be retrieved by means of graph-based anisotropic intensity interpolation. The method of interpolation used by the authors to enhance low dose CT images uses the combinatorial Laplace equation with Dirichlet boundary conditions given by the known values (Grady and Schwarz, 2003). Solving the Laplace equation allows one to "fill-in" missing values, as described in (Grady and Schwarz, 2003). The result of enhancement of a low-dose CT brain image is shown in Fig. 20.4. While Fig. 20.4a presents the original image, while Fig. 20.4b shows the image after graph-based intensity interpolation.

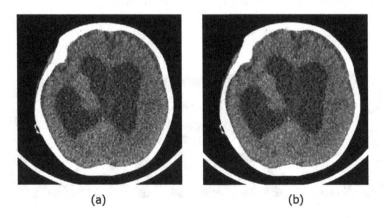

(a)                                      (b)

Fig. 20.4 Result of enhancement of low x-ray CT image (a) original image; (b) image after graph-based intensity interpolation.

### 20.3.2 *Image segmentation*

The purpose of image segmentation is to extract significant image information (relevant from the point of view of a certain application) from the entire information contained within a CT or MRI scan. The goal of segmentation is to simplify and change the representation of an image into binary segments (ROI's) which correspond to meaningful regions within the brain and are easier to analyze. In the case of brain image processing, image segmentation may involve both

(1) segmentation of intracranial brain;
(2) segmentation of the lesion;

The abovementioned steps are particularly important in the assessment of hydrocephalus in children. After the brain intracranial area and the brain ventricular system filled with cerebrospinal fluid (CSF) are properly segmented, the key characteristics of hydrocephalus can be determined based on the segmentation results (see Sect. 20.4).

The results of segmentation of the intracranial brain and lesion caused by hydrocephalus are presented in Fig. 20.5. In particular, Fig. 20.5a shows an exemplary CT input image and Fig. 20.5b presents the results of the segmentation of the intracranial brain, and Figure 20.5c shows the extracted hydrocephalus area.

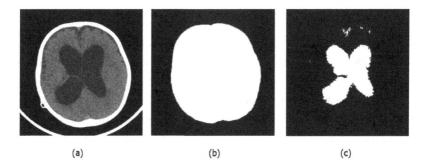

(a)                              (b)                              (c)

Fig. 20.5 Results of image segmentation (a) original image; (b) segmentation of intracranial brain; (c) segmentation of hydrocephalus.

The result of intracranial brain segmentation may be used as a binary mask that restricts the processing area. The background can easily be separated by suppressing (setting to zero) the pixels outside the mask. In particular, such boundaries are useful for segmentation by the traditional region-based methods. These methods are generally prone to over-segmentation. In addition, satisfactory results on the restricted intracranial brain area can be obtained using traditional thresholding algorithms (Węgliński and Fabijańska, 2011).

Brain lesion segmentation from volumetric CT or MRI images can be performed in both 2D and 3D manners. The choice of method depends on a particular application. The key factor that determines the selection of the algorithm is related to the expected accuracy of image segmentation and the processing time following image analysis. When the planar or linear analysis of the lesion is performed (as described in Section 4.1) 2D segmentation is sufficient. In such a case image segmentation can be performed on one – the most representative - slice of the brain scan. For the volumetric measurements (described in Section 4.2) both 2.5D (i.e. performed slice-by-slice) or 3D segmentation algorithms can be used. The results of the cerebrospinal fluid (CSF) segmentation from exemplary CT scans of hydrocephalic brains using various image segmentation algorithms are presented in Figures 20.6-20.8. In particular, Fig. 20.6 shows the results of planar segmentation provided by three popular thresholding algorithms compared to the results of the 3D segmentation algorithm based on region growing as described in (Węgliński and Fabijańska, 2011). In the case considered, of the several thresholding methods tested, namely: IsoData (Ridler and Calvard, 1978), Moments (Tsai, 1985) and MaxEntropy (Kapur et al., 1985), the MaxEntropy algorithm gives subjectively the best results (see Fig. 20.6).

In Fig. 20.7 the results provided by MaxEntropy thresholding are compared with the results of the 3D segmentation algorithm. Before the segmentation, the median filter was applied to the smooth contours of the brain lesion caused by hydrocephalus.

The image segmentation algorithms presented above represent the fully automatic approaches. However, in many cases automatic segmentation algorithms fail when applied for the extraction of complex structures within the brain. In these cases semi-automatic approaches often provide better results. These approaches require human interference. In particular, the operator (e.g. the radiologist) is expected to impose initial conditions on the lesion and the background by manually selecting the starting pixels (seeds) for each region being segmented.

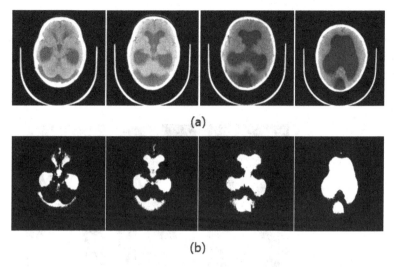

Fig. 20.6 Results CSF segmentation from brain using MaxEntropy tresholding (a) original images; (b) results of the MaxEntropy segmentation.

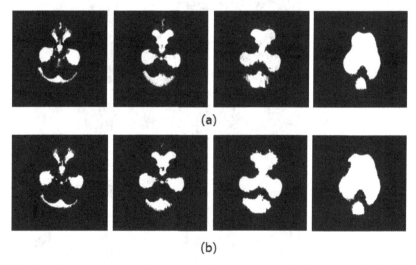

Fig. 20.7 Results of hydrocephalus segmentation using (a) MaxEntropy and (b) 3D region growing. Original images are shown in Fig. 20.6.

However, different algorithms have different requirements for the proper selection of seeds. For instance, random walk (Grady, 2006) and min-cut/max-flow (Boykov and Jolly, 2001) algorithms require the selection of both object and background seeds, whereas the level set approach is initialized only with object seeds. Fig. 20.8 shows the results of CSF segmentation using the approaches considered. Specifically, the first row presents the input images, the second row

contains the images with marked labels for the object and the background. The
third row presents the results of the random walk approach, the fourth row shows
the results of using the min-cut/max-flow segmentation algorithm. Finally, the
fifth row presents the results of level set segmentation.

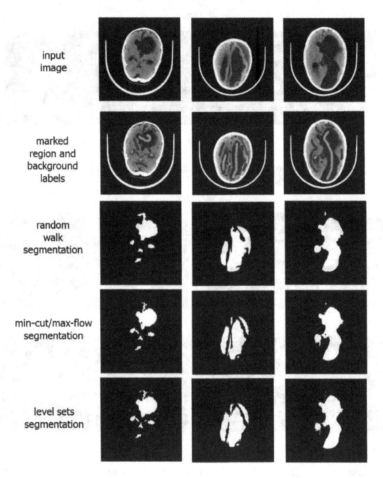

Fig. 20.8 Comparison of image segmentation results provided by the considered algorithms.

In the case of volumetric datasets, the results of image segmentation can later
be used to build a 3D model of the brain and the lesion. This is shown in Fig.
20.9, which presents the results of 3D segmentation of the intracranial brain from
the MRI image dataset, shown from four different angles. Then, in Fig. 20.10 the
results of the 3D tumor segmentation are compared with the whole brain.

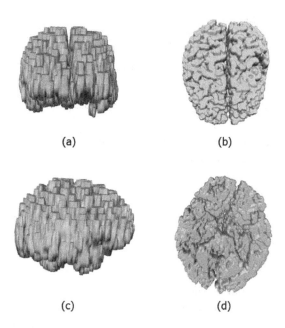

Fig. 20.9  3D view of segmented brain from four angles (a) front, (b) top, (c) left side and (d) bottom.

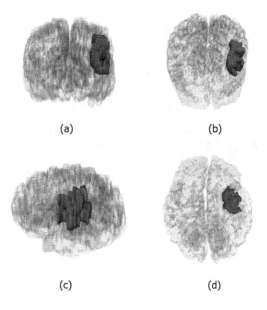

Fig. 20.10  3D view of segmented brain tumor compared to whole brain (a) front, (b) top, (c) side affected by a tumor, (d) bottom.

### 20.3.3 *Post-processing*

The goal of the post-processing step is similar to image enhancement. The difference is that it enhances the results of image segmentation and the input data is typically a binary result provided by the image segmentation algorithm. The quality improvement in the post-processing step generally aims at

(1) the removal of interconnected pixels and outliers;
(2) the smoothing of the edges;
(3) hole filling;

The post-processing step typically involves basic and complex morphological operations (e.g. erosion, dilatation, hole filling) and spatial filtering (i.e. median filter). The results of applying several post-processing morphological methods to exemplary binary images provided by segmentation are presented in Figures 20.11-20.13. In particular, Figure 20.11 presents the results of improving the quality of an exemplary image after segmentation by removing small outliers which are not connected with the main object (i.e. the lesion caused by hydrocephalus). The outliers are removed by morphological closing. Fig. 20.12 shows the binary images after hole filling. Finally, Fig. 20.13 presents the image with its edges smoothened by the median filter.

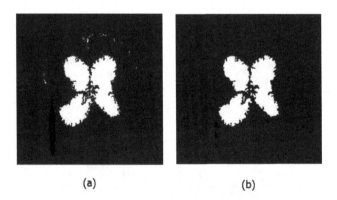

(a) (b)

Fig. 20.11 Result of removing outliers with morphological closing; a) input image; b) image after post processing.

The post-processing step is relevant to the accuracy of further analysis. The side effects of these methods generate errors, although the probable amount of lost information typically fluctuates around several percent. Hence, the post-processing step is desirable and important for most applications.

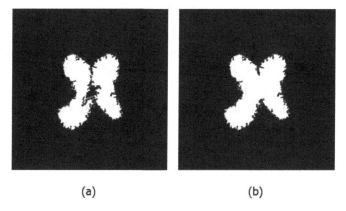

(a) (b)

Fig. 20.12 Result of hole filling in binary image after segmentation; a) input image; b) image after post processing.

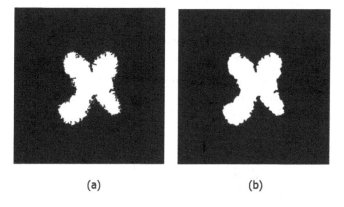

(a) (b)

Fig. 20.13 Result of edge smoothing in the binary image after segmentation using median filtering; a) input image; b) image after post processing.

## 20.4 Image analysis

The input data for image analysis is the binary image produced by image segmentation, optionally followed by the post-processing algorithm. This image is next used to determine specific measures describing the disease being assessed and typically involves the calculation of the key characteristics related to the lesion concerned.

The following sub-sections present a description of image analysis algorithms dedicated for a quantitative assessment and diagnosis of various brain lesions. The measured indices are particularly important for the assessment of disease progression. Depending on the application, image analysis can be performed linearly or volumetrically. In particular, planar or linear measurements are performed when one – the most representative slice is analyzed. But when the size or the volume of the lesion is concerned, volumetric calculations are made.

Because of a large number of indicators that can be used to analyze various brain diseases, in this section special attention is paid to the problem of quantitative analysis of the pathological changes associated with hydrocephalus.

### 20.4.1 *Planar and linear measurements*

The diagnosis of brain diseases is usually based on the clinical signs and symptoms and typical radiological characteristics related to particular lesion. However, in some cases, it is hard to clearly assess the dynamics of pathological changes. Hence, additional numerical calculations should be performed. The methods and results described in this section are used for the quantitative analysis of hydrocephalus, but they can also be applied for the analysis of other brain disorders, such as cerebral atrophy, intracranial hemorrhage or tumors. The major linear parameters used to classify hydrocephalus include (Barkovich, 2005)

(1) the Evans ratio (ER);
(2) the frontal and occipital horn ratio (FOHR);
(3) the ventricular angle (VA);
(4) the frontal horn radius (FHR);

All the above mentioned parameters used for the assessment of hydrocephalus are linear, which means that they are determined based on the most representative scan, selected automatically from the image dataset or manually by the radiologist. The determination of all the indices considered is performed on the binary images obtained from the segmentation process. The linear functions are determined based on a priori knowledge about the head shape.

The calculations of the Evans ratio and the Frontal and the occipital horn ratio are performed based on the measurements of the maximum widths within the considered regions. In turn, the frontal horn radius and the ventricular angle are calculated by measuring the diameters of the frontal horns. The following paragraphs briefly describe the considered parameters and present the methods proposed used to perform these measurements.

The Evans ratio (Synek et al., 1976) is the ratio between the maximum width of the anterior (frontal) horns to the maximum width of the inner skull (see Fig. 20.14a). Using the symbols shown in Fig. 20.14, it can be defined by Equation 20.6.

$$ER = \frac{E1}{E2} \tag{20.6}$$

The frontal and occipital horn ratio (FOHR) (O'Hayon, 1998) is the ratio between the sum of the maximum width of the anterior and the occipital horns and the width of the inner skull at the level of the minimum width of the ventricular system (see Fig. 20.14b). Using the symbols shown in Fig. 20.14, it can be defined by Equation 20.7.

$$FOHR = \frac{A + B}{2C} \tag{20.7}$$

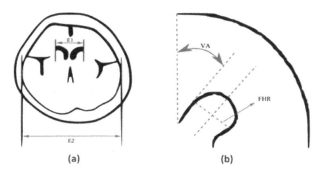

(a)  (b)

Fig. 20.14 Determination of Evans ratio (a) and Frontal Occipital Horn Ratio (b) (adapted from (Hammano et al., 1993) and (Lie et al., 2002)).

The widths considered must be measured in a direction perpendicular to the central sagittal plane of the head. The central sagittal axis of the head can be assumed to be a line which passes through the tip of the nose and the gravity center of intracranial region. To obtain reliable measurements, the characteristic dimensions must be measured in a direction perpendicular to the central sagittal axis of the head. These are the lengths of segments containing the brain characteristic dimensions as shown in Figure 20.15. In particular, the first row shows the results of determination of the maximum width of the inner skull. The second row presents the results of determination of the maximum width of the anterior horns and maximum width of the occipital horns.

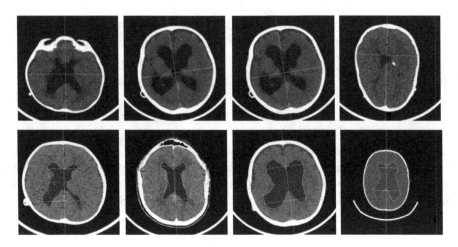

Fig. 20.15  Partial results of the calculation of Evans ratio and FOHR indices on the exemplary CT brain slices.

The ventricular angle (VA) is the angle made by the anterior or superior margins of the frontal horn at the level of the foramina of Monro (see Fig. 20.14). In turn, the frontal horn radius (FHR) is determined by measuring the widest diameter of the frontal horns (left and right) taken at a 90° angle to the long axis of the frontal horn (see Fig. 20.14). The increase in the FHR causes the narrowing of the ventricular angle.

The FHR may be found based on binary images of the segmented cerebrospinal fluid region as shown in Figure 20.16. First, the horn must be separated from the CSF region.

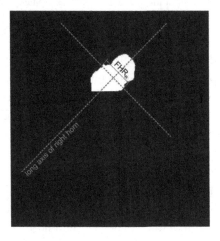

Fig. 20.16  Determination of long maximal width of FHR.

The value of VA may be determined as an angle between the central sagittal axis of the head and a line fitted into the upper edge of the horn, as shown in Figure 20.17.

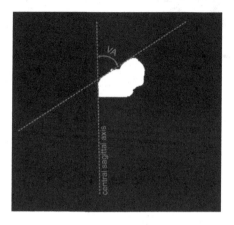

Fig. 20.17 Determination of VA.

### 20.4.2 *Volumetric measurements*

Image analysis based on the calculation of linear indices related to particular lesions ignores the volumetric information about the pathological changes inside the brain. However, in some applications volume measurements may provide relevant information about the activity of the disease process. Volumetric methods of disease assessment are currently not widespread in clinical practice because they require the manual tracing of lesion areas, sometimes across hundreds of scans covering the area of the brain. This solution is very inaccurate, time-consuming and too cumbersome to be performed in everyday clinical routine. However, such an analysis can easily be performed using the results of 3D image segmentation.

In the case of volumetric images stored in DICOM format, using the parameters stored in a header of each file, a single pixel can be regarded as a cuboid of the height equal to the slice thickness and the base of dimensions equal to pixel spacing X and pixel spacing Y, respectively. This idea is sketched in Figure 20.18.

Having the volume of a single pixel, approximate volumes of the brain and hydrocephalus can easily be calculated, simply by counting the number of the pixels qualified (by segmentation algorithms) to these regions and multiplying the result by the volume of a single pixel in accordance with Equations 20.8 and 20.9.

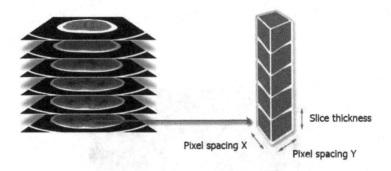

Fig. 20.18 The idea of brain volume calculation from binary images.

$$B = V_p \sum_{z=1}^{n} \sum_{x=1}^{M} \sum_{y=1}^{N} b(x, y, z) \qquad (20.8)$$

$$H = V_p \sum_{z=1}^{n} \sum_{x=1}^{M} \sum_{y=1}^{N} h(x, y, z) \qquad (20.9)$$

where:

$B$     – the volume of the brain;

$H$     – the volume of hydrocephalus;

$n$     – the number of images (brain cross sections) in the dataset;

$M,N$ – the dimensions of a single cross section (in pixels);

$h$     – the binary image of hydrocephalus;

$x,y,z$ – the pixel coordinates;

## 20.5 Conclusions

The notable growth of knowledge in contemporary medicine, especially in the field of radiology and medical imaging, exert influence on the significant progress in the computerized assessment and diagnosis of pathological conditions within the human brain. Medical image processing investigates various problems, from the image enhancement and reconstruction, to segmentation and image analysis. The main goal of applying image-processing algorithms in the computer-aided diagnostics is to increase the objectivity and repeatability of the assessment and diagnosis of various brain diseases.

In this chapter, the focus was on the description of the fundamentals of medical image processing and the authors' algorithms, used to process digital CT and MRI images. The results presented mainly come from the ongoing research performed by the authors, aimed at the detection and quantitative analysis of hydrocephalus in children. However, these methods make a universal set of tools, which can be extended to other problems and applications.

## References

Ahmed, M.N., Yamany, S.M., Mohamed, N., Farag, A.A., Moriarty, T. (2002). A modified fuzzy C-means algorithm for bias field estimation and segmentation of MRI data, *IEEE Transactions on Medical Imaging* 21 (3) , pp. 193-199

Artaechevarria, X., Muñoz-Barrutia, A., Ortiz-de-Solórzano, C. (2009). Combination strategies in multi-atlas image segmentation: Application to brain MR data, *IEEE Transactions on Medical Imaging*, 28 (8) , art. no. 4785214 , pp. 1266-1277

Barkovich A.J. (2005). Pediatric Neuroimaging, *Lippincott, Williams & Wilkins*, New York, USA

Boussion, N., Hatt, M., Lamare, F., Bizais, Y., Turzo, A., Rest, C.C.-L., Visvikis, D. (2005). Generating resolution-enhanced images for correction of partial volume effects in emission tomography: A multiresolution approach, *IEEE Nuclear Science Symposium Conference Record*, pp. 2423-2427

Boussion, N., Hatt, M., Lamare, F., Rest, C.C.L., Visvikis, D. (2008). Contrast enhancement in emission tomography by way of synergistic PET/CT image combination, *Computer Methods and Programs in Biomedicine* 90 (3) , pp. 191-201

Boussion, N., Hatt, M., Reilhac, A., Visvikis, D. (2007). Fully automated partial volume correction in PET based on a wavelet approach without the use of anatomical information, *IEEE Nuclear Science Symposium Conference Record*, pp. 2812-2816

Boykov Y., Jolly M.P. (2001). Interactive Graph Cuts for Optimal Boundary & Region Segmentation of Objects in N-D Images, *Proceedings of "Internation Conference on Computer Vision*, 1, pp. 105-112

Butman J.A., Linguraru M.G. (2008). Assessment of ventricle volume from serial MRI scans in communicating hydrocephalus, *5th IEEE International Symposium on Biomedical Imaging: From Nano to Macro*, pp. 49-52

Chen, J., Zhong, N. (2008). Data-Brain modeling based on Brain informatics methodology, *Proceedings - 2008 IEEE/WIC/ACM International Conference on Web Intelligence*, WI 2008, pp. 41-47

del Fresno, M., Vénere, M., Clausse, A. (2009). A combined region growing and deformable model method for extraction of closed surfaces in 3D CT and MRI scans, *Computerized Medical Imaging and Graphics*, 33 (5) , pp. 369-376

Grady L. (2006). Random Walks for Image Segmentation, *IEEE Transactions on Pattern Analysis and Machine Intelligence*, 28 (11), pp. 1768-1783

Grady L., Schwartz E. (2003). Anisotropic interpolation on graphs: The combinatorial dirichlet problem, *Tech. Report CAS/CNS-TR-03-014*, Boston University, Boston (MA)

Grady, L., Schwartz, E.L. (2003). The Graph Analysis Toolbox: Image processing on arbitrary graphs, *Technical Report TR-03-021*, Boston University, Boston (MA)

Hamano K., Iwasaki N., Takeya T. and Takita H. (1993). A Com-parative Study of Linear Measurements of The Brain and Three-Dimensional Measurement of Brain Volume Using CT Scans, *Pediatric Radiology*, 23, pp. 165–168

Kapur J.N., Sahoo P.K., Wong A.K.C. (1985), A new method for gray-level picture thresholding using the entropy of the histogram, *Computer Vision, Graphics, & Image Processing*, 29 (3), pp. 273-285

Lie W.N., Peng W.H. and Chuung C.H. (2002). Efficient Content-Based CT Brain Image Retrieval by Using Region Shape Features, *Proceedings of IEEE International Symposium on Circuits and Systems*, 4, pp. 157–160

Lo, M.-T., Hu, K., Liu, Y., Peng, C.-K., Novak, V. (2008). Multimodal pressure-flow analysis: Application of Hilbert Huang transform in cerebral blood flow regulation, *Eurasip Journal on Advances in Signal Processing*, 2008 , art. no. 785243

Maniadakis, M., Trahanias, P. (2005). A hierarchical coevolutionary method to support brain-lesion modelling, *Proceedings of the International Joint Conference on Neural Networks*, 1, pp. 434-439

Maniadakis, M., Trahanias, P. (2009). Agent-based brain modeling by means of hierarchical cooperative coevolution, *Artificial Life*, 15 (3) , pp. 293-336

Mohamed, Nevin.A., Ahmed, M.N., Farag, A. (1999). Modified fuzzy c-mean in medical image segmentation, ICASSP, IEEE International Conference on Acoustics, *Speech and Signal Processing – Proceedings*, 6 , pp. 3429-3432

Näppi, J., Yoshida, H. (2008). Adaptive correction of the pseudo-enhancement of CT attenuation for fecal-tagging CT colonography, *Medical Image Analysis* 12 (4) , pp. 413-426

Noda, N., Matsuura, H., Nemoto, T., Koide, K., Nakano, M. (2008). Two-dimensional simulation on liquid flows in vessels based on navier-stokes equations, *International Journal of Innovative Computing, Information and Control*, 4 (2) , pp. 255-262

O'Hayon B. B., Drake J. M., Ossip M. G., Tuli S. and Clarke M. (1998). Frontal and Occipital Horn Ratio: A Linear Esti-mate of Ventricular Size for Multiple Imaging Modalities in Pediatric Hydrocephalus, *Pediatric Neurosurgery*, 29 (5), pp. 245-249

Prastawa, M., Gilmore, J.H., Lin, W., Gerig, G. (2005). Automatic segmentation of MR images of the developing newborn brain, *Medical Image Analysis* 9 (5 SPEC. ISS.) , pp. 457-466

Pustkova R., Kutalek F., Penhaker M., Novak V. (2010). Measurement and calculation of cerebrospinal fluid in proportion to the skull, *9th RoEduNet IEEE International Conference*, RoEduNet, pp. 95-99

Ridler T.W., Calvard S. (1978). Picture thresholding using an iterative selection method, *IEEE Transactions on Systems*, Man and Cybernetics, 8, pp. 630-632

Sun, F., Morris, D., Lee, W., Taylor, M.J., Mills, T., Babyn, P.S. (2010), Feature-space-based fMRI analysis using the optimal linear transformation, *IEEE Transactions on Information Technology in Biomedicine*, 14 (5) , art. no. 5497155 , pp. 1279-1290

Synek V., Reuben J. R. and Du Boulay G. H. (1976). Comparing Evans Index and Computerized Axial Tomography in Assessing Relationship of Ventricular Size to Brain Size, *Neurology*, 26, pp. 231–233

Tsai W.-H. (1985), Moment-preserving thresholding: a new approach, *Computer Vision, Graphics & Image Processing*, 29 (3), pp. 377-393

Węgliński T., Fabijańska A. (2011), Image segmentation algorithms for diagnosis support of hydrocephalus in children, *Kwartalnik AGH Automatyka*, pp. 309-319

Xue, J.-H., Pizurica, A., Philips, W., Kerre, E., Van De Walle, R., Lemahieu, I. (2003). An integrated method of adaptive enhancement for unsupervised segmentation of MRI brain images, *Pattern Recognition Letters* 24 (15) , pp. 2549-2560

Zang, X., Wang, Y., Yang, J., Liu, Y. (2010). A novel method of CT brain images segmentation, *International Conference on Medical Image Analysis and Clinical Application*, MIACA 2010, pp. 109-112

Zhang, Y., Bazilevs, Y., Goswami, S., Bajaj, C.L., Hughes, T.J.R. (2007). Patient-specific vascular NURBS modeling for isogeometric analysis of blood flow, *Computer Methods in Applied Mechanics and Engineering*, 196 (29-30) , pp. 2943-2959

Zhang, Y., Brady, M., Smith, S. (2001). Segmentation of brain MR images through a hidden Markov random field model and the expectation-maximization algorithm, *IEEE Transactions on Medical Imaging* 20 (1) , pp. 45-57

# CHAPTER 21

## COMPUTER SYSTEMS FOR STUDYING DYNAMIC PROPERTIES OF MATERIALS AT HIGH TEMPERATURES

Marcin Bąkała, Rafał Wojciechowski, and Dominik Sankowski

*Institute of Applied Computer Science*
*Lodz University of Technology*
*90-924 Łódź, ul. Stefanowskiego 18/22*
*{m.bakala, r.wojciechowski, dsan}@kis.p.lodz.pl*

### 21.1 Introduction

Materials engineering is nowadays one of the most important and rapidly growing science area. Knowledge of the structure of materials, their physical-chemical properties and behavior in appropriate external conditions allows new materials to be designed and used according to the customer needs. The understanding and the ability to manipulate materials and their properties is a key factor in any industrial process or technology. In many processes in metallurgy, material and surface engineering areas, i.e. soldering technologies, composite material production, pulver sintering processes, etc., detailed information on the phenomena on solid and liquid phases boundary play an important role. Surface tension and liquid phase surface energy, wetting angle – including extreme basis wetting angle, adhesion energy and interphase tension are the parameters that are important in modern soldering technologies.

The introduction of EU directive RoHS (2002/95/WE), which specifies detailed requirements of use, in electrical and electronic devices, of certain substances which can have an effect on the natural environment during their activity and till the end of its usage, and the WEEE, concerning the use of electrical and electronic devices, are in force in many countries (ROHS, 2003; WEEE, 2003). The above-mentioned directives involve a production technology change, i.e. a need for the production of devices using non-lead soldering technologies. The main problems related to the new technology application are an increase in production costs, product quality and reliability improvement.

Nowadays, the research in materials engineering also focuses on modern, automatic measurement methods of joining process properties, i.e. wetting force and surface tension. The obtainment of quantitative results allows one to compare the physical-chemical properties of the solders and brazes examined as well as to check the cross-phase effects of appropriate materials. The article is concerned with computer systems for studying dynamic properties of materials at high temperatures – an integrated platform allowing the automatic determination of wettability and the surface tension of high-temperature brazes, and a concept of an automated device for surface tension determination at high temperatures (Wojciechowski, 2012).

## 21.2 Interphase phenomena in high-temperature processes

The driving force for the formation of an interface between two materials is a decrease in Gibbs energy that occurs when intimate contact is established between two material surfaces as the system strives towards the minimum total energy. If the liquid does not completely cover the solid, the liquid surface will intersect the solid surface at contact angle $\theta$. Young equation describes the equilibrium value of $\theta$, used to define the wetting behavior of the liquid.

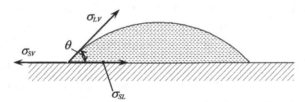

Fig. 21.1 A drop of liquid on the solid surface at equilibrium condition.

The condition of mechanical equilibrium of the system is:

$$|\bar{\sigma}_{LV}|\cos\theta + |\bar{\sigma}_{SL}| - |\bar{\sigma}_{SV}| = 0 \qquad (21.1)$$

For most liquid metals at high temperatures the numerical equality of surface tension and surface energy can be assumed. After the appropriate transformation, the following equation, known as Young's equation, is obtained:

$$\cos\theta = \frac{\omega_{SV} - \omega_{SL}}{\omega_{LV}} \qquad (21.2)$$

In the case of energy balance the shape of the drop on a horizontal surface is the result of two kinds of forces: surface tension forces, which try to give a spherical shape to the drop and the force of gravity by which the drop is flattened. Because no analytical solution of the shape of the surface in gravitational fields (differential equation of the second degree) is available in the sessile drop configuration, in further considerations a meniscus formed on a vertical plate was used.

$$\frac{\frac{\partial^2 z}{\partial x^2}[1+(\frac{\partial z}{\partial y})^2]+\frac{\partial^2 z}{\partial y^2}[1+(\frac{\partial z}{\partial x})^2]-2\frac{\partial z}{\partial x}\cdot\frac{\partial z}{\partial y}\cdot\frac{\partial^2 z}{\partial x\partial y}}{[1+(\frac{\partial z}{\partial x})^2+(\frac{\partial z}{\partial y})^2]^{3/2}}=\frac{z}{a^2}+c \qquad (21.3)$$

Young's equation can be used to describe the variation of the surface energy of the system consequent by a small displacement $dz$ of the S/L/V (point TL is three-phase contact point). The total free energy change can be obtained when the liquid surface initially in a horizontal position ($z^*$, $\theta = 90°$) is raised (or depressed) to form a meniscus of height $z^*$, corresponding to the contact angle $\theta$ (Kwok, D., 1995, Eustanthopoulos, N., 1999).

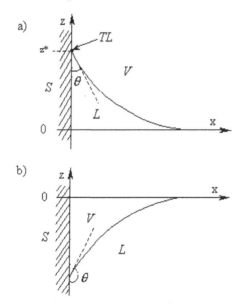

Fig. 21.2 Height of the meniscus on a vertical plate where $\theta$ is less than 90° (a), decrease when $\theta$ is greater than 90° (b).

For any rise $z^*$ of the meniscus, $z^*$ and $\theta$ are related by:

$$z^* = \pm\left(\frac{2\sigma_{LV}}{\rho g}\right)^{\frac{1}{2}}(1-\sin\theta)^{\frac{1}{2}} = \pm l_c(1-\sin\theta)^{\frac{1}{2}} \qquad (21.4)$$

## 21.3 Methodology of determination of dynamic properties of materials

### 21.3.1 *Surface tension determination using the maximum bubble pressure*

The measurement method for determining high-temperature surface tension is based on sinking the pipe in a liquid metal at an appropriate depth $h$. The experiment is carried out at a high temperature, in gas reductive-protective atmosphere. The first stage of the experiment flow is the initial process stabilizing thermodynamic conditions in the furnace chamber, where the chamber is heated in gas atmosphere. The main stage of the experiment starts with sinking a pipe in a liquid metal at an appropriate depth $h$. During the experiment, gas is delivered from outside the controlled system to the pipe, a bubble at the end of the pipe is formed and the pressure in the newly formed bubble is registered. Point A (Fig. 21.3) responds to a sweep gas pressure increase, which results in bubble formation. The shape of a gas bubble and the value of the registered pressure depend on the wettability of the pipe by the fluid metal. At point B, the pressure achieves the maximum $P_{max}$ and the bubble has a quasi-semicircular shape, the radius of which is equal to the geometrical radius of the pipe. At points B and C, the pressure decreases and the radius of the bubble curvature is still increasing. Finally, the bubble is released from the pipe and the cycle is repeated.

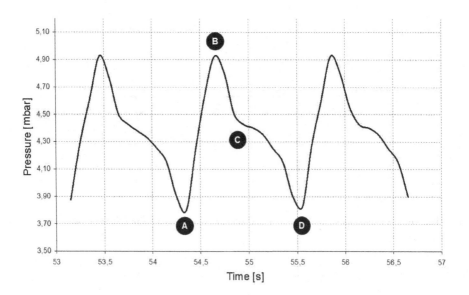

Fig. 21.3 Gas pressure changes during the bubbling process.

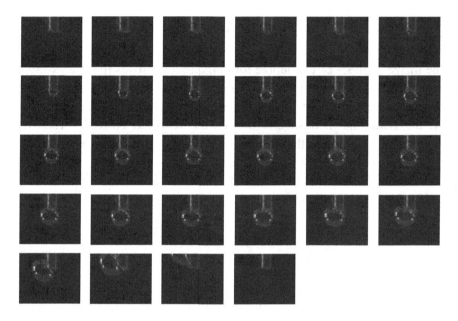

Fig. 21.4 Bubble forming process (pipe with 2mm diameter immersed in glycerol at a depth of 30mm).

In the equilibrium state, the bubble pressure $P_p$ equalizes the sum of pressures limiting the bubble volume increase. For every point A of the bubble surface:

$$P_p = P_r + P_h(A) + P_\sigma(A) \tag{21.5}$$

where: $P_p$ - the bubble pressure (not depending on point A), $P_r$ - the chamber pressure over the braze surface (not depending on point A), $P_h(A)$ - the hydrostatic pressure (function of the depth $h$ related to the braze surface), $P_\sigma(A)$ - the pressure related to the surface tension.

During the experiment, the pressure difference $\Delta P = P_p - P_r$ is measured; Eq. 21.5 can be transformed to:

$$\Delta P = \rho g h(A) + \sigma \left( \frac{1}{R_{1A}} + \frac{1}{R_{2A}} \right) \tag{21.6}$$

The assumption for surface determination using the maximum gas bubble pressure method is that the bubble shape near the pipe border slightly differs from the spherical surface with radius $r$, which allows one to simplify Eq. 21.6 to the final form, allowing the surface tension to be determined:

$$\sigma = (\Delta P_{max} - \rho gh) \cdot \frac{r}{2} \tag{21.7}$$

The image analysis of the bubble forming process proves that the assumption of semisphere bubble shape for the maximum pressure brings some method error, which increases accordingly to the factor $\rho / \sigma$ – the larger the value of the factor, the flatter the hemisphere towards the revolved ellipsoid. The ADSA analyzing method confirms that the maximum pressure occurs in a bubble with a bigger volume than the volume of hemisphere with radius $r$ (Fig. 21.5). Therefore, there exists another form of Eq. 21.7 formulated by Schroedinger (Miyazaki, M., 1997, Adamson, A., 1982):

$$\sigma = \frac{r \Delta P_{max}}{2} \left[ 1 - \frac{2}{3} gr^2 \omega - \frac{\left( r^2 g \omega \right)^2}{12} \right] \tag{21.8}$$

where: $\omega = \rho / \sigma$, $\sigma$ - the surface tension calculated using Eq. 21.7.

 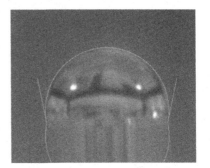

Fig. 21.5  ADSA analyzing method of gas bubble shape in a liquid.

### 21.3.2 Wettability determination using the immerse method

The measurement method for determining high-temperature surface tension is based on sinking the pipe in a liquid metal at an appropriate depth $h$. The experiment is carried out at a high temperature, in a gas reductive-protective atmosphere. The first stage of the experiment flow is the initial process stabilizing thermodynamic conditions in the furnace chamber, where the chamber is heated in gas atmosphere. The main stage of the experiment starts with contact of the specimen front surface with fluid braze (Fig. 21.6, a, point

A). In the middle of sinking the specimen (Fig. 21.6, a, stage A - C), the acting buoyancy force is rising linearly with an immersion depth. At the same time, the braze bath surface is deflected and a down-curved meniscus is formed. The wetting angle $\theta$ is changing until it reaches the minimum value, which is close to 180°. While the specimen is still immersing, the temperature of the specimen is growing up to the braze temperature. The bonds between the two-phase atoms and the SL-phase boundary are created. The end of the specimen immersion process is at point C (Fig. 21.6, a). The C - E stages correspond to the wetting progress. The wetting angle $\theta$ decreases to the equilibrium value $\theta_o$ at point E. The buoyancy force is constant at this stage; the capillary force is increasing, with a value equal to 0 at point D, where the wetting angle reaches 90°. From point D, the wetting angle forms an acute angle towards the limit value at point E. At point E, the specimen emergence process starts. The fluid meniscus is broken off at point F (Fig. 21.6, b). Point G corresponds the end of the experiment. The above-mentioned experiment flow is idealized, in fact, stages B - C can be very short or even put at stages A - B. If some interphases are created, the stage properties may change.

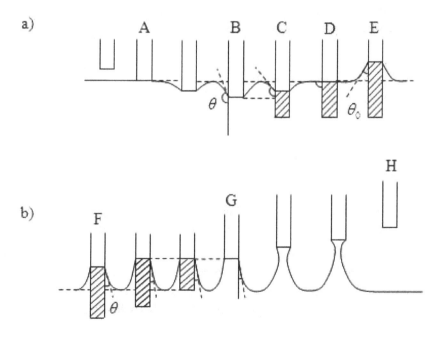

Fig. 21.6 Specimen immersed in fluid braze (a) and emergence (b) stages.

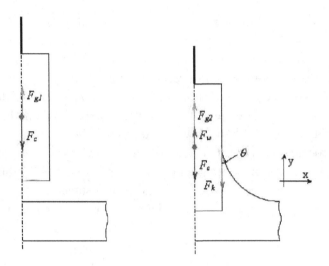

Fig. 21.7  Distribution of forces acting on the vertical specimen and the measurement unit.

The distribution of forces acting on the vertical specimen and the measurement unit (e.g. scales), before and after partial immersion, is presented in Fig. 21.7 (Park, J. 1999, Chang, H. 2003). The capillary force is determined by forces acting on the measurement device.

Before immersion:
  (1) the force registered by the measurement device before immersion, directed upwards: $F_{g1}$,
  (2) the gravity force, directed down $F_c$,

$$\sum_{i=1}^{n} F_{iy} = F_{g1} - F_c = 0 \qquad (21.9)$$

After immersion:
  (3) the force registered by the measurement device before immersion, directed upwards: $F_{g2}$, where $F_{g2} > F_{g1}$,
  (4) the buoyancy force directed upwards: $F_w$,
  (5) the capillary force: $F_k$ (directed upwards when $0 \leq \theta \leq 90°$, down when $90 \leq \theta \leq 180°$).

In the steady state ($\theta$ is an equilibrium angle $\theta_o$), when the specimen front surface is placed at the depth $h$ for the horizontal liquid surface, the force acting on the measurement device is denoted as:

$$F_{g2} - F_{g1} = F_k - F_w \qquad (21.10)$$

In that case, the capillary force $F_k$ is described by the formula:

$$F_k = O_p \sigma_{LV} \cos \theta_0 \qquad (21.11)$$

where: $O_p$ - the specimen perimeter, $\sigma_{LV}$ - the surface tension and $\theta_o$ - the equilibrium wetting angle.

And after the transformation:

$$Fk = \frac{F_{g2} - F_{g1} + P_p \rho g h}{O_p} \qquad (21.12)$$

where: $P_p$ - the cross-section area, $\rho$ - the braze density, $g$ - the gravity acceleration.

## 21.4 Integrated platform for the automatic determination of high-temperature braze properties

### 21.4.1 *Overview*

An integrated platform for testing properties of the solder surface is an automated stand enabling a comprehensive study of dynamic properties of solders and brazes – the surface tension and the wetting force, at temperatures up to 1000 °C with the use of various technological gas atmospheres (an overview of the integrated platform is presented in Fig. 21.8).

The integrated platform has the following features (Wojciechowski, 2012):

(1) the determination of braze parameters: wetting force - using the immerse method (specimen immersed in fluid braze), and surface tension - using the maximum bubble gas pressure (gas bubbles generated in fluid braze),

(2) a wide range of working temperatures of up to 1000 °C with a braze bath temperature control accuracy of 0.5 K,

(3) carrying out experiments in a gas reductive and protective atmosphere (commonly based on argonium, nitrogen and hydrogen, 5 N purity) controlled by gas MFCs , which ensure gas flow regulation with 1% of a set value accuracy,

(4) precise scale module measuring the force acting on scales' tie with a 0.1 mg accuracy,

(5) an automated measurement cycle, the control of process parameters and an experiment flow in real time,

(6) a wide scope of testing specimens in the form of flat specimens, pipes and cylinders limited in size to the furnace retort and ceramic pot dimensions,

(7) a user friendly graphical interface allowing one to monitor the device current state, analyzing and reporting the already made experiments,

(8) a possibility to be applied for industrial and laboratory purposes in new joining technologies and materials design, brazing/soldering process parameter optimization and the verification of quality of existing technologies.

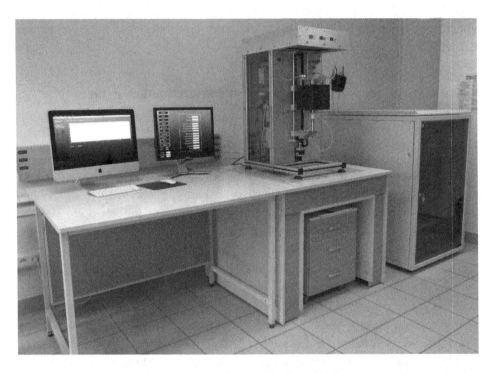

Fig. 21.8 Integrated platform for the automatic determination of high-temperature braze properties.

### 21.4.2 *System architecture*

The architecture of an integrated platform is based on a lazy coupled functional tier concept (Fig. 21.9).

**The hardware tier** is a separate autonomic device that allows experiments to be carried out according to given parameters. The main module of the hardware tier is a WAGO PLC controller running the Linux operating system, supervising all the hardware components in real-time, i.e. the execution units (furnace power controller, stepper driver, MFCs, scales) and the sensors (thermocouple, limit switches, optical sensors, pressure sensors). The communication with all the control and measurement devices is made through a set of extension I/O cards offering digital and analog current and voltage I/Os and serial ports in the RS232 and RS485 standard (Fig. 21.10) (WAGO, 2013). The PLC controller ensures also simple GUI with visualization of device's current state. The detailed structure of the hardware tier is presented in Fig. 21.11.

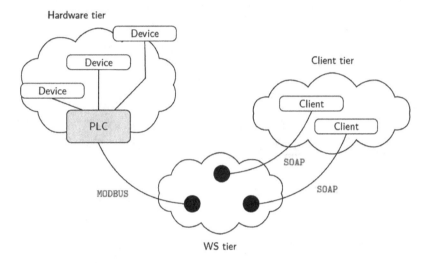

Fig. 21.9 General tier system architecture

**The web service tier** is an interface between the client tier, the testing device and the database repository system. The functionality of this tier consists of experiment template management methods (specimen-, materials- and experiment parameters definitions), testing device's current state monitoring and obtaining results of the experiments already carried out. The WS tier fulfills the SOA concept and exposes API for the client tier as a set of web service methods executed by HTTP/SOAP protocol. All the WSs were developed using the JAX-WS technology (JAX, 2013), the interaction with the database system, and necessary ORM mappings were made with the Hibernate framework (Hibernate, 2013); the communication with the PLC controller was based on the Modbus TCP protocol on common sockets. The application was integrated with the Spring framework and deployed on the Tomcat server (Spring, 2013).

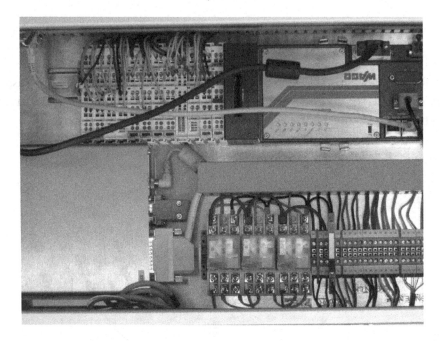

Fig. 21.10 Device controller and I/O modules setup.

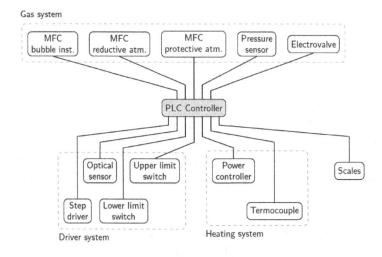

Fig. 21.11 Automation structure of wettability and surface tension analyzing system.

**The client tier** consists of software ensuring end-user interaction with all the other system components. The software offers a user friendly GUI, executes the WS methods of the neighbor tier and provides the information requested. The tier is also equipped with report- and experiment data analyzing engine. Due its graphical performance, the extensive component library and the possibility of execution in most operating systems and web browsers, the client tier was developed using the Adobe Flex technology and also deployed on the Tomcat server (Adobe, 2013). An example of GUI functionality – an experiment variable graph – is presented in Fig. 21.12.

Fig. 21.12 Experiment variable graph in the client's software.

The concept of platform architecture based on functional separated tiers allows easy system distribution (limited to network connectivity) and portability (platform independent programming technologies used for system development).

### 21.4.3 *Experiment flows*

The integrated platform for testing dynamic properties of the surface of solders at high temperatures allows determination of surface tension using the maximum gas pressure in a bubble released in fluid braze, and wettability using the immerse method. According to the above-mentioned measurement methodology (see chapter Methodology of determination of dynamic properties), the experiment flow can be organized as a set of a sequential action with conditional transitions between experiment stages (a sequence diagram of the immerse experiment is presented in Fig. 21.13 and a state diagram in Fig. 21.14).

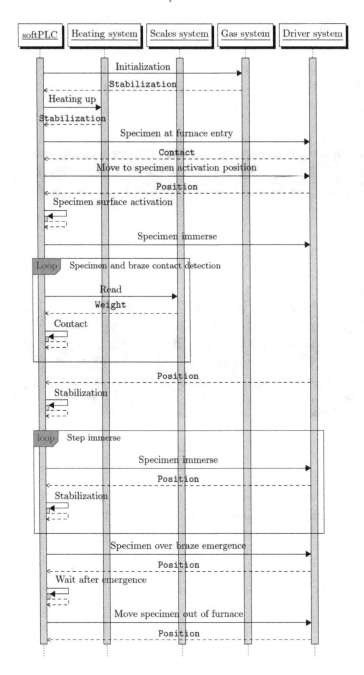

Fig. 21.13 Immerse experiment flow sequence diagram.

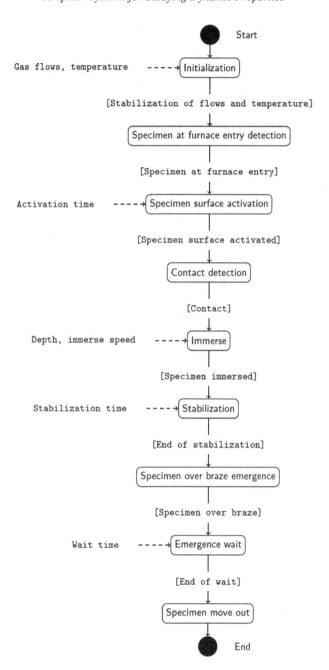

Fig. 21.14 Immerse experiment state diagram.

The experiment flow starts with a stage of stabilization of gas atmosphere and thermodynamic conditions in the furnace retort. The PLC controller sets appropriate gas flows on the MFCs according to the user defined settings (Fig. 21.14, gas flows and temperature) and begins to heat up the furnace. The stage ends, when the gas flows and the temperature in the furnace retort are stable (Fig. 21.14, stabilization of flows and temperature). The next step is the detection of the specimen by the furnace retort entrance while the heating system with the pot filled with the braze inside moves vertically to the specimen. The contact is detected by an optical sensor (Fig. 21.14, specimen at furnace entrance). In the next stage, the specimen is placed in the furnace retort, the surface of the specimen is activated and heated during the time set by the user (Fig. 21.14, activation time). After the specimen surface activation period, the specimen is immersed at an appropriate depth. The most important aspect at this stage is contact detection between the specimen front surface and the braze bath. The contact is detected by the scales system, where the registered specimen weight is rapidly changing (Fig. 21.13, specimen and braze contact detection). After the specimen is placed at an appropriate depth, it remains immersed for a given period of time (Fig. 21.14, stabilization time), and then is emerges over the braze bath. The specimen stays in this position during the time set by the user (Fig. 21.14, wait time), and then moves out of the furnace to the base position, which ends the experiment.

The bubble experiment has a similar flow, the difference is in immersing and remaining in immerse stages. The contact with the fluid braze during the immersion process is, in that case, detected by two electrodes, which are placed along the ceramic capillary, in an electrical manner – contact with a conductive braze closes the electrical circuit, which is registered by the PLC module. While it is at an immerse stage, the bubbles of gas are generated. The capillary is connected to an external gas source with a regulated pressure – the PLC opens the electrovalve and enables the gas to flow, which causes bubbling.

## 21.5. System for the automatic determination of high-temperature braze surface tension

### 21.5.1 *Overview*

The promising results from the image analysis of a bubble shape in the maximum gas bubble pressure method let one create a concept of an automated system for the automatic determination of high-temperature braze surface tension using the pendant drop method. The features and requirements should be compliant with the above-mentioned integrated platform for the automatic determination of high-temperature braze properties, in particular:

(1) the determination of the surface tension parameter using the pendant drop method,
(2) a wide range of working temperatures of up to 1000 °C with a temperature control accuracy of 0.5 K,
(3) carrying out experiments in a vacuum chamber,
(4) an automated measurement cycle, the control of the process parameters and the experiment flow in real time.

### 21.5.2 *System architecture*

The architecture of the system is based on the 3-tier concept presented in chapter 21.4.2. The tiers are extended with functionalities related to an image analyzing system.

**The hardware tier** consists of two main modules. The first one is a WAGO PLC controller running the Linux operating system, supervising all the hardware components in real-time, i.e. the execution units (power controller, vacuum pump, camera filters) and the sensors (thermocouple). The second one is the video grabbing unit consisting of a CCD video camera and a frame grabber supervised by dedicated drivers. The communication between the main modules can be done using the TCP MODBUS protocol, which should ensure turning on and off the grabber software and store the result images in an experiment repository. The detailed structure of the hardware tier is presented in Fig. 21.15.

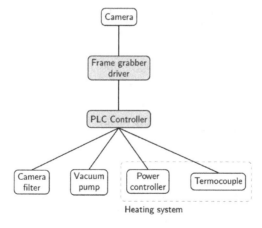

Fig. 21.15 Structure of the system for automatic determination of high-temperature braze surface tension of the hardware tier.

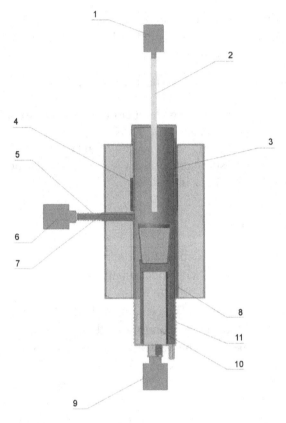

Fig. 21.16 Concept of a vacuum chamber integrated with a video system and a specimen installed inside.

**The web service tier,** besides all database and device monitoring activities, provides communication with the image result storage system allowing the client software to obtain a set of images of an appropriate experiment.

**The client tier** consists of image analysis methods based on ADSA methodology, described further in chapter 21.5.4.

The concept of a vacuum chamber is presented in Fig. 21.16. The central part of the chamber is the cylindrical retort space (3), where the specimen rod (2) hanging on the specimen grip (1) is placed. The heating system (4) is placed coaxially, round the specimen, just above the video system branch (5) creating a speculum (7) for camera and a filter set connection (6). The feedback for heating the system control comes from the thermocouple placed near the specimen. Below the specimen, on the ceramic pillar (8) filled with seal material (10), a pot for falling braze drops is placed. The pillar is connected with the loading system supported by a driver motor, which allows one to easily change the pot and caulk the chamber. The chamber is connected to the vacuum pump terminal (11).

### 21.5.3 *Experiment flow*

According to the above-mentioned measurement methodology (see chapter Methodology of determination of dynamic properties of materials), the pendant drop experiment flow can also be organized as a set of sequential actions with conditional transitions between experiment stages (Fig. 21.17).

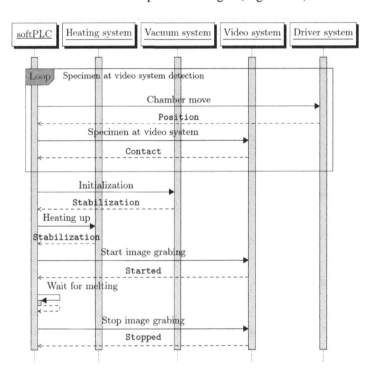

Fig. 21.17 Pendant drop experiment flow sequence diagram.

The experiment flow starts with the stage of the specimen position initialization – the specimen end has to be visible in the video system speculum. Then, the PLC controller waits for a vacuum stabilization signal from the vacuum pump system. The next step is the start of the specimen heat-up. When the temperature of the specimen rod is close to the expected temperature, the video system begins to register images. After a melting time specified by the user, the video grabber stops registering images and the experiment ends.

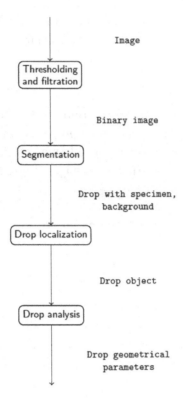

Fig. 21.18  Data analysis stages.

### 21.5.4 *Data analysis*

The pendant drop analysis is a sequence of image processing activities (Fig. 21.18). The starting point for the analysis is an image grabbed by the camera during the experiment. The first stage is image thresholding and filtering, based on the image histogram, which results in a binary image. The next step is image segmentation to obtain a drop with the specimen rod and background objects. At high temperatures the strong thermal radiation causes an aura phenomenon or a reflection effect in the areas around the drop and the specimen. This leads to the incorrect classification of the background pixels into the set of points representing the object, which results in a calculation inaccuracy. There is a large number of basic segmentation methods – Canny, Sobel, LOG and many others, which can be used for stage accomplishing (Kwok, D. 1995). Further, the stage of drop localization is performed. The result of this stage is a set of objects – the specimen rod, the drop and the background. The determination of surface tension is based on drop geometry calculations. The drop shape parameters described in chapter 21.3.2, can be estimated using the ADSA-P (Axisymmetric Drop Shape Analysis – Profile) developed by Neumann and coworkers, where the calculated theoretical Laplacian curve is compared to the actual drop profile until the specified overall is achieved. Finally, the surface tension is obtained from the gravitational parameter that results from the closest overlap between the theoretical curve and the actual drop profile.

## 21.6 Summary

Materials engineering is nowadays one of the most important and rapidly growing science area. It plays a key role in many different industrial processes. Therefore, the knowledge of the structure of materials, their physical and chemical properties, and behavior in appropriate external conditions is strongly required. The development of new technologies and new materials which meet customer requirements is also a very important part of materials engineering since it allows one to develop new solutions, which improves the quality of the technologies used.

It should also be borne in mind that all the experiments described in the present article were conducted at high temperatures. It is essential that an automatic system measures the wettability and the surface tension of high temperature brazes.

In this chapter two different solutions were presented: an integrated platform for the automatic determination of high-temperature braze properties, which allows full automatic determination of braze parameters: the wetting force – using the immerse method (specimen immersed in fluid braze), and the surface tension – using the maximum bubble gas pressure at a temperature of up to 1000 °C in a gas reductive- and protective atmosphere, and a concept of an automated system for the automatic determination of high-temperature braze surface tension using the pendant drop method, which allows one to carry out experiments in a vacuum chamber. Image processing and analysis plays an important role in both experimental stand concepts. In the first case it was used to verify a drop shape against the maximum pressure, which shows that a correction to Schroedinger's calculation of surface tension is needed. In the second case, the measurement methods are based on the grabbing of the specimen rod image during the experiment flow and image processing methods are used for drop shape geometrics determination, and indirectly for surface tension calculation.

# References

Adamson A.W.: *Physical Chemistry of Surfaces*. Wiley Int. Publ., New York, 1982

*Adobe Flex*, http://www.adobe.com, 2013.

Chang, H. Y., Chen, S. W., Wong, D. S. H., Hsu, H. F.: „Determination of reactive wetting properties of Sn, Sn–Cu, Sn–Ag, and Sn–Pb alloys using a wetting balance technique", *J. Mater. Res.*, 18, 1420–1428, 2003

Directive 2002/95/EC of the European Parliament and of the Council of 27 January 2003 on the restriction of the use of certain hazardous substances in electrical and electronic equipment (RoHS). Official Journal of the European Union, L37/19, 2003.

Directive 2002/96/EC of the European Parliament and of the Council of 27 January 2003 on the waste electrical and electronic equipment (WEEE). Official Journal of the European Union, L37/24, 2003.

Eustanthopoulos, N., Nicholas, M. G., Drevet, B.: „Wettability at High Temperatures", *Pergamon Press*, 1999.

*Hibernate*, http://www.hibernate.org, 2013

International Electrotechnical Commision: *IEC 61131-3*, PLC's programming languages, 2013.

*JAX-WS*, http://jax-ws.java.net, 2013.

Kwok D. Y.: Measurements of Static and Low Rate Dynamic Contact Angles by Means of an Automated Capillary Rise Technique", Journal of Colloid and interface Science, 173, 1995.

Miyazaki M., Mizutani M., Takemoto T., Mastunawa A., „Measurement of Surface Tension with Wetting Balance", Q. Japan, "Weld. Soc." Nr 15/1997, str. 681-687.

Park, J. Y., Kang, C. S., Jung, J. P.: „The analysis of the withdrawal force curve of the wetting curve using 63Sn–37Pb and 96.5Sn–3. 5Ag eutectic solders", *J. Electron. Mater.*, 28, 1256–1262, 1999.

*Spring*, http://www.springsource.org, 2013.

*WAGO Innovative Connections*, http://wago.com, 2013

Wojciechowski, R. (2012). Research grant no N 519 441839 of Polish National Research Committee: Integrated platform for automatic determination of wettability and surface tension of high-temperature brazes, Final report.

# INDEX

Printed in the United States
By Bookmasters